Instructor's Manual

for

The Nature of Mathematics

Tenth Edition

Karl J. Smith

Santa Rosa Junior College

THOMSON

BROOKS/COLE

Australia • Canada • Mexico • Singapore • Spain • United Kingdom • United States

Printed in the United States of America
1 2 3 4 5 6 7   07 06 05 04 03

Printer: Victor Graphics, Inc.

ISBN: 0-534-40026-4

For more information about our products,
contact us at:
Thomson Learning Academic Resource Center
1-800-423-0563

For permission to use material from this text,
contact us by:
Phone: 1-800-730-2214
Fax: 1-800-731-2215
Web: http://www.thomsonrights.com

Brooks/Cole—Thomson Learning
10 Davis Drive
Belmont, CA 94002-3098
USA

Asia
Thomson Learning
5 Shenton Way #01-01
UIC Building
Singapore 068808

Australia/New Zealand
Thomson Learning
102 Dodds Street
Southbank, Victoria 3006
Australia

Canada
Nelson
1120 Birchmount Road
Toronto, Ontario M1K 5G4
Canada

Europe/Middle East/South Africa
Thomson Learning
High Holborn House
50/51 Bedford Row
London WC1R 4LR
United Kingdom

Latin America
Thomson Learning
Seneca, 53
Colonia Polanco
11560 Mexico D.F.
Mexico

Spain/Portugal
Paraninfo
Calle/Magallanes, 25
28015 Madrid, Spain

INTRODUCTION

I have written this teacher's guide to help you with supplementary material when teaching from *The Nature of Mathematics, Tenth Edition.* I wrote this book as a textbook for a one or two-semester course for basic competencies, liberal arts, teacher training, or finite mathematics. I also wrote it to allow for a great deal of flexibility. As an instructor, one of the reasons I enjoy teaching this course so much is that it is one of the few courses in the mathematics curriculum that allows for different topics semester to semester.

You will, of course, build the course to suit your own class and style. I thought you might be interested in seeing the handout I use when I teach this book. It follows on pages 5-9.

I also assign a term project in which the student is to select one of the INDIVIDUAL or GROUP RESEARCH PROJECTS and develop it into a class project, paper, or demonstration. Suggestions for the INDIVIDUAL PROJECTS are given at the end of almost every section in the text, and the GROUP RESEARCH PROJECTS are listed in the index. I allow the student to set their own limits when doing their project. I have included the handout for the assignment of the term project on pages 11-12.

I would be happy to hear from you. Let me know both the positive and negative comments about my book.

Dr. Karl J. Smith
(707) 829-0606
email: SMITHKJS@mathnature.com

CONTENTS

MATH 10: NATURE OF MATHEMATICS
MWF 1:10 - 2:00 pm
PET 137

INSTRUCTOR: Dr. Karl J. Smith
PET 140
(707) 778-3942
e-mail: SMITHKJS@mathnature.com

OFFICE HOURS: MWF 11:10-12:00 noon
You can make an appointment if you prefer another time. Feel free to stop in anytime, an appointment is not necessary.

TEXT: *The Nature of Mathematics, Tenth Edition* by Karl J. Smith
Pacific Grove: Brooks/Cole, 2001
You are expected to read each section of the textbook as we cover that section in the class. You will find the course easier if you read it <u>before</u> we discuss it in class because you will be able to ask questions in class about troublesome material.

CALCULATOR: A calculator is required for class, homework, and for examinations.

COURSE DESCRIPTION

The main goal of this course is to create a positive attitude toward mathematics. Mathematics is not an endless procession of dull manipulations, theorems, proofs, and irrelevant topics. The purpose of the course is not to present the technical details needed to proceed to the next course, but to give insight as to what mathematics is, what it attempts to accomplish, and how mathematicians think.

As a mathematician, I frequently encounter people who relate their unpleasant mathematical experiences to me. I have a true sympathy for these people, and recall one of my elementary teachers who would assign additional arithmetic problems as punishment for misbehaving pupils. This can only create negative attitudes toward mathematics, which is indeed unfortunate. It is a vicious circle, and we must attempt to break it. If elementary teachers and parents have a positive attitude toward mathematics, their children cannot help but see some of the beauty of the subject. I want you to come away from this course with the feeling that mathematics can be pleasant, useful, and practical.

Mathematics is considered a **hard** subject. However, if you have the prerequisites for this course I believe it is my job as instructor to make sure you understand the material and that you succeed in this course. If you do not succeed, or if you do not pass this course, I feel I must accept the responsibility for your failure. I believe there is no such thing as a student failure, only an instructor failure. In order to accept this responsibility, however, I must have your assurance that you are willing to make a commitment to this course. There are two aspects to your commitment:

1. *You must make a commitment to attend each class.* Obviously, unforeseen circumstances can come up, but you must plan on being here for every class. There is no such thing as a class that is not important. If you cannot be in class, you must make up the material you missed before continuing with the class. Generally, the make up work for missing a class meeting is to outline the section you missed. Include in the outline each example worked in the text, and in addition make up an additional example of your own for each example in the book. Then work your own example. This outline should be attached when you turn in the assignment that was made the day of your absence. Please check with the instructor by phone or e-mail for any additional information about the class that you missed.

2. *Also you must make a commitment to daily work.* It takes one or two hours each day; you cannot save up and do 5 or 10 hours on the weekend or just before an exam. This works best if you set aside a fixed time for doing your math homework each day. If you do miss some assignment (for whatever reason) it should be made up as soon as possible. However, all assignments must be made up before you take an examination.

If you are not willing or able to make this kind of commitment you should drop the course now. If you are willing to make this kind of commitment, I will work with you, provide tutoring or extra help, or do whatever we need to do to make this course a success for you. However, it is your responsibility to let me know if you are having trouble.

COURSE REQUIREMENTS

Attendance

Attendance is required. I take roll each day. If you are not there for any part of the class (beginning or end), an absence will be recorded. I make no value judgments as to the nature of the absence or the reason for nonattendance, but simply record that you were not present. If you must be absent, the following policy will be used. If you come into class after the class begins, you will have been marked absent. It is your responsibility to see that this is absence erased. Please stop by my desk at the end of the hour and show me the assignment that was due that day. I will simply erase your recorded absence for that day. If you forget to inform me at the end of class that you were tardy (and not absent), it will become your responsibility to clear up the tardy in the same manner required for an absence.

1. *Policy on turning in homework when you must be absent.*
   a. Late papers receive half credit. Late papers are any papers turned in after the beginning of the class during which the paper is to be turned in.
   b. If you are ill or cannot make it to class and wish to receive full credit for a paper that is to be turned in, you may send it with someone else or deliver it to my office before it is late.

2. *Policy on class work when you are absent.*
   a. Class work cannot be made up. However, if you would like me to go over the material with you or if you want to borrow my notes, you only need to ask. Materials that were passed out in class can be picked up during my office hours. **It will probably take you at least 2 hours extra work for each hour that you miss in class.** You will need to turn in an outline for the class meeting that you missed.
   b. You are responsible to keep track of what happened while you were gone. You can call me or one of your classmates, or e-mail me, but you must keep up with all assignments (or changes in assignments) during your absence.

3. *Policy on excessive absences.*
   a. During the first two weeks, if there is a waiting list of students wishing to add the class you must notify me or leave a message the same day of your absence so that I know to hold a spot for you. If you do not contact me the day of your absence, I will assume you do not want this class and I will give your seat to the next person on the waiting list. This applies to all students, those enrolled, and those waiting to enroll.
   b. After the first two weeks, until the last day to drop a class, if you have more than three absences, I will attempt to call you. If you do not contact me, or if I cannot reach you by

phone, I will drop you from the class.

    c.   After the last day to drop from the class, if you have more than three absences, I will attempt to call you. If you do not contact me, or if I cannot reach you, I will lower your grade by one letter grade.

Homework

There will be 30 daily homework assignments of about 2-3 hours per night. Each assignment will be 10 points unless otherwise noted. You should plan on spending about 2 hours on the written assignment and about 1 hour on studying. Your homework will be collected and graded by a student assistant called a *reader.* The reader will keep track of your progress on homework scores, but I will grade your exams and keep track of your grades in the course.

1.   Scheduling homework
    I would like you to keep a daily log of the time of day you work on your math homework. You should SCHEDULE two to three hours for each hour in class. This scheduled time can be at different times each day, but you should set aside time for doing your math homework. Every time I make an assignment, you should enter the date assigned, the due date, and the date collected on your log. You should then enter on the log the time you actually spend working on the assignment, the time spent studying, and any comments you wish to make.

2.   Handing in homework
    The system I use allows you to ask questions on each assignment before it is turned in. The assignments are denoted by PROBLEM SET NUMBER. Even though I've given you a list of assignments, I will specifically make each assignment in class, so that you will be clear about the day of a particular assignment. The day I make an assignment is called the ASSIGNMENT DAY. This assignment is DUE the following class meeting. On the due date, the assignment should be completed and should be out on your desk when the class begins. If you do not have it out and completed on the due date, it will be marked for half credit by the instructor. You will have a chance to ask questions on the due assignment, and then you will be able to take it home and make corrections or changes. It will be turned IN at the beginning of the following class. If it is not turned in on the IN date, it will be marked half credit by the reader. When turning in your homework, fold your paper in half lengthwise and put your name, class, and problem set number on the outside. I will provide a daily log sheet and a calendar to help you keep track of this process.

3.   Extra Credit
    You can work any PROBLEM SOLVING problem on any homework assignment for extra credit. It will be scored and bonus points of 1 to 10 points given for each problem-solving problem (not otherwise assigned) worked correctly.

4.   Paper
    I would prefer you use engineering paper. It is available at the bookstore and is called "engineer's pad." Use only the face side of the tablet (the side without the printed lines) and do not write anything on the back.

5.   Answers alone are not sufficient.
    Show all of your work. You may do steps in your head or on a calculator (in fact, I encourage you to do so), but you need to show the set up of the problem, and enough so that I can follow your reasoning through the problem. Do not use scrap paper.

6. *Neatness*

All written work must be neat and legible. The margins should be kept clear and all work should be clearly shown and labeled.

7. *Scoring*

The emphasis on homework is on understanding, not on points. You should check your work with the answers in the back of the book. Ask questions in class, or work with fellow students in study groups. Not all problems you turn in will be graded, but those selected for grading will be indicated by a check mark if it is correct and by an x if it is wrong. Scores are recorded such as 8/10 which means that you received 8 points out of a 10 point assignment. Extra credit will be indicated by +2 EC which means that 2 extra points were recorded on your score.

Term Project

A term project will be assigned in this course. The exact details and scope of this project will be given later in the semester, but you will be expected to work on a major project during this course. This project might be a term paper, but more likely it will be a work or art or a demonstration that is related to mathematics. You will be judged not only by the instructor, but also by your classmates. There is an index of mathematical projects at the back of your text which will begin to give you some ideas about the type of project you might do. The important dates for this project are:

Project assignment: Sept. 27
Title or subject of your project due: Oct. 18
Paragraph description of project due: Oct. 30
Detailed description of project due: Nov. 20
Project due on or before: December 15

The penalty for late title, paragraph, or description of project will be 10 points. You should plan on turning in your project early since no late projects are accepted, and unforeseen circumstances may prevent you from turning it in on the last day. If you turn in your project early, you will be exempted from all remaining reports on the project.

> Late projects will not be accepted.

Tests

Quizzes

There may be daily quizzes. These will be worth 10 points each and will be short answer type questions over the previous day's material. The first quiz will be over the material in this handout and will be given at the beginning of the second class meeting. Daily recitation is required.

> No make up quizzes are given.

Tests

There will be four fifty-minute tests worth 100 points each. The three best scores will be counted. If you miss one test, for any reason, that is the test which will be deleted. Part of your preparation for each test is to make sure all of your outlines and homework are completed. You will not be permitted to take an examination if this preparation is not complete.

> No make up tests are given.

Final Exam There will be a final exam worth 300 points and will cover the entire course. Attendance for the final is mandatory. See the schedule of classes for the final examination schedule. If you must be absent for the final for any reason, you must contact me the day of the test about taking an incomplete grade for the course until you can make up the final exam. Otherwise, 0 points will be recorded and your grade computed accordingly.

GRADING

Grading is done on the basis of total points received in the course. The approximate weighting for the course will be:

> **HOMEWORK:** 300 points, about 35% of your grade
> **TERM PROJECT:** 100 points, about 15% of your grade
> **TESTS:** 300 points, about 30% of your grade
> **FINAL EXAM:** 300 points, about 20% of your grade

Rather than grade on a curve or a strict percentage, I use the following method for determining your grade. At the end of the course I total all of your points and make a distribution of total points for all members of the class, from the highest to the lowest. Next, I look for a "natural break" in the scores. *The philosophy I use is that similar scores will receive the same grade.* Generally the distribution is as follows:

> A: 90% - 100%
> B: 80% - 90%
> C: 65% - 80%
> D: 50% - 65%
> F: 0% - 50%

However, if there are a cluster of scores around one of the "breaking points," I will move that breaking point higher or lower to the closest or most reasonable break in the scores.

> "What is good teaching? Giving opportunity to the student to discover things by himself (or herself)."
> — Herbert Spencer

MATH 10   Karl J. Smith   Fall

A copy of this grade report will be filled out and given to you after each test so you can monitor your own progress through the course. I also want to make sure my records are correct. I'll be happy to discuss your grade with you at any time. If you think I've made a mistake, please let me know and I'll correct any mistake with apologies. If you do not make an attempt to correct this report, I'll assume my records are correct.

Name _____   **Address** _____   **Seat** _____

Social Security Number _____   Phone _____

Grades: *Final Grades*   *Possible*

TEST I: _____   HMK I: _____   PART I: _____ _____   Test Total _____ _____

TEST II: _____   HMK II: _____   PART II: _____ _____   HMK Total _____ _____

TEST III: _____ HMK III: _____ PART III: _____ _____ Subtotal _____ _____

POINT ADJUSTMENTS: Final Exam _____ _____

Comments:

Part I: TOTAL _____ _____

Part II:

Part III: Commitment: ☐

Missing Homework Papers: *Absences   Grading Scale: 90% - 100%   A
                                                                       80% - 90%   B
                                  _____   65% - 80%   C
                                  _____   50% - 65%   D
                                  _____   0% - 50%   F

Half Credit Homework Papers: More than 2 absences is   MIDTERM GRADE:
                                  considered excessive and   _____
                                  could result in an instructor   FINAL GRADE:
                                  drop or in a lowered grade.   _____

*When this reaches 3, please see me in my office. You should clear <u>each</u> absence in excess of two during my office hours. Generally after the 2nd absence, if you are doing satisfactorily **and are caught up**, I will require about two hours additional work for each hour of class you have missed.*

*An absence is defined as not being in class when roll is taken at the beginning of the hour, or not in class during any part of the hour. I make no value judgments as to the nature of the absence or the reason for nonattendance, but simply record that you were not present.

IF YOU ARE TARDY, you can have the absence removed if you come up to me after class or come into my office <u>that</u> <u>day</u>. Failure to do this will result in the tardy being counted as an absence.

In order to enroll in this class you must meet the prerequisites. You should also have sufficient time to devote to this class. This class meets for 3 hours per week and requires approximately 6 hours per week of preparation time. These hours should be scheduled throughout a week on a daily basis (1 or 2 hours per day, seven days a week).

Mathematics is considered a **hard** subject. However, if you have the prerequisites for this course, I believe it is my job as instructor to make sure you understand the material and that you succeed in this course. If you do not succeed, or if you do not pass this course, I feel I must accept the responsibility for your failure. I believe there is no such thing as a student failure, only an instructor failure. In order to accept this responsibility, however, I must have your assurance that you are willing to make a commitment to this course. There are two aspects to your commitment:

1. *You must make a commitment to attend each class.* Obviously, unforeseen circumstances can come up, but you must plan on being here for every class. There is no such thing as a class that is not important. I agree to turn in an outline of work that was covered in class on days that I could not attend. An outline should be submitted for each day missed and should include: the main ideas from the text; all the day's examples (copied from the text), as well as other topics that were covered that day. If you come in after class begins, it is your responsibility to see Dr. Smith after class to get that day's absent report removed. If you forget to get the absent report removed *that day*, the tardy will be treated as an absence.

2. *Also you must make a commitment to daily work.* It takes one or two hours each day; you cannot save up and do 5 or 10 hours on the weekend or just before an exam. This works best if you set aside a fixed time for doing your math homework each day. If you are not willing or able to make this kind of commitment you should drop the course now. If you are willing to make this kind of commitment, I will work with you, provide tutoring or extra help, or do whatever we need to do to make this course a success for you.

If you are willing to make this commitment please sign here. _____

MATH 10 TERM PROJECT
Karl J. Smith

Preparing a mathematics project can give you interesting and worthwhile experiences. In preparing a project you will get experience in using resources to find information, in doing independent work, in organizing your presentation, in communicating ideas orally, in writing, and in visual demonstrations. You will broaden your background in mathematics and contact new mathematical topics that you never before knew existed. In setting up an exhibit you will experience the satisfaction of demonstrating what you have accomplished. It may be a way of satisfying your curiosity and your desire to be creative. It is an opportunity for developing originality, craftsmanship, and new mathematical understandings.

Projects are judges on the following criteria: originality, completeness, clarity and craftsmanship, interest value, and mathematical thought. In order to organize a project that is a success, you may find the following suggestions helpful:

1. *Select a topic that has interest potential.* Do not do a project on a topic that does not interest you. Some ideas for projects will be given in class. Other ideas for projects can be found in the Index of Term Project at the back of your text. If you still do not see something that interests you, come to my office to talk about additional topics. My goal is to stimulate your interest to the point that you would want to do the project even if it were not assigned as part of this class.

2. *Find as much information about the topic as possible.* For those projects suggested in class or listed in the text, I've included a bibliography to get you started. In addition check the following sources:
   Periodicals: *The Mathematics Teacher, Arithmetic Teacher,* and *Scientific American;* each of these has their own cumulative index; also check the *Reader's Guide.*
   Source books: The World of Mathematics by Newman is a gold mine of ideas; *Mathematics,* a Time-Life book by David Bergimini may provide you with many ideas.
   Encyclopedias can be consulted after you have some project ideas; however, I do not have in mind that the term project necessarily be a term paper.
   Internet: Use one or more search engines on the internet for information on a particular topic. The more specific you can be in describing what you are looking for, the better the engine will be able to find your topic.

3. *Prepare and organize your material into a concise, interest report.* Include drawings in color, pictures, applications, examples that will get the reader's attention and add meaning to your report. Build models, mockups, or devices that add interest and understanding of the topics.

4. *Build an exhibit that will tell the story of your topic.* Remember the science projects in high school? That type of presentation is appropriate. Use models, applications, charts that lend variety. The Mobius strip in the center quad of the math building on the Santa Rosa Junior College Campus was a Math 10 project. If possible, prepare materials that viewers can manipulate. Give your exhibit a catchy, descriptive title. Label everything with brief captions or legends so that viewers will understand the principles involved. Make the display simple, but at the same time attractive and dramatic. Use color for emphasis. Write captions in a unique way such as with rope, pipe cleaners, plastics, or yarn. Show craftsmanship, creativeness, and diligence in arranging the exhibit. This is a **term** project, and cannot be done in one or two evenings.

5. *Grading.* The possible is 100 points. I will give you 50 points on your project and will grade it before it is seen by the class. I will use the following criteria in grading your project: Originality (10 pts), Completeness (10 pts), Clarity and craftsmanship (10 pts), interest value (10 pts), and mathematical thought (10 pts). Then, January 3, 20xx is PROJECT DAY and all of the projects will be on display (without names). I will ask everyone in the class to grade all of the projects on a 0-100 scale. I will take the average of all of the student evaluations and equate this average with a 50-point possible score. The sum of these scores (instructor + classmates) will be your score for the project.

6. *Deadlines.* The following time requirements must be met. *Late projects will not be accepted.*

    Project assignment: 9/27/xx
    Title due: 10/18/xx
    Paragraph due: 10/30/xx
    Detailed description due: 11/20/xx
    Project due on or before 12/15/xx

    If you turn in your project early, you will be exempted from all remaining reports on the project. Many students avoid the detailed description by turning in the project on November 20.

AUDIO-VISUALS

The use of films as part of the course is often desirable in a course of this type. Included below is a list of available video tapes and films that could fit nicely into the development of materials in this text. Of course, it is neither desirable nor recommended that a particular class use them all, but an instructor may select one or more films based on the particular class and availability of films and equipment.

VHS Video

For All Practical Purposes
This is a set of 26 half-hour lessons, 5 lesson per tape, $29.95 per tape with $350 for the full series. The Consortium for Mathematics and its Applications (COMAP) produced these lessons as an introduction to contemporary mathematics. The 26 lessons are divided into 5 categories: management science, statistics, social choice, geometry, and computer science. To order and get information or to preview tapes, call 1-800-539-7637.

Donald in Mathemagicland
20 min., available at many video stores. This 1959 classic film by Walt Disney is one of the best known mathematics films.

You Can Count on It
15 min., $8.75 fee includes UPS charge. It is a project of the Mathematical Association of America (MAA) and the National Council of Teachers of Mathematics (NCTM) with support from Texas Instruments. It is a nice film, but suitable for less able classes. It is available by calling 1-800-527-2156 (outside Texas) or 1-800-441-0145 (Texas) or from:
> Levco Marketing
> 2809 Ross Avenue
> Dallas, TX 75201

Math Anxiety: We Beat It, So Can You!
29 min., color, $35 fee for 3 days. It is also available as a 16mm film. It is available from:
> EDC Distribution Center
> 55 Chapel Street
> Newton, MA 02160

Powers of Ten
9 min. This film is an entirely new color remake of the original delightful film. It illustrates the relationship between the number ten and the size of the physical universe. It is available by calling 1-800-421-2304 or from:
> Pyramid Films
> Post Office Box 1048
> Santa Monica, CA 90406

Pythagorean Theorem
16 min., $29.95. This video presents an in-depth look at the Pythagorean Theorem. It is available by calling 1-800-262-8837 or from
> Teacher's Video Company
> P.O. Box MAW-4455
> Scottsdale, Arizona 85261

Chaos, Fractals, and Dynamics: Computers Experiments in Mathematics
1 hr., $59, order code VIDDEVANEY/NA. It is a clearly presented and richly illustrated instructional tool which is available by calling 1-800-321-4267 or from:
    American Mathematical Society
    Post Office Box 1571
    Annex Station
    Providence, RI 02901

The Beauty and Complexity of the Mandelbrot Set
1 hr., $59, order code VIDDHUBBARD/N. It is available by calling 1-800-321-4267 or from:
    American Mathematical Society
    Post Office Box 1571
    Annex Station
    Providence, RI 02901

The Challenge of the Unknown
7 20 min., about problem solving. These videos are available from:
    Karol Media
    22 Riverview Drive
    Wayne, NJ 07470

Math . . . Who Needs It?
58 minutes, $29.95. Jaime Escalante, Bill Crosby, and other celebrities get young people pumped up about math. It is available by calling 1-800-262-8837 or from
    Teacher's Video Company
    P.O. Box MAW-4455
    Scottsdale, Arizona 85261

Proof: Fermat's Last Theorem
60 minutes, $29.95. Nova presents the amazing story of how Princeton whiz Andrew Wiles solved one the most challenging mysteries of math. It is available by calling 1-800-262-8837 or from
    Teacher's Video Company
    P.O. Box MAW-4455
    Scottsdale, Arizona 85261

Code Breaking Spies
46 minutes, $29.95. In the secret world of spies, math experts become heroes as they break intricate codes. This is an intriguing application of math. It is available by calling 1-800-262-8837 or from
    Teacher's Video Company
    P.O. Box MAW-4455
    Scottsdale, Arizona 85261

Films

Flatland
12 min. It is an animated film based on the short novel by the same name by Edwin Abbott. It is available from:
    Moody Institute of Science
    1200 E. Washington Blvd.
    Whittier, CA 90606

Let Us Teach Guessing
61 min., color; $11 rental, $400 purchase. It presents a mathematical inquiry lesson on the process of making reasonable guesses to solve a problem or answer a question. Suggests that the reasonable guess should be based on testing extreme cases, testing the guess, and making analogies. This film is produced by the American Mathematical Association (MAA film number 194X1053) and should be used with a more able class. It is available from:

Modern Learning Aids
5100 West Henrietta Road
Rochester, NY 14692

The Seven Bridges of Königsberg
4 min., color; $6 rental, $65 purchase. This topological film recreates Leonhard Euler's analysis of the Königsberg Bridge Problem. This should be shown *after* the students have had an opportunity to work on the problem, since the film gives the solution. It is available from:

International Film Bureau, Inc.
332 S. Michigan Ave.
Chicago, IL 60604

Nim and Other Oriented Graph Games, 63 min., b/w
John von Neumann, A Documentary, 63 min., b/w
These films, along with many other 16-mm sound films are available for sale or rent from:

Wards Natural Science Establishment
(Modern Learning Aids)
5100 West Henrietta Road
Rochester, NY 14692

For more information call 716-359-2502

Mathematical Association of America Films (animated) available from Modern Learning Aids (address listed above).

Mr. Simplex Saves the Aspidistra (color, 33 min.)
What is a Set (color, 15 min.)
One-to-one Correspondence (color, 12 min.)
Counting (color, 10 min.)

Allendoefer Films (also available from Modern Learning Aids)
Area and pi (color, 10 min.)
Geometric concepts (color, 10 min.)
Geometric Transformations (color, 10 min.)
Cycloidal Curves or Tales from the Wanklenberg Woods (color, 22 min.)

Prologue, page P1

A theme of this book is problem solving, and this problem set was designed to give an overview of variety of ideas and concepts which will be considered in this book. Here is a list of where in the book the problem will be discussed:

Chapter 1:	10, 13, 14, 49, 55
Chapter 2:	22, 28, 52
Chapter 3:	26, 58
Chapter 4:	16, 21, 27, 31, 32
Chapter 5:	17, 30, 39
Chapter 6:	25, 29, 53
Chapter 7:	18, 33, 34, 35, 37, 38
Chapter 8:	19, 48, 51, 59
Chapter 9:	23, 24
Chapter 10:	11, 12, 50
Chapter 11:	52, 54, 60
Chapter 12:	36, 40
Chapter 13:	43, 56, 57
Chapter 14:	41, 46
Chapter 15:	15, 20, 47
Chapter 16:	44
Chapter 17:	42, 45

You can use this list of problems in many creative ways, but I've used it as a pretest at the beginning of the course and then as a post test at the end of the course.

PROLOGUE PROBLEM SET, page P17

1. Answers vary.
2. The seven chronological periods are:

Egyptian, Babylonian and Native American Periods 3000 BC to 601 BC	
Greek, Chinese, and Roman Periods	600 BC to 499 AD
Hindu and Arabian Period	500 to 1199
Transition Period	1200 to 1599
Century of Enlightenment	1600 to 1699
Early Modern Period	1700 to 1899
Modern Period	1900 to 2000

    Answers vary.

3-9. Answers vary.

10. A fence of 8 ft requires $\frac{8}{8} + 1 = 2$ poles; 16 ft needs $\frac{16}{8} + 1 = 3$ poles; 1,440 ft needs
$$\frac{1,440}{8} + 1 = 181 \text{ poles}$$

11. Must draw at least five cards. The worst-case scenario is to draw one from each suit for the first four draws; the 5th card must match one of those.

12. There must be 1,099 people. The worst-case scenario is for the first $3 \cdot 366$ people to have birthdays so that 3 have birthdays on each different day of the year (including leap years); thus would need to have $3 \cdot 366 + 1$ to be certain that there are four people with the same birthday.

13. $3^0 = 1$; $3^1 = 3$; $3^2 = 9$; $3^3 = 27$; $3^4 = 81$; $3^5 = 243$; $3^6 = 729$; $3^7 = 2,187, \cdots$

Thus,

$3^0, 3^4, 3^8, 3^{12}, \cdots$ have a last digit of 1

$3^1, 3^5, 3^9, 3^{13}, \cdots$ have a last digit of 3 $\leftarrow$ 3^{2001} is in here

$3^2, 3^6, 3^{10}, 3^{14}, \cdots$ have a last digit of 9

$3^3, 3^7, 3^{11}, 3^{15}, \cdots$ have a last digit of 7

Similarly, $2^0 = 1$; $2^1 = 2$; $2^2 = 4$; $2^3 = 8$; $2^4 = 16$; $2^5 = 32$; $2^6 = 64$; $2^7 = 128; \cdots$. For this pattern, we start after the first number, 2^0:

$2^1, 2^5, 2^9, 2^{13}, \cdots$ have a last digit of 2 $\leftarrow$ 2^{2001} is in here

$2^2, 2^6, 2^{10}, 2^{14}, \cdots$ have a last digit of 4

$2^3, 2^7, 2^{11}, 2^{15}, \cdots$ have a last digit of 8

$2^4, 2^8, 2^{12}, 2^{16}, \cdots$ have a last digit of 6

We see that 3^{2001} has a last digit of 3 and 2^{2001} has a last digit of 2, so $3^{2001} - 2^{2001}$ must have a last digit of 1.

14. February and March; February is the only month with exactly 28 days (exactly 4 weeks).

15. This is a problem which is equivalent to the famous Königsberg street problem (see Section 15.1). You will be able to trace out a path if there are no more than two vertices connecting an odd number of line segments. There are four such vertices (coffee shop, Santa Rosa Junior College, city hall, and the ice cream stand). It is, therefore, impossible to trace out this circuit.

16. $210 = 2 \cdot 3 \cdot 5 \cdot 7$ and $330 = 2 \cdot 3 \cdot 5 \cdot 11$ so the largest number that divides both 210 and 330 is $2 \cdot 3 \cdot 5 = 30$.

17. Ann's age = Brittany's age + 3; Brittany's age = Chelsea's age + 2;
Deidre's age = Elysse's age + 5; Elysse's age = Fawn's age + 1

There are relationships among Ann's age, Brittany's age, and Chelsea's age as well as relationships among Deidre's age, Elysse's age, and Fawn's age; however, nothing can be said about Deidre's age as compared to Chelsea's age.

18. The measurement is one of volume, not length. The bead forms a cylinder and the volume of a cylinder is $V = Bh$ where B is the area of the base and h is the length of the bead.

Tube A ($\frac{1}{4}$ in. bead)   Tube B ($\frac{1}{8}$ in. bead)

30 ft = 360 in.   96 ft = 1,152 in.

$B = \pi r^2$: $\pi(\frac{1}{8})^2 \approx 0.049$ $\pi(\frac{1}{16})^2 \approx 0.012$

$V = Bh$ $\pi(\frac{1}{8})^2 360 \approx 17.67$ in.^{3} $\pi(\frac{1}{16})^2 1,152 \approx 14.1$ in.^{3}

We see that the tube with the smaller bead actually contains less caulking, which confirms Marilyn's answer (which is not shown in the text).

19. Guesses vary. To calculate the date, first find the growth rate using the given information using the formula $P = P_0 e^{rt}$.

$$6.248 = 6e^{3r} \qquad \textit{The time, t, from Oct. 12, 1999 to Oct. 12, 2002 is 3.}$$

$$\frac{6.248}{6} = e^{3r} \qquad \textit{Divide both sides by 6.}$$

$$3r = \ln\left(\frac{6.248}{6}\right) \qquad \textit{Definition of logarithm.}$$

$$r = \frac{1}{3}\ln\left(\frac{6.248}{6}\right) \qquad \textit{Divide both sides by 3.}$$

$$\approx 0.0135 \qquad \textit{By calculator.}$$

Now use this value of r to calculate the date.

$$7 = 6.248 e^{rt} \qquad \textit{Use the population formula where P = 7 and P_0 = 6.248.}$$

$$rt = \ln\left(\frac{7}{6.248}\right) \qquad \textit{Definition of logarithm.}$$

$$t = \frac{1}{r}\ln\left(\frac{7}{6.248}\right) \qquad \textit{Divide both sides by r.}$$

$$\approx 8.418 \qquad \textit{By calculator.}$$

Subtract 8 (for 8 years), and then multiply by 12 (to obtain the nearest month), 5.016. Therefore, eight years five months from October 12, 2002 puts the predicted date at April, 2011.

20. **a.** Yes, there are many possibilities; for example, Washington, Oregon, California, and Arizona

b. No, it is not possible.

21. $(1, 2) = 1 \times 2 + 1 + 2 = 5$ and $(3, 4) = 3 \times 4 + 3 + 4 = 19$, so

$$((1, 2), (3, 4)) = (5, 19) = 5 \times 19 + 5 + 19 = 119$$

22. All Angelenos are Los Angeles residents.

23. If $2n - 1$ represents all odd numbers, then for $n = 1$ we have $2(1) - 1 = 1$; for $n = 2$ we have $2(2) - 1 = 3$; $\cdots$ for $n = 473$ we have $2(473) - 1 = 945$

24. $1 + 2 + 3 + \cdots + 100,000 = \dfrac{100,000(100,000 + 1)}{2} = 5,000,050,000$

25. After cutting, there are $4^3 = 64$ one-inch cubes; the corners are painted on 3 sides and the edges are painted on 2 sides; there are four other cubes on each face for a total of $6 \cdot 4 = 24$ cubes that are painted on one side only. The fraction is
$$\frac{24}{64} = \frac{3}{8}$$

26. This number system counts the first 20 numbers as: 0, 1, 2, 3, 4, 10, 11, 12, 13, 14, 20, 21, 22, 23, 24, 30, 31, 32, 33, and 34. The number we know as 18 would be written as 33.

27. $M(m(1, 2), m(2, 3)) = M(1, 2) = 2$

28. Draw a Venn Diagram; since there are 20 males and 50 people, there are 30 females. Thus, since there are 25 women there are 5 girls; this means that there are 7 boys (12 children) and consequently there are 13 men.

29. **a.** tetrahedron **b.** cube (hexahedron) **c.** octahedron **d.** icosahedron **e.** dodecahedron

30. We are looking for the least common multiple of 16 and 10; $16 = 2^4$ and $10 = 2 \cdot 5$, so the least common multiple is $2^4 \cdot 5 = 80$. It will require 80 minutes before they both return to the starting point at the same time.

31. Consider a pattern where we cross out multiples of 3:

$$1, 2, \cancel{3}, 4, 5, \cancel{6}, 7, 8, \cancel{9}, 10, 11, \cancel{12}, 13, 14, \cancel{15}, 16, 17, \cancel{18}, \cdots$$

Let n be the nth positive integer that in not divisible by 3.

The n is even, the nth number that is not divisible by 3 is $\frac{3}{2}n - 1$;

when n is odd, the nth number that is not divisible by 3 is $\frac{3}{2}n - \frac{1}{2}$.

If $n = 1$, then $\frac{3}{2}(1) - \frac{1}{2} = 1$ is the 1st number not divisible by 3.

If $n = 2$, then $\frac{3}{2}(2) - 1 = 2$ is the 2nd number not divisible by 3.

If $n = 3$, then $\frac{3}{2}(3) - \frac{1}{2} = 4$ is the 3rd number not divisible by 3.

If $n = 4$, then $\frac{3}{2}(4) - 1 = 5$ is the 4th number not divisible by 3.

$$\vdots$$

If $n = 1,000$, then $\frac{3}{2}(1,000) - 1 = 1,499$ is the 1000th number not divisible by 3.

32. We are looking for the least common multiple of 3, 5, and 7 which is 105. April has 30 days, May has 31 days, and June has 30 days. This is 91 days, so 105 days after March 31st is July 14th.

33. If you are checking the calculation for 450 dollar bills per pound, you find 1 lb $\approx$ 454 g, but there will be some waste in the process, so 450 is correct. Note, \$1 trillion would require $\frac{1 \times 10^{12}}{450}$ lb of wood or about 2.22×10^9 lb. To find the volume of a tree, we assume that it is a cylinder so that $V = Bh$. B is the area of the base (12 in. diameter is 1 ft diameter or 0.5 ft radius) and h is the height of the tree so that

$$V = 0.5^2\pi(50) \approx 39.26990817 \text{ ft}^3 \text{ or about } 1{,}963.5 \text{ lb.}$$

Thus, the number of trees necessary to print the money is

$$\frac{2.22 \times 10^9}{V} \approx 1{,}131{,}768$$

That is, more than a million trees must be cut down to print a trillion one-dollar bills.

34. Look at the cars in the front to estimate the height of each can. Estimates vary, but these cans are located at Heileman's Brewers in LaCrosse, Wisconsin. They are 53 ft high and each can has a diameter of 20 ft. $V = Bh = \pi r^2 h = \pi(10)^2(53) = 5{,}300\pi$ ft^3. A ft^3 is about 7.48 gal so each can has about 124,545 gal. The huge 6-pack is about 747,272 gal. This is almost 8,000,000 standard 12-oz cans of beer.

35. Instead of calculating the area of the shaded region directly, notice that the desired area is the area of the square less the area of the circle.

$$8^2 - \pi(4)^2 = 64 - 16\pi$$

This is an area of approximately 13.7 in.^{2}

36. Answers vary; the sample was not randomly selected.

37. Areas of the three circles:

$A = 2^2\pi$ (smallest area); $B = 3^2\pi$ (middle area); $C = 5^2\pi$ (largest area)

Area of shaded region: $C - (A + B) = 25\pi - (4\pi + 9\pi) = 12\pi$

Ratio of the smallest area to the shaded region is $\frac{4\pi}{12\pi} = \frac{1}{3}$.

38. Base area for the 1 in. hose is $\left(\frac{1}{2}\right)^2\pi = \frac{1}{4}\pi$

Base area for two $\frac{1}{2}$ in. hoses is $2 \cdot \left(\frac{1}{4}\right)^2\pi = \frac{1}{8}\pi$

Since $\frac{1}{4}\pi > \frac{1}{8}\pi$, the water will drain from the 1-in. diameter hose faster.

39. Let x be the amount she started with. Then at the end of the first day she had $2x - 30$. After the second day she had $3(2x - 30) - 20 = 6x - 110$. Thus,

$$6x - 110 = x$$
$$5x = 110$$
$$x = 22$$

She had \$22 at the start of the first day.

40. Answers vary. Deaths from HIV are rising faster than deaths from other causes.

41. Assume the cost of the charter is the same regardless of the number of travelers. Let x be the cost per traveler if 100 persons travel. Then,

$$100x = 125(x - 78)$$
$$100x = 125x - 9{,}750$$
$$25x = 9{,}750$$
$$x = 390$$

The cost is \$390 if 100 make the trip.

42. $\lim\limits_{n \to \infty} \left(1 + \frac{1}{n}\right)^n = e$; this is the definition of e.

43. Let x be the number of weeks enrolled, and y be the total cost. Then $y = 45(10 - x)$. The graph is shown.

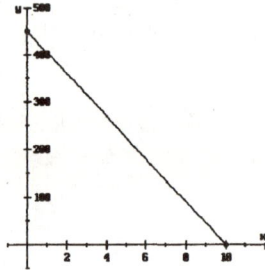

44. The vote is R is $38\% + 10\% = 48\%$, S is 29%, and T is 24%.

a. The plurality vote is for Rameriz (R).

b. Here is the Borda count (using percents):

	R	S	T		R	S	T
38(	3,	2,	1)		114	76	38
29(	2,	3,	1)		58	87	29
24(	1,	2,	3)		24	48	72
10(	3,	1,	2)		30	10	20
TOTALS:					226	221	159

The winner from a Borda count is Rameriz (R).

c.   Since Rameriz is the last choice of those in the 24% column, they could get a candidate more to their liking by changing their ranking to (S, T, R). The vote is now:

R is $38\% + 10\% = 48\%$, S is $29\% + 24\% = 53\%$, T is 0%

Smith (S) wins the majority vote.

45.   The rate at which alcohol is changing with respect to time is $\frac{dC}{dt} = 0.3(-\frac{1}{2})e^{-t/2} = 0.15e^{-t/2}$.

46.   Let $x =$ no. of units of food A and $y =$ no. of units of food B; minimize $C = 0.14x + 0.06y$

Subject to
$$\begin{cases} x \geq 0, \; y \geq 0 \\ 6x + 2y \geq 100 \\ 3x + 2y \geq 60 \\ x + 2y \geq 40 \end{cases}$$

Corner points are $(0, 50)$, $(12, 14)$, and $(20, 0)$. The minimum cost $C = \$2.52$ with 12 units of food A and 14 units of food B.

47.   Label each vertex; there are 33 paths.

48.   Start at 100 and halve the amount at each step until you arrive at an odd number; then subtract 1 and continue until you arrive at 0. The actual steps would then reverse this process. 100, 50, 25, 24, 12, 6, 3, 2, 1, 0; 9 steps

49. $\frac{5.7 \times 10^{12}}{2.749 \times 10^8} \approx 20{,}734.81$ days; divide by 365 to find 56.80770592 years.

This is in March 2057, and this is without paying off accrued interest in the meantime.

50.   The missing number is 31. The pattern is the representation of 16 in various bases is shown:

base 16:	10	base 9:	17
base 15:	11	base 8:	20
base 14:	12	base 7:	22
base 13:	13	base 6:	24
base 12:	14	**base 5:**	**31** missing entry
base 11:	15	base 4:	100
base 10:	**16**	base 3:	121
		base 2:	10000

51.
$$\log_2 x + \log_4 x = \log_b x$$

$$\log_2 x + \frac{\log_2 x}{\log_2 4} = \log_b x \qquad \text{\textit{Change base 4 to base 2.}}$$

$$\log_2 x + \frac{\log_2 x}{2} = \log_b x \qquad \text{\textit{$\log_2 4 = 2$ since $2^2 = 4$.}}$$

$$\frac{3}{2}\log_2 x = \log_b x \qquad \text{\textit{Common denominator.}}$$

$$\frac{3}{2}\log_2 x = \frac{\log_2 x}{\log_2 b} \qquad \text{\textit{Change base b to base 2.}}$$

$$\log_2 b = \frac{2\log_2 x}{3\log_2 x} \qquad \text{\textit{Multiply both sides by $2\log_2 b$ and divide both sides by $3\log_2 x$.}}$$

$$\log_2 b = \frac{2}{3}$$

$$b = \frac{2}{3}\ln 2 \qquad \text{\textit{Definition of logarithm.}}$$

52.   Consider the possibilities:

$$\boxed{1}\ \text{Ed};\ P(H) = \tfrac{1}{3}$$

$\boxed{2}\ \text{Red};\ P(H) = \tfrac{1}{2}$ $\qquad\qquad\qquad$ $\boxed{3}\ \text{Fred};\ P(H) = 1$

I.   If $\boxed{1}$ shoots at $\boxed{2}$ and wins, then he is dead. (Fred shoots him.)

If $\boxed{1}$ shoots at $\boxed{3}$ and wins, then he has a 50% chance of being dead.

If $\boxed{1}$ shoots and misses on purpose, then there are still three people. This is what Ed should do.

$P(\text{Ed survives}) = 1$.

II.   If $\boxed{2}$ shoots at $\boxed{1}$ and wins, then he is dead. (Fred shoots him.)

If $\boxed{2}$ shoots at $\boxed{1}$ and misses, then he is dead (Fred shoots him.)

Thus, $\boxed{2}$ shoots at $\boxed{3}$.

This is what Red should do.

III.   If $\boxed{3}$ shoots at $\boxed{1}$, then he has a $\frac{1}{2}$ chance in the second round.

If $\boxed{3}$ shoots at $\boxed{2}$, then he has a $\frac{1}{3}$ chance in the second round.

Thus, $\boxed{3}$ shoots at $\boxed{2}$. This is what Fred should do.

This means that Ed has the best chance of not ending up dead.

53. There are at least three cubes to be seen:

(1) A little cube nestled in the corner of a larger cube.

(2) A big cube with a cubical chunk removed from one corner.

(3) Two cubes meeting externally at a corner.

If you see any other configurations, please send them the author

(email: SMITHKJS@mathnature.com).

54. $_5P_5 = 5! = 120$, so the probability is $\frac{1}{120}$. The number of cards in the original deck does not matter.

55. This is what is called a magic square; the sum of the numbers in all of the rows, columns, and diagonals is 15. Wow!

56. Look for a pattern:

> Two distinct lines intersect in at most 1 point.
> **Three** lines intersect in at most 3 points; this is $1 + (3 - 1) = 3$.
> **Four** lines intersect in at most 6 points; this is $3 + (4 - 1) = 6$.
> **Five** lines intersect in at most 10 points; this is $6 + (5 - 1) = 10$.
> **Six** lines intersect in at most 15 points; this is $10 + (6 - 1) = 15$.

It looks like n lines would intersect in at most $n - 1$ added to the number of intersection points of $n - 1$ lines (the previous answer). That is, if s_n is the number of intersection points for n lines, then $s_n = s_{n-1} + (n - 1)$.

57. When $t = 0$ (corresponding to the year 2000), we find $P = 153{,}000e^0 = 153{,}000$. The graph is shown.

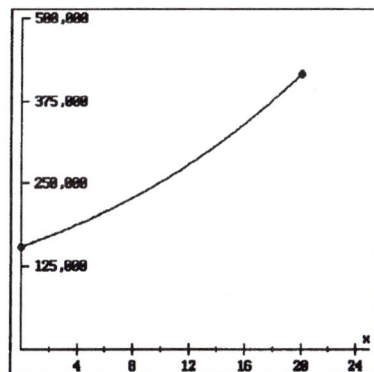

58. Answers vary.

59. The pattern is $1, $2, $4, $8, $16, $\cdots$ or 2^0, 2^1, 2^2, 2^3, $\cdots$ so:

1 day:	$1	This is $2^1 - 1$
2 days:	$1 + $2 = $3	This is $2^2 - 1$
3 days:	$1 + $2 + $4 = $7	This is $2^3 - 1$
4 days:	$1 + $2 + $4 + $8 = $15	This is $2^4 - 1 = 15$

For 30 days, this must be $2^{30} - 1$ or $1,073,741,823$

60. If you perform these experiments a large number of times, you will find that event E is slightly more likely. In particular

$$P(E) = 1 - \left(\tfrac{5}{6}\right)^4 \approx 0.52 \quad \text{and} \quad P(F) = 1 - \left(\tfrac{35}{36}\right)^{24} \approx 0.49$$

CHAPTER 1

THE NATURE OF PROBLEM SOLVING

1.1 Problem Solving, page 2

I use Transparency 1 (transparencies are found at the end of this *Instructor's Manual*) to introduce myself and the course.   You can write your name, office hours, and phone numbers on this transparency. Students taking this course often have a great deal of apprehension about mathematics. I use Transparency 2 to "break the ice."

The material presented in this section is not absolutely necessary for the remainder of the course. It is presented to set the style and the spirit of the course. There are three methods of approaching a given problem: (1) experimentation, (2) intuition, and (3) deduction. Although the latter is important, the first two are often neglected. In this and the next sections we examine some patterns and hopefully develop some intuitive ability. The students should realize that mathematicians do not always think in logically elegant ways. George Polya discusses this method in a film by the American Mathematical Association called *Let Us Teach Guessing*. You might also see Polya's book *Mathematical Discovery* (New York: Wiley, 1962) for a similar discussion.   We will use Polya's problem-solving theme throughout the book. I use Transparency 3 to introduce the ideas of problem-solving.

I want the first problem we solve to be one which does directly involve numbers, operations, or manipulations. To do this, I use Transparency 4 and ask the question, "Suppose I'm standing at the YWCA and want to walk to Macy's in San Francisco." (By the way, you can use a city close to your home.) I ask the student's to write out an answer, and then I have them read what they have written on their papers. We see that there are many possible answers. This breaks the myth that answers to math questions have only one right answer. Eventually, I ask them, "How many different ways can a person walk from YWCA to Macy's?" The answer leads us to Pascal's triangle (Transparency 5).

Throughout this book there are problems labeled IN YOUR OWN WORDS.   These are included to stimulate the students' writing skills and to help ensure understanding of important concepts.   The answers to these problems are, of course, individual and can vary considerably. In addition, there are problems labeled PROBLEM SOLVING. These problems may require additional insight, information, or effort to solve.

Level 1, page 12

1. Answers vary.   **2.** Answers vary.   **3.** Answers vary.   **4.** First, you have to *understand the problem*. Second, *devise a plan*. Find the connection between the data and the unknown. Look for patterns, relate to a previously solved problem or a known formula, or simplify the given information to give you an easier problem. Third, *carry out the plan*. Fourth, *look back*. Examine the solution obtained.   **5.** Answers vary.

6.   1 16 120 560 1,820   4,368   8,008   11,440   12,870   11,440 8,008   4,368   1,820 560 120 16 1

  1 17 136 680 2,380   6,188 12,376   19,448   24,310 24,310 19,448 12,376 6,188   2,380 680 136 17 1

7. first diagonal   **8.** third diagonal   **9.** Answers vary; yes; answers are the same because the triangle is symmetric.   **10.** 10   **11.** 20   **12.** 35   **13.** 56   **14.** 1 up, 2 over; 3 paths   **15.** 2 up, 3 over; 10 paths   **16.** 3 up, 8 over; 84 paths   **17.** 4 up, 6 over; 210 paths   **18.** 2 down, 3 over; 10 paths   **19.** 4 down, 3 over; 35 paths   **20.** 7 down, 4 over; 330 paths   **21.** 2 up, 4 over; 15 paths

Level 2, page 13

22.   15 penguins and 1 bear   **23.** Answers vary. Poor problem solvers will pick A (both deal with penguins and bears), better problem solvers will choose C (same type of question), but excellent problem solvers will choose B (structurally the same type of problem).

24.   WEIGHT OF TEN CRATES + WEIGHT OF WALNUTS = 410 LB

                    100 + WEIGHT OF WALNUTS = 410

                       WEIGHT OF WALNUTS = 310

The walnuts weigh 310 pounds.

25. 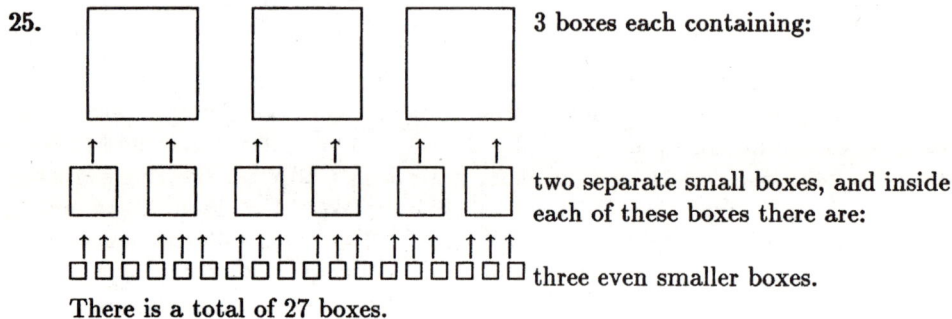 3 boxes each containing: two separate small boxes, and inside each of these boxes there are: three even smaller boxes.

There is a total of 27 boxes.

26. Label the tires A, B, C, D, and E. Record the mileage of each tire (in thousands of miles).

	A	B	C	D	E	
start:	0	0	0	0	0	Tire E is the spare.
12,500 mi:	12.5	12.5	12.5	12.5	**0**	Tire D now becomes the spare.
25,000 mi:	25.0	25.0	25.0	**12.5**	12.5	Tire C now becomes the spare.
37,500 mi:	37.5	37.5	**25.0**	25.0	25.0	Tire B now becomes the spare.
50,000 mi:	50.0	**37.5**	37.5	37.5	37.5	Tire A now becomes the spare.
62,500 mi:	**50.0**	50.0	50.0	50.0	50.0	Now, you need a set of new tires.

Rotate the tires every 12,500 miles for a total of 62,500 miles.

27. a. 2 **b.** 4 **c.** 8 **d.** 16   **28.** 2^n

29. There are intermediate paths, so number the vertices; 37 paths

30. 49 paths (by numbering the vertices)

31. 11 paths (by numbering the vertices; if you took a path down to Leavenworth, I think you would be backtracking when you reach McAllister, so basically your choices are to come down Taylor or Jones Street)

32. 284 paths (by numbering the vertices)

Level 3, page 13

33. Answers vary.   **34.** Answers vary.

35. Answers vary; the number of items in the grocery carts not in the express lane or the speed of the grocery clerks.

36. Answers vary; the number of people waiting in line inside the bank.

37. You need the formula $d = rt$; $r = 20$ mph and $t = 1$ hour (6 mph + 4 mph = 10 mph). Thus $d = 20(1) = 20$, so the fly flies 20 miles.

38. 4 mm (The bookworm needs only to chew through the covers.)

39. Draw a picture; there are two possible answers, 8 miles or 12 miles.

```
┌────── 10 mi (A to B) ──────┐   ┌───┐   ┐
A                          ↑   B   ↑
                    C (8 miles)     C (12 miles)
```

40. *dum cas*   **41.** 1,089; does not work for palindromes

42. 1.62   **43.** 6, 8, 10, 14, and 15

44. 12   **45.** A ton of anything is a ton; they are the same.

46. 7 cards

47. Draw a picture. The time between each bong is 6 sec. For 6 bongs, there are 5 time intervals ($5 \times 6 = 30$ sec) and for 12 bongs, there are 11 time intervals ($11 \times 6 = 66$ sec). It will take 66 seconds.

48. They don't bury the living.

49. The coins are a quarter and a nickel. (*One* is not a nickel, it is a quarter.)

50. triplets   **51.** 54 (6 outs per inning)

52. 11 (every 10 ft. plus one to start)

53. Answers vary. Take goose across; return and pick up the fox. Deliver the fox and take the goose back to the first side. Drop off the goose, pick up the grain and leave it with the fox. Return one last time to fetch the goose.

54. Answers vary; The first cup has one lump; the second cup has four lumps; and the third cup has five lumps. This accounts for the ten lumps, but the second cup does not have an odd number of lumps − not to worry; put the first cup (containing the one lump) inside the second cup. Now, the second cup has five lumps, so the conditions of the problem are satisfied.

55. Pour the contents of glass 5 into glass 2.

56. One mile per hour faster will take 10 hours to catch up. Since the fly flies at 20 miles per hour, the fly will fly $10 \times 20 = 200$ miles.

Level 3 Problem Solving, page 15

57. The answer is the number of routes from the start to the destination minus the number of routes passing through the barricade.

    a. $35 - (3 + 3 + 3) = 26$ **b.** $35 - (6 + 6) = 23$ **c.** $35 - (4 + 4) = 27$

58. Ahmes should tie one end of the rope onto the ankh at the outer edge of the funnel. Then walk around the outer perimeter of the funnel until he returns to the ankh. The rope will have wrapped around the ankh at the center so that if he ties off the rope he will have a rope bridge to the center. There is no reason for the chicken's skull.

59. There are 26 cards in each pile. The hidden card is 7 cards down in the pile. Three cards are removed from the second pile, leaving 23 cards. Look for a pattern:

    If the 3 cards are tens, then $23 + 7 = 30$ cards down to the hidden card.
    If the 3 cards total 29, then $22 + 7 = 29$ cards down.
    If the 3 cards total 28, then $21 + 7 = 28$ cards down.
    $$\vdots$$
    If the 3 cards total x, then $23 - (30 - x) + 7 = x$ cards down.

60. Answers vary. This solution uses some algebra. Let the given number be $10t + u$, where t is the tens digit and u is the units digit.

    $$10t + u \qquad \textit{Given number}$$
    $$70 + 6 \qquad \textit{Add 76}$$

    Since t is 5, 6, 7, 8, or 9, there is a carry of 1 into the hundreds column:

    $$100 + (10t + 70 - 100) + (u + 6)$$

    Add the *digit* in the hundreds place to the remaining two-digit number:

    $$(10t - 30) + (u + 6) + 1 = (10t - 30) + (u + 7)$$

    Subtract this from the original number:

    $$(10t + u) - [(10t - 30) + (u + 7)] = 30 - 7 = 23$$

1.2 Inductive and Deductive Reasoning, page 16

The general format I use for the course is to begin each class period with a "zinger." A zinger is something as contemporary and meaningful for the student as possible, but yet presents an unexpected result. I will give examples of some zingers you can use in this *Instructor's Manual.* For example, today's lesson is devoted to the idea of getting the student to make conjectures by looking for patterns. I put several patterns on transparencies or on the blackboard and ask the students for any observed patterns. I am also careful to accept any patterns they give and also give patterns of my own which are very trivial (to show that everyone can find patterns). For example, Transparency 6 (top) shows the multiplication table by nine. There are many patterns to discuss (most comes from the students), but eventually we are lead to the problem: Evaluate (find the value of)

$$\frac{999,999,999 \times 999,999,999}{1 + 2 + 3 + 4 + 5 + 6 + 7 + 8 + 9 + 8 + 7 + 6 + 5 + 4 + 3 + 2 + 1}$$

Students will find that they can not get an answer to this by calculator, so they are forced to look for patterns.

Level 1, page 21

1. Answers vary; *inductive reasoning* is looking at evidence or patterns to form a conclusion whereas deductive reasoning is based on proving results (called theorems) using laws of logic. With *deductive reasoning*, a set of unproved axioms and a set of undefined terms are accepted and then a theorem is derived using logic, axioms, terms, or other theorems.   **2.** The pattern of last digits are found by multiplication by consecutive integers by 7: 7, 5, 3, 1, 8, 6, 4, 2, 9 , ⋯.   **3.** First, perform any operations enclosed in parentheses. Next, perform multiplication and divisions as they occur by working from left to right. Finally, perform additions and subtractions as they occur by working from left to right.   **4.** Answers vary; the scientific method is a method of inductive reasoning which controls conditions, performs experiments, and then looks for patterns to form hypotheses about future occurrences under the same conditions. Based on these observations, conclusions are formed.   **5.** If $a = b$, then a may be **substituted** for b in any mathematical statement without affecting the truth or falsity of the given mathematical statement.

6. a. 17 **b.** 13   **7. a.** 32 **b.** 12   **8. a.** 45 **b.** 45   **9. a.** 7 **b.** 11   **10. a.** 14 **b.** 6

11. a. 505 **b.** 100   **12. a.** 50 **b.** 2   **13. a.** 29 **b.** 35   **14. a.** 38 **b.** 54   **15. a.** 23

b. 28   **16. a.** 10 **b.** 12   **17. a.** 13 **b.** 6   **18.** Inductive reasoning; answers vary.

19. Deductive reasoning; answers vary.   **20.** <u>3, 6, 9</u>, 3, 6, 9, . . .

21. <u>4, 8, 3, 7, 2, 6, 1, 5, 9</u>, 4, 8, 3, . . .   **22.** <u>5, 1, 6, 2, 7, 3, 8, 4, 9</u>, 5, 1, 6, 2, 7, 3, 8, 4, 9, 5, 1, 6, . . .

23. <u>6, 3, 9</u>, 6, 3, 9, 6, 3, 9, . . .   **24.** $25^2 = 625$ **25.** $50^2 = 2,500$ **26.** $1,000^2 = 1,000,000$

27. a. $3 + 2 \times 4$ **b.** $3(2 + 4)$ **28. a.** $\dfrac{3}{2 + 4}$ **b.** $10(4 + 3)$

29. a. $8 \times 9 + 10$ **b.** $8(9 + 10)$ **30. a.** $3^2 + 5^2$ **b.** $(3 + 5)^2$

31. a. $3^2 + 2^3$ **b.** $3^3 - 2^2$ **32. a.** $n^2 + 5$ **b.** $6n^3$

33. a. $4^2 + 9^2$ **b.** $(4 + 9)^2$

34. a. $n + 1 = 1 + n$ **b.** $2 + n = n + 2$

35. a. $3(n + 4) = 16$ **b.** $5(n + 1) = 5n + 5$

36. a. $6n + 12 = 6(n + 2)$ **b.** $n(7 + n) = 0$

37. a. $n^2 + 8n = 6$ **b.** $n^2 + 8n = n(8 + n)$

Level 2, page 22

38. $A = bh$ **39.** $A = \frac{1}{2}bh$ **40.** $A = \frac{1}{2}pq$ **41.** $A = \frac{1}{2}h(a + b)$ **42.** $V = s^3$ **43.** $V = \ell wh$

44. $V = \frac{1}{3}\pi r^2 h$ **45.** $V = \pi r^2 h$ **46.** $V = \frac{4}{3}\pi r^3$

Level 3, page 23

47. $9 \times 54{,}321 - 1$; $488{,}888$ **48.** $8{,}888{,}888{,}888$ **49.** $98{,}888{,}888{,}888$

50. $123{,}456{,}789 \times 36 = 4{,}444{,}444{,}404$ **51.** $1{,}111{,}111{,}101{,}000$ **52.** $9{,}999{,}999{,}909{,}000$

Level 3, Problem Solving, page 23

53. The figure has: 36 1-by-1 squares; 25 2-by-2 squares; 16 3-by-3 squares; 9 4-by-4 squares; 4 5-by-5 squares; and one 6-by-6 square for a total of 91 squares.   **54.** The figure has: 16 single triangles; 7 "four" triangles, 3 "nine" triangles, and 1 big triangle for a total of 27 triangles.   **55.** There are $4 \times 4 \times 4 = 64$ cubes; there are 6 faces and each face has 4 cubes painted on one side, so there are $6 \times 4 = 24$ cubes with one painted face. The fraction is $\frac{24}{64} = \frac{3}{8}$.

56. a.   2 $8\ (2 \times 2 \times 2)$ 8 $0\ (12 \times 0)$ $0\ (6 \times 0 \times 0)$ $0\ (0 \times 0 \times 0)$

    b.   3 $27\ (3 \times 3 \times 3)$ 8 $12\ (12 \times 1)$ $6\ (6 \times 1 \times 1)$ $1\ (1 \times 1 \times 1)$

    c.   4 $64\ (4 \times 4 \times 4)$ 8 $24\ (12 \times 2)$ $24\ (6 \times 2 \times 2)$ $8\ (2 \times 2 \times 2)$

    d.   5 $125\ (5 \times 5 \times 5)$ 8 $36\ (12 \times 3)$ $54\ (6 \times 3 \times 3)$ $27\ (3 \times 3 \times 3)$

    e.   6 $216\ (6 \times 6 \times 6)$ 8 $48\ (12 \times 4)$ $96\ (6 \times 4 \times 4)$ $64\ (4 \times 4 \times 4)$

    f.   7 $343\ (7 \times 7 \times 7)$ 8 $60\ (12 \times 5)$ $150\ (6 \times 5 \times 5)$ $125\ (5 \times 5 \times 5)$

g.   8   512 (8 × 8 × 8)   8   72 (12 × 6)   216 (6 × 6 × 6)   216 (6 × 6 × 6)

h.   9   729 (9 × 9 × 9)   8   84 (12 × 7)   294 (6 × 7 × 7)   343 (7 × 7 × 7)

57. Balance 3 with 3; if it balances, the heavier one is in the three not weighed; if it does not balance, then the pan balance will show which set of 3 is the heavier one. In any case, the first weighing narrows it down to a set of 3 coins, one of which contains the heavy coin. Take 2 from this set of heavy 3 and balance them. If they balance, the heavy coin is the one not weighed; if they do not balance, then the pan balance will show the heavier one. In any case, it takes two weighings.

58.
$$\begin{bmatrix} 2 & 7 & 6 \\ 9 & 5 & 1 \\ 4 & 3 & 8 \end{bmatrix}$$

59. Common sum is 4,440,084,513.

60. a. One diagonal sum is 38,307 and the other sums are 21,609.

   b. One diagonal sum is 13,546,875 and the other sums are 20,966,014.

If you find a 3 × 3 magic square with nine distinct square numbers, send me your solution and if you are correct I'll send you $100.

1.3 Scientific Notation and Estimation, page 24

I often begin this section by posing the problem, "What would happen if the *entire* population of the world moved to Florida?" (Transparency 7). It turns out that the amount of space allotted to each person is about 300 ft^2:

$$\frac{58{,}664 \times 5{,}280^2}{5.5 \times 10^9} \approx 297.3560832$$

The correct answer is D. A discussion of how we could go about computing something such as this (world population, 5,500,000,000 and the area of New Jersey, 7,836 square miles) leads us to the topic at hand.

Level 1, page 36

1. Answers vary; for any number b and any counting number n,

$$b^n = \underbrace{b \cdot b \cdot b \cdot \,\cdots\, \cdot b}_{n \text{ factors}}, \qquad b^0 = 1, \qquad b^{-n} = \frac{1}{b^n}$$

The number b is called the base, the number n in b^n is called the exponent, and the number b^n is called a power or exponential. **2.** The scientific notation for a number is that number written as a power of 10 times another number x, such that x is between 1 and 10, including 1; that is, $1 \le x < 10$. **3.** Answers vary. **4.** Answers vary; for exponents use $\boxed{y^x}$ or $\boxed{\hat{\ }}$, whereas for scientific notation use the $\boxed{\text{EXP}}$ or $\boxed{\text{EE}}$ keys. **5.** Assume a balloon is 1 ft^3 and that a typical classroom size of 30 ft × 50 ft × 10 ft = 15,000 ft^3. Since $10^6/(1.5 \times 10^4) \approx 67$ we see that it would take about 67 such classrooms to contain a million inflated balloons. **6.** Answers vary. **7.** A trillion is 10^{12}; this is a million times larger than one million, so (from Problem 5) it would take 67,000,000 classrooms to hold a trillion balloons.

8. **a.** 3.2×10^3; 3.2 03   **b.** 4×10^{-4}; 4 -04 **c.** 6.4×10^{10}; 6.4 10

9. **a.** 2.379×10^1; 2.379 01   **b.** 1×10^{-6}; 1 -06 **c.** 3.5×10^{10}; 3.5 10

10. **a.** 5.629×10^3; 5.629 03   **b.** 6.3×10^5; 6.3 05   **c.** 3.4×10^{-8}; 3.4 -08

11. **a.** 10^{100}; 1 100 (This number is too large for most calculators.)

    b. 1.2003×10^6; 1.2003 06   **c.** 1.23×10^{-6}; 1.23 -06

12. **a.** 49   **b.** 72,000,000,000   **c.** 4,560

13. **a.** 64   **b.** 0.0021   **c.** 40,700

14. **a.** 216   **b.** 0.00000 041   **c.** 0.00000 048

15. **a.** $0.02\overline{7}$ **b.** 32,170,000   **c.** 0.00000 00000 889

16. 3.0×10^{10} 17. 8.6×10^1; 6.1×10^3

18. 2.2×10^8; 2.5×10^{-6} 19. 6×10^{12}; 1.86×10^5

20. 3,600,000   21. 906

22. 0.00000 003   23. 38,000,000,000,000,000,000,000,000

Estimates in Problems 24-31 may vary.

24. Estimate $20 \times 400 = 8,000$ hr; $24 \times 365 = 8,760$

25. Estimate $2,000 \times 500 = 1,000,000$; $1,850 \times 487 = 900,950$

26. Estimate $1,000 - 250 = 750$; $\$1,025.66 - \$255.83 = \$769.83$

27. Estimate $1,500 \times 12 = 1,500 \times 10 + 1,500 \times 2 = 15,000 + 3,000 = 18,000$;

    $\$1,543 \times 12 = \$18,516$

28. Estimate $6 \times 8 \times 5 \times 50 = 30 \times 400 = 12,000$; $\$6.25 \times 8 \times 5 \times 52 = \$13,000$

29. Estimate $1,000,000$ people $\div 10,000$ people/yr $= 100$ yr; $1,000,000 \div 13,688 \approx 73$, so it will take her about 73 years.

30. Estimate $20 \times 15 = 300$; $23 \times 15 = 345$

31. Estimate $300 \div 10 = 30$; $280 \div 10.2 = 27.451$

Level 2, page 37

32. **a.** 1.2×10^9 **b.** 3×10^2 33. **a.** 10^4 **b.** 1.4×10^{-1} 34. **a.** 1.2×10^{-3} **b.** 2×10^7
35. **a.** x^2 **b.** $8xy^4$ 36. **a.** 4×10^{-8} **b.** 1.2   37. **a.** 3×10^9 **b.** 4×10^{-11} 38. Answers vary; 35 miles   39. Answers vary; 30 miles   40. Estimate 10/cm², so about 120 oranges
41. Estimate 15 persons/cm², so about 180 people   42. About 150 chairs   43. Estimate 8/cm², so

about 300 penguins   **44.** $72 \times 10^{12} \times 10^9 = 7.2 \times 10^{22}$ **45.** 5.157×10^{15}

46. $186,075,000 \approx 1.86 \times 10^8$; 1 box $\approx 10^2$ oranges, so the total is approximately 1.86×10^{10} oranges.

47. $1,000,000/233 \approx 4,292$ in. or 358 ft or 119 yd.   Thus, a million dollars would form a stack about 119 yd high.   **48.** There are approximately 17 pennies per inch. $1,000,000/17 \approx 58,823$ in. or 4,902 ft or 0.93 mi. Thus, a million pennies would be about a mile high.

49.   (grains per pound) $\cdot$ (pounds per ton) $\cdot$ (tons of sugar)

$$= (2.26 \times 10^6) \times (2.00 \times 10^3) \times (3 \times 10^7)$$
$$= (2.26)(2)(3)(10^{16})$$
$$= 1.356 \times 10^{17} \quad \text{(This is about 135,600,000,000,000,000 grains.)}$$

Level 3, page 38

50. a. Answers vary   **b.** 9^{9^9} **c.** This number is too large for your calculator; $9^{9^9} = 9^{387,420,489}$; the largest power of 9 most calculators will handle is $9^{104} \approx 1.7 \times 10^{99}$.

We can make an estimate by noting that a larger base 10 will have a smaller exponent so that
$$9^{387,420,489} \approx 10^{387,000,000}$$

In Chapter 8, we will learn how to find a better approximation:
$$9^{9^9} \approx 10^{369,692,902}$$

51. 8^{16}; if you use your calculator directly you obtain an answer in scientific notation, namely $2.814749767 \times 10^{14}$. To obtain an exact answer you might consider
$$8^{11} \times 8^5 = 8,589,934,592 \times 32,768 = 281,474,976,710,656$$

52. 23,451,789 yd $\times$ 5,642,732 yd $\times$ 54,465 yd $= 7.207471108 \times 10^{18}$ yd^3; we can also calculate that 1 yd^3 = 36 in. $\times$ 36 in. $\times$ 36 in. = 46,656 in.^{3}. Thus, $7.207471108 \times 10^{18} \times 46,656 = 3.36271772 \times 10^{23}$ (By the way, the answer Jedidiah gave was 336,271,772,009,375,336,113,920.)

53. Estimate $238,000 \approx 240,000$; $240,000 \div 4 = 60,000$ minutes; $60,000 \div 60 = 1,000$ hours; 1,000 hours $\div 24 \approx 1,200 \div 24 \approx 100 \div 2 = 50$ days.   Calculate: $238,000 \div 4 \div 60 \div 24 = 41.319444$ days. This is 41 days and $0.319444 \times 24 = 7.6666666667$ hours; this is 41 days, 7 hours, and 40 minutes.

54.   Answers vary; 2^{50} sheets high is $2^{50} \times 0.003$ in. $\approx 1.125899907 \times 10^{15} \times 3 \times 10^{-3}$

$= 3.377699721 \times 10^{12}$ in. Changing units, we obtain

$\qquad 2.814749767 \times 10^{11}$ ft   *Divide by 12*

$= \; 53,309,654.68$ mi   *Divide by 5,280*

This is more than halfway to the sun, which of course makes the task impossible.

Level 3 Problem Solving, page 38

55.

The units column will contain	1,000 digits
The tens column will contain	1,000 − 9 digits
The hundreds column will contain	1,000 − 99 digits
The thousands column will contain	1,000 − 999 digits
Sum	4,000 − 1,107 = 2,893 digits

Thus, the time required is 2,893 seconds or 2,893 ÷ 60 minutes = 48 minutes, 13 seconds.

56. Look for a pattern; consider each column separately:

The units column will contain	1,000,000 digits
The tens column will contain	1,000,000 − 9 digits
Hundreds column	1,000,000 − 99
Thousands column	1,000,000 − 999
Ten-thousands column	1,000,000 − 9,999
Hundred-thousands column	1,000,000 − 99,999
Millions column	1,000,000 − 999,999
Sum	7,000,000 − 1,111,104

Thus, the total number of digits is 5,888,896. The time required is about 68 days.

57. Look for a pattern; consider each column separately.

The units column will contain	100 zeros
The tens column will contain	100 − 9 zeros
Hundreds column	100 − 99
Sum	300 − 108 = 192 zeros

58. Look for a pattern; consider each column separately.

The units column will contain	100,000 zeros
The tens column will contain	100,000 − 9 zeros
Hundreds column	100,000 − 99
Thousands column	100,000 − 999
Ten-thousands column	100,000 − 9,999
Hundred-thousands column	100,000 − 99,999
Sum	600,000 − 111,105

Thus, the total number of zeros is 488,895.

59. Answers vary;

California $\approx 1.586 \times 10^5$ sq mi

$\approx 4.422 \times 10^{12}$ sq ft   (1 mi = 5,280 *ft, so multiply by* 5,2890^2)

Divide by world population 6.3×10^9 and obtain about 7.018×10^2 or about

700 sq ft per person; choose D.

60. $\dfrac{6.3 \times 10^9}{7.5} = 8.4 \times 10^8$ cubic feet of blood; to answer the question we need to know the size of an acre. A dictionary (or other source) provides that 1 acre is 43,560 ft^2 so 840 acres is

$840 \times 43{,}560$ ft$^2 = 3.65904 \times 10^7$ ft^2. The volume of the desired box of height h and base area of

3.65904×10^7 ft^2 is $V = 3.65904 \times 10^7 h$ ft^3 so

$$3.65904 \times 10^7 h = 8.4 \times 10^8 \text{ or } h = \frac{8.4 \times 10^8}{3.65904 \times 10^7} \approx 22.9568$$

The wall should be about 23 ft high.

1.4 Finite and Infinite, page 39

I ask the class "What is the largest number whose name you know?" and that discussion evolves from the finite to the infinite. In class I attempt to lead the class to a definition of what we mean by "infinite." In the process, the discussion allows us to define some of the sets of numbers that we will use in this course:

$\mathbf{N} = \{1,\ 2,\ 3,\ 4,\ \cdots\}$	Set of **natural**, or **counting, numbers**
$\mathbf{W} = \{0,\ 1,\ 2,\ 3,\ 4,\ \cdots\}$	Set of **whole numbers**
$\mathbf{Z} = \{\cdots,\ -2,\ -1,\ 0,\ 1,\ 2,\ \cdots\}$	Set of **integers**
$\mathbf{Q} = \{\ \frac{a}{b}\ \mid\ a \in \mathbf{Z},\ b \in \mathbf{Z},\ b \neq 0\}$	Set of **rational numbers**

The last definition for the rational numbers gives us the opportunity to use and mention set-builder notation.

Level 1, page 46

1. In any system of thought, there must be some starting point, or assumptions about the terminology. This means that *some* words must be assumed without definition, and set is one of those words which the mathematical world has decided to accept without proof. **2.** Sets are equal if they are identical and equivalent if they have the same cardinality. **3.** The universal set contains all the elements under consideration in a given discussion. **4.** The empty set is a set containing no elements; its cardinality is zero. **5.** Answers vary; set of people over 12 ft tall; set of living saber-tooth tigers; set of children who are U. S. Senators. **6.** Answers vary; the set of integers; the set of different routes that I can take to get from home to work; the set of possible temperatures between 0°C and 100°C. **7.** Well defined because it is possible to have a list of all such students. **8.** Not well defined since there might be disagreement about what we mean by "a grain of sand." Change it to be "The set of solid particles that are less than 1 mm in diameter." **9.** Well defined because the set is empty. **10.** Not well defined since there is no agreed definition for a "happy" person. Change it to be "The set of people who have smiled in the last 60 minutes."   **11.** Not well defined; change to, "The set of people with two ears."   **12.** Well defined because the set is empty.   **13.** Not well defined because a "good bet" is not a term that has universal meaning. Change to, "Bets over \$100 on the next race at Hialeah."   **14.** Not well defined because "bumper crop" does not have the same meaning to everyone. Change to "Years that the production of corn in Iowa exceeds the average production for the preceding ten years."   **15.** $\{m, a, t, h, e, i, c, s\}$ **16.** Answers may vary; {George W. Bush}   **17.** $\{1, 3, 5, 7, 9\}$ **18.** $\{3, 6, 9, \cdots \}$ **19.** $\{p, i, e\}$ **20.** $\{151, 152, 153, \cdots \}$ **21.** $\{6, 8, 10, 12, 14\}$ **22.** $\{1, 11, 111, 1111, \cdots \}$

23. The set of counting numbers less than 10.   **24.** The set of (nonnegative) powers of 11.

25. The multiples of 10 between 0 and 101.   **26.** The set of numbers 5×10^n for n a counting number.   **27.** The set of odd numbers between 100 and 170.   **28.** The set of letters in the word *Mississippi*.   **29.** The set of all x such that x is an odd counting number; $\{1, 3, 5, 7, \cdots \}$

30. The set of all x such that x is a natural number between 1 and 10; $\{2, 3, 4, 5, 6, 7, 8, 9\}$

31. The set of all x such that x is a natural number (except 8); $\{1, 2, 3, 4, 5, 6, 7, 9, 10, \cdots \}$

32. The set of all x such that x is whole number less than or equal to 8; $\{0, 1, 2, 3, 4, 5, 6, 7, 8\}$

33. The set of all x such that x is a whole number less than 8; $\{0, 1, 2, 3, 4, 5, 6, 7\}$

34. The set of all x such that x is a whole number that is not even; $\{0, 1, 3, 5, 7, \cdots \}$

35. $A \times B = \{(c, w), (c, x), (d, w), (d, x), (f, w), (f, x)\}$

36. $B \times A = \{(w, c), (x, c), (w, d), (x, d), (w, f), (x, f)\}$

37. $\{(1, a), (1, b), (1, c), (2, a), (2, b), (2, c), (3, a), (3, b), (3, c), (4, a), (4, b), (4, c), (5, a), (5, b), (5, c)\}$

38. $\{(a, 1), (b, 1), (c, 1), (a, 2), (b, 2), (c, 2), (a, 3), (b, 3), (c, 3), (a, 4), (b, 4), (c, 4), (a, 5), (b, 5), (c, 5)\}$

39. $|A \times B| = 26 \times 10 = 260$ **40.** $|A \times A| = 26 \times 26 = 676$ **41.** $|C \times D| = 100 \times 50 = 5{,}000$

42. $|M \times N| = 8 \times 8 = 64$

Level 2, page 46

43. a. $|A| = 3; |B| = 1; |C| = 3; |D| = 1; |E| = 1; |F| = 1$ **b.** $A \leftrightarrow C$; $B \leftrightarrow D \leftrightarrow E \leftrightarrow F$ **c.** $A = C; D = E = F$ **44. a.** $|A| = 1; |B| = 1; |C| = 2; |D| = 1; |E| = 2$ **b.** $A \leftrightarrow B \leftrightarrow D; C \leftrightarrow E$ **c.** $A = B$

Level 3, page 47

45. $\{m, a, t\}$; $\{m, a, t\}$; there are others.

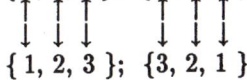

$$\{1, 2, 3\}; \quad \{3, 2, 1\}$$

46. $\{1, 2, 3, \cdots, n, \quad n+1, \cdots, 999, 1000\}$

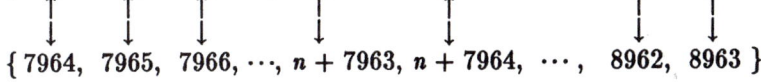

$$\{7964, 7965, 7966, \cdots, n+7963, n+7964, \cdots, 8962, 8963\}$$

47. $\{1, 2, 3, \cdots, n, \quad n+1, \cdots, 353, 354, 355, 356, \cdots 586, 587\}$

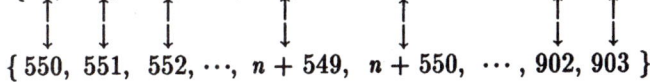

$$\{550, 551, 552, \cdots, n+549, n+550, \cdots, 902, 903\}$$

48. Notice $n \leftrightarrow 437$, so $784 \leftrightarrow 784 + 437 = 1,221$; since 784 is the last element listed in the first set, and since the second set has many more elements, we see that these sets do not have the same cardinality.

49. a. finite **b.** infinite **c.** finite **d.** finite　　**50.** $\left\{\frac{1}{2}, \frac{1}{3}, \frac{2}{3}, \frac{1}{4}, \frac{3}{4}, \frac{1}{5}, \frac{2}{5}, \frac{3}{5}, \frac{4}{5}, \frac{1}{6}, \frac{5}{6}, \frac{1}{7}, \cdots\right\}$

51. $\{1, 2, 3, \cdots, n, \cdots\}$

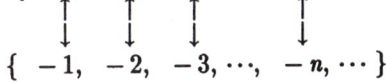

$$\{-1, -2, -3, \cdots, -n, \cdots\}$$

Since these sets can be put into a 1-1 correspondence, they have the same cardinality; namely, $\aleph_0$.

52. $\{1, 2, 3, \cdots, n, \cdots\}$

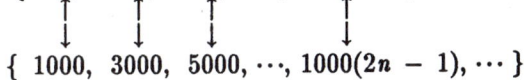

$$\{1000, 3000, 5000, \cdots, 1000(2n-1), \cdots\}$$

Since these sets can be put into a 1-1 correspondence, they have the same cardinality; namely, $\aleph_0$.

53. $\{1, 2, 3, \cdots, n-1, n, \cdots\}$

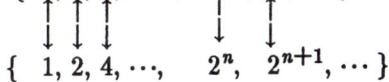

$$\{1, 2, 4, \cdots, 2^n, 2^{n+1}, \cdots\}$$

Since these sets can be put into a 1-1 correspondence, they have the same cardinality; namely, $\aleph_0$.

54. $\{1, 2, 3, \cdots, n, \cdots\}$

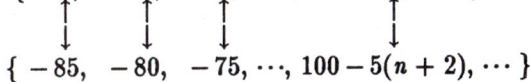

$$\{-85, -80, -75, \cdots, 100 - 5(n+2), \cdots\}$$

Since these sets can be put into a 1-1 correspondence, the have the same cardinality; namely, $\aleph_0$.

55. $W = \{0, 1, 2, 3, \cdots, n, \cdots\}$

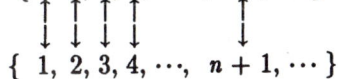

$$\{1, 2, 3, 4, \cdots, n+1, \cdots\}$$

56. $N = \{\ 1,\ 2,\ 3,\ 4,\ \cdots,\quad n\,,\ \cdots\ \}$

$\{\ 2,\ 3,\ 4,\ 5,\ \cdots,\quad n+1,\ \cdots\ \}$

57. $\{\ 12,\ 14,\ 16,\ \cdots,\quad n\,,\ \cdots\ \}$

$\{\ 14,\ 16,\ 18,\ \cdots,\quad n+2,\ \cdots\ \}$

58. $\{\ 4,\ 44,\ 444,\ \cdots,\quad$ integer with n 4s $,\ \cdots\ \}$

$\{\ 44,\ 444,\ 4444,\ \cdots,\quad$ integer with $n+1$ 4s, $\cdots\ \}$

Level 3 Problem Solving, page 47

59. Answers vary; the set, E, of even integers has cardinality $\aleph_0$; the set, O, of odd integers has cardinality $\aleph_0$. If we add the cardinality of the even integers to the cardinality of the odd integers, we have $\aleph_0 + \aleph_0$. However, if we put the even integers together with the odd integers, we have the set of counting numbers, which has cardinality $\aleph_0$. Thus, our example illustrates why $\aleph_0 + \aleph_0 = \aleph_0$.

60. Draw lines through the left and right endpoints, respectively. Label the point of intersection of these new lines, P, as shown:

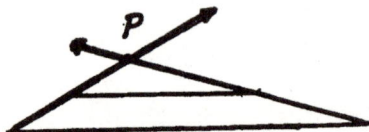

Now, to show the two segments have the same number of points (that is, the same cardinality), choose some point x_1 on the shorter segment. Draw the line through P and x_1; this line will correspond to exactly one point on the longer segment. Similarly, choose some point y_1 on the longer segment. Draw the line through P and y_1; this line will correspond to exactly one point on the shorter segment. This means, that for every point on the shorter segment, there is exactly one corresponding point on the longer segment, and for every point on the longer segment, there is exactly one corresponding point on the shorter segment.

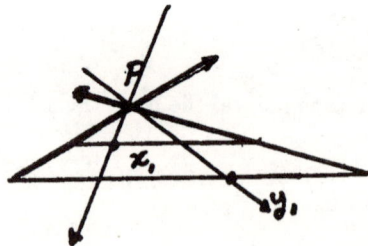

Chapter 1 Review Questions, Page 48

1. Understand the problem, devise a plan, carry out the plan, and then look back.

2. Use Pascal's triangle; look at 5 blocks down and 4 blocks over to find 126.

3. Turn the chessboard so that it forms a triangle with the rook at the top. To get the appropriate square, look at 7 blocks down and 4 blocks over using Pascal's triangle to find 330 paths.

4. By patterns:
$$1 \times 1 = 1$$
$$11 \times 11 = 121$$
$$111 \times 111 = 12{,}321$$
$$\vdots$$
$$111{,}111{,}111^2 = 12{,}345{,}678{,}987{,}654{,}321$$

5. Order of operations: (1) First, perform any operations enclosed in parentheses. (2) Next, perform multiplications and divisions as they occur by working from left to right. (3) Finally, perform additions and subtractions as they occur by working from left to right.

6. The scientific notation for a number is that number written as a power of 10 times another number x such that $1 \le x < 10$.

7. Inductive reasoning; the answer was found by looking at the pattern of questions.

8. Answers vary.

    a. $\boxed{2}$ $\boxed{y^x}$ $\boxed{63}$ $\boxed{=}$ or $\boxed{2}$ $\boxed{\wedge}$ $\boxed{63}$ $\boxed{\text{ENTER}}$ 9.223372037E18

    b. $\boxed{9.22}$ $\boxed{\text{EE}}$ $\boxed{18}$ $\boxed{\div}$ $\boxed{6.34}$ $\boxed{\text{EE}}$ $\boxed{6}$ $\boxed{=}$ 1.454258675E12

9. A. \$50/person times 263 million is about \$13 billion; not even close.

    B. $\$1 \times 60$ sec $\times 60$ min $\times 24$ hr $\times 365.25$ days $\times 1{,}000$ years is about \$3.1 billion; not even close.

    C. 1 million people $\times \$80{,}000 + 1$ billion people $\times 200 \approx 280$ billion; choice C comes closest.

10. $C = \frac{\ell w}{15}$ where C is the capacity of the boat, ℓ and w are the length and width of the boat.

11. $\frac{3}{4}$ hr $\times 365 = 273.75$ hr. This is approximately $2\frac{3}{4}$ hr/book. Could not possibly be a complete transcription of each book.

12. An ice cube is about 1 in.^{3} and a cubic foot has 12^3 in.^{3}. A classroom 30 ft $\times$ 50 ft $\times$ 10 ft would hold about 2.592×10^7 ice cubes. $\dfrac{7 \times 10^{16}}{2.592 \times 10^7} \approx 2.7 \times 10^9$ classrooms; this is not meaningful, but is shown here because it is typical of first attempts. It is important to look for a meaningful comparison. Let's try again. Lake Mead is about 35,154,000 m^3 (**www.infoplease.com**). To covert this to "ice cubes," we look up a cubic meter (in a dictionary) to find it is 35.314667 ft^3. Thus, we have: $(3.5154 \times 10^7)(35.314667)(12^3) \approx 2.145 \times 10^{12}$ in.^{3}. Finally, we divide the

reported size of the iceberg by the capacity of Lake Mead to find:

$(7 \times 10^{16}) \div (2.145 \times 10^{12}) \approx 32{,}630$. It would take more than 32,600 dams the size of the Hoover Dam (which forms Lake Mead) to hold the capacity of the iceberg. This is, no doubt, larger than the capacity of all the lakes behind all the world's dams, even the world's largest one to be completed in 2009 on the Yangzi River in China.

13. 6.3×10^{12}; $(6.3 \times 10^{12}) \div (2.9 \times 10^{8}) \approx 2.1724 \times 10^{4}$; each person's share is about \$21,724.

14. Arrange the cards as $\begin{bmatrix} 2 & 7 & 6 \\ 9 & 5 & 1 \\ 4 & 3 & 8 \end{bmatrix}$.

15. Measure a dollar bill (2.5 in. by 6 in.). The floor is 240 in. by 360 in. This is 96 bills by 60 bills, or 5,760 bills on the floor. The ceiling is 10 ft = 120 in. with 233/in., so the total number of bills in the room is

$$5{,}760 \times 120 \times 233 = 161{,}049{,}600$$

The national debt is \$6,300,000,000,000 so

$$\frac{6.3 \times 10^{12}}{161{,}049{,}600} \approx 39{,}118 \text{ classrooms}$$

This number of classrooms is still hard to comprehend, so we will fill them with \$100 bills instead. Now we can estimate it will take 392 classrooms. How many classrooms are in your city? Imagine all of them *filled* with \$100 bills.

16.   **a.** The set of rational numbers is the set of all numbers of the form $\frac{a}{b}$ such that a is an integer and b is a counting number.   **b.** $\frac{2}{3}$; answers vary.

17. $\{5, 10, 15, \cdots, 5n, \cdots\}$
$$\{10, 20, 30, \cdots, 10n, \cdots\}$$

Since the second set is a proper subset of the first set, we see that the set F is infinite.

18.   **a.** $\{ n \mid (-n) \in \mathsf{N}\}$; answers vary.

  b. $\{ 1, 2, 3, \cdots, n, \cdots \}$
$$\{-1, -2, -3, \cdots, -n, \cdots\}$$

Since the first set is the set of counting numbers, it has cardinality $\aleph_0$; so the given set also has cardinality $\aleph_0$ since it can be put into a one-to-one correspondence with the set of counting numbers.

19. Answers vary; 20 is unhappy because $2^2 + 0^2 = 4$, which is unhappy. 100 is happy because $1^2 + 0^2 + 0^2 = 1$, which is happy.

20. By patterns: $7^1 = 7$; $7^2 = 49$ ends in 9; $7^3 = 343$ ends in 3; $7^4 = 2{,}401$ ends in 1; 7^5 ends in 7; 7^6 ends in 9; 7^7 ends in 3; 7^8 ends in a 1, $\cdots$. Looking ahead, we see that 7^{1000} must end in 1.

Group Research Problems, page 50

Generally, complete answers to these problems will not be given in this *Instructor's Manual*. But notice that these problems introduce the student to some of the more important literature in mathematics. Some good general references, in addition to those listed as references in this section, are:

> *Mathematics*, Life Science Library. This is an excellent book to introduce mathematics to the layperson.
> *The World of Mathematics* by Newman. This classic is full of ideas.
> *Mathematics and the Imagination* by Kasner and Newman.

These three books should be on the shelf of every instructor teaching this course. They will provide endless ideas for discussion. These problems will also introduce the students to some of the mathematical journals: *The Arithmetic Teacher*, *The Mathematics Teacher*, and *The Journal of Recreational Mathematics*. There are a multitude of links for research problems at the website for this book, **mathnature.com**.

G1. Answers vary.

G2. a. Successive powers of 11 represent the rows of Pascal's triangle. That is,
$$11^0 = 1, \ 11^1 = 11, \ 11^2 = 121, \ 11^3 = 1331, \ 11^4 = 14641, \ 11^5 = 1 \ 5 \ 10 \ 10 \ 5 \ 1 \text{ where two}$$
digits numbers require a "carry" for the answer: 161051.

b. The natural numbers are found in diagonal 1.

c. The triangular numbers are found in diagonal 2.

d. The tetrahedral numbers are those numbers of billiard balls necessary to build a tetrahedron, and these numbers are found in diagonal 3.

e. The pattern labeled "multiples of 2" is found by making a dot for those numbers in Pascal's triangle that are even and an x otherwise; The pattern labeled "multiples of 3" is found by making a dot for those numbers in Pascal's triangle that are multiples of three, and an x otherwise. As an extension, you might provide other "multiple" patterns.

Individual Research Problems, pages 51-52

P1.1. They were all math majors in college.

P1.2. Answers vary.

P1.3.   Answers vary.

P1.4.   A magic square is an arrangement of numbers in the shape of a square with the sums of each vertical column, each horizontal row, and each diagonal all equal. The magic square by Dürer in 1514 (notice that the date appears in the magic square) is called *Melancholia*.

$$\begin{bmatrix} 16 & 3 & 2 & 13 \\ 5 & 10 & 11 & 8 \\ 9 & 6 & 7 & 12 \\ 4 & 15 & 15 & 1 \end{bmatrix}$$

All the rows, columns, and diagonals add up to 34, so it is a magic square. Here are some other properties: (1) The four corners add up to 34. (2) The four center squares (10, 11, 6, 7) add up to 34. (3) Opposite pairs of squares (3, 2, 15, 14) add up to 34. (4) Slanting squares (2, 8, 15, 9) add up to 34. (5) The sum of the first eight numbers equal the sum of the second eight numbers. (6) The sum of the squares of the first eight numbers equals the sum of the squares of the second eight numbers.

P1.5.   Answers vary.

P1.6.

$$\begin{bmatrix} \text{eight} & \text{nineteen} & \text{eighteen} \\ \text{twenty-five} & \text{fifteen} & \text{five} \\ \text{twelve} & \text{eleven} & \text{twenty-two} \end{bmatrix}$$

The alpha magic squares in this problem, as well as a method of constructing them is given in the "Mathematical Recreations" department of *Scientific American*, January 1997, by Ian Stewart.

P1.7.   Answers vary.

P1.8.   Answers vary; Fermat's last theorem and the four-color problem are discussed in this text.

CHAPTER 2

THE NATURE OF LOGIC

2.1 Deductive Reasoning, page 54

I generally begin this section with a discussion and examples of deductive vs. inductive reasoning. The discussion sometimes includes the scientific method, scientific predictions, psychic predictions, Star Trek, proving results, and winning everyday arguments. I allow the class to lead the discussion to various popular usages of the word "proof." In this section the groundwork for the rest of the chapter is presented. We introduce the important operators *and, or,* and *not.* It should be made clear that they are defined in terms of the given tables, not according to everyday usage (even though the tables and everyday usage usually agree). In this section the student also determines the truth values of the variables. In the next section we will find the truth value of a given statement for all values of the variable.

Level 1, page 60

1. A *logical operator* is a way of connecting simple statements. The fundamental operators are *and, or,* and *not.* Other operators include *because, either ... or, neither... nor,* and *unless.*

2. *Conjunction* is the operator defined by Table 2.1. It is defined so it is true only when both of the simple propositions are true, and it is false otherwise. **3.** *Disjunction* is the operator defined by Table 2.2. It is defined so it is false only when both of the simple propositions are false, and it is true otherwise. **4.** *Negation* is the operator defined by Table 2.3. It is defined so it has the opposite truth value from the given proposition. **5.** Begin by listing all the possibilities of truth and falsity for the propositions; 2 possibilities for 1 proposition, 2^2 for 2 propositions, 2^3 for 3 propositions, $\cdots$, 2^n for n propositions. Then use Tables 2.1. 2.2, and 2.3 to fill in the truth values of pairs of propositions until the entire list of propositions is filled in. **6.** a, b, and d are statements **7.** a, b, and d are statements **8.** a, b, and d are statements **9.** a and b are statements **10.** Some mathematicians are not ogres. **11.** Some dogs do not have fleas. **12.** No integer is negative. **13.** All people pay taxes. **14.** Some even integers are divisible by 5. **15.** Some triangles are squares. **16.** Some squares are not rectangles. **17.** Some counting numbers are not divisible by 1. **18.** No apples are rotten. **19.** All integers are odd. **20. a.** $p \lor \sim q$ **b.** $p \land \sim q$ **c.** $p \land q$ **d.** $\sim p \land q$ **21. a.** T **b.** F **c.** T **d.** F **22. a.** T **b.** T **c.** F **d.** F **23. a.** Prices or taxes will rise.   **b.** Prices will not rise and taxes will rise.   **c.** Prices will rise or taxes will not rise. **d.** Prices will not rise or taxes will not rise.   **24. a.** $p \wedge q$ **b.** $q \vee p$ **c.** $q \wedge \sim p$ **d.** $\sim q \wedge \sim p$ **25.** Given q is T and p is F. **a.** F **b.** T **c.** T **d.** F   **26.** Given q is F and p is F. **a.** F **b.** F **c.** F **d.** T   **27.** *Answers may vary.* **a.** Paul is peculiar and likes to read math texts. **b.** Paul is not peculiar and likes to read math texts. **c.** Paul is peculiar or does not like to read math texts. **d.** Paul is not peculiar or does not like to read math texts.   **28.** *Answers may vary.* **a.** Today is Friday and there is homework tonight. **b.** Today is not Friday and there is homework tonight. **c.** Today is Friday or there is no homework tonight. **d.** Today is not Friday or there is no homework tonight.   **29. a.** T **b.** F **c.** T **d.** F   **30. a.** F **b.** T **c.** F **d.** T **31. a.** T **b.** F   **32. a.** T **b.** T   **33. a.** F **b.** T   **34. a.** T **b.** T   **35. a.** T **b.** T **36. a.** F **b.** F   **37. a.** T **b.** F   **38. a.** F **b.** T

Level 2, page 61

39. $e \wedge d \wedge t$ where e: W. C. Fields is eating; d: W. C. Fields is drinking;
                                  t: W. C. Fields is having a good time.

40. $\sim s \wedge \sim a$ where s: Sam will seek the nomination;
                                    a: Sam will accept the nomination.

41. $\sim t \wedge \sim m$ where t: Jack will go tonight;
                                    m: Rosamond will go tomorrow.

42. $\ell \wedge \sim e$ where ℓ: Fat Albert lives to eat; e: Fat Albert eats to live.

43. $(j \vee i) \wedge \sim p$ where j: The decision will depend on judgment;
                                    i: The decision will depend on intuition;
                                    m: The decision will depend on who paid the most.

44. $a \vee p$ where a: The successful applicant for the job has a B.A.
                                        in liberal arts;
                                    p: The successful applicant for the job has a B.A. in psychology.

45. $d \vee e$ where d: The winner must have an A.A. degree in drafting;
                                    e: The winner has three years professional experience.

46. Let p: Dinner includes soup; d: Dinner includes salad; v: Dinner includes the vegetable of the day.   **a.** $(p \wedge d) \vee v$ **b.** $p \wedge (d \vee v)$ **47.** Let s: Marsha finished the sign; t: Marsha finished the table; c: Marsha finished a pair of chairs.   **a.** $(s \wedge t) \vee c$ **b.** $s \wedge (t \vee c)$ **48.** F

49. T   **50.** F   **51.** T   **52.** T   **53.** F   **54.** T   **55.** F   **56.** T   **57.** T

Level 3 Problem Solving, page 61

58. If p is true, then $\sim(\sim p)$ is $\sim(\sim \text{T})$ or $\sim \text{F}$ which is T;
if p is false, then $\sim(\sim p)$ is $\sim(\sim \text{F})$ or $\sim \text{T}$ which is F. In either
case, p has the same truth value as $\sim(\sim p)$.

59. Each phrase is equivalent to a negation:
Dr. Smith, I wish to explain that...

I was	I didn't			
joking...	mean...	reconsider...	not...	change mind
$\sim$	$\sim$	$\sim$	$\sim$	

By the law of double negation (Problem 58 used twice) this is equivalent to changing her mind.
Yes, Melissa did change her mind.

60. The words "there" and "errors" are misspelled. The third error is that there
were only two errors in the sentence.

2.2 Truth Tables and the Conditional, page 61

The conditional is an extremely important operator in logic and mathematics. The students may have
trouble accepting the F → T or F → F as true. Care should be exercised when presenting this concept.
Ultimately, however, you may simply have to resort to the fact that, regardless of our intuitive notions
about the conditional, the relationship is *defined* by Table 2.6.

The students should now have Tables 2.5 and 2.6 (conjunction, disjunction, negation, and
conditional) committed to memory (Transparency 8, top). This is essential before continuing.

Level 1, page 67

1. A device used in logic which lists all the possible truth and false conditions for a given argument.

2. The statement $p \rightarrow q$ is translated as "if p, then q" and is called the conditional. It is defined by
Table 2.6. It is false when the antecedent is true and the conclusion is false; otherwise it is true.

3. The *law of double negation* is: $\sim(\sim p)$ may be replaced by p in any logical expression.   **4.** A
conditional may be replaced by its contrapositive without having its truth value affected.

5.

p	q	$\sim p$	$\sim p \vee q$
T	T	F	T
T	F	F	F
F	T	T	T
F	F	T	T

6.

p	q	$\sim p$	$\sim q$	$\sim p \wedge \sim q$
T	T	F	F	F
T	F	F	T	F
F	T	T	F	F
F	F	T	T	T

7.

p	q	$p \wedge q$	$\sim (p \wedge q)$
T	T	T	F
T	F	F	T
F	T	F	T
F	F	F	T

8.

r	s	$\sim r$	$\sim s$	$\sim r \vee \sim s$
T	T	F	F	F
T	F	F	T	T
F	T	T	F	T
F	F	T	T	T

9.

r	$\sim r$	$\sim (\sim r)$
T	F	T
F	T	F

10.

r	s	$r \wedge s$	$\sim s$	$(r \wedge s) \vee \sim s$
T	T	T	F	T
T	F	F	T	T
F	T	F	F	F
F	F	F	T	T

11.

p	q	$\sim q$	$p \wedge \sim q$
T	T	F	F
T	F	T	T
F	T	F	F
F	F	T	F

12.

p	q	$\sim p$	$\sim q$	$\sim p \vee \sim q$
T	T	F	F	F
T	F	F	T	T
F	T	T	F	T
F	F	T	T	T

13.

p	q	$\sim p$	$\sim p \wedge q$	$\sim q$	$(\sim p \wedge q) \vee \sim q$
T	T	F	F	F	F
T	F	F	F	T	T
F	T	T	T	F	T
F	F	T	F	T	T

14.

p	q	$\sim q$	$p \wedge \sim q$	$(p \wedge \sim q) \wedge p$
T	T	F	F	F
T	F	T	T	T
F	T	F	F	F
F	F	T	F	F

15.

p	q	$p \rightarrow q$	$p \vee (p \rightarrow q)$
T	T	T	T
T	F	F	T
F	T	T	T
F	F	T	T

16.

p	q	$p \wedge q$	$(p \wedge q) \rightarrow p$
T	T	T	T
T	F	F	T
F	T	F	T
F	F	F	T

17.

p	q	$p \vee q$	$p \wedge (p \vee q)$	$[p \wedge (p \vee q)] \rightarrow p$
T	T	T	T	T
T	F	T	T	T
F	T	T	F	T
F	F	F	F	T

18.

p	q	$p \wedge q$	$p \vee (p \wedge q)$	$[p \vee (p \wedge q)] \to p$
T	T	T	T	T
T	F	F	T	T
F	T	F	F	T
F	F	F	F	T

19.

p	q	r	$p \vee q$	$(p \vee q) \vee r$
T	T	T	T	T
T	T	F	T	T
T	F	T	T	T
T	F	F	T	T
F	T	T	T	T
F	T	F	T	T
F	F	T	F	T
F	F	F	F	F

20.

p	q	r	$p \wedge q$	$\sim r$	$(p \wedge q) \wedge \sim r$
T	T	T	T	F	F
T	T	F	T	T	T
T	F	T	F	F	F
T	F	F	F	T	F
F	T	T	F	F	F
F	T	F	F	T	F
F	F	T	F	F	F
F	F	F	F	T	F

21.

p	q	r	$p \vee q$	$\sim r$	$(p \vee q) \wedge (\sim r)$	$[(p \vee q) \wedge \sim r] \wedge r$
T	T	T	T	F	F	F
T	T	F	T	T	T	F
T	F	T	T	F	F	F
T	F	F	T	T	T	F
F	T	T	T	F	F	F
F	T	F	T	T	T	F
F	F	T	F	F	F	F
F	F	F	F	T	F	F

22.

p	q	r	$\sim p$	$q \vee \sim p$	$p \wedge (q \vee \sim p)$	$[p \wedge (q \vee \sim p)] \vee r$
T	T	T	F	T	T	T
T	T	F	F	T	T	T
T	F	T	F	F	F	T
T	F	F	F	F	F	F
F	T	T	T	T	F	T
F	T	F	T	T	F	F
F	F	T	T	T	F	T
F	F	F	T	T	F	F

23. Statement: $\sim p \to \sim q$.   Converse: $\sim q \to \sim p$.
Inverse: $p \to q$.   Contrapositive: $q \to p$.

24. Statement: $\sim r \to t$.   Converse: $t \to \sim r$.
    Inverse: $r \to \sim t$.   Contrapositive: $\sim t \to r$.

25. Statement: $\sim t \to \sim s$.   Converse: $\sim s \to \sim t$.
    Inverse: $t \to s$.   Contrapositive: $s \to t$.

26. Statement: If you break the law, then you will go to jail.
    Converse: If you go to jail, then you have broken the law.
    Inverse: If you do not break the law, then you will not go to jail.
    Contrapositive: If you do not go to jail, then you did not break the law.

27. Statement: If I get paid, then I will go Saturday.
    Converse: If I go Saturday, then I will get paid.
    Inverse: If I do not get paid, then I will not go Saturday.
    Contrapositive: If I do not go Saturday, then I do not get paid.

28. Statement: If you brush your teeth with Smiles toothpaste, then you will have fewer cavities.
    Converse: If you have fewer cavities, then you brush your teeth with Smiles toothpaste.
    Inverse: If you do not brush your teeth with Smiles toothpaste, then you will not have fewer cavities.
    Contrapositive: If you do not have fewer cavities, then you do not brush your teeth with Smiles toothpaste.

Level 2, page 68

29. If it is a triangle, then it is a polygon.

30. If it is a prime number greater than two, then it is an odd number.

31. If you are a good person, then you will go to heaven.

32. If there is time enough, then everything happens to everybody sooner or later.

33. If we make a proper use of those means which the God of Nature has placed in our power, then we are not weak.

34. If it is a useless life, then there is an early death.

35. If it is work, then it is noble.

36. If only you can find it, then everything's got a moral. (Don't confuse "if only" with the "only if" operator.)

37. $F \to F$ is true   38. F only if F is true   39. $F \to F$ is true   40. T only if F is false

41. p is T and q is T. a. T b. F c. F   42. p is T and q is F.   a. T b. T c. T

43. p is F and q is T.   a. F b. T c. T   44. p is T and q is F.   a. T b. T

45. $(q \wedge \sim d) \to n$, where q: The qualifying person is a child; d: This child is your dependent; n: You enter your child's name.   46. $(a \wedge c) \to p$, where a: The amount on line 31 is less than $26,673; c: A child lives with you; p: You turn to page 27.   47. $(b \vee c) \to n$, where b: The amount on line 32 is $86,025; c: The amount on line 32 is less than $86,025; n: You multiply the

number of exemptions on line 6e by $2,500.   **48.** $(m \wedge j) \to s$, where m: You are married; j: You are filing a joint return; s: You enter your spouse's earned income.   **49.** $(m \vee d) \to s$, where m: You are a student; d: You are a disabled person; s: You see line 6 of instructions.   **50.** $[(i \wedge s) \to t] \wedge c$, where i: The income of line 1 as reported on line 1 of Form W-2; s: The "Statutory employee" box on form W-2 was checked; t: You see instructions for Schedule C, line 1; c: you will check here.

51. Assume d, c, and b are true, and w is false: $(\sim T \wedge T) \to (F \wedge T)$ is true.

52. Construct a truth table:

d	c	w	b	$\sim d$	$\sim d \wedge c$	$w \wedge b$	$(\sim d \wedge c) \to (w \wedge b)$
T	T	T	T	F	F	T	T
T	T	T	F	F	F	F	T
T	T	F	T	F	F	F	T
T	T	F	F	F	F	F	T
T	F	T	T	F	F	T	T
T	F	T	F	F	F	F	T
T	F	F	T	F	F	F	T
T	F	F	F	F	F	F	T
F	T	T	T	T	T	T	T
F	T	T	F	T	T	F	F
F	T	F	T	T	T	F	F
F	T	F	F	T	T	F	F
F	F	T	T	T	F	T	T
F	F	T	F	T	F	F	T
F	F	F	T	T	F	F	T
F	F	F	F	T	F	F	T

The statement is true except when d is false c is true and $(b \wedge w)$ is false.

Level 3, page 68

53. **a.** $a \to (e \vee f)$ **b.** $(a \wedge e) \to q$ **c.** $(a \wedge f) \to q$ **54.** **a.** $(\sim m \wedge \sim i) \to \sim q$
b. $(\sim m \wedge i) \to q$ **c.** $[m \wedge (i \wedge b)] \to q$ **55.** $[(m \vee t \vee w \vee h) \wedge s] \to q$ **56.** $(t \wedge d) \to \sim c$
57. $t \to (m \wedge s \wedge p)$ **58.** $t \to [u \wedge \sim (w \vee g)]$

Level 3 Problem Solving, page 69

59. This is not a statement, since "Apollo can do anything" gives rise to a paradox and is neither true nor false.   **60.** It is always true, since the antecedent is false.

2.3 Operators and Laws of Logic, page 69

As an introduction to today's lesson, I ask several students to give the meaning of the everyday connectives shown at the bottom of Transparency 8.

Level 1, page 74

1. Answers vary; the conditional has the form $p \rightarrow q$, whereas the biconditional $p \leftrightarrow q$ not only means $p \rightarrow q$ but also $q \rightarrow p$.   2. To find the negation of a compound statement use De Morgan's law as well as the law of double negation.   3. The symbol $\leftrightarrow$ is used to symbolize the biconditional, whereas the symbol $\Leftrightarrow$ is used to symbolize a tautology (a statement that is always true).   4. The symbol $\rightarrow$ is used to symbolize the conditional, whereas the symbol $\Rightarrow$ is used to symbolize an implication (a conditional which is always true).   5. Answers vary.   6. yes   7. no   8. no   9. no   10. no   11. yes   12. yes   13. no

14.

p	q	$p \vee q$	$p \wedge q$	$\sim (p \wedge q)$	$(p \vee q) \wedge \sim (p \wedge q)$
T	T	T	T	F	F
T	F	T	F	T	T
F	T	T	F	T	T
F	F	F	F	T	F

15.

p	q	$p \vee q$	$\sim (p \vee q)$
T	T	T	F
T	F	T	F
F	T	T	F
F	F	F	T

16.

p	q	$\sim q$	$\sim q \rightarrow p$
T	T	F	T
T	F	T	T
F	T	F	T
F	F	T	F

17.

p	q	$\sim q$	$p \rightarrow \sim q$
T	T	F	F
T	F	T	T
F	T	F	T
F	F	T	T

18. Let s: Smoking is good for your health; d: Drinking is good for your health.
Neither s nor d: $\sim (s \vee d)$.

19. Let h: I will buy a new house; p: All provisions of the sale are clearly understood.
Not h unless p: $\sim p \rightarrow \sim h$.

20. Let ℓ: I obtain the loan; i: I have an income of \$85,000 per year. If ℓ, then i: $\ell \rightarrow i$

21. Let r: I am obligated to pay the rent; s: I signed the contract. r because s: $(r \wedge s) \wedge (s \rightarrow r)$

22. Let g: I can go with you; e: I have a previous engagement.
Not g because e: $(\sim g \wedge e) \wedge (e \rightarrow \sim g)$.

23. Let m: It is a man; i: It is an island. No m is i: $m \rightarrow \sim i$

24. Let i: I will invest my money in stocks; s: I will put my money in a savings account.
Either i or s: $(i \vee s) \wedge \sim (i \wedge s)$.

25. Let n: You are nice to people on your way up; m: You will meet people on your way down.
n because m: $(n \wedge m) \wedge (m \rightarrow n)$.

26. Let h: One has once heartily laughed; w: One has once wholly laughed; b: One is altogether irreclaimably bad.
No $(h \wedge w)$ is b: $(h \wedge w) \rightarrow \sim b$.

27. Let *f*: The majority, by mere force of numbers, deprives a minority of a clearly written constitutional right; *r*: Revolution is justified. If *f* then *r*: $f \rightarrow r$.

Answers to Problems 28-33 may vary.

28. I did not go or I paid $100.   **29.** The cherries have not turned red or they are ready to be picked.   **30.** If we visit New York, then we will visit the Statue of Liberty.   **31.** If Melissa watches Jay Leno, then she watches the NBC late-night orchestra.   **32.** If the sun is not shining, then I will not go to the park.   **33.** If the money is not available, then I will not take my vacation.

Level 2, page 74

34.

p	q	$\sim p$	$\sim q$	$p \vee q$	$\sim p \wedge \sim q$	$\sim(p \vee q)$
T	T	F	F	T	F	F
T	F	F	T	T	F	F
F	T	T	F	T	F	F
F	F	T	T	F	T	T

↑_____↑
same

35. See Example 2, Section 2.2.

p	$\sim p$	$\sim(\sim p)$	$p \leftrightarrow \sim(\sim p)$
T	F	T	T
F	T	F	T

Thus, $p \Leftrightarrow \sim(\sim p)$.

36.

p	q	$p \rightarrow q$	$\sim q$	$\sim p$	$\sim q \rightarrow \sim p$	$(p \rightarrow q) \leftrightarrow (\sim q \rightarrow \sim p)$
T	T	T	F	F	T	T
T	F	F	T	F	F	T
F	T	T	F	T	T	T
F	F	T	T	T	T	T

Thus, $(p \rightarrow q) \Leftrightarrow (\sim q \rightarrow \sim p)$.

37.

p	q	$p \vee q$	$\sim(p \vee q)$	$\sim p$	$\sim q$	$\sim p \wedge \sim q$	$\sim(p \vee q) \leftrightarrow (\sim p \wedge \sim q)$
T	T	T	F	F	F	F	T
T	F	T	F	F	T	F	T
F	T	T	F	T	F	F	T
F	F	F	T	T	T	T	T

Thus, $\sim(p \vee q) \Leftrightarrow \sim p \wedge \sim q$.

38.

p	q	$p \rightarrow q$	$\sim(p \rightarrow q)$	$\sim q$	$p \wedge \sim q$	$\sim(p \rightarrow q) \leftrightarrow (p \wedge \sim q)$
T	T	T	F	F	F	T
T	F	F	T	T	T	T
F	T	T	F	F	F	T
F	F	T	F	T	F	T

Thus, $\sim(p \rightarrow q) \Leftrightarrow (p \wedge \sim q)$.

39.

p	q	$p \to q$	$\sim p$	$\sim p \vee q$	$(p \to q) \leftrightarrow (\sim p \vee q)$
T	T	T	F	T	T
T	F	F	F	F	T
F	T	T	T	T	T
F	F	T	T	T	T

Thus, $(p \to q) \Leftrightarrow (\sim p \vee q)$.

40. $p \wedge \sim q$ **41.** $p \wedge q$ **42.** $\sim p \wedge \sim q$ **43.** $\sim p \wedge q$ **44.** John did not go to Macy's and he did not go to Sears.   **45.** Jane did not go to the basketball game and she did not go to the soccer game.   **46.** Tim is here or he is at home.   **47.** Sally is on time or she did not miss the boat.

48. I can't go with you, and I will not go with Bill.   **49.** You are out of Schlitz and you have beer.

50. $x + 2 = 5$ and $x \neq 3$.   **51.** $x - 5 = 4$ and $x \neq 1$.   **52.** $x = -5$ and $x^2 \neq 25$.

53. $x = 1$ and $y = 2$, and $2x + 3y \neq 8$.

Level 3 Problem Solving, page 74

54. Let s: The person is single (assume $\sim s$: The person is married); a: The person has assets of at least \$50,000; c: The person has a gross income of \$72,000; d: The person has a combined income (with spouse) of \$100,000; q: The person qualifies for a loan of \$200,000. Symbolic statement: $[(s \wedge a \wedge c) \vee (\sim s \wedge a \wedge d)] \to q$; For this problem, s is T, a is T, c is F, and d is F, so that $[(T \wedge T \wedge F) \vee (\sim T \wedge T \wedge F)] \to [F \vee F] \to F$ so that Liz does not qualify.

55. Let d: You purchase your ticket between January 5 and February 15; f: You fly round trip between February 20 and May 3; m: You depart on Monday; t: You depart on Tuesday; w: You depart on Wednesday; h: You return on Tuesday; i: You return on Wednesday; j: You return on Thursday; s: You stay over a Saturday night; e: You obtain 40% off regular fare. Symbolic statement: $[d \wedge f \wedge (m \vee t \vee w) \wedge (h \vee i \vee j) \wedge s] \to e$

56. Let a: An alteration is made to the building; r: The building is redecorated; t: tacks are inserted onto the building; n: Nails are hammered into the building; p: Permission is obtained; Symbolic statement: $\sim (a \vee r \vee t \vee m)$ unless p. This means: $\sim p \to \sim (a \vee r \vee t \vee m)$

57. Let ℓ: The tenant lets the premises; m: The tenant lets a portion of the premises; s: The tenant sublets the premises; t: The tenant sublets a portion of the premises; p: Permission is obtained. Symbolic statement: $\sim [(\ell \vee m) \vee (s \vee t)]$ unless p. This means:

$\sim p \to \sim [(\ell \vee m) \vee (s \vee t)]$ or $\sim p \to [\sim (\ell \vee m) \wedge \sim (s \vee t)]$. This can also be written (using De Morgan's law) as $\sim p \to [(\sim \ell \wedge \sim m) \wedge (\sim s \wedge \sim t)]$.

58. Every person would say that he or she is a member of the Veracious party. Thus the second person's assertion was true, and the third person was lying. Hence the third was a member of the Deceit party.

59. Let a: Alfie is afraid to go; b: Bogie lied; c: Clyde lied.

Either $(\sim a)$ or $(b \wedge c)$. Symbolic statement: $[(\sim a) \vee (b \wedge c)] \wedge \sim [(\sim a) \wedge (b \wedge c)]$.

60. Let p: Number 1 is the hot seat. q: You are telling the truth. Question: Is it true that number 1 is the hot seat if and only if you are telling the truth?

p	q	$q \leftrightarrow p$	*How the jailer would answer*
T	T	T	YES (q is T so he tells the truth)
T	F	F	YES (q is F so he lies)
F	T	F	NO (q is T so he tells the truth)
F	F	T	NO (q is F so he lies)

If the answer is yes, then number 1 is the hot seat. If the answer is no, then 2 is the hot seat.

2.4 The Nature of Proof, page 75

When finished with this section, the student should be able to distinguish among direct reasoning, indirect reasoning, and transitive reasoning, as well as give examples of each.

Level 1, page 81

1. *Direct reasoning* consists of two *premises*, or *hypotheses*, and a *conclusion*. The argument form for direct reasoning is $[(p \rightarrow q) \wedge p] \rightarrow q$.

2. *Indirect reasoning* consists of two *premises*, or *hypotheses*, and a *conclusion*. The argument form for indirect reasoning is $[(p \rightarrow q) \wedge \sim q] \rightarrow \sim p$.

3. *Transitive reasoning* consists of two *premises*, or *hypotheses*, and a *conclusion*. The argument form for transitive reasoning is $[(p \rightarrow q) \wedge (q \rightarrow r)] \rightarrow (p \rightarrow r)$.

4. A *logical fallacy* is an invalid argument form. The words usually refer to a common fallacy. In this book we considered three such logical fallacies:

Fallacy of	*Fallacy of*	*False Chain*
the Converse	*the Inverse*	*Pattern*
(assuming the consequent)	*(denying the antecedent)*	
$p \rightarrow q$	$p \rightarrow q$	$p \rightarrow q$
q	$\sim p$	$p \rightarrow r$
$\therefore p$	$\therefore \sim q$	$\therefore q \rightarrow r$

5. A *syllogism* is a form of reasoning in which two statements or premises are made and a logical conclusion is drawn from them. In this book, we considered three types of syllogisms: direct reasoning, indirect reasoning, and transitive reasoning.

6. Look at a truth table:

p	q	$p \to q$	$\sim q$	$(p \to q) \land \sim q$	$\sim p$	$[(p \to q) \land \sim q] \to \sim p$
T	T	T	F	F	F	T
T	F	F	T	F	F	T
F	T	T	F	F	T	T
F	F	T	T	T	T	T

Since all of the entries in the last column are Ts, it is proved. It is called indirect reasoning.

7. We show this is a fallacy by constructing a truth table:

p	q	r	$p \to q$	$p \to r$	$q \to r$	$(p \to q) \land (p \to r)$	$[(p \to q) \land (p \to r)] \to (q \to r)$
T	T	T	T	T	T	T	T
T	T	F	T	F	F	F	T
T	F	T	F	T	T	F	T
T	F	F	F	F	T	F	T
F	T	T	T	T	T	T	T
F	T	F	T	T	F	T	F
F	F	T	T	T	T	T	T
F	F	F	T	T	T	T	T

        ↑
     Not all Ts

Since the result does not show all Ts, it is a fallacy. This is the false chain pattern.

8. Look at a truth table:

p	q	$p \to q$	$\sim p$	$(p \to q) \land q$	$\sim q$	$[(p \to q) \land \sim p] \to \sim q$
T	T	T	F	T	F	T
T	F	F	F	F	T	T
F	T	T	T	T	F	F
F	F	T	T	F	T	T

Since the result does not show all Ts, it is a fallacy. This is the fallacy of the inverse.

9.

Problem-solving language	*Detective language*
Understand the problem.	*Understand the case.*
What is the unknown?	What are you looking for?
Devise a plan.	*Investigate the case.*
Do you know a related problem?	Have you solved a similar case?
Can you simplify the problem?	What are the facts?
Carry out the plan.	*Analyze the facts/data.*
What information is important?	What information is important?
What information is not important?	What information is not important?
What pieces of information fit together logically?	What pieces of information do not seem to fit together logically?
Which information is consistent with the given information?	Which data are inconsistent with the given information?
Look back.	*Reexamine the facts.*
Examine the solution obtained.	Do the facts support the solution?
Does it make sense?	Can we obtain a conviction?

10. a. valid; indirect **b.** invalid; fallacy of the converse   **11. a.** valid; by truth table (law of the excluded middle) **b.** valid; law of the excluded middle   **12. a.** invalid; fallacy of the inverse **b.** valid; direct   **13. a.** valid; indirect **b.** valid; direct   **14.** valid; direct   **15.** valid; indirect **16.** valid; transitive   **17.** valid; indirect   **18.** valid; direct   **19.** valid; direct   **20.** valid; direct **21.** invalid; fallacy of the inverse   **22.** invalid; false chain   **23.** invalid; fallacy of the inverse **24.** valid; indirect   **25.** valid; indirect (along with the law of double negation)   **26.** valid; indirect **27.** invalid; fallacy of the converse   **28.** invalid; fallacy of the converse   **29.** valid; direct **30.** valid; transitive   **31.** invalid; a statement is not equivalent to its converse.   **32.** valid; transitive   **33.** valid; indirect   **34.** valid; indirect   **35.** valid; indirect   **36.** valid; contrapositive and transitive (twice)

Level 2, page 82

37. If you learn mathematics, then you understand human nature. (*transitive*)   **38.** I am lazy. (*direct*)   **39.** We do not go to the concert. (*indirect*)   **40.** If you climb the highest mountain, then you are happy. (*transitive*)   **41.** $b = 0$ (*excluded middle*)   **42.** $a = 0$ or $b = 0$ (*direct*)   **43.** We do not interfere with the publication of false information. (*indirect*)   **44.** If a nail is lost, then the kingdom is lost. (*transitive*)   **45.** I will not eat that piece of pie. (*indirect*)   **46.** Two does not divide a positive integer N or N is not greater than 2. (*indirect*)

47. Let p: We win first prize; e: We will go to Europe; i: We are ingenious.

Given argument: (1) $p \rightarrow e$
                   (2) $i \rightarrow p$
                   (3) i

Rearrange the premises:

(2)	$i \rightarrow p$	*given*
(1)	$p \rightarrow e$	*given*
	$i \rightarrow e$	*transitive law*
(3)	i	*given*
	e	*direct reasoning*

Conclusion: We will go to Europe.

48. Let t: I am tired; f: I finish my homework; u: I understand the homework.

Given argument: (1) $t \rightarrow \sim f$ *given*
                   (2) $u \rightarrow f$ *given*

Then we have: (2′) $\sim f \rightarrow \sim u$ *contrapositive of (2)*
                       $t \rightarrow \sim u$ *transitive law*

Conclusion: If I am tired, then I did not understand the homework.

Level 3, page 83

49. Let b: This is a baby; l: This is an illogical creature; d: This is a despised creature; c: This is one who can manage a crocodile.

Given argument:
(1) $b \rightarrow l$
(2) $d \rightarrow \sim c$
(3) $l \rightarrow d$

Then we have:

(1)	$b \rightarrow l$	*given*
(3)	$l \rightarrow d$	*given*
	$b \rightarrow d$	*transitive law*
(2)	$d \rightarrow \sim c$	*given*
	$b \rightarrow \sim c$	*transitive law*

Conclusion: If it is a baby, then it cannot manage a crocodile. Or, babies cannot manage crocodiles.

50. Let m: This is a hummingbird; c: This is a richly richly colored creature; b: This is a large bird; h: this bird lives on honey.

Given argument:
(1) $m \rightarrow c$
(2) $b \rightarrow \sim h$
(3) $\sim h \rightarrow \sim c$

Then we have:

(2)	$b \rightarrow \sim h$	
(3)	$\sim h \rightarrow \sim c$	
	$b \rightarrow \sim c$	*transitive law*
(1′)	$\sim c \rightarrow \sim m$	*contrapositive of* (1)
	$b \rightarrow \sim m$	*transitive law*

Conclusion: If it is a large bird, then it is not a hummingbird.

51. Let d: It is a duck; w: It waltzes; f: It is an officer; p: It is my poultry.

Given argument:
(1) $d \rightarrow \sim w$
(2) $f \rightarrow w$
(3) $p \rightarrow d$

Then we have:

(3)	$p \rightarrow d$	*given*
(1)	$d \rightarrow \sim w$	*given*
	$p \rightarrow \sim w$	*transitive law*
(2′)	$\sim w \rightarrow \sim f$	*contrapositive of* (2)
	$p \rightarrow \sim f$	*transitive law*

Conclusion: None of my poultry are officers. Here is another possibility: My poultry are not officers.

52. Let s: This is a sane person (assume $\sim s \leftrightarrow$ lunatic); l: This is a person who can do logic; j: This person is fit to serve on a jury; y: This is your son.

Given argument:
(1) $s \rightarrow l$
(2) $\sim s \rightarrow \sim j$
(3) $y \rightarrow \sim l$

Then we have:

(2′)	$j \rightarrow s$	*contrapositive of* (2)
(1)	$s \rightarrow l$	*given*
	$j \rightarrow l$	*transitive law*
(3′)	$l \rightarrow \sim y$	*contrapositive of* (3)
	$j \rightarrow \sim y$	*transitive law*

Conclusion: No one fit to serve on a jury is your son.

53. Let g: The government awards the contract to Airfirst Aircraft Company;

s: Senator Firstair stands to earn a great deal of money;

f: Airsecond Aircraft Company suffers financial setbacks.

Given argument:
(1) $g \to s$
(2) $\sim f \to \sim s$
(3) g

Then we have:
(1)	$g \to s$	given
(3)	g	given
	s	direct reasoning
(2)	$\sim f \to \sim s$	given
	s	
	f	indirect reasoning

Conclusion: Airsecond Aircraft Company suffers financial setbacks.

54. Let c: You go to college; j: You obtain a good job; m: You will make a lot of money; l: You obey the law.

Given argument:
(1) $c \to j$
(2) $j \to m$
(3) $\sim l \to \sim m$
(4) c

Then we have:
(1)	$c \to j$	given
(2)	$j \to m$	given
	$c \to m$	transitive law
(3')	$m \to l$	contrapositive of (3)
	$c \to l$	transitive law
	c	given
	l	direct reasoning

Thus, the valid conclusion is: You will obey the law.

Level 3 Problem Solving, page 83

55. The janitor could not have taken the elevator because the building's fuses were blown.

56. The window was facing east and it was a westerly wind, so it was probably his brother.

57. Let a: This is a person who appreciates Beethoven; s: This is a person who keeps silent while the *Moonlight Sonata* is played; p: This person is a guinea pig; i: This person is hopelessly ignorant of music.

Given argument:
(1) No $a \to \sim s$ or $a \to s$
(2) $p \to i$
(3) No $i \to s$ or $i \to \sim s$

Then we have: (2) $p \rightarrow i$ *given*

                (3) $i \rightarrow\ \sim s$ *given*

                    $p \rightarrow\ \sim s$ *transitive law*

             $(1')$ $\sim s \rightarrow\ \sim a$ *contrapositive*

                    $p \rightarrow\ \sim a$ *transitive law*

Thus, the valid conclusion is: Guinea pigs do not appreciate Beethoven.

58.   Let f: This kitten loves fish; t: This kitten is teachable; l: This kitten has a tail; g: This kitten plays with a gorilla; w: This kitten has whiskers; e: This kitten has green eyes.

Given argument: (1) $f \rightarrow\ \sim (\sim t)$

                    (2) $l \rightarrow\ \sim g$

                    (3) $w \rightarrow f$

                    (4) $t \rightarrow\ \sim e$

                    (5) $\sim w \rightarrow l$

Then we have: (5) $\sim w \rightarrow l$ *given*

                (2) $l \rightarrow\ \sim g$ *given*

                (6) $\sim w \rightarrow\ \sim g$ *transitive*

                (7) $g \rightarrow w$ *contrapositive of* (6)

                (3) $w \rightarrow f$ *given*

             $(1')$ $f \rightarrow t$ *double negative on* (1)

                (8) $w \rightarrow t$ *transitive*

                (9) $g \rightarrow t$ *transitive on* (7) *and* (8)

                (4) $t \rightarrow\ \sim e$ *given*

             (10) $g \rightarrow\ \sim e$ *transitive*

           $(10')$ $e \rightarrow\ \sim g$ *contrapositive of* (10)

Conclusion: No kitten with green eyes plays with a gorilla.

59.   Let g: I work a logic problem without grumbling   (1) $g \rightarrow u$

                                               (2) $\sim r$

         u: It is a logic problem I can understand   (3)   no e is h or $e \rightarrow\ \sim h$

         r: These problems are arranged in   (4) $\sim r \rightarrow\ \sim u$

             regular order like the problems I am $(4')$ $u \rightarrow r$ (contrapositive)

             used to   (5) g unless h or $\sim h \rightarrow g$

         e: It is an easy problem

         h: The problem makes my head ache

Argument:

       (3) $e \rightarrow\ \sim h$

       (5) $\sim h \rightarrow g$

             $e \rightarrow g$ *transitive*

       (1) $g \rightarrow u$

             $e \rightarrow u$ *transitive*

      $(4')$ $u \rightarrow r$

             $e \rightarrow r$ *transitive*

       (2) $\sim r$

             $\sim e$ *indirect*   Therefore, this is not an easy problem.

60. Let *s*: This idea of mine can be
expressed as a syllogism.

d: This idea of mine is really
ridiculous.

r: This idea of mine is about
rock stars.

w: This idea of mine is worth
writing down.

f: This idea of mine fails to come
true.

l: This idea of mine is referred to
my lawyer.

m: This idea of mine is a dream.

(1) $\sim s$ is d
(2) no r is w
(3) no f is s
(4) no d is $\sim l$
(5) All m is r
(6) $\sim l$ unless w

or

(1) $\sim s \rightarrow d$
(2) $r \rightarrow \sim w$
(3) $f \rightarrow \sim s$
(4) $d \rightarrow \sim (\sim l)$
(5) $m \rightarrow r$
(6) $\sim w \rightarrow \sim l$

Argument:

(3) $f \rightarrow \sim s$
(1) $\dfrac{\sim s \rightarrow d}{f \rightarrow d}$ *transitive*

(4′) $\dfrac{d \rightarrow l}{f \rightarrow l}$ *double negative*
   transitive

(6′) $\dfrac{l \rightarrow w}{f \rightarrow w}$ *contrapositive*
   transitive

(2′) $\dfrac{w \rightarrow \sim r}{f \rightarrow \sim r}$ *contrapositive*
   transitive

(5′) $\dfrac{\sim r \rightarrow \sim m}{f \rightarrow \sim m \text{ or } m \rightarrow \sim f}$ *contrapositive*

Therefore, no dream of mine ever fails to come true.

2.5 Problem Solving using Logic, page 84

Most students enjoy doing the problems presented in this section. A good procedure is to work through the examples in the text and then assign the problems of this section as extra credit problems. Since the methods of solution can vary greatly, we give the answers only.

Level 3 Problem Solving, page 89

1. Moe constructed a truth table of all possibilities (underscore means hand is raised):

Harry: <u>B</u> <u>B</u> <u>B</u> <u>W</u> <u>W</u> <u>W</u> B W

Larry: <u>B</u> <u>B</u> W <u>B</u> <u>W</u> B <u>W</u> W

Moe: <u>B</u> <u>W</u> <u>B</u> <u>B</u> <u>B</u> <u>W</u> <u>W</u> W

Moe sees two hands and two black hats and concludes his hat must be black. If Moe's hat were white, then Harry and Larry would each have a solution because they would each see one white hat and one black hat with two hands raised. Thus, Moe knows that his hat must be black.

2.   Consider all possibilities:

Answers of:	Alice	Bob	Cole	Analysis:
Assume Alice did it:	T, F	T, F	T, T	Not possible, because one does not always lie.
Assume Bob did it:	T, T	F, F	T, F	Truth teller is Alice, Bob is the liar, and Cole told the truth half the time.
Assume Cole did it:	F, F	T, T	F, F	Not possible.

Bob did it.

3.   White, since this could happen only at the North Pole.

4.   Gary was 17 and had 15 marbles; Harry was 10 and had 12 marbles; Iggy was 3 and had 12 marbles; Jack was 18 and had 9 marbles. They shot in the order of Gary, Jack, Harry, and then Iggy.

5.   There can be only one yellow flower, so there must be 49 red flowers.

6.   Need to choose 3 socks.

7.   Curly committed the murder.

8.   Three weighings

9.   Only one question is necessary. Ask Connie (who falsely claimed to have the mixed bag) to pull out one fruit. Suppose she pulls out a peach; this means she has the bag containing two peaches. Then, Alice, who falsely claimed two peaches, must have two plums. This leaves Betty with the mixed bag. Suppose she pulls out a plum; this means she has the bag containing two plums. Then, Betty, who falsely claimed two plums, must have two peaches. This leaves Alice with the mixed bag.

10.   a. 7 b. 8 c. 2 d. 6 e. 6 f. 1 g. 5 h. 4 i. 5 j. 1 k. 3 ℓ. 5 m. 4 n. 2
      p. 2 q. 3 r. 2 s. 1

2.6 Logic Circuits, page 90

Level 1, page 93

1. The light is on when when the circuit is complete (does not have a break).

2. a. The light is on when both switches are on.

   b. The light is on when either of the switches are on.

3. a. battery (or power source) b. light c. switch   4. a. AND-gate b. OR-gate c. NOT-gate

5. $p \wedge q$

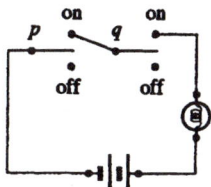

6. $p \vee q$

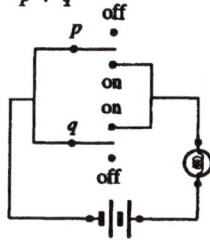

7. $\sim p \wedge q$

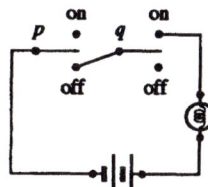

8. $p \wedge \sim q$

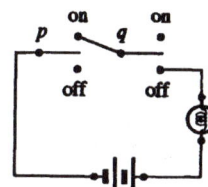

9. $\sim (p \vee q)$

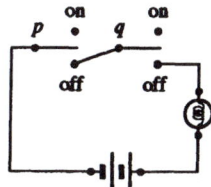

10. $\sim (p \wedge q)$

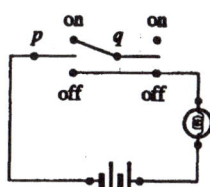

11. $p \rightarrow q$

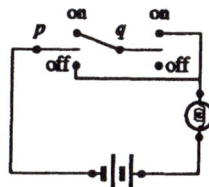

12. $q \rightarrow p$

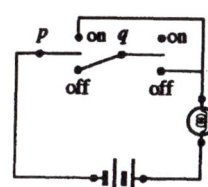

13. $p \rightarrow \sim q$

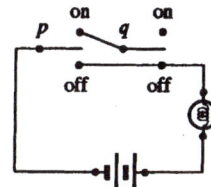

14. $p \leftrightarrow q$

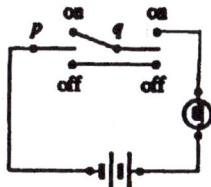

15. $p \leftrightarrow \sim q$

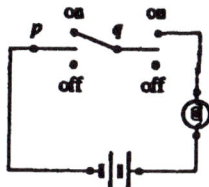

16. $\sim p \leftrightarrow q$

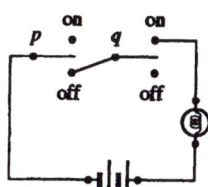

Level 2, page 93

17.

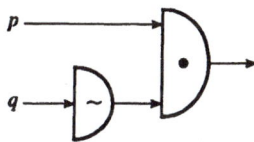

18.

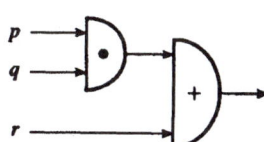

19.

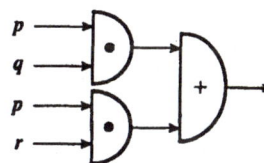

20.

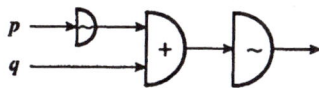

21.

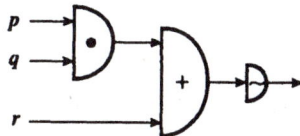

22.

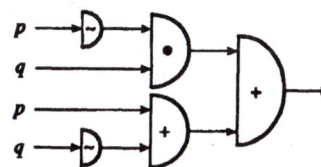

Level 3, page 93

23.

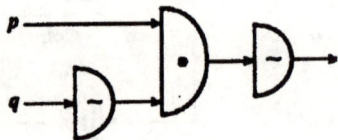

24.

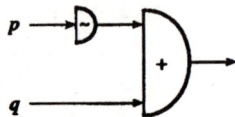

p	q	$p \rightarrow q$	$\sim p$	$\sim p \vee q$
T	T	T	F	T
T	F	F	F	F
F	T	T	T	T
F	F	T	T	T

$(p \rightarrow q) \Leftrightarrow (\sim p \vee q)$

25. Notice $\sim p \rightarrow q \Leftrightarrow p \vee q$ **26.** $p \rightarrow \sim q \Leftrightarrow \sim p \vee \sim q$ **27.** $\sim q \rightarrow \sim p \Leftrightarrow \sim p \vee q$

   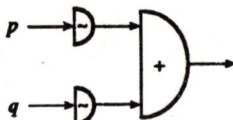   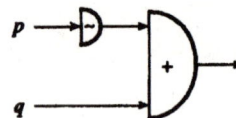

Level 3 Problem Solving, page 94

28. a.

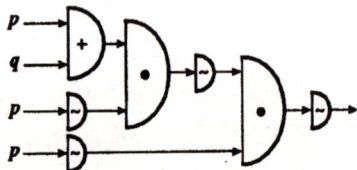

b. $140,000

c.

p	q	$p \vee q$	$\sim p$	$(p \vee q) \wedge \sim p$	$\sim [(p \vee q) \wedge \sim p]$	$\sim [(p \vee q) \wedge \sim p] \wedge \sim p$ $\downarrow$	ANS
T	T	T	F	F	T	F	T
T	F	T	F	F	T	F	T
F	T	T	T	T	F	F	T
F	F	F	T	F	T	T	F

ii.   *i.*

iii. They are the same.

d.

e. $20,000; savings $120,000

29. Let the committee members be *a*, *b*, and *c*, respectively. Light *on* represents a majority. The circuit is shown at the right. Light is on for a yea majority and light is off otherwise.

30. Let the committee members be *a*, *b*, *c*, *d*, and *e*.

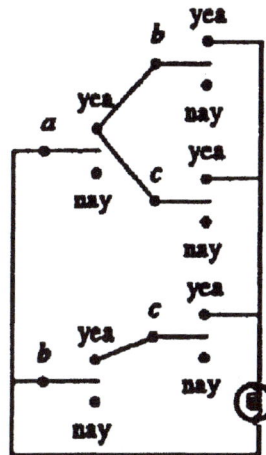

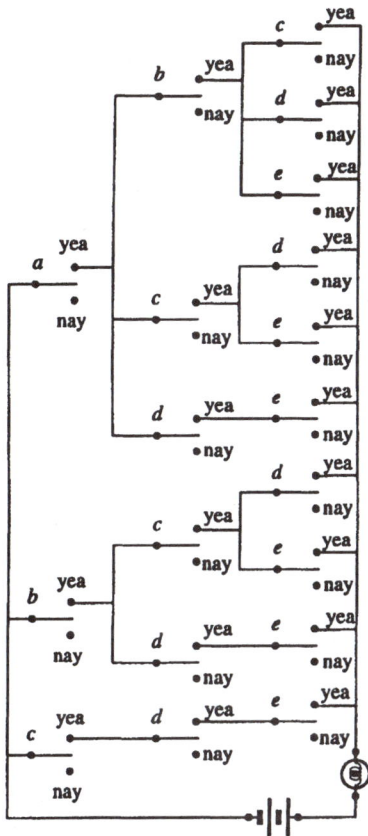

Chapter 2 Review Questions, page 95

1. **a.** A *logical statement* is a declarative sentence that is either true or false.

   b. A *tautology* is a logical statement in which the conclusion is equivalent to its premise.

   c. $(p \rightarrow q) \leftrightarrow (\sim q \rightarrow \sim p)$ or, in words: A conditional may always be replaced by its contrapositive without having its truth value affected.

2.

p	q	$\sim p$	$p \wedge q$	$p \vee q$	$p \to q$	$p \leftrightarrow q$
T	T	F	T	T	T	T
T	F	F	F	T	F	F
F	T	T	F	T	T	F
F	F	T	F	F	T	T

3.

p	q	$p \wedge q$	$\sim (p \wedge q)$
T	T	T	F
T	F	F	T
F	T	F	T
F	F	F	T

4.

p	q	$\sim q$	$p \vee \sim q$	$\sim p$	$(p \vee \sim q) \wedge \sim p$	$[(p \vee \sim q) \wedge \sim p] \to \sim q$
T	T	F	T	F	F	T
T	F	T	T	F	F	T
F	T	F	F	T	F	T
F	F	T	T	T	T	T

5.

p	q	r	$p \wedge q$	$(p \wedge q) \wedge r$	$[(p \wedge q) \wedge r] \to p$
T	T	T	T	T	T
T	T	F	T	F	T
T	F	T	F	F	T
T	F	F	F	F	T
F	T	T	F	F	T
F	T	F	F	F	T
F	F	T	F	F	T
F	F	F	F	F	T

6.

p	q	$\sim p$	$\sim q$	$p \wedge q$	$\sim (p \wedge q)$	$\sim p \vee \sim q$	$\sim (p \wedge q) \leftrightarrow (\sim p \vee \sim q)$
T	T	F	F	T	F	F	T
T	F	F	T	F	T	T	T
F	T	T	F	F	T	T	T
F	F	T	T	F	T	T	T

7.

$p \to q$
$\dfrac{p}{\therefore q}$

p	q	$p \to q$	$(p \to q) \wedge p$	$[(p \to q) \wedge p] \to q$
T	T	T	T	T
T	F	F	F	T
F	T	T	F	T
F	F	T	F	T

8. Answers vary. If you study hard, then you will get an *A*. You do not get an *A*. Therefore, you did not study hard.

9. Answers vary; fallacy of the converse, fallacy of the inverse, or false chain pattern

    For example: If you study hard, then you will get an A. You do not study hard, so you do not

    get an A. (Fallacy of the inverse.)

10. Yes; it is indirect reasoning.

11. **a.** Some birds do not have feathers. **b.** No apples are rotten. **c.** Some cars have two wheels.

    d. All smart people attend college. **e.** You go on Tuesday and you can win the lottery.

12. **a.** T $(F \to F)$ **b.** T $(F \to T)$ **c.** T $(T \to T)$ **d.** T $(F \to F)$ **e.** T $(F \to F)$

13. **a.** If P is a prime number, then $P + 2$ is a prime number.

    b. Either P or $P + 2$ is a prime number.

14. **a.** Let p: This machine is a computer; q: This machine is capable of self-direction. $p \to \sim q$

    b. Contrapositive: $\sim (\sim q) \to \sim p$; $q \to \sim p$. If this machine is capable of self-direction, then

    it is not a computer.

15. Let p: There are a finite number of primes; q: There is some natural number, greater than 1,

    that is not divisible by any prime. Then the argument in symbolic form is:

    $$p \to q$$
    $$\underline{\sim q}$$
    $$\therefore \quad \sim p$$

    Conclusion: There are infinitely many primes (assuming that "not a finite number"

    is the same as "infinitely many"). This is indirect reasoning.

16. **a.**   **b.**

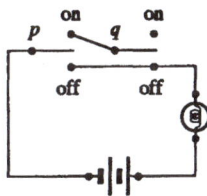   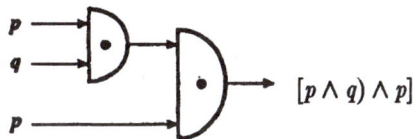

$$[p \wedge q) \wedge p]$$

17. Let d: I attend to my duties; r: I am rewarded; ℓ: I am lazy. Symbolic argument:

    (1) $d \to r$ (2) $\ell \to \sim r$ (1) $d \to r$

    (2) $\ell \to \sim r$ (3) $\underline{\ell}$ (4) $\underline{\sim r}$

    (3) ℓ (4) $\therefore \sim r$ *Direct reasoning* $\therefore \quad \sim d$ *Indirect reasoning*

    Conclusion: I do not attend to my duties.

18. Let o: This is organic food; h: This is healthy food; s: This is an artificial sweetener; p: This

    is a prune. Symbolic argument:

(1) $o \to h$ (3) $p \to o$

(2) $s \to \sim h$ (1) $o \to h$

(3) $p \to o$ (4) $\overline{p \to h}$ *Transitive*

                  (2′) $\underline{h \to \ \sim s}$ *Law of contraposition*

                      $p \to \sim s$ *Transitive*

Conclusion: No prune is an artificial sweetener.

19.   Let s: This is a square; r: This is a rectangle; q: This is a quadrilateral; This is a polygon.

Symbolic argument:

(1) $s \to r$

(2) $\underline{r \to q}$

∴ $s \to q$ *Transitive*

(3) $\underline{q \to p}$

∴ $s \to p$ *Transitive*

Conclusion: All squares are polygons.

20.

              Maureen (baker)

Warren   Terry

(surfer)   (hiker)

              Josie (gardener)

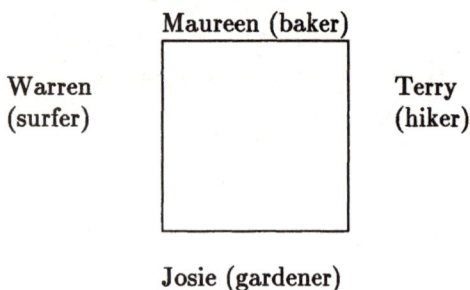

Group Research Problems, pages 97-98

G3.   (1)   Let

     d: Donna likes Elmer.

     e: Elmer is bald.

     f: Donna likes Frank.

     g: Donna likes George.

     t: Frank and George are twins.

Either [$\sim d$ because e] or [$(f \wedge g)$ because t]

Either [$(\sim d \wedge e) \wedge (e \to \sim d)$] or [$(f \wedge g \wedge t) \wedge (t \to (f \wedge g))$]

Translation:

$\{[(\sim d \wedge e) \wedge (e \to \ \sim d)] \ \vee \ [(f \wedge g \wedge t) \wedge (t \to (f \wedge g))]\}$

$\wedge \sim \{[(\sim d \wedge e) \wedge (e \to \sim d)] \wedge [(f \wedge g \wedge t) \ \wedge \ (t \to (f \wedge g))]\}$

If you needed to know when this statement is true, there would be 32 possibilities in the truth table.

(2) Let

        y: You are honest.

        i: I am honest.

        h: Hank is a liar.

        g: Iggy had the money.

We assume that "honest" is the same as "not a liar."

Either [neither y nor i] or [h because $\sim g$]

Either [$\sim y \wedge \sim i$] or [$(h \wedge \sim g) \wedge (\sim g \rightarrow h)$]

Translation:

$$\{[\sim y \wedge \sim i] \vee [(h \wedge \sim g) \wedge (\sim g \rightarrow h)]\} \wedge \sim \{[\sim y \wedge \sim i] \wedge [(h \wedge \sim g) \wedge (\sim g \rightarrow h)]\}$$

G4. Georgia did. For a detail of solution, see "Ask Marilyn" *Parade Magazine*, October 31, 1993.

G5. Answers vary. See "Proof by Contradiction and the Electoral College," by Charles Redmond, Michael P. Federici, and Donald M. Platte in *The Mathematical Teacher*, November 1998.

G6. The baker is Tom Brown. Burt's last name is Evans. Ulysses Baurger is going to Baker Street.

Cab Number:	1	2	3	4	5
First name:	Sam	Tom	Ulysses	Victor	Winston
Last name:	Adams	Brown	Camp	Duncan	**Evans**
Profession:	lawyer	**accountant**	banker	teacher	stock broker
Wife's name:	Donna	Alice	Barbara	Connie	Eve
Distinction:	Achor St.	Denver St.	**Elm St.**	Burbon St.	Camp St.

G7. The reward can be obtained with six balls as follows: 1, 2 (or 3), 4, 5, 8, and 11 (or 12).

Individual Research, pages 98-99

P2.1 They were all math majors.

P2.2 See **www.mathnature.com**.

P2.3 p is sufficient for q means $p \rightarrow q$; q is necessary for p means $p \rightarrow q$

P2.4 *Contradictory* statements are related so that if either of the two is true, then the other is false and if either is false, the other must be true. *Contrary* statements are related so that both cannot both be true (even though they could both be false). *Inconsistent* statements have incompatible elements.

P2.5 Answers vary.

P2.6 If the statement on door 1 is true, then the statement on door 2 is also true; not possible. On the other hand, if the statement on door 2 is true, then the statement on door 2 is false. This means that door 2 is the one that leads to freedom.

P2.7 To say that at least one of the two would be male is another way of saying that no two could be female. If there were two or more women, a pair might include two women. Thus, the only possibility is that there is only one woman. Now this means that there must be 49 males, so 2% of the people attending the convention are female.

P2.8 The answer is yes. There could not be two of any one color (say red) in the garden, because if there were, then one could pick two red, one white, and one yellow, thus contradicting one of the observations that one flower must be blue. If each student's observations is true, the only possibility is that there were just four flowers in the entire garden — one of each color. If all four were picked, then, of course, at least one had to be white.

P2.9 Pitcher: Jones; catcher: Smith; first base: Brown; second base: White; third base: Adams; shortstop: Miller; left field: Green; center field: Hunter; right field: Knight

P2.10 Activity: build a device to add numbers.

CHAPTER 3

THE NATURE OF NUMERATION SYSTEMS

3.1 Early Numeration Systems, page 102

In this section, we examine different numeration systems in order to more fully understand our own numeration (decimal) system and to see how the ideas of these systems have been incorporated into our numeration system. We wish to illustrate two basic ideas: (1) a simple grouping system and (2) a simple positional system. Our goal here is not to gain a proficiency in Egyptian, Roman, or Babylonian systems, but to achieve an understanding of why we need a numeration system. Understanding what is desirable in a numeration system will give us a point of reference when we are discussing the decimal numeration system we use every day. The chapter could be supplemented by a discussion of other numeration systems, such as the Greek, Mayan, or Chinese numeration systems.

Level 1, page 108

1. A *number* is a concept, whereas a *numeral* is the symbol used to represent a number.   2. Answers vary.   3. Answers vary; no zero symbol, difficult and cumbersome to write larger numbers.

4. Answers vary; no zero symbol, additive and subtractive properties are difficult to use.

5. Answers vary; only two symbols is difficult, but positional idea makes it easier to write large numbers.   6. Answers vary.   7. a. Egyptian, Roman   b. Roman (because IX and XI are represent different numbers), Babylonian   c. Egyptian, Roman, Babylonian   d. Egyptian, Roman, Babylonian   e. Roman, Babylonian   f. Roman   8. 1,256   9. $\frac{1}{200}$ 10. $\frac{2}{3}$ 11. 100,010

12. $123\frac{1}{2}$ 13. $\frac{1}{12}$ 14. $24\frac{1}{2}$ 15. $\frac{1}{100}$ 16. 11,352   17. 1,000,001   18. 1,100   19. 1,997

20. 48   21. 709   22. 1,997   23. 2,001   24. 550,000   25. 400,000   26. 7,600   27. 9,712

28. 261   29. 24   30. 71   31. 671   32. 99   33. 28   34. 47   35. 25   36. 7 houses; 7^2 cats; 7^3 mice killed; 7^4 ears of spelt; 7^5 kehat of grain. Thus, the correct solution is $7^5 = 16,807$. (Note: The solution given on the Rhind papyrus is $19,607 = 7 + 7^2 + 7^3 + 7^4 + 7^5$, which is not correct or is not translated correctly.)   37. One; the only one known to be going to St. Ives is myself.

Level 2, page 109

38. **a.** ∩∩∩∩||||| || **b.** XLVII   **c.** ◁ ◁ ◁ ◁ ▽ ▽ ▽ ▽ ▽ ▽ ▽

39. **a.** ∩∩∩∩∩∩ ||||| **b.** LXXV   **c.** ▽ ◁ ▽ ▽ ▽ ▽

40. **a.** ∂∂∩∩∩∩∩||||| ||| **b.** CCLVIII   **c.** ▽ ▽ ▽ ▽ ◁ ▽ ▽ ▽ ▽ ▽ ▽ ▽ ▽

41. **a.** ∂∂∂∂∂∩ ∩| **b.** DXXI   **c.** ▽ ▽ ▽ ▽ ▽ ▽ ▽ ▽ ◁ ◁ ◁ ◁ ▽

42. **a.** ∂∂∂∂∂∂∂∂∩ ∩ ∩ ∩ ∩|| **b.** DCCCLII   **c.** ◁ ▽ ▽ ▽ ▽ ◁ ▽ ▽

43. **a.** Υ Υ | **b.** MMI   **c.** ◁ ◁ ◁ ▽ ▽ ▽ ◁ ◁ ▽

44. $143 - 15 - 36 - 14 = 78$ 45. $45 + 34 + 54 = 133$

46. Υ∂∂∂∩ ||||| ||| 47. ∂∩∩∩∩∩∩∩∩||||| ||

48. ∩∩∩||||| || 49. ◁ ◁ ◁ ◁ ◁

50. ▽ ▽ ▽ ▽ ▽ ▽ ▽ ▽   51. ◁ ◁ ▽ ▽ ▽ ▽ ▽

Level 3 Problem Solving, page 109

52. Choice A

53. **a.** MDCLXVI = 1,666   **b.** MCDXLIV = 1,444

54. MDCCCLXXXVIII = 1888

55. **a.** IX = 9   **b.** VIII = 8   **c.** XCIX = 99   **d.** LXXXIX = 89   **e.** CMXCIX = 999

56.

$x + 7$ (top), x (side)

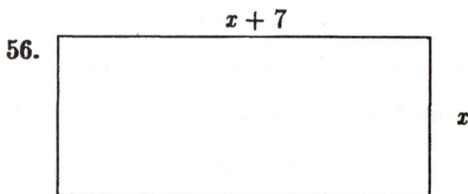

$$x(x + 7) = 60$$
$$x^2 + 7x - 60 = 0$$
$$(x + 12)(x - 5) = 0$$
$$x = -12, 5$$
↑
Reject

The dimensions are 5 by 12; that is,
▽ ▽ ▽ ▽ ▽ by ◁ ▽ ▽.

57.

$x + 10$ (top), x (side)

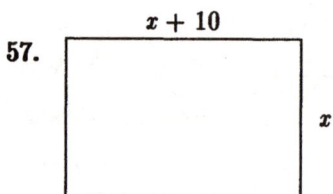

$$x(x + 10) = 600$$
$$x^2 + 10x - 600 = 0$$
$$(x + 30)(x - 20) = 0$$
$$x = 20, -30$$
↑
Reject

The dimensions are 20 by 30; that is
◁ ◁ by ◁ ◁ ◁ .

58. Answers vary.   □, △, ▽, □o, □□, □△, □▽, □o, □oo, □o□, □o△, □o▽,

1,   2,   3,   4,   5,   6,   7,   8,   9,   10,   11,   12,

□□o, □□□, □□△, □□▽, △oo, △o□, △o△, △o▽,

13,   14,   15,   16,   17,   18,   19,   20

In this system, □ + ▽ = □o, ▽ × ▽ = □oo.

59. Answers vary;   — = ≡ ≡ ≡ |— |= |≡ |≡ $\frac{o}{-}$ $\frac{o}{-}$

1   2   3   4   5   6   7   8   9   10 ··· 20

60. Answers vary;   α, β, γ, δ, ε, ϛ, ζ, η, θ, ι, ια, ιβ, ιγ, ιε, ιϛ, ιζ, θη, ιθ, κ, κα, ···

3.2 Hindu-Arabic Numeration System, page 110

The Hindu-Arabic numeration system is introduced in this section, and what the notation means is discussed. Exponential notation for any integer is defined and then used to reinforce the notions of a positional numeration system via expanded notation. The ideas presented in this section will be used in the development of other numeration systems. The words "Hindu-Arabic," "decimal," and "base ten" will all be used synonymously in reference to the Hindu-Arabic numeration system.

I ask the students to count the number of people at the top of Transparency 9. Answers: **a.** Hindu-Arabic, 13 **b.** Egyptian: ∩||| **c.** Babylonian: ≺ ▽ ▽ ▽ **d.** Roman: XIII   This leads (in Section 3.3) to other bases as shown at the bottom of Transparency 9.

Level 1, page 113

1.

2.

3.   Let $\boxed{X}$ represent $\boxed{\bullet\ \bullet\ \bullet\ \bullet\ \bullet\ \bullet\ \bullet\ \bullet\ \bullet\ \bullet}$ of the larger groups.

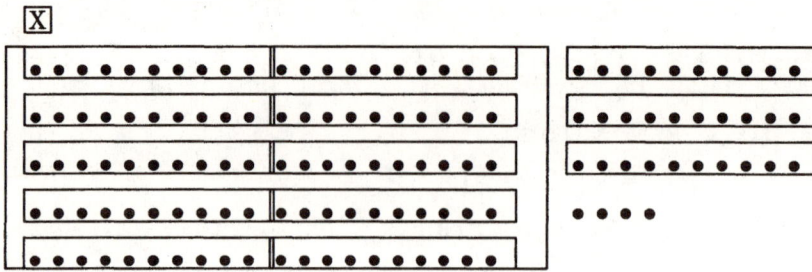

4. Expanded notation is writing a number so that you can see the meaning of each digit.

5. $b^n = b \cdot b \cdot \cdots \cdot b$ where there are n factors of b (for any counting number n)

6. $b^{-n} = \dfrac{1}{b^n}$ (for any counting number n)   **7.** 5 units   **8.** 5 hundreds   **9.** 5 thousandths

10. 5 ten-millions   **11. a.** 100,000   **b.** 1,000   **12. a.** 1,000,000   **b.** 10,000   **13. a.** 0.0001

b. 0.001   **14. a.** 0.01   **b.** 0.00000 1   **15. a.** 5,000   **b.** 500   **16. a.** 0.0008   **b.** 0.007

17. a. 0.06   **b.** 0.00009   **18. a.** 0.00000 5   **b.** 0.00000 0002   **19.** 10,234   **20.** 65.089

21. 521,658   **22.** 60,004,001   **23.** 7,000,000.03   **24.** 6,000,000,000.002   **25.** 500,457.34

26. 3,028.5402   **27.** 20,600.40769   **28.** $7 \times 10^2 + 4 \times 10^1 + 1 \times 10^0$

29. $7 \times 10^5 + 2 \times 10^4 + 8 \times 10^3 + 4 \times 10^2 + 7 \times 10^0$

30. $9 \times 10^{-2} + 6 \times 10^{-3} + 4 \times 10^{-4} + 2 \times 10^{-5} + 1 \times 10^{-6}$

31. $2 \times 10^1 + 7 \times 10^0 + 5 \times 10^{-1} + 7 \times 10^{-2} + 2 \times 10^{-3}$

32. $4 \times 10^1 + 7 \times 10^0 + 2 \times 10^{-3} + 1 \times 10^{-4} + 5 \times 10^{-5}$

33. $5 \times 10^2 + 2 \times 10^1 + 1 \times 10^0$ **34.** $6 \times 10^3 + 2 \times 10^2 + 4 \times 10^1 + 5 \times 10^0$

35. $2 \times 10^6 + 3 \times 10^5 + 5 \times 10^3 + 6 \times 10^2 + 8 \times 10^1 + 1 \times 10^0$

36. $4 \times 10^2 + 2 \times 10^1 + 8 \times 10^0 + 3 \times 10^{-1} + 1 \times 10^{-2}$

37. $5 \times 10^3 + 2 \times 10^2 + 4 \times 10^1 + 5 \times 10^0 + 5 \times 10^{-1}$

38. $5 \times 10^{-6} + 2 \times 10^{-7} + 7 \times 10^{-8}$ **39.** $1 \times 10^5 + 1 \times 10^{-3}$

40. $8 \times 10^2 + 9 \times 10^1 + 3 \times 10^0 + 1 \times 10^{-4}$ **41.** $8 \times 10^0 + 5 \times 10^{-5}$

42. $6 \times 10^5 + 7 \times 10^4 + 8 \times 10^3 + 1 \times 10^{-2}$

43. $5 \times 10^4 + 7 \times 10^3 + 2 \times 10^2 + 8 \times 10^1 + 5 \times 10^0 + 9 \times 10^{-1} + 3 \times 10^{-2} + 6 \times 10^{-3} + 1 \times 10^{-4}$

Level 2, page 113

44. $30 + 1 = 31$ **45.** $3,000 + 200 + 1 = 3,201$ **46.** $10,000 + 500 + 400 + 5 = 10,905$

47. $5,000,000 + 1,000 + 5 = 5,001,005$ **48.** $1,000,000 + 50,000 + 1,000 + 4 = 1,051,004$

49. $5,000,000 + 3,000,000 + 5,000 + 4,000 + 20 + 5 + 1 = 8,009,026$

50. 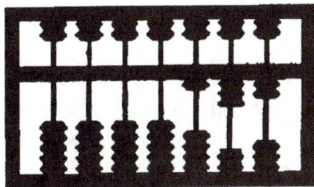   **51.**    **52.**

53.    **54.**    **55.**

56.    **57.**

Level 3 Problem Solving, page 114

58.   0, 1, 2, 10, 11, 12, 20, 21, 22, 100, **101, 102, 110, 111, 112,** ⋯.

59.   0, 1, 2, 3, 4, 10, 11, 12, 13, 14, 20, 21, **22, 23, 24, 30, 31, 32, 33, 34, 40,** ⋯.

60.

$$
\begin{array}{r}
Bb \\
\times\ 11 \\
\hline
\end{array}
\qquad
\begin{array}{r}
(B \times 10^1) + (b \times 10^0) \\
\times \quad (1 \times 10^1) + (1 \times 10^0) \\
\hline
(B \times 10^1) + (b \times 10^0) \\
(B \times 10^2) + (b \times 10^1) \\
\hline
(B \times 10^2) + [(B+b) \times 10^1] + (b \times 10^0)
\end{array}
$$

first digit   sum of the digits   last digit

3.3 Different Numeration Systems, page 114

If time permits, I have the class invent its own numeration system. I ask the students to invent a system to count the number of people in class. I lead them to some names for numbers:

   fe, fi, fo, fum, ...

If the number of students in the class is not too large, we might continue to invent new sounds for larger numbers of elements. How do we count these students? We put them into a one-to-one correspondence with certain sounds having specific meaning to us. For example, such a counting procedure might look like the following (at this point I move across a row of students in the class counting out each one):

fe, fi, fo, fum, fu, fa, de, di, do, dum,..

I stop here and notice that to have a workable numeration system, we should have symbols for each number and should also limit the number of symbols. This means we must invent a method for combining these symbols to represent large numbers. Suppose we count *fe, fi, fo,* and *fum.* If we continue as did the Egyptians, we could add one more and use the additive principle to say *fum fe* (which is the same as *fe fum*). Using this principle we would count:

fe, fi, fo, fum, fum fe, fum fi, fum fo, fum fum, fum fum fe, ...

Is this additive system very practical for large numbers? Could we improve the situation? We could invent some additional symbols for larger numbers, as did the Egyptians, or we could invent a positional or grouping system as did the Babylonians. Let's create a group called a *fiddle:*

/	*fe*
//	*fi*
///	*fo*
////	*fum*
/////̄	*fe fiddle*
/////̄ /	*fe fiddle fe*
/////̄ //	*fe fiddle fi*
/////̄ ///	*fe fiddle fo*
/////̄ ////	*fe fiddle fum*
/////̄ /////̄	*fi fiddle*

This method of counting seems a little better than simply using the additive principle. Count some things using this system. If you practice before class, you can have a lot of fun with your students doing this. Someone usually suggests the need for a placeholder; when they do I use □ and call it a *null.* What if the class had 25 or more students?

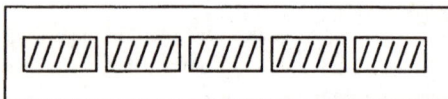

This is a *fe-fiddle* of *fiddles.* What about calling this a *fe-fiddle-fiddle.* That is, a *fiddle-fiddle* is

If you guide your class in inventing this numeration system, you can have an enjoyable class, make them think for themselves, and give a nice introduction to a base 5 system.

Level 1, page 118

1. Write the base eight number in expanded notation, and then carry out the arithmetic.

2. Write the base sixteen number in expanded notation, and then carry out the arithmetic.

3. Do repeated division by eight, and then read the remainders down for the base 8 numbers.

4. Do repeated division by sixteen, and then read the remainders down for the base 8 numbers.

5. Change from base two to base ten and then from base ten to base eight. Problems 49-59 of Section 3.4 use a simpler method: group the digits in the base two number into groups of three; then convert each group of three base two numbers into a base eight number using the following correspondence: $000 \leftrightarrow 0$, $001 \leftrightarrow 1$, $010 \leftrightarrow 2$, $011 \leftrightarrow 3$, $100 \leftrightarrow 4$, $101 \leftrightarrow 5$, $110 \leftrightarrow 6$, $111 \leftrightarrow 7$.   **6.** Answers vary.

7. **a.** 13 **b.** 23_{five} **c.** $10_{thirteen}$ **d.** 15_{eight} **e.** 1101_{two} **f.** 11_{twelve} **8. a.** 9 **b.** 14_{five}
c. 100_{three} **d.** 11_{eight} **e.** 1001_{two} **f.** 10_{nine}

9. **a.** $6 \times 8^2 + 4 \times 8^1 + 3 \times 8^0$ **b.** $5 \times 12^3 + 3 \times 12^2 + 8 \times 12^1 + 7 \times 12^0 + 9 \times 12^{-1}$

10. **a.** $1 \times 2^8 + 1 \times 2^7 + 1 \times 2^6 + 1 \times 2^4 + 1 \times 2^3$ **b.** $7 \times 8^2 + 5 \times 8$

11. **a.** $1 \times 2^5 + 1 \times 2^4 + 1 \times 2^2 + 1 \times 2^1 + 1 \times 2^0 + 1 \times 2^{-1} + 1 \times 2^{-4}$
    b. $5 \times 6^3 + 4 \times 6^2 + 1 \times 6^1 + 1 \times 6^0 + 1 \times 6^{-1} + 2 \times 6^{-3} + 3 \times 6^{-4}$

12. **a.** $6 \times 8^7 + 4 \times 8^6 + 2 \times 8^5 + 5 \times 8^1 + 1 \times 8^0$
    b. $1 \times 3^3 + 2 \times 3^1 + 1 \times 3^0 + 2 \times 3^{-1} + 2 \times 3^{-2} + 1 \times 3^{-3}$

13. **a.** $3 \times 4^5 + 2 \times 4^4 + 3 \times 4^3 + 2 \times 4^{-1}$ **b.** $2 \times 5^5 + 3 \times 5^4 + 4 \times 5^3$

14. **a.** $3 \times 5^0 + 4 \times 5^{-1} + 2 \times 5^{-3} + 3 \times 5^{-4} + 1 \times 5^{-5}$
    b. $2 \times 4^3 + 3 \times 4^1 + 3 \times 4^0 + 1 \times 4^{-1}$

15. 343   **16.** 751   **17.** 13.75   **18.** 4,307   **19.** 116   **20.** 53   **21.** 11.625   **22.** 1,862

23. 807   **24.** 2,250   **25.** 66   **26.** 582   **27.** 351.125   **28.** 2,001   **29.** 3,042   **30.** 377

31. 21310_{four} **32.** 10344_{five} **33.** $2E7_{twelve}$ **34.** 100000000_{two} **35.** 1147_{eight}

36. 3122_{five} **37.** 1001100111_{two} **38.** $2E79_{twelve}$ **39.** 2214_{seven} **40.** 1000000000_{two}

41. $28T3_{twelve}$ **42.** 1221_{three} **43.** 1030_{four} **44.** 1132_{eight} **45.** 3720_{eight}

46. 11111010001_{two}

Level 2, page 119

47. 7 weeks, 3 days   **48.** 6 days, 14 hours   **49.** 4 ft, 7 in.   **50.** 2 lb, 7 oz   **51.** 3 gross, 5 doz, 8 units   **52.** 18 quarters, 1 nickel, 4 pennies   **53.** $84 = 314_{five}$ so you would need 8 coins

54. 242_{five} **55.** $954_{twelve} = 1,360$ **56.** 33 quarters, 1 nickel and 4 pennies   **57.** $44 = 62_{seven}$; 6 weeks and 2 days   **58.** $54 = 46_{twelve}$; 4 years and 6 months   **59.** $29 = 15_{twenty-four}$; 1 day and 5 hours

Level 3 Problem Solving, page 119

60. Answers vary.

3.4 Binary Numeration System, page 119

This section introduces some general ideas about computers. The student is not expected to achieve any proficiency in understanding how a computer "works," but rather should gain an understanding of how data and programs can be stored in a computer. The binary numeration system is introduced in this section. The binary numeration system has a direct application and tie-in to computers, and is usually sufficient to give the student an appreciation of other number bases.

The way I introduce the ideas of this section is to ask for five or six "volunteers." After the volunteers are obtained, we make the following agreements: (1) Each person represents a "switch" in our binary counter. (2) A person is "off" when he holds his arms down at his sides, as shown at the left below.† (See Transparency 10.)

(3) A person is "on" when he holds his right arm up as shown at the right above. (4) An "on" switch represents "1" and an "off" represents "0." (5) A person can change state (*on* to *off*; *off* to *on*) in only one way: by receiving an impulse via his left shoulder. The only thing a "switch" must remember is to change state *every* time he receives an impulse to his left shoulder. In order to change state, the "switch" simply lowers his arm with a sweeping downward pattern.

The "switches" are then arranged so they are standing at arm's length apart. When a "switch" changes from an "on" position to an "off" position, the sweeping downward pattern is such that his fingertips give an "impulse" to the "switch" at his right as shown above.

---

†My thanks to Pat Boyle for this art work.

We are now ready to begin. Suppose we use 5 "switches" and we turn "on" the first switch by giving an impulse:

The number of impulses given to the first switch will now be recorded for display. After 6 impulses our "counter" is shown below:

After 31 impulses all switches will be "on."

If we wish, we can obtain another person to represent "overflow." For example, if we do this and give another impulse, the overflow switch will go "on."

After a little practice, we can use our binary counter to do arithmetic. Suppose, for example, we wish to add 12 plus 7. In binary notation (some students in the class work to "code") this problem becomes:

$$1100_{two}$$
$$111_{two}$$

First, clear the switches by returning them all to an "off" position. Next, give impulses to represent 1100_{two}.

Next, give impulses to represent 111_{two}. The answer is now displayed.

EXAMPLES:

$$13 \rightarrow 1101_{two}$$
$$\underline{15} \rightarrow \underline{1111}_{two}$$
$$11100_{two}$$

$$11 \rightarrow 1011_{two}$$
$$\underline{6} \rightarrow \underline{110}_{two}$$
$$10001_{two}$$

$$5 \rightarrow 101_{two}$$
$$3 \rightarrow 11_{two}$$
$$10 \rightarrow 1010_{two}$$
$$\underline{13} \rightarrow \underline{1101}_{two}$$
$$11111_{two}$$

Level 1, page 123

1. Figure 3.4 is an example of a two-state device which is used to illustrate how computers operate. In particular, if a light is on it represents a "1" and if it is off it represents a "0". 2. Answers vary. Divide the given decimal numeral to find a quotient and a remainder. Then divide the quotient by two to find the second quotient and save the remainder. Repeat this process until the quotient is zero. The base two numeral is found by listing the remainder in the reverse order from which they were saved.

3. 39	**4.** 42	**5.** 167	**6.** 170
7. 13	**8.** 9	**9.** 11	**10.** 15
11. 29	**12.** 23	**13.** 27	**14.** 31
15. 99	**16.** 119	**17.** 184	**18.** 255
19. 1101_{two}	**20.** 1111_{two}	**21.** 100011_{two}	**22.** 101110_{two}
23. 110011_{two}	**24.** 111111_{two}	**25.** 1000000_{two}	**26.** 100000000_{two}
27. 10000000_{two}	**28.** 1001100111_{two}	**29.** 1100011011_{two}	**30.** 1100100011_{two}

Level 2, page 123

31. 68 79	**32.** 80 82 73 78 84	**33.** 69 78 68	**34.** 83 65 86 69
35. HAVE	**36.** FUN	**37.** STUDY	**38.** HARD
39. 101_{two}	**40.** 10010_{two}	**41.** 1101_{two}	**42.** 11001_{two}
43. 10_{two}	**44.** 101_{two}	**45.** 100011_{two}	**46.** 11000110_{two}

Level 3, page 123

47. If we assume that war and piece are complements, then one or the other must be true.

Thus, the computer is correct.

48. Answers vary; changing from octal to decimal use expanded notation and changing from decimal to octal use repeated division by eight.

49. a. 101_{two} **b.** 110_{two} **50. a.** $001\ 100_{two}$ **b.** $101\ 110_{two}$

51. a. $001\ 110\ 111_{two}$ **b.** $110\ 010\ 100_{two}$

52. a. $111\ 000\ 100\ 101_{two}$ **b.** $011\ 000\ 110\ 010_{two}$

53. a. $101\ 111\ 000\ 000_{two}$ **b.** $000\ 100\ 011\ 010\ 000_{two}$

54. a. 5_{eight} **b.** 4_{eight}

55. a. 3_{eight} **b.** 1_{eight} **56.** 007750_{eight} **57.** 400567_{eight} **58.** 773521_{eight} **59.** 453127_{eight}

Level 3 Problem Solving, page 124

60. The trick works because card 1 represents each number from 1 to 31 (inclusive) having a one in the units column of its binary representation; card 2, each number having a one in the twos column of its binary representation; cards 3, 4, and 5 similarly. The final result leaves only one number showing through the remaining hole; this number is the chosen number.

3.5 History of Calculating Devices, page 124

The students usually enjoy the novelty of finger multiplication. I begin with the examples given in this section of the text. Then I show them the following finger multiplication, which works only for the multiplication table for 6-10. Number the fingers of both hands as shown:

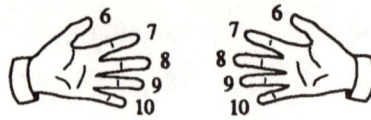

To multiply 8 × 6, place finger 8 on one hand against finger 6 on the other hand. Bend down all of the fingers below the touching fingers as shown:

The tens digit of the answer is determined by adding the open fingers on both hands (3 + 1 for this example), and the units digit is obtained by multiplying the number of closed fingers on one hand by the number of closed fingers on the other hand (2 × 4 for this example). We see that the result is 48. (See if you can explain how this works with a "carry" in the units digit; 7 × 6, for example.)

Level 1, page 134

1. Answers vary; fingers, Napier's rods, abacus, slide rule, Pascal's calculator, Leibniz' reckoning machine, Babbage's calculating machines   2. Hardware refers to the actual computer and peripherals, whereas software refers to the programs that cause the computer to operate.   3. The CPU is the microprocessor the computer uses; for example, INTEL Pentium Pro.   4. Input devices are means for inputting information into a computer; for example, keyboard, modem (Internet), CD, zip disk, or even an old floppy disk.   Output devices are means of getting information from a computer; for example, monitor, printer, modem (Internet).   5. ROM is read-only memory, and RAM is random-access memory.   6. Answers vary; data processing, census records; information retrieval, internet search; pattern recognition, television; simulation, weather forecasting

7. Answers vary; writing term papers, using the internet, chat rooms, shopping, playing games

8. Answers vary; bookkeeping, record keeping, inventory control, cash control, personnel records

9. Answers vary.   10. Answers vary.

11. a. 27 **b.** 63 **c.** 54   **12. a.** 45 **b.** 72 **c.** 81   **13. a.** 243 **b.** 432 **c.** 504   **14. a.** 315 **b.** 423 **c.** 612   **15.** ENIAC (1942), UNIVAC (1951), Cray (1958), Altair (1975), and Apple (1976) *Answers to Problems 16-33 vary.*

16. Allen was a partner with Bill Gates in beginning the Microsoft company.   **17.** Aristophanes developed a finger-counting systems about 500 B.C.   **18.** Atanasoff built the first computer in the 1940s (but he could not get it to work properly).   **19.** Babbage built a calculating machine in the 19th century.   **20.** Baran was the first to envision the internet.   **21.** Berry helped Atanasoff built the first computer.   **22.** Brand first used the word "personal computer."   **23.** Cray developed the first supercomputer.   **24.** Eckert helped build the first working computer, the ENIAC.   **25.** Engelbart invented the computer mouse.   **26.** Gates is the founder of Microsoft.   **27.** Jobs was the cofounder of Apple Computer and codesigned the Apple II computer.   **28.** Berners-Lee developed the World Wide Web.   **29.** Leibniz built a calculating device in 1695.   **30.** Mauchley helped built the first working computer, the ENIAC.   **31.** Napier invented a calculating device to do multiplication (in 1617).   **32.** Pascal built a calculating machine in 1642.   **33.** Wozniak codesigned the Apple II computer.

Level 2, page 134

34. yes; speed, complicated computations   **35.** yes; speed, complicated computations   **36.** yes; repetition   **37.** yes; speed, repetition   **38.** yes; repetition   **39.** yes; repetition   **40.** yes; repetition   **41.** yes; ability to make corrections easily   **42.** yes; tutorial   **43.** yes (but not completely); it can help with some of the technical aspects   **44.** no, although written or descriptive directions might be stored on a computer or found on the internet   **45.** yes; speed, complicated computations, repetition   **46.** B   **47.** E   **48.** D   **49.** D   **50.** E

Level 3, page 135

51. Answers vary.   **52.** Answers vary.   **53.** Answers vary.

54.
  a. Using this criterion, computers can think. In fact, they are superior to man in this category.

  b. Again, the answer is yes.

  c. In this category, computers are not too advanced. However, engineers at the Stanford Research Institute are working on a computer that can make perceptions using a video eye that recognizes objects and patterns.

d. In this category, computers also fall short of the designation of "thinking." However, it is possible to program a computer so that it can recognize a human face, or one's handwriting.

e. Computers utterly fail as thinking machines in this category, since they have absolutely no feeling or consideration. However, they can be programmed to simulate all the appropriate responses and reactions that emotions require. They can respond as if they were happy, sad, worried, or angry.

f. According to this criterion, many would say that computers cannot think. However, the answer here is not that clear-cut. Computers can write songs, play original music, write original poems, and draw original designs and paintings. For example, at Stuttgart's Technical College in Germany, a computer was programmed to write original poetry. One of the first tries with the Stuttgart computer produced a staggering result. Fed style and content in the manner of Franz Kafka, the computer turned out a manuscript so Kafkaesque that it fooled language experts. It was, they said, Kafka's writing. In the category of creating, although work is currently being done, we would have to say that a computer cannot (yet!) compete with artists.

g. There is a lot of current research on programming computers to learn by experience. Many computers today do learn by experience, however such efforts are still in the early stages of developments.

55. Answers vary.   **56.** Answers vary.   **57.** Answers vary.

Level 3 Problem Solving, page 135

58. Answers vary.   **59.** Answers vary.   **60.** Answers vary.

Chapter 3 Review Questions, page 137

1. Answers vary. The position in which the individual digits are listed is relevant; examples will vary.

2. Answers vary. Addition is easier in a simple grouping system; examples will vary.

3. Answers vary. It uses ten symbols; it is positional; it has a placeholder symbol (0); and it uses 10 as its basic unit for grouping.

4. Answers vary. Should include finger calculating, Napier's rods, Pascal's calculator, Leibniz' reckoning machine, Babbage's difference and analytic engines, ENIAC, UNIVAC, Atanasoff's and

Eckert and Mauchly's computers, and the dispute they had in proving their position in the history of computers. Should also include the role and impact of the Apple and Macintosh computers, as well as the supercomputers (such as the Cray).

5.　Answers vary. Should include illegal (breaking into another's computer, adding, modifying, or destroying information; copying programs without authorization or permission) and ignorance (assuming that output information is correct, or not using software for purposes for which it was intended).

6.　Answers vary. **a.** the physical components (mechanical, magnetic, electronic) of a computer system **b.** the routine programs, and associated documentation in a computer system **c.** the process of creating, modifying, deleting, and formatting text and materials **d.** a device connected to a computer that allows the computer to communicate with other computers using electronic cables or phone lines **e.** electronic mail — that is, messages sent along computer modems **f.** Random-Access Memory or memory where each location is uniformly accessible, often used for the storage of a program and data being processed **g.** an electronic place to exchange information with others, usually on a particular topic

7. 10^9　　**8.** $4 \times 10^2 + 3 \times 10^1 + 6 \times 10^0 + 2 \times 10^{-1} + 1 \times 10^{-5}$　　**9.** $5 \times 8^2 + 2 \times 8^1 + 3 \times 8^0$

10. $1 \times 2^6 + 1 \times 2^3 + 1 \times 2^2 + 1 \times 2^1$　　**11.** $4,020,005.62$　　**12.** $1 \times 2^4 + 1 \times 2^3 + 1 \times 2^2 + 1 \times 2^0 = 29$

13. $1 \times 2^6 + 1 \times 2^5 + 1 \times 2^4 + 1 \times 2^3 + 1 \times 2^1 + 1 \times 2^0 = 123$

14. $1 \times 3^2 + 2 \times 3^1 + 2 \times 3^0 = 17$　　**15.** $8 \times 12^2 + 2 \times 12^1 + 1 \times 12^0 = 1,177$

16.
```
        0    r. 1
   2) ‾‾1‾   r. 1
   2) ‾‾3‾   r. 0
   2) ‾‾6‾   r. 0
   2) ‾12‾            12 = 1100_two
```

17.
```
        0    r. 1
   2) ‾‾1‾   r. 1
   2) ‾‾3‾   r. 0
   2) ‾‾6‾   r. 1
   2) ‾13‾   r. 0
   2) ‾26‾   r. 0
   2) ‾52‾
              52 = 110100_two
```

18.
```
        0      r. 1
   2) ‾‾‾1‾    r. 1
   2) ‾‾‾3‾    r. 1
   2) ‾‾‾7‾    r. 1
   2) ‾‾15‾    r. 1
   2) ‾‾31‾    r. 0
   2) ‾‾62‾    r. 1
   2) ‾125‾    r. 0
   2) ‾250‾    r. 0
   2) ‾500‾    r. 1
   2)‾1001‾    r. 1
   2)‾2003‾
         2,003 = 11111010011_two
```

19. $11110100001001000000_{two}$

INSTRUCTOR'S MANUAL FOR *The Nature of Mathematics, Tenth Edition*

20. a.

```
        0    r. 9
  12) 9     r. 2
  12) 110   r. E
  12) 1,331    1,331 = 92E_twelve
```

b.

```
        0    r. 4
  5) 4      r. 0
  5) 20     r. 0
  5) 100        100 = 400_five
```

Group Research Problems, page 138

G8. Answers vary.　**G9.** Answers vary.　**G10.** Answers vary.　**G11.** Answers vary.

G12. Answers vary.

Individual Research Problems, page 139

P3.1 They all have Ph.D.'s in mathematics.　**P3.2** Answers vary.　**P3.3** Answers vary.

P3.4 Answers vary.　**P3.5** Answers vary.　**P3.6** Answers vary.　**P3.7** Answers vary.

P3.8 Answers vary.　**P3.9** Answers vary.　**P3.10** Answers vary.　**P3.11** Answers vary.

P3.12 Answers vary.

CHAPTER 4
THE NATURE OF NUMBERS

4.1 Natural Numbers, page 142

The ideas of a mathematical system and the closure, associative, commutative, and distributive properties are presented in this section. Rather than work with familiar sets of numbers, an abstract system is used so that the students can focus their attention on the *properties*. If we used sets of familiar numbers with ordinary operations, the students might know many of the results *without investigating the properties*. The student should also be encouraged to invent some mathematical systems and then to investigate their properties. The student should be able not only to describe each property but also to check whether a given set satisfies each property for a particular given operation.

The *Leprechaun Puzzle* is the best example I've found for illustrating the commutative property (or lack of it). It is shown in Problem 60, but to be effective you should have your own copy cut out and enlarged. I have mine mounted on large pieces of cardboard.

Level 1, page 146

Answers to Problems 1-9 may vary.   **1.** $N = \{1, 2, 3, 4, 5, \cdots\}$ **2.** Multiplication is repeated addition: $m \times n = n + n + n + \cdots + n$ where there are m terms of n.   **3.** Subtraction is defined in terms of addition: $m - n = x$ means $m = n + x$.   **4.** If a and b are members of a set S, then the sum $(a + b)$ is also a member of the set S.   **5.** For an operation $\circ$ and elements a and b in a set S, $a \circ b = b \circ a$.   **6.** For an operation $\circ$ and elements a, b, and c in a set S, $(a \circ b) \circ c = a \circ (b \circ c)$ **7.** For operations $\circ$ and $\otimes$ and elements a, b, and c in S, $a \otimes (b \circ c) = (a \otimes b) \circ (a \otimes c)$ **8.** If a and b are members of a set S, then the product ab is also a member of the set S.   **9.** Answers vary; associative property changes the grouping and the commutative property changes order.

10. a. $3 + 3$ **b.** $2 + 2 + 2$ **11. a.** $4 + 4 + 4$ **b.** $3 + 3 + 3 + 3$ **12. a.** $2 + 2 + 2 + 2 + 2$ **b.** $5 + 5$ **13. a.** $184 + 184$ **b.** $2 + 2 + \cdots + 2$ (a total of 184 terms)   **14. a.** $145 + 145 + 145$ **b.** $3 + 3 + 3 + \cdots + 3$ (a total of 145 terms)   **15. a.** $y + y + y + \cdots + y$ (a total of x terms) **b.** $x + x + x + \cdots + x$ (a total of y terms)   **16.** commutative   **17.** commutative   **18.** associative   **19.** associative   **20.** both   **21.** commutative   **22.** commutative   **23.** commutative   **24.** commutative   **25.** commutative   **26.** both   **27.** commutative   **28.** no   **29.** None are associative.   **30.** Answers vary; purple people eater, red hot mama, free air pump, no frill tour, bare bones budget, high flying plane

Level 2, page 147

31. Answers vary; dictionary definitions are often circular. **32. a.** 4 **b.** 4 **c.** 1 **d.** 7 **e.** 4 **f.** 4

g. 1 **h.** 7   **33. a.** $-i$ **b.** -1 **c.** 1 **d.** $-i$ **e.** i **f.** -1 **34.** Answers vary; yes

35. Answers vary; yes   **36.** Answers vary; **a.** yes **b.** yes   **37.** Answers vary; **a.** yes **b.** yes

38. a. no; $b \star c \neq c \star b$ **b.** yes; $a \star (b \star c) = (a \star b) \star c$ **39.** yes; $a \leftarrow b = a$, which, by

definition, is an element of N.   **40. a.** yes; answers vary **b.** no; give a counterexample:

$5 \leftarrow 3 \neq 3 \leftarrow 5$ **41. a.** $\square$ **b.** $\circ$ **c.** yes; no, $\circ \bullet \triangle \neq \triangle \bullet \circ$ **d.** yes

42. Check: $a \rightarrow (b \downarrow c) = (a \rightarrow b) \downarrow (a \rightarrow c)$. Try several examples; it holds.

43. Check: $a \downarrow (b \rightarrow c) = (a \downarrow b) \rightarrow (a \downarrow c)$. Try several examples; it holds.

44. a. $\begin{aligned} 6 \times (80 + 2) &= 480 + 12 \\ &= 492 \end{aligned}$ **b.** $\begin{aligned} 8 \times (40 + 1) &= 320 + 8 \\ &= 328 \end{aligned}$

c. $\begin{aligned} 7 \times (50 - 1) &= 350 - 7 \\ &= 343 \end{aligned}$ **45. a.** $\begin{aligned} 5 \times (100 - 1) &= 500 - 5 \\ &= 495 \end{aligned}$

b. $\begin{aligned} 4 \times (90 - 2) &= 360 - 8 \\ &= 352 \end{aligned}$ **c.** $\begin{aligned} 8 \times (50 + 2) &= 400 + 16 \\ &= 416 \end{aligned}$

46. yes; $0 \times 1 = 1 \times 0$

47. yes; try all possibilities

48. yes; $(-1) \times 0 = 0 \times (-1)$; $1 \times 0 = 0 \times 1$; $1 \times (-1) = (-1) \times 1$

49. yes; even + even = even   **50.** yes; even $\times$ even = even

51. yes; odd $\times$ odd = odd   **52.** no; odd + odd = even

Level 3, page 147

53. yes; let a and b be any elements in S. Then, $a \odot b = 0 \cdot a + 1 \cdot b = b$ and we know that b is an

element of S.   **54.** It is neither commutative nor associative: $2 \oplus 3 = 7$ but $3 \oplus 2 = 8$;

also $(2 \oplus 3) \oplus 4 \neq 2 \oplus (3 \oplus 4)$.   **55.** It is commutative but not associative:

$2 \otimes 3 = 10$ and $3 \otimes 2 = 10$; try other examples. $(2 \otimes 3) \otimes 4 \neq 2 \otimes (3 \otimes 4)$

Level 3 Problem Solving, page 148

56.

$\star$	ℓ	r	a	f
ℓ	a	f	r	ℓ
r	f	a	ℓ	r
a	r	ℓ	f	a
f	ℓ	r	a	f

57. yes   **58. a.** yes **b.** yes

59. Answers vary; $0 + 1 + 2 - 3 - 4 + 5 - 6 + 7 + 8 - 9 = 1$

60. I will answer this question by working a similar problem. Suppose you take a \$1 bill and cut it $\frac{1}{9}$ of the way from the end; another $\frac{2}{9}$; another $\frac{3}{9}$; $\cdots$, as shown:

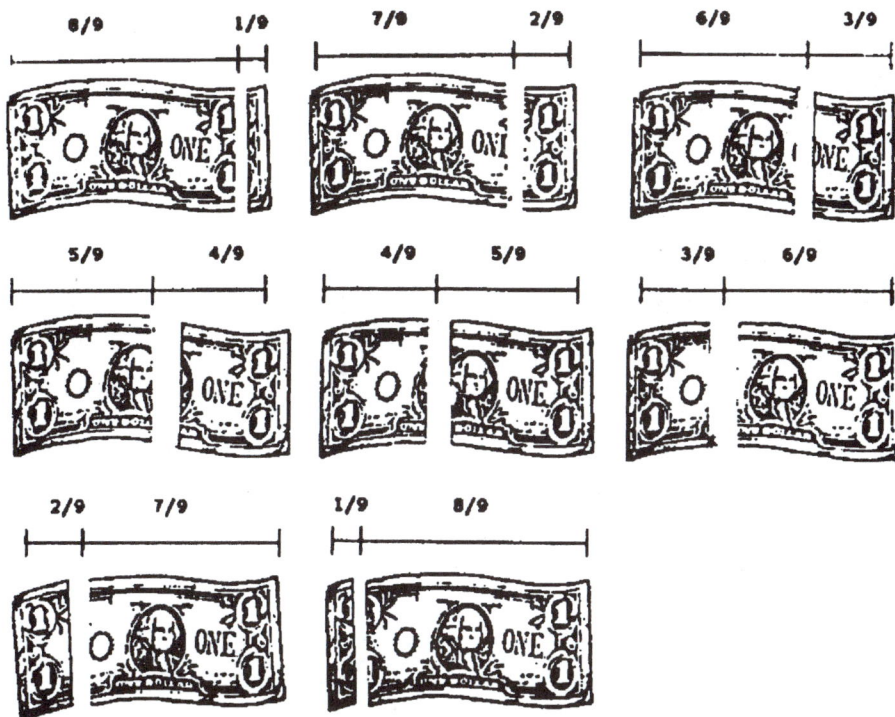

Now suppose you take the $\frac{1}{9}$ piece and tape it to the $\frac{7}{9}$ piece; the $\frac{2}{9}$ piece and tape it to the $\frac{6}{9}$ piece; and so on. Notice that you have 9 \$1 bills even though you started with only 8! (*The Vanishing Dollar Bill?*) However, the total length of the 8 original bills and the new bill is the same. The difficulty (other than with the Treasury agents) is that the bills would obviously be defective. However, if we do it with more bills as shown below, the defect is less obvious.

Level 3 Problem Solving, page 184

54.

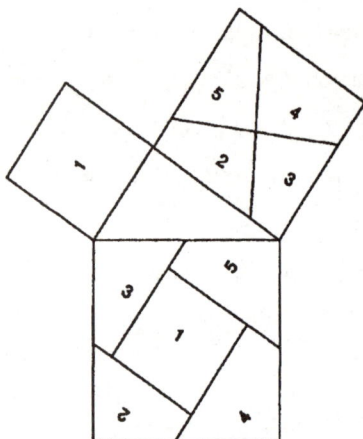

55.

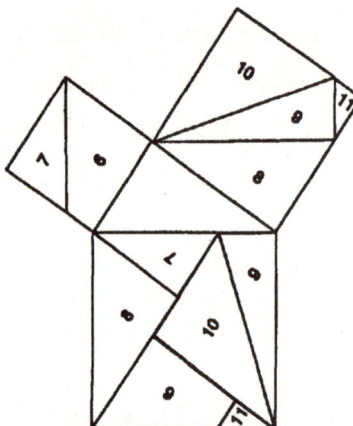

56. He can put it in a box along the diagonal since $3^2 + 4^2 = 5^2$.

57. Answers vary; any number can be written as the sum of three triangular numbers. He must have meant 3 or fewer triangular numbers, because to represent 2, for example, we write $1 + 1$.

58. Answers vary; any square number can be written as the sum of two consecutive triangular numbers.

59. Consider the second diagonal: 1, 3, 6, 10, 15, 21, $\cdots$. If we take the sum of any 2 adjacent of these terms, we obtain square numbers: 4, 9, 16, 25, $\cdots$. (*Note*: Don't forget that we start counting with the 0th diagonal.)

60. Entries in the second diagonal: 1, 3, 6, 10, 15, $\cdots$.

4.6 Groups, Fields, and Real Numbers, page 184

I introduce this section by having the students construct a table of symmetries of a square as shown on Transparency 13. They begin in class and finish at home. After they are completed, I ask them to verbalize patterns they notice. *After* they have verbalized a particular property, I give it a name. Invariably, students identify most of the field properties, but if they miss one or two, then I lead them to a discovery of those they miss. At the end, we have discussed all of the important properties, after which I use Transparency 14 to review this section. The statement of properties is written too small for the students to see. This is intentional, because it gives you a chance to call on students and ask them to talk about each of these properties.

Level 1, page 192

1. Rational numbers have repeating or terminating decimal representations, and irrational numbers do not.

The idea with the leprechauns is the same. Each of the 14 leprechauns contributes a small part in making up the 15 leprechauns, and also each of the 14 leprechauns contributes a small part in making up the leprechaun. If you want to verify this in a rather tedious way, measure the total height of the 14 leprechauns and the fifteen leprechauns. These totals are the same as the total lengths of the 14 leprechauns and 15 leprechauns, respectively.

4.2 Prime Numbers, page 149

This is one of the richest topics in all of mathematics. You could develop an entire course around the ideas of primes (see any elementary number theory book). There is a story told by Norm Shaumburger about a student who came to class after hearing a lecture on prime numbers, and was asked what the previous day's lecture was all about. The student replied, "Chuck numbers."

"Chuck numbers," retorted the professor. "Last time we talked about prime numbers!"

"Oh yah," said the student, "I knew it had something to do with meat."

Level 1, page 158

1. A prime number is a counting number with exactly two factors.   **2.** Write out a factor tree.

3. It is writing a prime factorization of a number so that the factors are written in exponential form in the order of smallest to largest.   **4.** Answers vary; g.c.f. stands for greatest common factor which means the largest number that divides evenly into all of the given numbers. To find the g.c.f. factor each of the numbers and then take one representative of each different factor, and choose as a representative the one with the smallest exponent.   **5.** Answers vary; l.c.m. stands for least common multiple which means the smallest number that all of the given numbers divides into evenly. To find the l.c.m. factor each of the numbers and then take one representative of each different factor, and choose as a representative the one with the largest exponent.   **6.** Answers vary; see the previous two answers.

7.   **a.** prime; use sieve   **b.** not prime; $3 \cdot 19$

   c. not prime; only 1 divisor   **d.** prime; check primes under 45

8.   **a.** not prime; $3^2 \cdot 7$ **b.** prime; use sieve

   c. prime; use sieve   **d.** prime; check primes under 45

9.   **a.** prime; use sieve   **b.** prime; use sieve

   c. not prime; $3^2 \cdot 19$ **d.** not prime; $3 \cdot 23 \cdot 29$

10.   **a.** not prime; $7 \cdot 13$ **b.** not prime; $3 \cdot 29$

      c. not prime; $3 \cdot 37$ **d.** prime; check primes under 45

11.   **a.** T   **b.** F   **c.** T   **d.** F

12.   **a.** T   **b.** F   **c.** F   **d.** F

13.   **a.** F   **b.** T   **c.** T   **d.** T

14.   **a.** F   **b.** T   **c.** F   **d.** F

15. Use the sieve of Eratosthenes; you only need to cross out the multiples of primes up to 17. The list of primes is: 2, 3, 5, 7, 11, 13, 17, 19, 23, 29, 31, 37, 41, 43, 47, 53, 59, 61, 67, 71, 73, 79, 83, 89, 97, 101, 103, 107, 109, 113, 127, 131, 137, 139, 149, 151, 157, 163, 167, 173, 179, 181, 191, 193, 197, 199, 211, 223, 227, 229, 233, 239, 241, 251, 257, 263, 269, 271, 277, 281, 283, and 293.

16.   **a.** 13   **b.** 19   **c.** 31   **d.** 997

17.   **a.** $2^3 \cdot 3$ **b.** $2 \cdot 3 \cdot 5$ **c.** $2^2 \cdot 3 \cdot 5^2$ **d.** $2^4 \cdot 3^2$

18.   **a.** $2^2 \cdot 3^3$ **b.** $2^2 \cdot 5 \cdot 37$ **c.** $3 \cdot 233$ **d.** $3 \cdot 41$

19.   **a.** $2^3 \cdot 3 \cdot 5$ **b.** $2 \cdot 3^2 \cdot 5$ **c.** $3 \cdot 5^2$ **d.** $3 \cdot 5^2 \cdot 13$

20.   **a.** $2 \cdot 5 \cdot 7^2$ **b.** $2^4 \cdot 3^3 \cdot 11$ **c.** $11 \cdot 13$ **d.** $3 \cdot 17$

21. prime   22. prime   23. prime   24. prime

25. $13 \cdot 29$ 26. prime   27. $3 \cdot 5 \cdot 7$ 28. $11 \cdot 17$

29. prime   30. prime   31. $3^2 \cdot 5 \cdot 7$ 32. $3 \cdot 37$

33. $3^4 \cdot 7$ 34. $2^3 \cdot 71$ 35. $19 \cdot 151$ 36. $13 \cdot 61$

37. 12; 360   38. 95; 1,425   39. 3; 2,052   40. 1; 3,289

41. 1; 252   42. 3; 270   43. 15; 1,800   44. 5; 20,400

Level 2, page 159

45. The least common multiple of 6 and 8 is 24, so the next night off together is in 3 weeks and 3 days.

46. The least common multiple of 75 and 90 is 450; 2:30 A.M. (450 minutes later)

47. Answers vary.

48. Answers vary; all primes except 2 and 3 are one more or one less than a multiple of 6.

49. 1, 3, 7, 9, 13, 15, 21, 25, 31, 33, 37, 43, 49, 51, 63, 67, 69, 73, 75, 79, 87, 93, and 99

Level 3, page 159

50. Answers vary; see answer to Problem 10 for possible pairs.

51. Let $2n + 1$, $2n + 3$, and $2n + 5$ be any 3 consecutive odd numbers. If $n = 1$, we have the prime triplets 3, 5, and 7. If $n > 1$, then one of the three numbers must be divisible by 3. Suppose the first is divisible by 3, then they are not prime triplets. If the first is not divisible by 3, then dividing it by 3 leaves a remainder of 1 or 2. If it leaves a remainder of 1, then the middle number is divisible by 3. If it leaves a remainder of 2, then the last one is divisible by 3. In all cases, the numbers will not be prime triplets.

52. Answers vary; $10 = 5 + 5$; $12 = 7 + 5$; $14 = 11 + 3$; $16 = 13 + 3$; $18 = 13 + 5$; $20 = 17 + 3$; $40 = 37 + 3$; $80 = 73 + 7$; $100 = 97 + 3$

53. **a.** yes   **b.** yes   **c.** yes

54. Suppose 23 is the largest prime. Consider
$$M = 2 \cdot 3 \cdot 5 \cdot 7 \cdot 11 \cdot 13 \cdot 17 \cdot 19 \cdot 23 + 1$$
which is either prime or composite. If M is prime, then 23 is not the largest prime. If M is composite, then it has a prime divisor. Try all prime divisors: 2, 3, 5, 7, 11, 13, 17, 19, and 23. None of these divide into M, so M must have a prime divisor that is larger than 23. In either case, 23 is not the largest prime.

Level 3 Problem Solving, page 160

55. $2 \cdot 3 \cdot 5 \cdot 7 \cdot 11 \cdot 13 + 1 = 30{,}031$, which is not prime, since $30{,}031 = 59 \cdot 509$

56. Write out the prime factorizations for the numbers 1 to 20 to find:
$$2^4 \cdot 3^2 \cdot 5 \cdot 7 \cdot 11 \cdot 13 \cdot 17 \cdot 19 = 232{,}792{,}560$$

57. Answers vary; some possibilities are: 2, 5 (included as part of the question), 17, 37, 101, 197, 257, 401, 577, 677, 1297, 1601, 2917, 3137, 4357, 5477, 7057, 8101, 8837. A computer could also be used to answer this question.

58. There is only one possibility, the one shown. This can be shown by factoring
$$p = n^2 - 1 = (n - 1)(n + 1)$$
which is prime only when $n = 2$.

59. In $N = 5$, then $2^{5-1}(2^5 - 1) = 2^4(2^5 - 1) = 16(31) = 496$. The proper divisors of 496 are 1, 2, 4, 8, 16, 31, 62, 124, and 248. The sum of these numbers is 496.

60. 1,184 has the following proper divisors: 1, 2, 4, 8, 16, 32, 37, 74, 148, 296, and 592; 1,210 has the proper divisors: 1, 2, 5, 10, 11, 22, 55, 110, 121, 242, and 605

Checking:

$$1 + 2 + 4 + 8 + 16 + 32 + 37 + 74 + 148 + 296 + 592 = 1{,}210$$
$$1 + 2 + 5 + 10 + 11 + 22 + 55 + 110 + 121 + 242 + 605 = 1{,}184$$

It might be of interest to know that there are at least 1,095 pairs of friendly numbers, and the largest known pair is:

4,522,265,534,545,208,537,974,785 and 4,539,801,326,233,928,286,140,415

4.3 Integers, page 160

Level 1, page 166

1. If $x = 0$, then $x + 0 = 0 + x = x$. To add nonzero integers x and y, look at the signs of x and y:

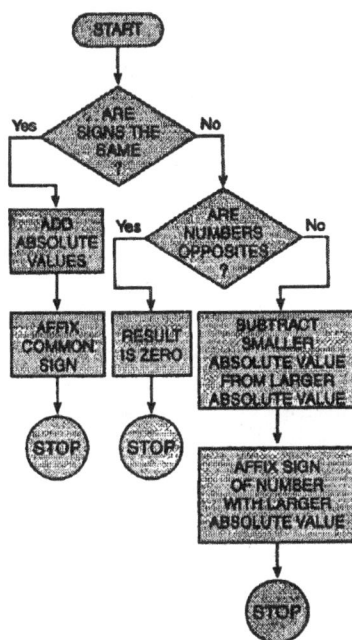

START

ARE SIGNS THE SAME ?
— Yes → ADD ABSOLUTE VALUES → AFFIX COMMON SIGN → STOP
— No → ARE NUMBERS OPPOSITES ?
— Yes → RESULT IS ZERO → STOP
— No → SUBTRACT SMALLER ABSOLUTE VALUE FROM LARGER ABSOLUTE VALUE → AFFIX SIGN OF NUMBER WITH LARGER ABSOLUTE VALUE → STOP

2. $x - y = x + (-y)$; that is, to subtract, add the opposite.

3.　　If $x = 0$, then $x \cdot 0 = 0 \cdot x = 0$. To multiply nonzero integers x and y, look at the signs of x and y:

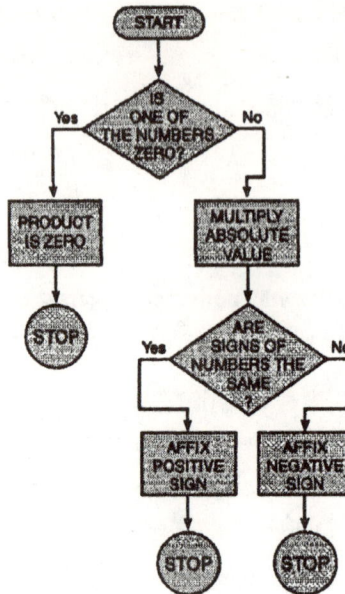

START

IS ONE OF THE NUMBERS ZERO? — Yes / No

PRODUCT IS ZERO — STOP

MULTIPLY ABSOLUTE VALUE

ARE SIGNS OF NUMBERS THE SAME? — Yes / No

AFFIX POSITIVE SIGN — STOP

AFFIX NEGATIVE SIGN — STOP

4.　　$x \div y = \frac{x}{y}$ means that we seek a number m so that $x = my$.

5.　　$0 \div 5$ means $\frac{0}{5}$ which is equal to 0; on the other hand, $5 \div 0$ is not defined.

6.　　If we use the definition of division $\frac{x}{0} = m$, then $0 \cdot m = x$. If $x \neq 0$, then there is no solution because $0 \cdot m = 0$ for all numbers m. Also, if $\frac{0}{0} = m$, then $0 = 0 \cdot m$ which is true for *every* number m. Thus, $\frac{0}{0} = 0$ checks, and $\frac{0}{0} = 1$ checks, and since two numbers equal to the same number must also be equal gives us the statement $0 = 1$.

7. a. 30　b. 30　c. 0　d. -30　e. 0　　8. a. 18　b. 18　c. -18　d. 18　e. 36

9. a. 8　b. -2　　10. a. -3　b. -6　　11. a. 10　b. -4　　12. a. -6　b. -18　　13. a. -7

b. 12　　14. a. 2　b. -52　　15. a. 3　b. -3　　16. a. 10　b. 15　　17. a. -18　b. -20

18. a. 2　b. 2　　19. a. -70　b. 70　　20. a. 20　b. -52　　21. a. -3　b. 7　　22. a. -3

b. 42　　23. a. 132　b. -1　　24. a. 1　b. 40　　25. a. -56　b. -75　　26. a. 30　b. -14

27. a. -4　b. 4　　28. a. 9　b. -9　　29. a. 150　b. -19　　30. a. 0　b. 10　　31. a. 5

b. 10　　32. a. 11　b. 174　　33. a. -12　b. 15　　34. a. -35　b. -4　　35. a. 0　b. -8

36. a. -37　b. -9　　37. a. 0　b. -23

Level 2, page 167

38. **a.** 15 **b.** -15 **39. a.** -7 **b.** 0 **40. a.** -26 **b.** 2 **41. a.** 8 **b.** -2

42. **a.** -9 **b.** -2 **43. a.** 6 **b.** 7 **44. a.** 7 **b.** 1 **45. a.** 1 **b.** 22 **46. a.** 2 **b.** 18

47. **a.** 4 **b.** 16 **48. a.** 11 **b.** -43 **49. a.** 44 **b.** 16 **50. a.** $-3,628,800$ **b.** -5

Level 3, page 167

51. **a.** 1 **b.** 1 **c.** 1 **d.** 1 **e.** 1 **52. a.** 4 **b.** 8 **c.** 16 **d.** 32 **e.** positive

53. **a.** 1 **b.** -1 **c.** -1 **d.** 1 **e.** -1 **54. a.** 4 **b.** -8 **c.** 16 **d.** -32 **e.** negative

55. Answers vary.

    a.   An operation $\star$ is commutative for a set S if $a \star b = b \star a$ for all elements a and b in S.

    b.   Yes, addition is commutative for the integers (try several examples).

    c.   Not commutative for subtraction since $7 - 5 = 2$, but $5 - 7 \neq 2$.

    d.   Yes, multiplication is commutative for integers.

    e.   Not commutative for division since $8 \div 4 = 2$, but $4 \div 8 \neq 2$.

56. **a.**   An operation $\star$ is associative for a set S if $(a \star b) \star c = a \star (b \star c)$ for all elements a, b, and c in S.

    b.   Yes, addition is associative for the integers.

    c.   Not associative for subtraction since $(2 - 3) - 5 \neq 2 - (3 - 5)$.

    d.   Yes, multiplication is associative for the integers.

    e.   Not associative for division since $20 \div (10 \div 5) \neq (20 \div 10) \div 5$.

57. Use the fact that the integers are closed for addition and that the opposite of any integer is also an integer. From this and the definition of subtraction, the desired result follows.

Level 3 Problem Solving, page 167

58. Answers vary; the set $\{-1, 0, 1\}$ is closed for multiplication.

59. **a.** 12345668765433; *note:* most calculators will show scientific notation: 1.234566877 E13.
    b. Search for patterns:

$$1 \times 9 = 9$$
$$12 \times 99 = 1188$$
$$123 \times 999 = 122877$$
$$1234 \times 9999 = 12338766$$

    By patterns: 12345668765433

60. Answers are not unique. $4 = 4 + (4 - 4) \div 4$; $5 = (4 \cdot 4 + 4) \div 4$; $6 = 4 + (4 + 4) \div 4$; $7 = (44 \div 4) - 4$; $8 = 4 + 4 + 4 - 4$; $9 = 4 + 4 + (4 \div 4)$; $10 = (44 - 4) \div 4$. This problem could be extended to finding the first 25, or the first 100 numbers using only four 4s. (See the group project G14 at the end of this chapter.)

4.4 Rational Numbers, page 168

Level 1, page 175

1. A fraction is reduced if the numerator and denominator have no common factors (except 1 or -1).

2. $\frac{a}{b} \times \frac{c}{d} = \frac{ac}{bd}$ **3.** $\frac{a}{b} \div \frac{c}{d} = \frac{ad}{bc}$ $(c \neq 0)$ **4.** $\frac{a}{b} + \frac{c}{d} = \frac{ad}{bd} + \frac{bc}{bd} = \frac{ad + bc}{bd}$

5. $\frac{a}{b} - \frac{c}{d} = \frac{a}{b} + (-1)\frac{c}{d} = \frac{a}{b} + \frac{-1}{1} \cdot \frac{c}{d} = \frac{a}{b} \cdot \frac{d}{d} + \frac{-c}{d} \cdot \frac{b}{b} = \frac{ad}{bd} + \frac{-bc}{bd} = \frac{ad}{bd} - \frac{bc}{bd} = \frac{ad - bc}{bd}$

6. a. $\frac{1}{3}$ **b.** $\frac{2}{3}$ **7. a.** $\frac{1}{5}$ **b.** $\frac{1}{4}$ **8. a.** $\frac{1}{3}$ **b.** $\frac{1}{2}$ **9. a.** 2 **b.** 2   **10. a.** $\frac{23}{5}$ **b.** $\frac{24}{5}$ or $4\frac{4}{5}$

11. a. 3 **b.** $\frac{2}{3}$ **12. a.** $\frac{3}{5}$ **b.** $\frac{2}{3}$ **13. a.** $\frac{1}{8}$ **b.** $\frac{1}{3}$ **14. a.** $\frac{3}{20}$ **b.** $\frac{1}{4}$ **15. a.** $\frac{6}{35}$ **b.** $\frac{3}{20}$

16. a. $\frac{5}{14}$ **b.** $\frac{7}{15}$ **17. a.** $\frac{17}{21}$ **b.** $\frac{7}{8}$ **18. a.** $\frac{13}{9}$ **b.** $\frac{13}{21}$ **19. a.** $\frac{-92}{105}$ **b.** $\frac{5}{6}$ **20. a.** $\frac{10}{3}$ **b.** $\frac{5}{2}$

21. a. $\frac{1}{9}$ **b.** $\frac{71}{63}$ **22. a.** $\frac{1}{15}$ **b.** -2 **23. a.** $\frac{20}{7}$ **b.** $\frac{-7}{27}$ **24. a.** 6 **b.** $\frac{10}{3}$ **25. a.** $\frac{8}{15}$ **b.** $\frac{7}{10}$

26. a. $\frac{31}{30}$ **b.** $\frac{31}{30}$ **27. a.** $\frac{10}{21}$ **b.** $\frac{1}{20}$ **28. a.** $\frac{2}{3}$ **b.** 21   **29. a.** 1 **b.** -1 **30. a.** 1 **b.** -1

31. a. 36 **b.** -25 **32. a.** $\frac{-1}{25}$ **b.** $\frac{-1}{10}$ **33. a.** $\frac{18}{35}$ **b.** $\frac{4}{45}$ **34. a.** $\frac{4}{5}$ (Note: Use distributive property) **b.** $\frac{4}{5}$ **35. a.** $\frac{1}{12}$ **b.** $\frac{-53}{48}$ **36. a.** $\frac{80}{27}$ **b.** $\frac{-15}{56}$ **37. a.** 20 **b.** 6

Level 2, page 175

38. a. $\frac{21}{20}$ **b.** $-5\frac{23}{40}$ **39. a.** $\frac{-5}{43}$ **b.** $\frac{28}{33}$ **40. a.** $\frac{11}{15}$ **b.** $\frac{39}{5}$ **41. a.** $\frac{-3}{4}$ (Note: Use distributive property) **b.** $\frac{-3}{4}$ **42. a.** $\frac{31}{9}$ **b.** $\frac{131}{25}$ **43.** $\frac{2,137}{10,800}$ **44.** $\frac{1,403}{10,800}$ **45.** $\frac{971}{3,060}$ **46.** $\frac{47}{900}$

47. $\frac{10,573}{13,020}$ **48.** $\frac{409,489}{2,831,400}$

Level 3, page 175

49. Given any two rationals $\frac{a}{b}$ and $\frac{c}{d}$; to show that $\frac{a}{b} - \frac{c}{d}$ is rational. Now,

$$\frac{a}{b} - \frac{c}{d} = \frac{ad - bc}{bd}$$

by the definition of subtraction. Also, ad and bc are integers and bd is a nonzero integer since a, c are integers and b, d are nonzero integers and the set of integers is closed for multiplication. Finally, $ad - bc$ is an integer because the integers are closed for subtraction. Therefore, $\frac{ad - bc}{bd}$ is a rational by the definition of a rational number.

50. Given any two rationals $\frac{a}{b}$ and $\frac{c}{d}$ where $c \neq 0$; to show $\frac{a}{b} \div \frac{c}{d}$ is rational. Now, $\frac{a}{b} \div \frac{c}{d} = \frac{ad}{bc}$ by the definition of division. ad is an integer because a and d are integers; also, bc is a nonzero integer because b and c are nonzero integers. Therefore $\frac{ad}{bc}$ is a rational number by the definition of a rational number.

51. The set of integers, $\mathbf{Z}$, is closed for addition, subtraction, and multiplication, but not for division since $4 \div 5$ is not an integer.

52. Yes; answers vary.   53. Yes; answers vary.

54. Yes; the set $\mathbf{Q}$ is both associative and commutative for multiplication.

Level 3 Problem Solving, page 175

55. Answers vary; $\frac{3}{4} = \frac{1}{2} + \frac{1}{4}$

56. Answers vary; $\frac{47}{60} = \frac{1}{3} + \frac{1}{5} + \frac{1}{4}$ or $\frac{1}{2} + \frac{1}{4} + \frac{1}{30}$

57. Answers vary; $\frac{67}{120} = \frac{1}{3} + \frac{1}{8} + \frac{1}{10}$ or $\frac{1}{2} + \frac{1}{20} + \frac{1}{120}$

58. Answers vary; $\frac{7}{17} = \frac{1}{3} + \frac{1}{15} + \frac{1}{85}$ or $\frac{1}{3} + \frac{1}{17} + \frac{1}{51}$

59. Answers vary; solve
$$x + \frac{2}{3}x + \frac{1}{2}x + \frac{1}{7}x = 33$$
$$\frac{97}{42}x = 33$$
$$x = \frac{1{,}386}{97}$$

Check papyrus answer: $14 + \frac{1}{4} + \frac{1}{56} + \frac{1}{97} + \frac{1}{194} + \frac{1}{388} + \frac{1}{679} + \frac{1}{776} = \frac{1{,}386}{97}$;

it checks.

60. There is only one example: $2 + 2 = 2^2$ and $2 - 2 = 0^2$

4.5 Irrational Numbers, page 175

Irrational numbers are presented in this section. The students should be able to distinguish an irrational from a rational number. The argument in this section that $\sqrt{2}$ is irrational is one that the students should be able to follow, but will probably not be able to do on their own.

The Pythagorean theorem is an important result and can be used to motivate irrational numbers. One of the very best zingers is the Railroad Problem as shown on Transparency 11. I do not give the answer (shown on Transparency 12) until the next class. Allow the students to work at this problem a little.

Level 1, page 182

1. For a right $\triangle ABC$, with sides of length a, b, and hypotenuse c, $a^2 + b^2 = c^2$. Also, if $a^2 + b^2 = c^2$ for a triangle with sides a, b, and c, then $\triangle ABC$ is a right triangle.   2. The symbol $\sqrt{4}$ is used to indicate a *number* which when it is squared is equal to 4. For this example the number $\sqrt{4}$ is the same as the number 2. On the other hand, $\sqrt{4}$ is also used to indicate an operation of finding an approximation (to a given accuracy) for the number whose square is 4. Another example is $\sqrt{3}$ is a number so that $\sqrt{3} \cdot \sqrt{3} = 3$, and as an operation $\sqrt{3} \approx 1.73$ (to two decimal place accuracy).

3. An irrational number cannot be written as a terminating or a repeating decimal. In particular, an irrational number is usually approximated by 10 or 12 decimal places when using a calculator. The symbol $\sqrt{2}$ is considered an exact representation for the number square root of 2, whereas, if I press $\sqrt{2}$ on my calculator I see a display of 1.414213562 which is considered an approximate representation for the number square root of 2.   4. A square root is *simplified* if: The radicand (the number under the radical sign) has no factor with an exponent larger than 1 when it is written in factored form. The radicand is not written as a fraction or by using negative exponents. There are no square root symbols used in the denominators of fractions.   5. The square root of 2 is a number which, when squared, is equal to 2. If we look at the last digits of the given number, we see it is 4 and $4 \cdot 4 = 16$, which does not have a last digit of 2 or 0. Consequently, it is not the square root of 2.   6. Answers vary; if you take a pile of pennies and form squares, you would find that 1, 4, 9, 16, $\cdots$ would be the number of pennies necessary to form squares. In a similar fashion, 1, 3, 6, 10, $\cdots$ would be the number of pennies necessary to form triangles. These numbers are called triangular numbers.

7. $\dfrac{12 + 144 + 20 + 3\sqrt{4}}{7} + 5 \cdot 11 = 9^2 + 0$ is a true statement.   8. a. 6 b. 7 c. 9 d. 14

9. a. 30 b. 36 c. 807 d. 169   10. a. 400 b. 2.5 c. 2.4 d. 0.25   11. a. a b. xy c. $4b$ d. $40w$ 12. a. rational; 3 b. rational; 5 c. irrational; 2.718 d. irrational; 3.142

13. a. irrational; 3.162 b. irrational; 5.477 c. irrational; 9.870 d. irrational; 7.389

14. **a.** rational; 6   **b.** irrational; 7.071   **c.** irrational; 1.649   **d.** irrational; 1.571

15. **a.** rational; 13   **b.** rational; 20   **c.** irrational; 23.141   **d.** irrational; 22.459

16. **a.** irrational; 22.361   **b.** irrational; 31.623   **c.** irrational; 0.524   **d.** irrational; 8.540

17. **a.** rational; 32   **b.** rational; 44   **c.** irrational; 1.772   **d.** irrational; 1.253   **18. a.** -4
b. -12 **c.** $5\sqrt{5}$ **d.** $4\sqrt{6}$ **19. a.** $10\sqrt{10}$ **b.** $20\sqrt{7}$ **c.** $8\sqrt{35}$ **d.** $21\sqrt{10}$ **20. a.** $15\sqrt{3}$
b. $6\sqrt{10}$ **c.** $20\sqrt{3}$ **d.** $12\sqrt{6}$ **21. a.** $\frac{1}{2}\sqrt{2}$ **b.** $\frac{1}{3}\sqrt{3}$ **c.** $\frac{1}{5}\sqrt{15}$ **d.** $\frac{1}{7}\sqrt{21}$ **22. a.** $-0.1\sqrt{10}$
b. $-0.2\sqrt{10}$ **c.** $0.5\sqrt{3}$ **d.** $0.1\sqrt{5}$ **23. a.** $\frac{1}{2}\sqrt{2}$ **b.** $-\frac{1}{3}\sqrt{3}$ **c.** $\frac{2}{5}\sqrt{5}$ **d.** $\frac{1}{2}\sqrt{10}$ **24. a.** $a+b$
b. simplified   **25. a.** simplified   **b.** $x+2$ **26. a.** 1   **b.** 3   **27. a.** $10\sqrt{2}$ **b.** 0

28. **a.** $2\sqrt{15}$ **b.** $2\sqrt{2}$ **29. a.** $3+\sqrt{5}$ **b.** $2-\sqrt{3}$ **30. a.** $\frac{4-\sqrt{2}}{2}$ **b.** $\frac{3-\sqrt{5}}{2}$

31. **a.** $1-3\sqrt{x}$ **b.** $-3-\sqrt{x}$ **32. a.** $\frac{3\sqrt{x}}{x}$ **b.** $\frac{-7\sqrt{y}}{y}$ **33. a.** $\frac{2x}{5y}\sqrt{y}$ **b.** $\frac{1}{4x}\sqrt{5xy}$

34. $\dfrac{-7+\sqrt{25}}{4} = -\dfrac{1}{2}$ **35.** $\dfrac{2-\sqrt{76}}{12} = \dfrac{2-2\sqrt{19}}{12} = \dfrac{1-\sqrt{19}}{6}$

36. $\dfrac{-10-\sqrt{28}}{6} = \dfrac{-10-2\sqrt{7}}{6} = \dfrac{-5-\sqrt{7}}{3}$

37. $\dfrac{12+\sqrt{148}}{2} = \dfrac{12+2\sqrt{37}}{2} = 6+\sqrt{37}$

Level 2, page 183

38. 24 ft   **39.** 24 ft   **40.** 13 ft   **41.** 10 ft   **42.** $\sqrt{325}$ or $5\sqrt{13}$; 18 ft; 55 ft   **43.** $\sqrt{8}$ in. or
$2\sqrt{2}$ in.   **44.** $\sqrt{18}$ ft or $3\sqrt{2}$ ft   **45.** 200 ft   **46.** $\sqrt{464}$ or $4\sqrt{29}$; 22 ft   **47.** 15 ft

Level 3, page 183

48. **a.** All three figures have the same total length of segments; it is 2 in. (one unit up, and one unit over).   **b.** The diagonal has length $\sqrt{1+1} = \sqrt{2}$ in. (by the Pythagorean theorem).   **c.** No matter how many stairs there are, the two answers of 2 and $\sqrt{2}$ are not changing. Another way of saying this is to say that, with an appropriate scale change, the steps will always look like the one shown in Figure 4.9.   **49.** Answers vary; $1.2323323332\cdots$ **50.** Answers vary; $0.53353335\cdots$ **51.** Answers vary;
$0.0919919991\cdots$ **52.** Answers vary; $0.232332\cdots$ **53.** Solution comes from Pythagorean theorem.
a. 2-in. square   **b.** same   **c.** 7-in. square   **d.** 9-in. square   **e.** 15-in. square

2. If you look at Figure 4.4, you will see that the segment with length $\sqrt{2}$ is constructed by drawing a right triangle with legs of length 1. Then, by the Pythagorean theorem, the length of the hypotenuse is $\sqrt{1^2 + 1^2} = \sqrt{2}$. Now, with that diagram, use the hypotenuse of length $\sqrt{2}$ as one of the legs of *another* right triangle constructed so that the other leg has length 1, as shown below. The length of the hypotenuse of this new triangle is $\sqrt{(\sqrt{2})^2 + 1^2} = \sqrt{2+1} = \sqrt{3}$.

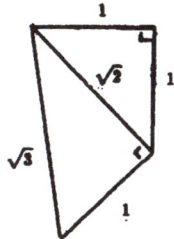

3. Answers vary; The set **S** satisfies the *identity* for ∘: There exists a number I ∈ **S** so that $x \circ \text{I} = \text{I} \circ x = x$ for *every* $x \in$ **S**. You might also mention the identity for addition in the set of real numbers: There exists in **R** a number 0, called **zero**, so that $0 + a = a + 0 = a$ for any $a \in$ **R**. Finally, you might recall the identity for multiplication in the set of real numbers: There exists in **R** a number 1, called **one**, so that $1 \times a = a \times 1 = a$ for any $a \in$ **R**.

4. Answers vary;   The set **S** satisfies the *inverse* for ∘: For *each* $x \in$ **S**, there exists a corresponding $x^{-1} \in$ **S** so that $x \circ x^{-1} = x^{-1} \circ x = \text{I}$ where I is the identity element in **S**. You might also mention the inverse property for addition in the set of real numbers: For each $a \in$ **R**, there is a unique number $(-a) \in$ **R**, called the **opposite** (or additive inverse) of a, so that $a + (-a) = -a + a = 0$. Finally, you might recall the inverse property for multiplication in the set of real numbers. For *each* number $a \in$ **R**, $a \neq 0$, there exists a number $a^{-1} \in$ **R**, called the **reciprocal** (or **multiplicative inverse**) of a, so that $a \times a^{-1} = a^{-1} \times a = 1$.

5. Let **S** be any set, let ∘ be any operation, and let a, b, and c be any elements of **S**. We say that **S** is a **group** for the operation of ∘ if the following properties are satisfied:

1. The set **S** is *closed* for ∘: $(a \circ b) \in$ **S**.

2. The set **S** is *associative* for ∘: $(a \circ b) \circ c = a \circ (b \circ c)$.

3. The set **S** satisfies the *identity* for ∘: There exists a number I ∈ **S** so that $x \circ \text{I} = \text{I} \circ x = x$ for *every* $x \in$ **S**.

4. The set **S** satisfies the *inverse* for ∘: For *each* $x \in$ **S**, there exists a corresponding $x^{-1} \in$ **S** so that $x \circ x^{-1} = x^{-1} \circ x = \text{I}$, where I is the identity element in **S**.

6. A **field** is a set $\mathbb{R}$, with two operations $+$ and $\times$ satisfying the following properties for any elements a, b, $c \in \mathbb{R}$:

	Addition, $+$	*Multiplication,* $\times$
Closure:	1. $(a + b) \in \mathbb{R}$	2. $ab \in \mathbb{R}$
Associative:	3. $(a + b) + c = a + (b + c)$	4. $(a \times b) \times c = a \times (b \times c)$
Identity:	5. There exists $0 \in \mathbb{R}$ so that $0 + a = a + 0 = a$ for every element a in $\mathbb{R}$.	6. There exists $1 \in \mathbb{R}$ so that $1 \times a = a \times 1 = a$ for every element a in $\mathbb{R}$.
Inverse:	7. For each $a \in \mathbb{R}$, there is a unique number $(-a) \in \mathbb{R}$ so that $a + (-a) = (-a) + a = 0$	8. For each $a \in \mathbb{R}$, $a \neq 0$, there is a unique number $\frac{1}{a} \in \mathbb{R}$ so that $a \times \frac{1}{a} = \frac{1}{a} \times a = 1$
Commutative:	9. $a + b = b + a$	10. $ab = ba$

Distributive for

     multiplication over addition: 11. $a \times (b + c) = a \times b + a \times c$

7. a. $\mathbb{N}, \mathbb{Z}, \mathbb{Q}, \mathbb{R}$ **b.** $\mathbb{Q}, \mathbb{R}$ **c.** $\mathbb{Q}', \mathbb{R}$ **d.** $\mathbb{Q}', \mathbb{R}$ **e.** $\mathbb{Q}, \mathbb{R}$ **f.** $\mathbb{Q}, \mathbb{R}$ **g.** $\mathbb{Q}, \mathbb{R}$ **h.** $\mathbb{N}, \mathbb{Z}, \mathbb{Q}, \mathbb{R}$

i. $\mathbb{Z}, \mathbb{Q}, \mathbb{R}$ **8. a.** $\mathbb{N}, \mathbb{Z}, \mathbb{Q}, \mathbb{R}$ **b.** $\mathbb{Q}, \mathbb{R}$ **c.** $\mathbb{Q}, \mathbb{R}$ **d.** $\mathbb{Q}', \mathbb{R}$ **e.** $\mathbb{Q}', \mathbb{R}$ **f.** $\mathbb{Q}, \mathbb{R}$

g. $\mathbb{N}, \mathbb{Z}, \mathbb{Q}, \mathbb{R}$ **h.** $\mathbb{Q}', \mathbb{R}$ **i.** $\mathbb{Q}, \mathbb{R}$ **9. a.** 1.5 **b.** 0.7 **c.** 0.6 **d.** 1.8 **10. a.** $0.8\overline{3}$

b. $0.\overline{285714}$ **c.** $0.1\overline{2}$ **d.** $2.1\overline{6}$ **11. a.** $0.\overline{6}$ **b.** $2.\overline{153846}$ **c.** 5 **d.** $1.\overline{09}$ **12. a.** -0.8

b. $-0.\overline{6}$ **c.** $-2.8\overline{3}$ **d.** -2.8 **13. a.** $\frac{1}{2}$ **b.** $\frac{4}{5}$ **14. a.** $\frac{1}{4}$ **b.** $\frac{3}{4}$ **15. a.** $\frac{9}{20}$ **b.** $\frac{117}{500}$

16. a. $\frac{111}{1,000}$ **b.** $\frac{13}{25}$ **17. a.** $\frac{987}{10}$ **b.** $\frac{63}{100}$ **18. a.** $\frac{6}{25}$ **b.** $\frac{329}{20}$ **19. a.** $\frac{153}{10}$ **b.** $\frac{139}{20}$

20. a. $\frac{16}{25}$ **b.** $\frac{349}{50}$ **21. a.** 3.179 **b.** -5.504 **c.** 1.901 **d.** -6.31 **22. a.** -5.121 **b.** 8.225

c. 27.738 **d.** 45.063018 **23. a.** -0.13112 **b.** -65.415 **c.** 12.4 **d.** 0.85 **24. a.** 653

b. 1.9990005 **c.** 4.8 **d.** 0.25625 **25. a.** 7.46 **b.** -5.15 **c.** 72.6 **d.** 45.96

26. commutative   **27.** commutative   **28.** inverse   **29.** distributive   **30.** associative

31. commutative   **32.** commutative   **33.** associative   **34.** inverse   **35.** identity

Level 2, page 193

36. $\mathbb{N}$ is not a group for $+$ since there is no identity.   **37.** $\mathbb{N}$ is not a group for $-$ since it is not closed.   **38.** $\mathbb{W}$ is not a group for $+$ since the inverse property is not satisfied.   **39.** $\mathbb{W}$ is not a group for $\times$ since the inverse property is not satisfied.   **40.** $\mathbb{Z}$ is a group for $+$.   **41.** $\mathbb{Z}$ is not a group for × since the inverse property is not satisfied.   **42. Q** is a group for +.   **43. Q** is not a group for × since there is no multiplicative inverse for the rational number 0.   **44. Q** is not a group for ÷ since it is not associative. Also, 0 does not have an inverse for the operation of division.

45. a.

×	1	2	3	4
1	1	2	3	4
2	2	4	6	8
3	3	6	9	12
4	4	8	12	16

b.

*	1	2	3	4
1	2	2	2	2
2	4	4	4	4
3	6	6	6	6
4	8	8	8	8

c. Operation ×: Not closed; associative; identity is 1; no inverse property; commutative

Operation *: Not closed; not associative; no identity; no inverse property; not commutative

Level 3, page 194

Answers for Problems 46-60 may vary.   **46.** 2; π **47.** 2.5; $\frac{2\pi}{3}$ **48.** 3.5; π

49. $4.\overline{5}$; $4.545545554\cdots$ **50.** 8.005; $8.005000500005\cdots$

51. $\left\{\frac{1}{2}, \frac{1}{3}, \frac{2}{3}, \frac{1}{4}, \frac{3}{4}, \frac{1}{5}, \frac{2}{5}, \frac{3}{5}, \frac{4}{5}, \frac{1}{6}, \frac{5}{6}, \frac{1}{7}, \frac{2}{7}, \cdots\right\}$ **52.** $a \circ b = a + b$ **53.** $a \circ b = ab$ **54.** $a \circ b = a - b$

55. $a \circ b = a + b + 1$ **56.** $a \circ b = ab + 1$ **57.** $a \circ b = 2(a + b)$ **58.** $a \circ b = 2a - b$

59. $a \circ b = a^2 + 1$

Level 3 Problem Solving, page 194

60. a.

★	A	B	C	D	E	F	G	H
A	B	C	D	A	H	G	E	F
B	C	D	A	B	F	E	H	G
C	D	A	B	C	G	H	F	E
D	A	B	C	D	E	F	G	H
E	G	F	H	E	D	B	A	C
F	H	E	G	F	B	D	C	A
G	F	H	E	G	C	A	D	B
H	E	G	F	H	A	C	B	D

b. yes **c.** yes **d.** no; $A \star E \neq E \star A$ **e.** yes; identity is D **f.** inverse of A is C; inverse of B is B; inverse of C is A; inverse of D is D; inverse of E is E; inverse of F is F; inverse of G is G; and inverse of H is H. **g.** Yes; closure, associative, identity, and inverse properties are satisfied.

4.7 Discrete Mathematics, page 194

Our purposes for studying a finite modular system are twofold. First, we wish to emphasize the nature of different types of algebras. The students can contrast a finite algebra with an infinite one. Second, we are able to again define the operations of subtraction, multiplication, and division. Notice that they are consistent with our previous definitions, but the problems of this section will require that the students use and understand *what* these operations mean. If we were to define them in terms of ordinary arithmetic, it would be possible for the students to answer the questions on the basis of past experience and intuition, without really using the definitions of these operations. We feel that a better understanding of the *operation* will result from discussing them in a context other than ordinary arithmetic. This section will probably take two or three days of class time.

Level 1, page 202

1. Answers vary.   **2.** Addition: $a + b$ is found on a 24-hr clock by starting at midnight and then counting out a hours, followed by b hours. The answer is the time found on the clock after those moves. Multiplication: $a \times b$ means $b + b + \cdots + b$ where there are a addends defined as 24-hr clock addition.   **3.** Answers vary.   **4.** The real numbers a and b are congruent modulo m, written $a \equiv b$, (mod m) if a and b differ by a multiple of m.   **5. a.** 3 **b.** 10   **6. a.** 3 **b.** 2   **7. a.** 5 **b.** 10   **8. a.** 6 **b.** 5   **9. a.** 12 **b.** 8   **10. a.** 8 **b.** 3   **11. a.** 8 **b.** 6   **12. a.** no values **b.** 4   **13. a.** 8 **b.** 1   **14. a.** 11 **b.** 12   **15. a.** T **b.** F   **16. a.** T **b.** F   **17. a.** T **b.** T   **18. a.** F **b.** F   **19. a.** T **b.** T   **20. a.** F **b.** F   **21. a.** 0, (mod 5) **b.** 8, (mod 12)   **22. a.** 2, (mod 5) **b.** 3, (mod 5)   **23. a.** 3, (mod 4) **b.** 3, (mod 5)   **24. a.** 4, (mod 8) **b.** 2, (mod 9)   **25. a.** 2, (mod 5) **b.** 2, (mod 8)   **26. a.** 3, (mod 7) **b.** 0, (mod 121)   **27. a.** 3, (mod 5) **b.** 10, (mod 11)   **28. a.** 0, (mod 2) **b.** 6, (mod 13) *Note: You must try all possibilities in doing these problems; there may be more than one solution.*   **29. a.** 4, (mod 7) **b.** 4, (mod 5)   **30. a.** 6, (mod 9) **b.** no values **31. a.** 5, (mod 6) **b.** 3, (mod 7)   **32. a.** 6, (mod 7) **b.** 5, (mod 11)   **33. a.** 1, 3, (mod 4) **b.** 2, (mod 9)   **34. a.** 1, 4, (mod 5) **b.** 5, (mod 13)   **35. a.** 0, (mod 4) **b.** 2, (mod 4)   **36.** 3, 4, (mod 7)   **37.** all x's (mod 2); namely, 0, 1

Level 2, page 202

38. a. 3, (mod 7); Thursday **b.** 1, (mod 7); Tuesday **c.** 1, (mod 7); Tuesday **d.** 2, (mod 7); Wednesday   **39. a.** 2, (mod 7); Sunday **b.** 6, (mod 7); Thursday **c.** 5, (mod 7); Wednesday **d.** 3, (mod 7); Monday   **40.** On a 24-hour clock, $8 + 8k \equiv 0, 8, 16$, (mod 24) for any value of k; thus, you will never have to take your medication between midnight and 8 A.M. The medication would be taken at 8:00 A.M., 4:00 P.M., and 12 midnight. Therefore, you will not need to take it between midnight and 7:00 A.M.   **41.** Let $x =$ the number of miles to aunt's house. Then the total mileage for the six round trips is $12x \equiv 8$, (mod 10). Solving, $x \equiv 4, 9$, (mod 10). The possible distances are 4, 14, 24, 34, $\cdots$ or 9, 19, 29, 39, $\cdots$.   **42.** You must buy $12x$ inches of rope, where x is the number of multiples you will buy. Also, you need $7k + 80$ inches where k is the number of pieces you will obtain. This means that $12x \equiv 80 \pmod 7$. Solving, $x \equiv 2 \pmod 7$. Thus, $x = 2, 9, 16, 23, \cdots$. If $x = 16$ you will have one piece 80 inches long and 16 pieces 7 inches long with no waste. Thus, you must buy 192 inches of rope (or 16 ft of rope). **43.** She lives 14 miles from your house. **44.** Answers vary; it is a group for addition modulo 6. **45.** Answers vary; it is a group for addition modulo 7.

46. Answers vary; it is not a field because the inverse property for multiplication is not defined.

47.

+	0	1	2	3	4	5	6	7	8	9	10
0	0	1	2	3	4	5	6	7	8	9	10
1	1	2	3	4	5	6	7	8	9	10	0
2	2	3	4	5	6	7	8	9	10	0	1
3	3	4	5	6	7	8	9	10	0	1	2
4	4	5	6	7	8	9	10	0	1	2	3
5	5	6	7	8	9	10	0	1	2	3	4
6	6	7	8	9	10	0	1	2	3	4	5
7	7	8	9	10	0	1	2	3	4	5	6
8	8	9	10	0	1	2	3	4	5	6	7
9	9	10	0	1	2	3	4	5	6	7	8
10	10	0	1	2	3	4	5	6	7	8	9

×	0	1	2	3	4	5	6	7	8	9	10
0	0	0	0	0	0	0	0	0	0	0	0
1	0	1	2	3	4	5	6	7	8	9	10
2	0	2	4	6	8	10	1	3	5	7	9
3	0	3	6	9	1	4	7	10	2	5	8
4	0	4	8	1	5	9	2	6	10	3	7
5	0	5	10	4	9	3	8	2	7	1	6
6	0	6	1	7	2	8	3	9	4	10	5
7	0	7	3	10	6	2	9	5	1	8	4
8	0	8	5	2	10	7	4	1	9	6	3
9	0	9	7	5	3	1	10	8	6	4	2
10	0	10	9	8	7	6	5	4	3	2	1

48. yes   **49.** No, because 0 does not have a multiplicative inverse.   **50.** yes   **51.** No, because it is not a group (see Problem 49), it cannot be a commutative group.   **52.** yes

Level 3 Problem Solving, page 203

53. a. The sum is 168, so the check (last) digit is 8.   **b.** The sum is 182, so the check digit is 5.

54. a. The sum is 164, so the check digit is 1.   **b.** The sum is 332, so the check digit is 9.

55. The sum is 100, so the check digit is 10.

56. When dividing by 24, look for a pattern on the remainders:

$99,999,999,999 \equiv 15 \pmod{24}$; thus, the answer is 5 A.M.

57. Assign eleven teams the numbers 1 to 11. On the first day, all teams whose sum is 1 (mod 11) play; Day 2, matches those whose sum is 2 (mod 11); ... The last team plays the left over numbered team so each team plays everyday. The schedule follows:

Day 1:	1-11;	2-10;	3-9;	4-8;	5-7;	6-12
Day 2:	2-11;	3-10;	4-9;	5-8;	6-7;	1-12
Day 3:	3-11;	4-10;	5-9;	6-8;	1-2;	7-12
Day 4:	4-11;	5-10;	6-9;	7-8;	1-3;	2-12
Day 5:	5-11;	6-10;	7-9;	1-4;	2-3;	8-12
Day 6:	6-11;	7-10;	8-9;	1-5;	2-4;	3-12
Day 7:	7-11;	8-10;	1-6;	2-5;	3-4;	9-12
Day 8:	9-10;	1-7;	2-6;	3-5;	8-11;	4-12
Day 9:	9-11;	1-8;	2-7;	3-6;	4-5;	10-12
Day 10:	10-11;	1-9;	2-8;	3-7;	4-6;	5-12
Day 11:	1-10;	2-9;	3-8;	4-7;	5-6;	11-12

Note: The answer is not unique. In fact, there are 39,916,800 possible solutions.

58. 5 roosters, 1 hen, and 94 chicks (answer is not unique)

Problems 59 and 60 can be solved using Diophantine equations, but I used a simple computer program to do them by brute force.   **59.** The smallest result is 785; at first, there were 46 coins for each pirate plus 3 for the cook. After 6 were killed, there are 71 coins for each of the 11 pirates plus 4 for the cook. Finally, the 6 remaining pirates received 130 coins each with 5 for the cook.   **60.** The second smallest result is 1907; the successive "share" for each pirate is 112, 173, and 317 coins each.

4.8 Cryptography, page 203

To introduce this day's lesson, I use Transparency 15 and invite the students to encode a message. Then I ask them to pass their message to someone else for decoding. This gives rise to the consideration that the idea of a coded message is to code it so that it cannot be decoded. I ask them to try again, this time trying to make decoding more difficult. This leads to a discussion of the topics of this section.

Level 1, page 207

1. 14-5-22-5-18-29-19-1-25-29-14-5-22-5-18-28

2. 20-8-5-29-5-1-7-12-5-29-8-1-19-29-12-1-14-4-5-4-28

3. 25-15-21-29-2-5-20-29-25-15-21-18-29-12-9-6-5-28

4. 13-25-29-2-1-14-11-29-2-1-12-1-14-3-5-29-9-19-29-14-5-7-1-20-9-22-5-28

5. ARE WE HAVING FUN YET   **6.** I LOVE MATHEMATICS

7. FAILURE TEACHES SUCCESS   **8.** SQUARE MEALS MAKE ROUND PEOPLE

9. Divide by 8.   **10.** Multiply by 6.   **11.** Subtract 2 and divide by 20.   **12.** Add 3 and multiply by 4.   **13.** Divide by 2, then subtract 2, then divide by 4.   **14.** Multiply by 2, add 3 and then divide by 3.

Level 2, page 207

15. 14-5-22-5-18-29-19-1-25-29-14-5-22-5-18-28   *Original message*

42-15-66-15-54-87-57-3-75-87-42-15-66-15-54-84   *Encode (multiply by 3)*

13-15-8-15-25-29-28-3-17-29-13-15-8-15-25-26   *Modulate*

MOHOY .CQ MOHOYZ   *Coded message*

16. 20-8-5-29-5-1-7-12-5-29-8-1-19-29-12-1-14-4-5-4-28   *Original*

100-40-25-145-25-5-35-60-25-145-40-5-95-145-60-5-70-20-25-20-140   *Encode*

13-11-25-29-25-5-6-2-25-29-11-5-8-29-2-5-12-20-25-20-24   *Modulate*

MKY YEFBY KEH BELTYTX   *Coded*

17. 25-15-21-29-2-5-20-29-25-15-21-18-29-12-9-6-5-28   *Original*

55-35-47-63-9-15-45-63-55-35-47-41-63-29-23-17-15-61   *Encode*

26-6-18-5-9-15-16-5-26-6-18-12-5-29-23-17-15-3   *Modulate*

ZFREIOPEZFRLE WQOC   *Coded message*

18. 13-25-29-2-1-14-11-29-2-1-12-1-14-3-5-29-9-19-29-14-5-7-1-20-9-22-5-28   *Original*

Encode: 42-90-106-27-23-46-34-106-27-23-38-23-46-2-10-106-26-66-106-46-10-18-
23-70-26-78-10-102

Modulate: 13-3-19-27-23-17-5-19-27-23-9-23-17-2-10-19-26-8-19-17-10-18-23-12-
26-20-10-15

Coded message: MCS,WQES,WIWQBJSZHSQJRWLZTJO

19. HUMPTY DUMPTY IS A FALL GUY.

20. A PROFESSOR IS ONE WHO TALKS IN SOMEONE ELSES SLEEP

21. MIDAS HAD A GILT COMPLEX.

22. TMMRU.Y,SQ,RVJRQT BY.RZ.,GY.RKQSFM.ZC.P

23. D UPEOTEHEQAUXUJGEWAUOEHEQHGHEXAUNTDDCRZEDN
UULEHRQECEPCLLED UPEVUJEHERCRNUOXJGTAB

Level 3 Problem Solving, page 208

24. I LOATHE PEOPLE WHO KEEP DOGS. THEY ARE COWARDS WHO
HAVEN'T GOT THE GUTS TO BITE PEOPLE THEMSELVES.

25. ANYONE WHO SLAPS CATSUP, MUSTARD, AND RELISH ON HIS HOT
    DOG IS TRULY A MAN FOR ALL SEASONINGS.

26. RIVAL TV NETWORKS, SEEKING HIGH RATINGS, CONSTANTLY SCHEDULE BEST
    SHOWS AGAINST EACH OTHER.

27. CRYSTAL-CLEAR AIR OF ROCKY MOUNTAINS PROVIDES IDEAL ENVIRONMENT FOR
    WEATHER STATION.

Answers for Problems 28-30 are not unique.

28.   9567
    + 1085
    ‾‾‾‾‾‾
     10652

29. 3915
       15
    + 4826
    ‾‾‾‾‾‾
     8756

30.   636
     8526
    + 1095
    ‾‾‾‾‾‾
    10257

The 3 and the 9 are interchangeable.

Chapter 4 Review Questions, page 210

1. $-4 + 5(-3) = -4 - 15 = -19$ 2. $\frac{4}{7} \cdot \frac{9}{9} + \frac{5}{9} \cdot \frac{7}{7} = \frac{36}{63} + \frac{35}{63} = \frac{71}{63}$

3.
$$30 = 2^1 \cdot 3^1 \cdot 5^1 \cdot 7^0 \cdot 11^0$$
$$42 = 2^1 \cdot 3^1 \cdot 5^0 \cdot 7^1 \cdot 11^0$$
$$99 = 2^0 \cdot 3^2 \cdot 5^0 \cdot 7^0 \cdot 11^1$$
$$\text{l.c.m.} = 2^1 \cdot 3^2 \cdot 5^1 \cdot 7^1 \cdot 11^1 = 6,930$$

$$\frac{7}{2 \cdot 3 \cdot 5} \cdot \frac{3 \cdot 7 \cdot 11}{3 \cdot 7 \cdot 11} = \frac{1,617}{2 \cdot 3^2 \cdot 5 \cdot 7 \cdot 11}$$
$$\frac{5}{2 \cdot 3 \cdot 7} \cdot \frac{3 \cdot 5 \cdot 11}{3 \cdot 5 \cdot 11} = \frac{825}{2 \cdot 3^2 \cdot 5 \cdot 7 \cdot 11}$$
$$\frac{5}{3^2 \cdot 11} \cdot \frac{2 \cdot 5 \cdot 7}{2 \cdot 5 \cdot 7} = \frac{350}{2 \cdot 3^2 \cdot 5 \cdot 7 \cdot 11}$$
$$\overline{}$$
$$\frac{2,792}{2 \cdot 3^2 \cdot 5 \cdot 7 \cdot 11} = \frac{1,396}{3^2 \cdot 5 \cdot 7 \cdot 11} = \frac{1,396}{3,465}$$

4. $-\sqrt{10} \cdot \sqrt{10} = -10$ 5. $\left(\frac{11}{12} + 2\right) + \frac{-11}{12} = 2 + \left(\frac{11}{12} + \frac{-11}{12}\right) = 2$

6. $\dfrac{3^{-1} + 4^{-1}}{6} = \dfrac{\frac{1}{3} + \frac{1}{4}}{6} = \dfrac{\frac{7}{12}}{6} = \dfrac{7}{12} \times \dfrac{1}{6} = \dfrac{7}{72}$

7. $\dfrac{-7}{9} \cdot \dfrac{99}{174} + \dfrac{-7}{9} \cdot \dfrac{75}{174} = \dfrac{-7}{9}\left(\dfrac{99}{174} + \dfrac{75}{174}\right) = \dfrac{-7}{9}(1) = -\dfrac{7}{9}$

8. $\dfrac{-3 + \sqrt{3^2 + 4(2)(3)}}{2(2)} = \dfrac{-3 + \sqrt{33}}{4}$ 9. $\frac{8}{3}$ (Note; $2\frac{2}{3}$ is mixed number form, but $\frac{8}{3}$ is reduced.)

10. $\frac{16}{18} = \frac{2 \cdot 8}{2 \cdot 9} = \frac{8}{9}$ 11. $\frac{100}{825} = \frac{25 \cdot 4}{25 \cdot 33} = \frac{4}{33}$ 12. $\frac{184}{207} = \frac{23 \cdot 8}{23 \cdot 9} = \frac{8}{9}$ 13. $\frac{1,209}{2,821} = \frac{3 \cdot 13 \cdot 31}{7 \cdot 13 \cdot 31} = \frac{3}{7}$

14. 0.375, rational   15. $2.\overline{3}$, rational   16. $0.\overline{428571}$, rational   17. 6.25, rational

18. $0.\overline{230769}$, rational   **19.** 89 is prime   **20.** 101 is prime   **21.** 349 is prime (check primes up to 17)

22. $1,001 = 7 \cdot 11 \cdot 13$ (use a factor tree)   **23.** $6,825 = 3 \cdot 5^2 \cdot 7 \cdot 13$ (use a factor tree)

24. $\frac{x}{5} \equiv 2$, (mod 8) means $x = 5 \cdot 2 = 10 \equiv 2$, (mod 8)

25. $2x \equiv 3$, (mod 7); consider the set $\{0, 1, 2, 3, 4, 5, 6\}$ and try each, one at a time to find

$2 \cdot 5 = 10 \equiv 3$, (mod 7), so $x \equiv 5$, (mod 7).

26. $2x^2 + 7x + 1 \equiv 0$, (mod 2); consider the set $\{0, 1\}$ and try each: $x \equiv 1$, (mod 2)

27.
$$49 = 7^2$$
$$1,001 = 7^1 \cdot 11^1 \cdot 13^1$$
$$2,401 = 7^4$$
$$\text{g.c.f.} = \overline{7^1 = 7}$$

28. l.c.m. $= 7^4 \cdot 11^1 \cdot 13^1 = 343,343$

29. $1 \odot 2 = 1 \times 2 + 1 + 2 = 2 + 1 + 2 = 5$

$3 \odot 4 = 3 \times 4 + 3 + 4 = 12 + 3 + 4 = 19$

$5 \odot 19 = 5 \times 19 + 5 + 19 = 95 + 5 + 19 = 119$

30. $(1 \downarrow 2) \uparrow (2 \downarrow 3) = 1 \uparrow 2 = 2$

31. For $a \neq 0$, then multiplication is defined as: $a \times b$ means $\underbrace{b + b + b + \cdots + b}_{a \text{ addends}}$. If $a = 0$, then

$0 \times b = 0$.

32. $a - b = a + (-b)$

33. $\frac{a}{b} = m$ means $a = bm$ where $b \neq 0$.

34. If we use the definition of division, $\frac{x}{0} = m$, then $0 \cdot m = x$. If $x \neq 0$, then there is no solution because $0 \cdot m = 0$ for all numbers m. Also, if $\frac{0}{0} = m$, then $0 = 0 \cdot m$ which is true for *every* number m. Thus, $\frac{0}{0} = 0$ checks, and $\frac{0}{0} = 1$ checks, and since two numbers equal to the same number must also be equal, we obtain the statement $0 = 1$; thus we say $\frac{0}{0}$ is indeterminate.

35. Answers vary; $34.1011011101111011\cdots$

36. A **field** is a set $\mathbb{R}$, with two operations $+$ and $\times$ satisfying the following properties for any elements a, b, $c \in \mathbb{R}$:

	Addition, +	*Multiplication, ×*
Closure:	1. $(a + b) \in \mathbb{R}$	2. $ab \in \mathbb{R}$
Associative:	3. $(a + b) + c = a + (b + c)$	4. $(a \times b) \times c = a \times (b \times c)$
Identity:	5. There exists $0 \in \mathbb{R}$ so that $0 + a = a + 0 = a$ for every element a in $\mathbb{R}$.	6. There exists $1 \in \mathbb{R}$ so that $1 \times a = a \times 1 = a$ for every element a in $\mathbb{R}$.

Inverse:

7. For each $a \in \mathbb{R}$, there is a unique number $(-a) \in \mathbb{R}$ so that

8. For each $a \in \mathbb{R}$, $a \neq 0$, there is a unique number $\frac{1}{a} \in \mathbb{R}$ so that

$$a + (-a) = (-a) + a = 0 \qquad a \times \tfrac{1}{a} = \tfrac{1}{a} \times a = 1$$

Commutative:

9. $a + b = b + a$ 10. $ab = ba$

Distributive for

  multiplication over addition: 11. $a \times (b + c) = a \times b + a \times c$

37. Let a, b, and c be any elements in $\mathbb{N}$.

Closure: $a \not\!\!D\, b \in \mathbb{N}$ because the g.c.f. is the product of factors and $\mathbb{N}$ is closed for multiplication.

Associative: $(a \not\!\!D\, b) \not\!\!D\, c = a \not\!\!D\, (b \not\!\!D\, c)$ because the g.c.f. is the product of factors and $\mathbb{N}$ is associative for multiplication.

Identity: Look for I so that $a \not\!\!D\, I = I \not\!\!D\, a = a$, but there is no such number.

Inverse: Since there is no identity, there can be no inverse property.

$\mathbb{N}$ is not a group for $\not\!\!D$, and therefore cannot be a commutative group.

38. $\text{BMI} = \dfrac{703w}{h^2}$ **a.** $\dfrac{703(165)}{65^2} \approx 27.5$ **b.** $\dfrac{703(185)}{72^2} \approx 25.1$

 c. Solve $25 = \dfrac{703w}{(5 \times 12 + 6)^2}$ $25 = \dfrac{703w}{(5 \times 12 + 6)^2}$

$$25(5 \times 12 + 6)^2 = 703w$$

$$w = \frac{25(5 \times 12 + 6)^2}{703}$$

$$\approx 154.9$$

The desired weight is 155 lb.

39. You can use a sieve or prime factorizations. A door will be opened or closed by a tenant only if the door number can be divided evenly by the number of that tenant. For example, door 9 will be touched (opened or closed) by tenants 1, 3, and 9; door 10 by tenants 1, 2, 5, and 10. Thus, the only doors left open are those with an odd number of divisors. The open doors are the perfect squares: 1, 4, 9, 16, 25, 36, $\cdots$, 841, 900, 961. There are 31 doors left open.

40. 8 ft = 96 in. and $\frac{96}{8} = 12$, so 12 stairs are necessary. The total length of the segments is $12 + 8 = 20$ ft, and the length of the diagonal is $\sqrt{12^2 + 8^2} = \sqrt{208} = 4\sqrt{13}$.

Group Research Problems, page 212

G13.

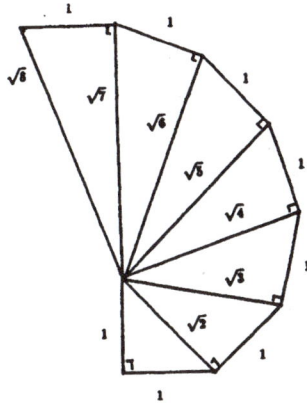

$$\sqrt{2} \approx 1.4$$
$$\sqrt{3} \approx 1.7$$
$$\sqrt{4} = 2$$
$$\sqrt{5} \approx 2.2$$

G14. Answers vary; the most difficult numbers seems to be 73. Other very difficult numbers to write with four 4s are 77, 81, 83, 87, 89, and 93. We did not use 44 or 4.4, although these would be permissible representations.

$0 = 4 + 4 - 4 - 4$ $1 = \left(\frac{4}{4}\right)\left(\frac{4}{4}\right)$ $2 = \left(\frac{4}{4}\right) + \left(\frac{4}{4}\right)$ $3 = \frac{4+4+4}{4}$

$4 = 4 + 4 - \sqrt{4} - \sqrt{4}$ $5 = \frac{4(4)+4}{4}$ $6 = 4 + 4 - 4 + \sqrt{4}$ $7 = 4 + 4 - \frac{4}{4}$

$8 = 4 + 4 + 4 - 4$ $9 = 4 + 4 + \frac{4}{4}$ $10 = 4(4) - 4 - \sqrt{4}$ $11 = 4 + \frac{4!+4}{4}$

$12 = 4(4) - \sqrt{4} - \sqrt{4}$ $13 = \frac{4!+4!+4}{4}$ $14 = 4 + 4 + 4 + \sqrt{4}$ $15 = 4(4) - \frac{4}{4}$

$16 = 4 + 4 + 4 + 4$ $17 = \frac{4}{4} + 4(4)$ $18 = 4(4) + 4 - \sqrt{4}$ $19 = 4! - 4 - \frac{4}{4}$

$20 = 4\left(4 + \frac{4}{4}\right)$ $21 = 4! - 4 + \frac{4}{4}$ $22 = 4(4) + 4 + \sqrt{4}$ $23 = 4! - \sqrt{4} + \frac{4}{4}$

$24 = 4(4) + 4 + 4$ $25 = 4! + \sqrt{4} - \frac{4}{4}$ $26 = 4(4 + \sqrt{4}) + \sqrt{4}$ $27 = 4! + \sqrt{4} + \frac{4}{4}$

$28 = 4! + 4 + 4 - 4$ $29 = 4! + 4 + \frac{4}{4}$ $30 = 4! + \sqrt{4} + \sqrt{4} + \sqrt{4}$ $31 = 4! + \sqrt{4} + \frac{\sqrt{4}}{.4}$

$32 = 4(4) + 4(4)$ $33 = 4! + 4 + \frac{\sqrt{4}}{.4}$ $34 = 4(4)\sqrt{4} + \sqrt{4}$ $35 = 4! + \frac{4! - \sqrt{4}}{\sqrt{4}}$

$36 = 4! + 4 + 4 + 4$ $37 = 4! + \frac{4! + \sqrt{4}}{\sqrt{4}}$ $38 = 4! + 4(4) - \sqrt{4}$ $39 = 4! + \frac{4 + \sqrt{4}}{.4}$

$40 = 4(4)(4) - 4!$ $41 = \frac{4(4) + .4}{.4}$ $42 = 4! + 4! - 4 - \sqrt{4}$ $43 = \frac{4! - 4}{.4} - \sqrt{4}$

$44 = 4! + 4! - \sqrt{4} - \sqrt{4}$ $45 = \dfrac{4(4) + \sqrt{4}}{.4}$ $46 = 4! + 4! - 4 + \sqrt{4}$ $47 = 4! + 4! - \dfrac{4}{4}$

$48 = 4! + 4! + 4 - 4$ $49 = 4! + 4! + \dfrac{4}{4}$ $50 = 4! + 4! + 4 - \sqrt{4}$ $51 = \dfrac{4! - \sqrt{4}}{.4} - 4$

$52 = 4! + 4! + \sqrt{4} + \sqrt{4}$ $53 = 4! + 4! + \dfrac{\sqrt{4}}{.4}$ $54 = 4! + 4! + 4 + \sqrt{4}$ $55 = \dfrac{4!}{.4} - \dfrac{\sqrt{4}}{.4}$

$56 = 4! + 4! + 4 + 4$ $57 = \dfrac{4! - \sqrt{4}}{.4} + \sqrt{4}$ $58 = 4! + 4! + \dfrac{4}{.4}$ $59 = \dfrac{4!}{.4} - \dfrac{4}{4}$

$60 = 4(4)(4) - 4$ $61 = \dfrac{4!}{.4} + \dfrac{4}{4}$ $62 = 4(4)(4) - \sqrt{4}$ $63 = \dfrac{4^4 - 4}{4}$

$64 = (4 + 4)(4 + 4)$ $65 = \dfrac{4^4 + 4}{4}$ $66 = 4(4)(4) + \sqrt{4}$ $67 = \sqrt{4} + \dfrac{4! + \sqrt{4}}{.4}$

$68 = 4(4)(4) + 4$ $69 = 4 + \dfrac{4! + \sqrt{4}}{.4}$ $70 = 4! + 4! + 4! - \sqrt{4}$ $71 = \dfrac{4! + 4 + .4}{.4}$

$72 = 4(4 \cdot 4 + \sqrt{4})$ $73 = \sqrt{\sqrt{\sqrt{4^{4!}}}} + \dfrac{4}{.4}$ $74 = 4! + 4! + 4! + \sqrt{4}$ $75 = \dfrac{4! + 4 + \sqrt{4}}{.4}$

$76 = 4! + 4! + 4! + 4$ $77 = \left(\dfrac{4}{.4}\right)^{\sqrt{4}} - 4$ $78 = (4! - 4)4 - \sqrt{4}$ $79 = \dfrac{4! - \sqrt{4}}{.4} + 4!$

$80 = 4(4 \cdot 4 + 4)$ $81 = \left(4 - \dfrac{4}{4}\right)^4$ $82 = 4(4! - 4) + \sqrt{4}$ $83 = \dfrac{4! - .4}{.4} + 4!$

$84 = 4(4! - 4) + 4$ $85 = \dfrac{4! + .4}{.4} + 4!$ $86 = 4(4! - \sqrt{4}) - \sqrt{4}$ $87 = 4(4!) - \dfrac{4}{.4}$

$88 = 4(4!) - 4 - 4$ $89 = \dfrac{4! + \sqrt{4}}{.4} + 4!$ $90 = 4(4!) - 4 - \sqrt{4}$ $91 = 4(4!) - \dfrac{\sqrt{4}}{.4}$

$92 = 4(4!) - \sqrt{4} - \sqrt{4}$ $93 = 4(4!) - \sqrt{\dfrac{4}{.4}}$ $94 = 4(4!) + \sqrt{4} - 4$ $95 = 4(4!) - \dfrac{4}{4}$

$96 = 4! + 4! + 4! + 4!$ $97 = 4(4!) + \dfrac{4}{4}$ $98 = 4(4!) + 4 - \sqrt{4}$ $99 = \dfrac{4!\sqrt{4} - 4}{.4}$

$100 = 4\left(4! + \dfrac{4}{4}\right)$ **G15.** Answers vary   **G16.** Answers vary   **G17. a.** yes   **b.** $|b| < a^2$ is the

appropriate condition   **c.** Answers vary

Individual Research Problems, pages 213-214

P4.1 Even though they did not major in mathematics, they did study mathematics in college or they received a minor in mathematics.

P4.2 Answers vary; the value given is a value for which the expression is not a prime.
    a. 41   **b.** 1,601   **c.** 29   **d.** 163   **e.** 8

P4.3 First, look at a pattern:

$$11 \cdot 14^1 + 1 = 155$$
$$11 \cdot 14^2 + 1 = 2{,}157$$
$$11 \cdot 14^3 + 1 = 30{,}185$$
$$11 \cdot 14^4 + 1 = 422{,}577$$
$$11 \cdot 14^5 + 1 = 5{,}916{,}065$$

If n is even, the number is divisible by 3. If n is odd, the number is divisible by 5. Conjecture: none are prime.

P4.4 Answers vary; if $n = 1$, $\sqrt{1 + 24(1)} = 5$, a prime; if $n = 2$, $\sqrt{1 + 24(2)} = 7$, a prime; if $n = 3$, $\sqrt{1 + 24(3)} = \sqrt{73}$ not a prime. If you read the problem as "all primes are contained in the list," then the formula is also a fraud because "n" is also such a formula.

P4.5 Answers vary.   **P4.6** Answers vary.

P4.7 $333{,}333{,}331 = 17 \cdot 19{,}607{,}843$ (See *International Mathematics Magazine*, February 1962, for a method of solution that does not involve calculators; otherwise, use a calculator to divide by consecutive primes until you reach 17.)

P4.8 a. Since $5 \cdot 2 = 10$ and 10 times any other number ends in a zero, the desired product must end in 0.

   b. Consider $1 \cdot 2 \cdot 3 \cdot 4 \cdot 5 \cdot 6 \cdot 7 \cdot 8 \cdot 9 \cdot 10$ and cross out the even factors:
$$1 \cdot \cancel{2} \cdot 3 \cdot \cancel{4} \cdot \cancel{5} \cdot \cancel{6} \cdot 7 \cdot \cancel{8} \cdot 9 \cdot \cancel{10} = 189$$
There are 200 products like this (all ending in 9). Since $9 \cdot 9 = 81$, and since the number of products is even, the last digit of their product is 1.

P4.9 See "Egyptian Fractions," by Bernhardt Wohlegmuth, *Journal of Recreational Math*, Vol. 5, No. 1 (1972), pp. 55-58

P4.10 Answers vary.   **P4.11** Answers vary.   **P4.12** Answers vary.   **P4.13** Answers vary.

P4.14 Answers vary.   **P4.15** Answers vary.   **P4.16** Answers vary.   **P4.17** Answers vary.

CHAPTER 5

THE NATURE OF ALGEBRA

5.1 Polynomials, page 216

Because of the prerequisites for this course, your students will have had some exposure to the material of this chapter. In fact, if your students have recently had intermediate algebra, then this chapter can easily be omitted. When I cover this material, I have found that it is best to treat it differently from the way they might have seen it in a previous algebra course. I motivate the material using the concept of area. At first there is some resistance, but after spending a little time with this tie between algebra and geometry, you will come to appreciate using it as an introduction for this material.

Level 1, page 222

1. A *polynomial* is a term or a sum of terms. **2.** The *degree* of a polynomial is the same as its largest degree term. **3.** Answers vary; look for similar terms. **4.** Answers vary. **5.** Answers vary. **6.** Answers vary. **7.** For any positive integer n,

$$(a+b)^n = \binom{n}{0}a^n + \binom{n}{1}a^{n-1}b + \binom{n}{2}a^{n-2}b^2 + \cdots + \binom{n}{n-1}ab^{n-1} + \binom{n}{n}b^n$$

where $\binom{n}{r}$ is the number in the nth row, rth diagonal of Pascal's triangle.

8. $6x - 4$; 1st degree binomial **9.** $-x - 8$; 1st degree binomial **10.** $3x + 6y - 2z$; 1st degree trinomial **11.** $3x - 6y - 4z$; 1st degree trinomial **12.** $8x^2 - 3x + 2$; 2nd degree trinomial **13.** $5x^2 - 7x - 7$; 2nd degree trinomial **14.** $7y - 7z$; 1st degree binomial **15.** $-x^2 - 5x + 3$; 2nd degree trinomial **16.** $-x^2 + 6x - 13$; 2nd degree trinomial **17.** $x - 31$; 1st degree binomial **18.** $4x + 4$; 1st degree binomial **19.** $16x^2 + 9x - 3$; 2nd degree trinomial **20.** $5x^2 + 9x + 28$; 2nd degree trinomial **21.** $-3x^2 + 15x - 17$; 2nd degree trinomial **22.** $-2x^2 - 10x - 53$; 2nd degree trinomial **23. a.** $x^2 + 5x + 6$ **b.** $y^2 + 6y + 5$ **c.** $z^2 + 4z - 12$ **d.** $s^2 + s - 20$

24. a. $x^2 - x - 2$ **b.** $y^2 - y - 6$ **c.** $a^2 - 8a + 15$ **d.** $b^2 - b - 12$

25. a. $c^2 - 6c - 7$ **b.** $z^2 + 2z - 15$ **c.** $2x^2 - x - 1$ **d.** $2x^2 - 5x + 3$

26. a. $3x^2 + 4x + 1$ **b.** $3x^2 + 5x + 2$ **c.** $6a^2 + 5a - 6$ **d.** $6a^2 + 13a + 6$

27. a. $x^2 + 2xy + y^2$ **b.** $x^2 - 2xy + y^2$ **c.** $x^2 - y^2$ **d.** $a^2 - b^2$

28. a. $25x^2 - 16$ **b.** $9y^2 - 4$ **c.** $a^2 + 4a + 4$ **d.** $b^2 - 4b + 4$

29. a. $x^2 + 8x + 16$ **b.** $y^2 - 6y + 9$ **c.** $s^2 + 2st + t^2$ **d.** $u^2 - 2uv + v^2$

Level 2, page 222

30. $15x^3 - 22x^2 + 5x + 2$ **31.** $6x^3 + x^2 - 12x + 5$ **32.** $3x^3 + 8x^2 - 9x + 2$

33. $5x^4 - 9x^3 + 13x^2 + 3x$ **34.** $14x^3 - 18x^2 + 4x + 6$ **35.** $-4x^3 + 25x^2 - 19x + 22$

36. $x^2 - 3x + 11$ **37.** $7x^2$

38.

$(x + 2)(x + 4) = x^2 + 6x + 8$

39.

$(x + 1)(x + 4) = x^2 + 5x + 4$

40.

$(x + 2)(x + 5) = x^2 + 7x + 10$

41.

$(x + 3)(x + 4) = x^2 + 7x + 12$

42.

$(2x + 3)(3x + 2) = 6x^2 + 13x + 6$

43.

$(2x + 1)(2x + 3) = 4x^2 + 8x + 3$

44. $(x + 1)^3 = x^3 + 3x^2 + 3x + 1$ **45.** $(x - 1)^3 = x^3 - 3x^2 + 3x - 1$

46. $(x + y)^5 = x^5 + 5x^4y + 10x^3y^2 + 10x^2y^3 + 5xy^4 + y^5$

47. $(x + y)^6 = x^6 + 6x^5y + 15x^4y^2 + 20x^3y^3 + 15x^2y^4 + 6xy^5 + y^6$

48. $(x - y)^7 = x^7 - 7x^6y + 21x^5y^2 - 35x^4y^3 + 35x^3y^4 - 21x^2y^5 + 7xy^6 - y^7$

49. $(x - y)^8 = x^8 - 8x^7y + 28x^6y^2 - 56x^5y^3 + 70x^4y^4 - 56x^3y^5 + 28x^2y^6 - 8xy^7 + y^8$

50. $$(5x - 2y)^3 = (5x)^3 + 3(5x)^2(-2y) + 3(5x)(-2y)^2 + (-2y)^3$$
$$= 125x^3 - 150x^2y + 60xy^2 - 8y^3$$

51. $$(2x - 3y)^4 = (2x)^4 + 4(2x)^3(-3y) + 6(2x)^2(-3y)^2 + 4(2x)(-3y)^3 + (-3y)^4$$
$$= 16x^4 - 96x^3y + 216x^2y^2 - 216xy^3 + 81y^4$$

Level 3, page 223

52. $x^{12} + 12x^{11}y + 66x^{10}y^2$ **53.** $91x^2y^{12} + 14xy^{13} + y^{14}$ **54.** $(5d + 2)(2d - 1) = 10d^2 - d - 2$

55. $(6x + 2)(51x - 7) = 306x^2 + 60x - 14$ **56.** $(800 - d)(6d + 12) = 9{,}600 + 4{,}788d - 6d^2$

57. $(6b + 15)(10 - 2b) = 150 + 30b - 12b^2$

Level 3 Problem Solving, page 223

58. a. 4,216 **b.** 3,021 **c.** 4,221 **d.** 9,025 **e.** 5,625

59. $$(10x + y)(10x + z) = 100x^2 + 10xz + 10xy + yz$$
$$= 100x^2 + 10x(z + y) + yz$$
$$= 100x^2 + 10x(10) + yz$$
$$= 100x^2 + 100x + yz$$
$$= 100(x^2 + x) + yz$$
$$= 100[x(x + 1)] + yz$$

60.

$$\overline{12 \times 13}$$
$$\downarrow \downarrow$$
$$124 \times 126 = 156\ 24$$
$$\uparrow\underline{\uparrow}\uparrow$$
$$4 \times 6$$

Same procedure as illustrated in Problem 58; it is just a little more difficult to multiply 12 and 13 mentally.

5.2 Factoring, page 223

Level 1, page 228

1. Answers vary; (1) look for common factor, then (2) factor a difference of squares, and then (3) use FOIL to factor if it is a trinomial.   **2.** $a^2 - b^2 = (a - b)(a + b)$ **3.** $2x(5y - 3)$ **4.** $5(x + 1)$

5. $2x(4y - 3)$ **6.** $2(3x - 1)$ **7.** $(x - 3)(x - 1)$ **8.** $(x - 3)(x + 1)$ **9.** $(x - 3)(x - 2)$

10. $(x - 3)(x + 2)$ **11.** $(x - 4)(x - 3)$ **12.** $(x - 8)(x + 1)$ **13.** $(x - 6)(x + 5)$ **14.** $(x + 7)(x + 2)$

15. $(x - 7)(x + 5)$ **16.** $(x - 8)(x + 2)$ **17.** $(3x + 10)(x - 1)$ **18.** $(2x - 3)(x + 5)$

19. $(2x - 1)(x - 3)$ **20.** $(3x - 1)(x - 3)$ **21.** $(3x + 1)(x - 2)$ **22.** $(6y - 1)(y - 1)$

23. $(2x + 1)(x + 4)$ **24.** $(7x - 3)(x + 1)$ **25.** $(3x - 2)(x + 1)$ **26.** $(3x + 1)(x + 2)$

27. $x(5x^2 + 7x - 6) = x(5x - 3)(x + 2)$ **28.** $4x(2x^2 + 3x + 1) = 4x(2x + 1)(x + 1)$

29. $x^2(7x^2 - 11x - 6) = x^2(7x + 3)(x - 2)$ **30.** $3x^2(x^2 + x - 12) = 3x^2(x + 4)(x - 3)$

31. $(x - 8)(x + 8)$ **32.** $(x - 13)(x + 13)$ **33.** $25(x^2 + 2)$ **34.** $(4x - 5)(4x + 5)$

35. $(x^2 - 1)(x^2 + 1) = (x - 1)(x + 1)(x^2 + 1)$

36. $(x^4 - 1)(x^4 + 1) = (x^2 - 1)(x^2 + 1)(x^4 + 1) = (x - 1)(x + 1)(x^2 + 1)(x^4 + 1)$

Level 2, page 228

37.

$(x + 3)(x + 2)$

38.

$(x + 4)(x + 3)$

39.

$(x + 3)(x + 1)$

40.

$(x + 4)(x + 1)$

41.

$(x + 2)(x + 4)$

42.

Cannot form a rectangle with these pieces.

Not factorable

In Problems 43-48, cut out the shaded portions.

43.

Move this (unshaded) piece to the side of the upper rectangle and rotate it 90° to form the area: $(x-1)(x+1)$

44.

Move this (unshaded) piece to the right of the upper rectangle (after rotating it 90°) to form the area: $(x-2)(x+2)$

45.

Move this (unshaded) piece to the right of the upper rectangle (after rotating it 90°) to form the area: $(x-1)(x+2)$

46.

Move this (unshaded) piece (after rotating it 90°) to form the area: $(x-2)(x+3)$

47.

← Move this (unshaded) piece and rotate it 90° to form the area: $(x-2)(x+1)$

48.

← Move unshaded piece (and rotate it 90°) to form the area: $(x-3)(x+2)$

Level 3, page 228

49. The dimensions are $x-13$ feet by $x+11$ feet. **50.** The dimensions are $x-15$ feet by $x+11$ feet. **51.** The time is $2x-1$ hours. **52.** There are $x+10$ rows of $x-60$ seats (or $x-60$ rows of $x+10$ seats). **53.** $x^2 + 6x + 8 + 5x^2 + x - 6 = 6x^2 + 7x + 2 = (3x+2)(2x+1)$

54. $6x^2 + 7x + 2 + 3x^2 - x - 10 = 9x^2 + 6x - 8 = (3x-2)(3x+4)$

55. $x^2(x^4 - 13x^2 + 36) = x^2(x^2 - 9)(x^2 - 4) = x^2(x-3)(x+3)(x-2)(x+2)$

56. $x^2(x^4 - 26x^2 + 25) = x^2(x^2 - 25)(x^2 - 1) = x^2(x-5)(x+5)(x-1)(x+1)$

57. $x^2(20y^2 + 17yz - 10z^2) = x^2(4y + 5z)(5y - 2z)$

58. $2x^2(6y^2 + 5yz - 6z^2) = 2x^2(3y - 2z)(2y + 3z)$

Level 3 Problem Solving, page 228

59. Let $x-1$, x, and $x+1$ be the three integers. Then $(x-1)(x+1) = x^2 - 1$ so the square of the middle integer is 1 more than the product of the first and third.

60.

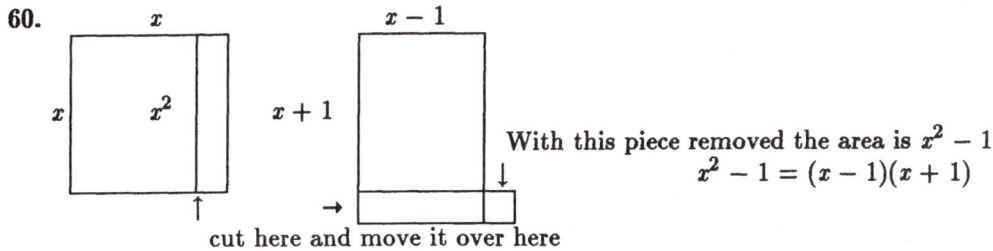

With this piece removed the area is $x^2 - 1$

$$x^2 - 1 = (x - 1)(x + 1)$$

cut here and move it over here

5.3 Evaluation, Applications, and Spreadsheets, page 228

Level 1, page 237

1. A *variable* is a symbol used to represent an unspecified member of some set (called the domain). A variable is a "place holder" for the name of some member of the set.   **2.** A *spreadsheet* is a computer program used to manipulate data and carry out calculations, or chains of calculations.   **3.** Answers vary.   **4.** Answers vary.   **5. a.** $+(2/3)*A1^2$ **b.** $+5*A1^2 - 6*A2^2$ **6. a.** $+3*A1^2 - 17$
b. $+14*A2^2 + 12*A1^2$ **7. a.** $+12*(A1^2 + 4)$ **b.** $+(15*A1 + 7)/2$ **8. a.** $+(3*A1 + 1)/12$
b. $+3*A1 + 1/12$ **9. a.** $+(5 - A1)*(A1 + 3)^2$ **b.** $+6*(A1 + 3)*(2*A1 - 7)^2$
10. a. $+(A1 + 1)*(2*A1 - 3)*(A1^2 + 4)$ **b.** $+(2*A1 - 3)*(3A*1^2 + 1)$
11. a. $+(1/4)*A1^2 - (1/2)*A1 + 12$ **b.** $+(2/3)*A1^2 + (1/3)*A1 - 17$ **12. a.** $+1 - A1/(A2*A3)$
b. $+(1 - A1)/(A2*A3)$ **13. a.** $4x + 3$ **b.** $5x^2 - 3x + 4$ **14. a.** $36z^2 - 13z + 2$ **b.** $13y^2 + \frac{15}{2}$
15. a. $\frac{5}{4}x + 14^2$ **b.** $(\frac{5}{4}x + 14)^2$ **16. a.** $5x^2 - 3y + 4$ **b.** $4(x^2 + 5)(3x^2 - 3)^2$ **17. a.** $\frac{x}{y}(z)$
b. $\frac{x}{yz}$ **18. a.** $b + ac$ **b.** $x + b^2$ **19.** $A = 1 + 4 + 8 = 13$ **20.** $B = 5(1) + 2 - 4 = 3$
21. $C = 10 - 2 = 8$ **22.** $D = 3(4) = 12$ **23.** $E = 25 - 2^2 = 25 - 4 = 21$
24. $F = 2(2 - 1 + 2 \cdot 4) = 2(2 - 1 + 8) = 2(9) = 18$ **25.** $G = 5(1) + 3(4) + 2 = 5 + 12 + 2 = 19$
26. $H = 3(1) + 2(2) = 3 + 4 = 7$ **27.** $I = 5(2) - 2(4) = 10 - 8 = 2$
28. $J = 2(2) - 4 = 4 - 4 = 0$ **29.** $K = 2(1)(2) = 4$ **30.** $L = 1 + 2^2 = 1 + 4 = 5$
31. $M = (1 + 2)^2 = 3^2 = 9$ **32.** $N = 1^2 + 2(1)(2) + 2^2 + 1 = 1 + 4 + 4 + 1 = 10$
33. $P = 2^2 + 4^2 = 4 + 16 = 20$ **34.** $Q = 2(1 + 2) = 2(3) = 6$
35. $R = 4^2 - 2^2 - 1^2 = 16 - 4 - 1 = 11$ **36.** $S = (1 + 2 + 4)^2 = 7^2 = 49$
37. $T = 1^2 + 2^2(4) = 1 + 16 = 17$ **38.** $U = \frac{2 + 2}{4} = \frac{4}{4} = 1$ **39.** $V = \frac{(3)(2)(2)(4)}{1} = 48$
40. $W = \frac{3(2) + 6(4)}{(1)(2)} = \frac{6 + 24}{2} = \frac{30}{2} = 15$ **41.** $X = [1^2(4) + 1]^2(4) = [5]^2(4) = 25(4) = 100$
42. $Y = [(2)(2)]^2 + 2^2(2) + 3(1) = 4^2 + 4(2) + 3 = 16 + 8 + 3 = 27$

Level 2, page 237

43. ALGEBRA_IS__THE_GREATEST LABOR SAVING DEVICE_EVER_INVENTED.

44. a. 2   **b.** 0   **c.** −12   **d.** 7   **45. a.** 8   **b.** 5   **c.** 9   **d.** 3.5

46.

  a.

  b.

	A	B	C	D	E
1	1	3	4	7	11
2	1	3	-2	5	-7
3					
4					
5					

47.

  a.

  b.

	A	B	C	D	E
1	1	3	4	5	6
2	1	3	3	3	3
3					
4					
5					

48. a.

	A	B	C
1	8	1	6
2	3	5	7
3	4	9	2
4			
5			

b.

	A	B	C
1	-3	-10	-5
2	-8	-6	-4
3	-7	-2	-9
4			
5			

49. a.

	A	B	C
1	7	0	5
2	2	4	6
3	3	8	1
4			
5			

b.

	A	B	C
1	107	100	105
2	102	104	106
3	103	108	101
4			
5			

50.

	A	B	C
1	Deposit	Year	Amount
2	1000	1	+A2*1.08
3		+B2+1	+C2*1.08
4		+B3+1	+C3*1.08
5		replicate	replicate

51.

	A	B	C	D	E
1	NAME	SALES	COST	PROFIT	COMMISSION
2				+B2 - C2	+.08*D2
3	Replicate Row 2 for Rows 3 to 21				
4					
5					

52. Genotype: brown eyes, 56.25%; brown eyes (recessive blue), 37.5%; blue eyes, 6.25%
Phenotype: brown eyes, 93.75%; blue eyes, 6.25%

53. Genotype: black, 42.25%; black (recessive brown), 45.5%; brown, 12.25%
Phenotype: black, 87.75%; brown, 12.25%

54. Genotype: tall, 25%; tall (short recessive), 50%; short, 25%
Phenotype: tall, 75%; short, 25%

55. Genotypes and phenotypes are the same: red, 4%; pink, 32%; white, 64%

Level 3 Problem Solving, page 238

56. They are magic squares.

57. $(B + b)^2 = B^2 + 2Bb + b^2$; if $b^2 = 0.25$, then $b = 0.5$, so $B = 1 - 0.5 = 0.5$. Thus,
$B = 50\%$ and $b = 50\%$

58. $(T + t)^2 = T^2 + 2Tt + t^2$; if $t^2 = 0.36$, then $t = 0.6$, so $T = 1 - 0.6 = 0.4$. Thus,
$T = 40\%$ and $t = 60\%$

59. Answers vary. For the general population, FF is 49%, Ff is 42%, and ff is 9%; Free hanging is
91% and attached is 9%. In other words, F is 70% and f is 30%.

60. $p = 20\%$, $q = 30\%$, and $r = 50\%$; $(p + q + r)^2 = p^2 + 2pr + q^2 + 2qr + 2pq + r^2$
AA: $p^2 = (0.20)^2 = 4\%$; AO: $2pr = 2(0.2)(0.5) = 20\%$; AB: $2pq = 2(0.2)(0.3) = 12\%$;
BO: $2qr = 2(0.3)(0.5) = 30\%$; BB: $q^2 = (0.3)^2 = 9\%$; OO: $r^2 = (0.5)^2 = 25\%$

5.4 Equations, page 239

After introducing the general process of equation solving, and after the students have had some practice with the general techniques, I have the class consider the following problem.

Rhonda, a very small person, needs to move a very large, heavy object.
She does not know what to do, so she asks her friend, Jane, for help.
Jane tells Rhonda, "Let's get a big board. I remember from school that
you can move a big object with a lever."

"I get it," Rhonda cried loudly. "I have just the right stuff out
back. Let's see.... Look at how I've put this together."

"What have you placed on the other end of the lever?"
asked Jane.

"Well, it is a can of Coke. Will that not work?"
asked Rhonda.

"That is really dumb, Rhonda," said Jane sarcastically.

Rhonda quickly responded, "Not if the weights of the can of Coke and trunk are the same. You do not know how much each of these containers weigh, and I'm prepared to show you that they weight the same!"

Polya's method leads the way.

Understand the Problem. In some sort of a silly game, Rhonda wants to show Jane that the weight of the can of Coke is the same as the weight of the heavy trunk.

Devise a Plan. Rhonda asserts: "There must be some weight w (probably very large) so that

$$T = c + w$$

where T is the weight of trunk, and c is the weight of the can of Coke. The plan is to use the elementary properties of equations presented in this chapter to conclude that $T = c$.

Carry Out the Plan.

$T = c + w$	*Given*
$T(T - c) = (c + w)(T - c)$	*Multiply both sides by $T - c$.*
$T^2 - Tc = cT + wT - c^2 - wc$	*Distributive property*
$T^2 - Tc - wT = cT - c^2 - wc$	*Subtract wT from both sides.*
$T(T - c - w) = c(T - c - w)$	*Use the distributive property to factor both sides.*
$T = c$	*Divide both sides by $T - c - w$.*

"Thus," says Rhonda, "The weight of the trunk is the same as the weight of the can of Coke! Since their weights are the same, the two objects will balance on the lever."

I give this much to the class and challenge them to tell me what is wrong with Rhonda's reasoning. Here is the correct answer: The step that is left off of Polya's method is *Look back*. The error in the reasoning is in the last step when both sides are divided by $T - c - w$. Since $T = c + w$ (first step), we see $T - c - w = 0$, and you cannot divide both sides by 0 in the last step.

Level 1, page 245

1. The procedure for solving a first-degree equation is to isolate the variable on one side of the equation. We do this by using the equation properties and the idea of opposite operations to decide which one of these properties to use.   **2.** The procedure for solving a second-degree equation is to use the equation properties to get a zero on one side. Then factor, if possible, using the zero product rule. If the polynomial does not factor, then use the quadratic formula.

3. a. 15   **b.** 8   **4. a.** -4 **b.** -8 **5. a.** 32   **b.** -44 **6. a.** 3   **b.** -5 **7. a.** 0   **b.** 0

8. a. $A = 5$ **b.** $B = 2$ **9. a.** $X = \frac{1}{5}$ **b.** $C = 10$ **10. a.** $D = 4$ **b.** $E = \frac{1}{3}$ **11. a.** $F = 1$

b. $G = 6$ **12. a.** $H = 0$ **b.** $I = 7$ **13. a.** $J = 15$ **b.** $K = 9$ **14. a.** $L = \frac{20}{3}$ **b.** $M = -9$

15. a. $N = -6$ **b.** $P = -8$ **16. a.** $Q = -2$ **b.** $R = \frac{11}{5}$ **17.** $S = 8$ **18.** $T = -3$

19. $U = 3$ **20.** $V = -4$ **21.** $W = -15$ **22.** $Y = -28$ **23.** $Z = 14$

Level 2, page 245

24. __THIS__ PROBLEM__ __ __ WILL__ UNDOUBTEDLY__ASSIST__ YOU __ __ IN__ FINDING __ ERRORS!!

25. $0, 10$ **26.** $0, 14$ **27.** $11, -6$ **28.** $\frac{2}{5}, -\frac{2}{3}$ **29.** $0, 2, -2$ **30.** $0, 1, -1$ **31.** $-\frac{3}{2}, \frac{3}{2}$

32. $-\frac{2}{3}, \frac{2}{3}$ **33.** $\dfrac{-7 \pm \sqrt{41}}{2}$ **34.** $\dfrac{3 \pm \sqrt{5}}{2}$ **35.** $\dfrac{5 \pm \sqrt{37}}{2}$ **36.** 3 **37.** $3 \pm \sqrt{2}$

38. $3 \pm \sqrt{3}$ **39.** $\dfrac{-5 \pm \sqrt{73}}{6}$ **40.** no real value   **41.** no real value   **42.** $\frac{3}{2}, \frac{2}{3}$ **43.** $4, -\frac{1}{3}$

44. $-\frac{1}{6}, 3$ **45.** $0, \frac{5}{6}$ **46.** $0, \frac{2}{9}$ **47.** $1.41, 2.83$ **48.** $1.73, 5.20$ **49.** $0.91, -2.33$

50. $1.46, -0.34$ **51.** $-41.54, -0.01$ **52.** $1.73, -1.67$ **53.** $7.98, 3.02$ **54.** $0.02, -4.11$

Level 3, page 246

55. **a.** CHILD'S DOSE $= \dfrac{10}{10 + 12} \times (100 \text{ mg}) \approx 45.5 \text{ mg}$

  b. $10 \text{ mg} = \dfrac{12}{12 + 12} \times$ ADULT DOSE; ADULT DOSE $= 20 \text{ mg}$

56. **a.** INFANT'S DOSE $= \dfrac{10}{150} \times 50 \text{ mg} \approx 3.3 \text{ mg}$

  b. $7.5 \text{ mg} = \dfrac{15}{150} \times$ ADULT DOSE; ADULT DOSE $= 75 \text{ mg}$

Level 3 Problem Solving, page 246

57. On the last step, dividing by $a + b - c$ is not allowed because $a + b = c$ so $a + b - c = 0$; can not divide by 0.

58. **a.** $\pi = \sqrt{6(1 + \frac{1}{2^2} + \frac{1}{3^2} + \frac{1}{4^2} + \cdots)}$ **b.** 3.094669524

  c. $\frac{22}{7} \approx 3.142857143$; this is a much better approximation of π

59. **a.**
$$x^2 + 10x = 39$$
$$x^2 + 10x - 39 = 0$$
$$(x + 13)(x - 3) = 0$$
$$x = -13, 3$$

  b. It is a geometric proof (which, by the way, gives only one of the answers).

Al-Khwârizmî starts with a square of side x, which therefore represents x^2, as shown at the right. To the square we must add $10x$ and this is done by adding four rectangles each of breadth 10/4 the length x to the square. The figure how has area $x^2 + 10x$ which is equal to 39. Now complete the square by adding four little squares each of area $(5/2)^2 = 25/4$.

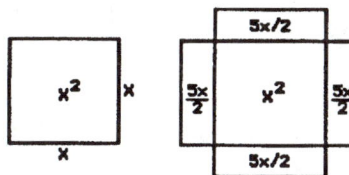

Thus, the outside square has area $4(25/4) + 39 = 64$.

The side of the square is therefore 8. But the length $5/2 + x + 5/2$, so $x + 5 = 8$, which gives $x = 3$.

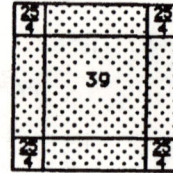

60. a. This is a research activity;

$$x^2 + 10x = 39$$

$$x^2 + 10x + 5^2 = 39 + 25$$

$$(x + 5)^2 = 64$$

$$x + 5 = \pm 8$$

$$x = -5 \pm 8$$

$$= -13, 3$$

   b. Problem 59a

5.5 Inequalities, page 247

Level 1, page 250

1. For any two numbers x and y, exactly one of the following is true: 1. $x = y$ (x is equal to y — the same as) 2. $x > y$ (x is greater than y — larger than) 3. $x < y$ (x is less than y — smaller than).   **2.** The procedure for solving inequalities is the same as the procedure for solving equations except that, if you multiply or divide by a negative number, you reverse the order of the inequality.

3.

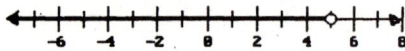

4.

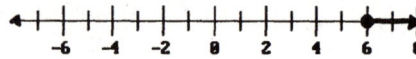

5.

6.

7.

8.

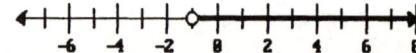

9.

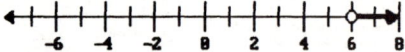

10.

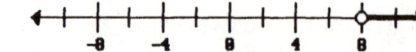

11.

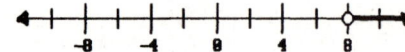

12.

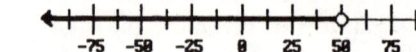

13. 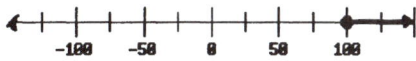   **14.**

15. 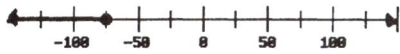   **16.**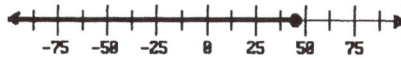

17.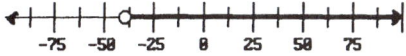

18. $x \geq -4$ **19.** $x \leq 7$ **20.** $x \geq -2$ **21.** $y > 5$ **22.** $y > -6$ **23.** $y > -8$

24. $s > -2$ **25.** $t \geq 3$ **26.** $m < 5$ **27.** $y \leq -1$ **28.** $x > -1$ **29.** $w \geq -4$

30. $x \leq 1$ **31.** $y \geq 5$ **32.** $s < -6$ **33.** $s > -3$ **34.** $a \geq 2$ **35.** $b < 8$ **36.** $s > 2$

37. $t \geq 3$ **38.** $u \leq 2$ **39.** $v > 2$ **40.** $w < -1$ **41.** $x \leq -2$

Level 2, page 251

42. $A > 0$ **43.** $B < -\frac{3}{2}$ **44.** $C > \frac{19}{2}$ **45.** $D > -4$ **46.** $E < -1$ **47.** $F > \frac{3}{5}$

48. $G > 7$ **49.** $H < -\frac{17}{2}$ **50.** $I > \frac{5}{4}$ **51.** $J \leq \frac{15}{4}$

52.
$$7(\text{NUMBER}) + 35 > 0$$
$$7N + 35 > 0$$
$$7N > -35$$
$$N > -5$$

The number is any number greater than -5.

53.
$$-(\text{NUMBER}) > 2(\text{NUMBER})$$
$$-N > 2N$$
$$-3N > 0$$
$$N < 0$$

The number is any number less than 0.

54.
$$3N + 12 < 0$$
$$3N < -12$$
$$N < -4$$

The number is any number less than -4.

55.
$$-x < 5$$
$$x > -5$$

The number is any number greater than -5.

Level 3 Problem Solving, page 251

56. $x = -x + 4$
$$2x = 4$$
$$x = 2$$

The number is 2.

57. $x + 6 = 2(-x)$
$$3x + 6 = 0$$
$$x = -2$$

The number is -2.

58. $x < (-x) + 4$
$$2x < 4$$
$$x < 2$$

The number is less than 2.

59. $x < 6 - 2(-x)$
$$x < 6 + 2x$$
$$-x < 6$$
$$x > -6$$

The number is greater than -6.

60.
$$\text{LENGTH} + \text{WIDTH} + \text{HEIGHT} \leq 72$$
$$4(\text{HEIGHT}) + \text{HEIGHT} + \text{HEIGHT} \leq 72$$
$$4h + h + h \leq 72$$
$$6h \leq 72$$
$$h \leq 12$$

The height and width must each be 12 in. or less and the length 48 in. or less.

5.6 Algebra in Problem-Solving, page 251

Level 1, page 259

1. Answers vary. First: You have to *understand the problem*. This means read the problem and note what it is all about. Focus on processes rather than numbers. You cannot work a problem you do not understand. A sketch may help in understanding the problem. Second: *Devise a plan.* Write down a verbal description of the problem using operation signs and an equal or inequality sign. Third: *Carry out the plan.* In the context of word problems, we *translate, evolve, solve,* and *answer.* Fourth: *Look back.* Be sure your answer makes sense by checking it with the original question in the problem. **Remember to answer the question that was asked.** **2.** *Translate* means to begin with a verbal statement and then let it *evolve* (which means use substitution) into a symbolic statement. The word *solve* applies to finding the replacements that make an equation true.

3. $2(\text{NUMBER}) + 7 = 17$ **4.** $2(\text{NUMBER}) - 12 = 6$

5. $5(\text{NUMBER}) - (-10) = -30$ **6.** $3(\text{NUMBER}) - 6 = 2(\text{NUMBER})$

7. $\text{INTEGER} + \text{NEXT INTEGER} = 117$

8. $\text{EVEN INTEGER} + \text{NEXT EVEN INTEGER} = 94$

9. $\text{INTEGER} + \text{NEXT INTEGER} + \text{THIRD CONSECUTIVE INTEGER} = 105$

10. $\text{INTEGER} + \text{NEXT INTEGER} + \text{3RD CONSECUTIVE INTEGER} + \text{4TH CONSECUTIVE INTEGER} = 74$

11. $\text{VALUE OF HOUSE} + \text{VALUE OF LOT} = 212{,}400$

12. $\text{PRICE OF FIRST CABINET} + \text{PRICE OF SECOND CABINET} = 4{,}150$

13. $\text{AMOUNT PAID FOR CORRECT PROBLEMS} + \text{AMOUNT OF FINES} = 0$

14. $4(\text{AMOUNT OF MONEY TO START}) - 72 = 48$

15. $10^2 + 15^2 = (\text{LENGTH OF GUY WIRE})^2$ **16.** $8^2 + 14^2 = (\text{LENGTH OF BRACE})^2$

17. $\text{DIST FROM J TO O} + \text{DIST FROM O TO P} + \text{DIST FROM P TO M} = \text{TOTAL DISTANCE}$

18. $\text{DIST FROM N TO M} + \text{DIST FROM M TO C} + \text{DIST FORM C TO D} = \text{TOTAL DISTANCE}$

19. $\text{DIST FROM S TO A} + \text{DIST FROM A TO W} + \text{DIST FROM W TO D} = \text{TOTAL DISTANCE}$

20. DIST OF SLOWER RUNNER + HEAD START = DIST OF FASTER RUNNER

21. DIST SHUTTLECRAFT TRAVELS + HEAD START = DIST ENTERPRISE TRAVELS

22. DIST OF CAR + HEAD START = DIST OF POLICE CAR

23. DIST OF SLOWER WALKER + HEAD START = DIST OF FASTER WALKER

24. DIST OF 1ST JOGGER + DIST OF 2ND JOGGER = TOTAL DISTANCE

25. DIST OF 1ST JOGGER + DIST OF 2ND JOGGER = TOTAL DISTANCE

Level 2, page 260

26. (See Problem 3) The number is 5.   **27.** (See Problem 4) The number is 9.   **28.** (See Problem 5) The number is -8.   **29.** (See Problem 6) The number is 6.   **30.** (See Problem 7) The integers are 58 and 59.   **31.** (See Problem 8) The integers are 46 and 48.   **32.** (See Problem 9) The integers are 34, 35, and 36.   **33.** (See Problem 10) The integers are 17, 18, 19, and 20.

34. (See Problem 11) The house is worth $177,000.   **35.** (See Problem 12) The prices of the cabinets are $830 and $3,320.   **36.** (See Problem 13) The daughter correctly solved 10 problems.

37. (See Problem 14) The gambler started with $30; note, all that is necessary is to solve: $4(\text{ORIG AMT}) - 72 = 48$.   **38.** (See Problem 15) The exact length of the brace is $5\sqrt{13} \approx 18.03$ ft; purchase at least 73 ft.   **39.** (See Problem 16) The exact length of the brace is $2\sqrt{65}$, or approximately 16 ft.   **40.** (See Problem 17) It is 150 miles from Jacksonville to Orlando.

41. (See Problem 18) The Cincinnati-Detroit leg of the trip is 250 miles.   **42.** (See Problem 19) It is 90 miles from Waco to Dallas.   **43.** (See Problem 20) It will take 12.5 seconds.   **44.** (See Problem 21) It will take 1.5 hours.   **45.** (See Problem 22) It will take 6 minutes (or 0.1 hour).

46. (See Problem 23) The faster person walked 8 miles and the slower walked 7 miles.   **47.** (See Problem 24) Their rates are 6 mph and 8 mph.   **48.** (See Problem 25) Their rates are 7.75 mph and 9.25 mph.

Level 3, page 261

49. Area $= \frac{1}{2}bh = 17.5$ cm^2. Let $x =$ length of shortest side.

$$\frac{1}{2}(x + 2)x = 17.5$$
$$x^2 + 2x = 35$$
$$x^2 + 2x - 35 = 0$$
$$(x + 7)(x - 5) = 0$$
$$x = -7, 5$$

The length of the shorter side is 5 cm.

50. Let $x =$ length of shortest side.

$$x^2 + (x + 6)^2 = 13^2$$

$$x^2 + x^2 + 12x + 36 = 169$$

$$2x^2 + 12x - 133 = 0$$

$$x = \frac{-12 \pm \sqrt{12^2 - 4(2)(-133)}}{2(2)}$$

$$\approx 5.689, -11.689 \qquad \textit{Reject} -11.689 \textit{ since } x \textit{ is a distance.}$$

The sides have length 5.7, 11.7, and 13.0.

51. Area $= \frac{1}{2}bh$; let $x =$ height of the triangle.

$$3 = \tfrac{1}{2}(x + 2)x$$

$$6 = x^2 + 2x$$

$$x^2 + 2x - 6 = 0$$

$$x = \frac{-2 \pm \sqrt{2^2 - 4(1)(-6)}}{2}$$

$$\approx 1.646, -3.646 \qquad \textit{Reject} -3.646 \textit{ since } x \textit{ is a distance.}$$

The base is 3.6 ft and the height is 1.6 ft.

52. Area $= \frac{1}{2}bh$; let $x =$ height of the triangle.

$$75 = \tfrac{1}{2}(x + 10)x$$

$$150 = x^2 + 10x$$

$$x^2 + 10x - 150 = 0$$

$$x = \frac{-10 \pm \sqrt{10^2 - 4(1)(-150)}}{2}$$

$$\approx 8.229, -18.229 \qquad \textit{Reject} -18.229 \textit{ since } x \textit{ is a distance.}$$

The base is 18.2 in. and the height is 8.2 in.

53. Find r when $A = 2$, so solve:

$$(1 + r)^2 = 2$$

$$1 + r = \sqrt{2} \qquad \textit{Disregard negative values.}$$

$$r = \sqrt{2} - 1$$

$$\approx 0.4142$$

Need to obtain about 41.4% interest.

54.
$$A = P(1 + r)^2$$

$$1,500 = 1,000(1 + r)^2$$

$$1.5 = (1 + r)^2$$

$$\sqrt{1.5} = 1 + r \qquad \textit{Disregard negative values.}$$

$$r \approx 0.2247 \qquad \text{Need to obtain about 22.5\% interest.}$$

Level 3 Problem Solving, page 262

55. Recall, 1 mile = 5,280 ft = 63,360 in.; $\frac{1}{2}$ mi = 31,680 in.

$$(\text{SIDE})^2 + (\text{HEIGHT})^2 = (\text{HYPOTENUSE})^2$$
$$(31{,}680)^2 + (\text{HEIGHT})^2 = (0.5 \text{ mi} + \frac{1}{4} \text{ in.})^2$$
$$31{,}680^2 + (\text{HEIGHT})^2 = (31{,}680 + 0.25)^2 \qquad \textit{Let h} = \text{HEIGHT}$$
$$31{,}680^2 + h^2 = 31{,}680.25^2$$
$$h^2 = 15{,}840.063$$
$$h \approx 125.86$$

The height is approximately 10.5 ft.

56. Let the depth of the water be x; then, $x^2 + 2^2 = (x+1)^2$

$$x^2 + 4 = x^2 + 2x + 1$$
$$3 = 2x$$
$$\frac{3}{2} = x$$

The water is $1\frac{1}{2}$ cubits.

57.
$$x - (\tfrac{1}{3}x + \tfrac{1}{5}x + \tfrac{1}{6}x + \tfrac{1}{4}x) = 6$$
$$60x - (20x + 12x + 10x + 15x) = 360$$
$$3x = 360$$
$$x = 120 \qquad \text{There are 120 lilies.}$$

58.
$$\tfrac{1}{5}x + \tfrac{1}{3}x + 3(\tfrac{1}{3}x - \tfrac{1}{5}x) + 1 = x$$
$$3x + 5x + 15x - 9x + 15 = 15x$$
$$15 = x \qquad \text{There are 15 bees.}$$

59.

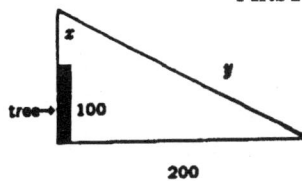

FIRST MONKEY'S DISTANCE = SECOND MONKEY'S DISTANCE

$$100 + 200 = x + y$$
$$300 = x + \sqrt{(100 + x)^2 + 200^2}$$
$$300 - x = \sqrt{(100 + x)^2 + 200^2}$$
$$300^2 - 600x + x^2 = (100 + x)^2 + 200^2$$
$$300^2 - 600x + x^2 = 100^2 + 200x + x^2 + 200^2$$
$$-800x = 100^2 + 200^2 - 300^2$$
$$x = 50$$

The second monkey jumped 50 cubits into the air.

60. Let g be the number of geese.

$$10\sqrt{g} + \tfrac{1}{8}g + 2 \cdot 3 = g$$

$$80\sqrt{g} = 7g - 48$$

$$49g^2 - 7{,}072g + 2{,}304 = 0$$

$$g = \frac{7{,}072 \pm \sqrt{(-7{,}072)^2 - 4(49)(2{,}304)}}{2(49)}$$ Note: Remember, the people

to whom this problem was originally directed did not have calculators; some problem!

$$g = \frac{7{,}072 \pm 7{,}040}{2(49)} = \underset{\underset{\text{reject}}{\uparrow}}{.327},\ 144$$

There were 144 geese.

5.7 Ratios, Proportions, and Problem Solving, page 262

Level 1, page 270

1. A *ratio* is a quotient of two numbers; a *proportion* is a statement of equality between ratios.

2. The property of proportions is equivalent to multiplying both sides of the equation $\frac{a}{b} = \frac{c}{d}$ by the number bd.   **3.** It saves a few steps; with equations it is necessary to multiply both sides by the least common denominator, and then to simplify, whereas the property of proportions does the same thing in one step.   **4.** First, find the product of the means or the product of the extremes, whichever does not contain the unknown term. Then, divide this product by the number that is opposite the unknown term.   **5.** 20 to 1   **6.** 1 to 20   **7.** 53 to 50   **8.** 50 to 53   **9.** 18 to 1   **10.** 37 to 2

11. a. yes **b.** yes **c.** yes   **12. a.** no **b.** yes **c.** yes   **13. a.** yes **b.** no **c.** no

14. a. no **b.** no **c.** yes   **15. a.** $>$ **b.** $<$ **c.** $>$ **16. a.** $=$ **b.** $=$ **c.** $>$

17. a. $<$ **b.** $>$ **c.** $>$ **d.** $<$ **18. a.** $<$ **b.** $>$ **c.** $>$ **d.** $=$ **19.** $A = 30$

20. $B = 36$ **21.** $C = 10$ **22.** $D = 56$ **23.** $E = 8$ **24.** $F = 25$ **25.** $G = 21$ **26.** $H = 20$

27. $I = 16$ **28.** $J = \frac{15}{4}$ or 3.75   **29.** $K = \frac{15}{2}$ or 7.5   **30.** $L = 15$ **31.** $M = 4$ **32.** $N = 2$

33. $P = 3$ **34.** $Q = \frac{63}{2}$ or 31.5   **35.** $R = 9$ **36.** $S = 32$ **37.** $T = \frac{1}{4}$ or 0.25   **38.** $U = 1$

39. $V = \frac{5}{2}$ or 2.5   **40.** $X = \frac{4}{5}$ or 0.8   **41.** $Y = 22$ **42.** $Z = \frac{5}{6}$ or $0.8\overline{3}$

Level 2, page 270

43. Seven melons cost $0.91.   **44.** The distance is 286 miles.   **45.** The family needs 2 gallons.

46. The amount of paint needed is 24 quarts.   **47.** It will take Jack 27 minutes.   **48.** It will take Jill 38 minutes.   **49.** A 165-lb person needs 2,475 calories per day.

50. _ _THIS_PROBLEM_ _WILL_UNDOUBTEDLY_ASSIST_YOU_ _IN_FINDING_ERRORS!!

Level 3, page 271

51. 159.25   **52.** 4   **53.** The pitch is 2-ft to 3-ft or 2/3.   **54.** The pitch is 1 ft to 2 ft or 1/2.

55. The longer side should be $3\frac{1}{3}$ ft or 3 ft 4 in.   **56.** The longer side is $8\frac{1}{3}$ in.   **57.** The tax is $780.

Level 3 Problem Solving, page 271

58. $\dfrac{\text{PERCENT OF DILUTE SOLUTION}}{\text{PERCENT OF ORIGINAL SOLUTION}} = \dfrac{\text{AMOUNT OF STRONG SOLUTION NEEDED}}{\text{AMOUNT OF DILUTED SOLUTION WANTED}}$

$$\frac{0.02}{0.10} = \frac{x}{500}$$

$$x = 100$$

Need 100 units of a 10% solution.

59. Since she started with 32 gallons and ended with 1 pt (of pure soft drink) the amount of soft drink served was 31 gal, 3 qt, and 1 pt.

60. Let x be the weight of the unpeeled banana. Then $x = \frac{7}{8}x + \frac{7}{8}$. Solve for x to find that the banana weighs 7 oz.

5.8 Percents, page 272

Level 1, page 279

1. $\frac{3}{4}$; 75%   **2.** $\frac{1}{5}$; 20%   **3.** $\frac{2}{5}$; 0.4   **4.** 1; 1   **5.** $0.\overline{3}$; $33\frac{1}{3}$%   **6.** 0.2; 20%   **7.** $\frac{17}{20}$; 85%

8. $\frac{3}{5}$; 0.6   **9.** 0.375; 37.5%   **10.** $\frac{9}{20}$; 0.45   **11.** $\frac{6}{5}$; 1.2   **12.** $0.\overline{6}$; $66\frac{2}{3}$%   **13.** $\frac{1}{20}$; 5%

14. $\frac{13}{200}$; 0.065   **15.** $0.1\overline{6}$; $16\frac{2}{3}$%   **16.** $0.8\overline{3}$; $83\frac{1}{3}$%   **17.** $\frac{2}{9}$; $0.\overline{2}$ **18.** $\frac{7}{20}$; 35%   **19.** $\frac{7}{40}$; 17.5%

20. $0.08\overline{3}$; $8\frac{1}{3}$%   **21.** $\frac{1}{400}$; 0.25%   *Estimates in Problems 22-28 may vary.* **22. a.** 1,000 **b.** 100

23. a. 9,500 **b.** 8.56   **24. a.** 5,000 **b.** 1,200   **25. a.** 200 **b.** 200   **26. a.** 750 or 800 **b.** 75

27. a. 40 **b.** 8,100   **28. a.** 6,000 **b.** 24   **29.** $\frac{15}{100} = \frac{A}{64}$; 9.6   **30.** $\frac{120}{100} = \frac{A}{16}$; 19.2

31. $\frac{14}{100} = \frac{21}{W}$; 150   **32.** $\frac{40}{100} = \frac{60}{W}$; 150   **33.** $\frac{P}{100} = \frac{10}{5}$; 200%   **34.** $\frac{P}{100} = \frac{1.2}{20}$; 6%

35. $\frac{P}{100} = \frac{4}{5}$; 80%   **36.** $\frac{P}{100} = \frac{2}{5}$; 40%   **37.** $\frac{P}{100} = \frac{9}{12}$; 75%   **38.** $\frac{P}{100} = \frac{25}{5}$; 500%

39. $\frac{35}{100} = \frac{49}{W}$; 140   **40.** $\frac{12}{100} = \frac{3}{W}$; 25   **41.** $\frac{120}{100} = \frac{16}{W}$; $13\frac{1}{3}$ **42.** $\frac{66\frac{2}{3}}{100} = \frac{21}{W}$; 31.5

43. $\frac{33\frac{1}{3}}{100} = \frac{12}{W}$; 36   **44.** $\frac{8}{100} = \frac{A}{2,425}$; $194   **45.** $\frac{6}{100} = \frac{A}{8,150}$; $489   **46.** $\frac{400}{100} = \frac{150}{W}$; 37.5

Level 2, page 280

47. Approximately 19.8 million Americans live in poverty.   **48.** Approximately 11.16 million Americans are unemployed.   **49.** The tax is $10.86.   **50.** The tax is $37.40.   **51.** The tax rate is 5%.   **52.** The increase is 25%.   **53.** The decrease is 40%.   **54.** The necessary number is 59 (you can't employ half a person).   **55.** The tax withheld is $2,624.   **56.** The necessary score is 94 points.   **57.** The percentage correct is 90%.   **58.** The test had 20 questions.   **59.** The old wage was $1,250 and the new wage is $1,350.   **60.** The expected gas mileage is 29.3 mpg.

5.9 Modeling Uncategorized Problems, page 281

Level 1, page 287

1. Answers vary.   **2.** Answers vary. GUIDELINES FOR PROBLEM SOLVING: First, you have to *understand the problem.* Second, *devise a plan.* Find the connection between the data and the unknown. Look for patterns, relate to a previously solved problem or a known formula, or simplify the given information to give you an easier problem. Third, *carry out the plan.* Fourth. look back. Examine the solution obtained.   **3.** The numbers are 20 and 22.   **4.** The numbers are 15 and 17. **5.** The number is 2.   **6.** There are two numbers, 1 and 2.

Level 2, page 287

7.

Bus and driver:	$150.00
10 children @ $5.00 each:	$50.00
6 teens @ $6.00 each:	$36.00
3 adults @ $7.00 each:	$21.00
20 people for lunch @ $2.50 each:	$50.00
TOTAL:	$307.00

Answer is A.

8. Since $\frac{7}{8}$ are mongrels, $\frac{1}{8}$ are purebreads: $\frac{1}{8}(48) = 6$. Answer is D.

9.

One hour (minimum):	$20.00
30 sheets @ $0.10 each:	$3.00
bundling fee:	$1.25
Additional copy @ 7¢/pg:	$2.10
TOTAL:	$26.35

Answer is A.

10.

23 cones @ $1.25 each:	$28.75
tips:	$3.50
TOTAL:	$32.25
less costs: $4.95 + $0.35:	$5.30
TOTAL:	$26.95

Answer is B.

11. 15% less is 0.85($125.00): $106.25
      Plus sales tax, 0.07($106.25): $7.44
                      TOTAL: $113.69   Answer is A.

12. 28%(SALARY) = $560

$$\text{SALARY} = \frac{\$560}{28\%}$$

$$= \$2,000 \quad \text{Answer is A.}$$

13.   35% of LOAN PAYMENT = $180

$$\text{LOAN PAYMENT} = \frac{\$180}{35\%}$$

$$\approx \$514.29$$

Combined total is $514.29 + $180.00 = $694.29. Answer is B.

14. Value of guitar = 0.90($1,500) = $1,350

Amount of loan = 0.80($1,350) = $1,080

The down payment is $1,500 − $1,080 = $420. Answer is A.

15. Mortgage payment: $2,320 + 0.03($2,320) = $2,320(1 + 0.03)

$$= \$2,389.60$$

Increase is: $2,389.60 − $2,320.00 = $69.60. Answer is D.

16. Rebate: 0.05($1,280) = $64.00

COST OF TV + TX − REBATE = $1,280 + 0.07($1,280) − $64

$$= \$1,305.60 \qquad \text{Answer is B.}$$

17. 2 hoses @ 2.50 gal/min = $5.00 gal/min. This is 60(5) gal/hr = 300 gal/hr

Thus, the number of hours to fill the pool is

$$\frac{25,500 \text{ gal}}{300 \text{ gal/hr}} = 85 \text{ hr} \qquad \text{Answer is C.}$$

18. 4 manifolds = 4(3 sprinkler heads) = 12 sprinkler heads

Installation: 4($85) = $340.00

Sprinkler heads: 12($9.50) = $114.00

TOTAL: $454.00

OVERHEAD: $ 45.40

$499.40 Answer is B.

19. Votes in favor: $\frac{1}{4}(40) = 10$

Votes against: $\frac{3}{5}(40) = 25$

TOTAL: 34

Since there are 40 voters, we see then there a 6 abstentions. Answer is A.

20. A grades: $\frac{1}{7}(28) = $ 4

B grades: $\frac{1}{4}(28) = $ 7

C grades: 11

TOTAL: 22

Since there are 28 students, we see, then, there are 6 D's and F's. Answer is C.

21. Alice's fare: $350

Bobbie's fare: $278

Cal's face: $322

TOTAL: $950 Answer is A.

22. Doug's first raise: 0.05($1,250) = $62.50

Doug's salary after 1st raise: $1,250 + $62,50 + $1,312.50

Doug's 2nd raise: 0.02($1,312.50) = $26.25

Doug's salary after 2nd raise: $1,312.50 + $26.25 = $1,338.75

Annual salary is: 12($1,338.75) = $16,065.00 Answers is D.

23. TAMIKA'S SCORE + NEXT PLAYER'S SCORE + THIRD PLAYER'S SCORE

$$= \text{T'S SCORE} + (\text{T'S SCORE} - 13) + (\text{T'S SCORE} - 14)$$

$$= 3(\text{T'S SCORE}) - 27$$

This is the total of 360, so

$$3T - 27 = 360$$

$$3T = 387$$

$$T = 129$$

Tamika's score was 129, with the next two player's scores: $129 - 13 = 116$ and 115.

Answer is B.

24. $194 + 16 = 210$ passes incomplete or intercepted. Thus,

$$551 - 210 = 341 \text{ completed passes.}$$

The percent completed: $\dfrac{341}{551} = 61.9\%$ Answer is A.

25. Total salary: $143(\$104,895) \approx \$15,000,000$ Answer is D.

26. Income from 2.5% investment: $0.025(\$2,500) = \$\ 62.50$

Income from 3.2% investment: $0.032(\$21,300) = \$681.60$

Income form 3.8% investment: $0.038(\$8,540) = \$324.52$

TOTAL: $\$1,068.62$

Taxes: $0.38(\$1,068.62) \approx \$406.08$

Income (after taxes): $\$1,068.62 - \$406.08 = \$662.54$ Answer is C.

27. Annual salary: $\underset{\text{weeks/yr}}{52} \times \underset{\text{hr/wk}}{40} \times \underset{\text{hr wage}}{25} = \$52,000$ Answer is B.

28. Hourly salary: $\$83,000 \div 52 \div 40 \approx \$39.90$ Answer is C.

29. There is only one even prime number, so the answer is B.

30. The prime numbers less than 100 are: 2, 3, 5, 7, 11, 13, 17, 19, 23, 29, 31, 37, 41, 43, 47, 53, 59, 61, 67, 71, 73, 79, 83, 89, and 97.   Answer is C.

31. Let x be the distance the ladder reaches up the wall. From the Pythagorean theorem,

$$3^2 + x^2 = 10^2$$

$$x^2 = 100 - 9$$

$$= 91$$

$$x = \sqrt{91} \approx 9.54$$ Answer is C.

32. Let x be the length of one guy wire. From the Pythagorean theorem,

$$10^2 + 12^2 = x^2$$

$$244 = x^2$$

$$15.62 \approx x$$

The guy wires are $3x \approx 46.86$.   Answer is B.

33. The (distinct) factors of 100 are: 1, 2, 4, 5, 10, 20, 25, 50, and 100. Answer is D.

34. The (distinct) prime factors of 100 are: 2 and 5.   Answer is A.

35. Answers vary.   **36.** Answers vary.   **37. a.** $0.38(\$7,415) = \$2,817.70$
b. $0.062(\$7,415) = \$459.73$ **c.** $0.10(\$7,415) = \$741.50$ **d.** $0.45(\$7,415) = \$333.68$ **e.** Total deductions are $\$4,352.61$, so the take home pay is $\$7,415 - \$4,35.61 = \$3,062.39$

38. a. $\$251,000 - \$112,950 = \$138,050$ so the percent loss is $\dfrac{\$138,050}{\$251,000} = 0.55$ or 55%   **b.** To gain back $\$138,050$ the percent gain is $\dfrac{\$138,050}{\$112,950} \approx 1.222$ or about 122%   **39.** Standard Oil bldg., 1,136 ft; Sears, 1,454 ft.   **40.** Calif., 156,361 mi.^{2}; New York, 47,831 mi.^{2} **41.** The sum is 10,100.

42. There are 49 people who do not participate.   **43.** There are 70 nonbusiness degrees.   **44.** It will take 56 hours to fill the pool.   **45.** If the drain is not opened, it will take 18 hours past the point where the pool is half full (since the total is 36 hours). However, if the drain is opened when the pool is half full, it will take an additional 180 hours for total time of 198 hours to fill the pool.

46. $x = 2\ell - 4$ **47.** $x = 2\ell - \frac{1}{30}c\ell - 4$ **48.** The drainage rate is 45 ft^3/min.

Level 3 Problem Solving, page 289

49. The selling price of the house was $\$224,000$.   **50.** $\$64,000$ in savings and $\$36,000$ in annuities.

51. Cut off 2 in.; the box is $8 \times 9 \times 12$ in.   **52.** The rate of the current is 1.5 mph.   **53.** The wind speed is 21 mph.   **54.** The plane's speed is 304.5 mph.   **55.** It will cost $2,350 per person with 30 additional persons.   **56.** There are 488,895 zeros.   **57.** The reading would not change. Think of fish in an aquarium. As the fish swims around in the water would the scale change? Of course not. Air is a medium with weight just as water is; the bird flying is the same as a fish swimming.

58. The product is 0; at least one person in NYC is totally bald.   **59. a.** 2, 5, 8, 10, 13, $\cdots$ **b.** No; $2 \cdot 8 = 16$ and 16 is not sacred.   **60.** Diophantus is 84 years old.

Chapter 5 Review Questions, Page 293

1. **a.** Algebra refers to a structure as a set of axioms that forms the basis for what is accepted and what is not.

   b. The four main processes are *simplify* (carry out all operations according to the order of operations agreement and write the result in a prescribed form), *evaluate* (replace the variable(s) with specified numbers and then simplify), *factor* (write the expression as a product), and *solve* (find the replacement(s) for the variable(s) that make the equation true).

2. **a.**
$$(x - 1)(x^2 + 2x + 8) = (x - 1)(x^2) + (x - 1)(2x) + (x - 1)(8)$$
$$= x^3 - x^2 + 2x^2 - 2x + 8x - 8$$
$$= x^3 + x^2 + 6x - 8$$

   b.
$$x^2(x^2 - y) - xy(x^2 - 1) = x^4 - x^2y - x^3y + xy$$
$$= x^4 - x^3y - x^2y + xy$$

3. **a.** $x^3 - y^3 = 2^3 - 3^3 = -19$ **b.** $(x - y)(x^2 + xy + y^2) = (2 - 3)(2^2 + 2 \cdot 3 + 3^2) = -19$

4. **a.** $3x^2 - 27 = 3(x^2 - 9) = 3(x - 3)(x + 3)$ **b.** $x^2 - 5x - 6 = (x - 6)(x + 1)$

5. **a.** $8x - 12 = 0$ **b.** $8x - 12 = -2x^2$

$$8x = 12 \qquad\qquad 2x^2 + 8x - 12 = 0$$

$$x = \frac{12}{8} = \frac{3}{2} \qquad\qquad x^2 + 4x - 6 = 0$$

$$x = \frac{-4 \pm \sqrt{16 - 4(1)(-6)}}{2}$$

$$= \frac{-4 \pm \sqrt{40}}{2}$$

$$= \frac{-4 \pm 2\sqrt{10}}{2}$$

$$= \frac{2(-2 \pm \sqrt{10})}{2}$$

$$= -2 \pm \sqrt{10}$$

6.

$$x^2 + 5x + 6 = (x + 2)(x + 3)$$

7.   **a.** $2x + 5 = 13$ **b.** $3x + 1 = 7x$

$$2x = 8 \qquad\qquad 1 = 4x$$

$$x = 4 \qquad\qquad \tfrac{1}{4} = x$$

8.   **a.** $\dfrac{2x}{3} = 6$ **b.** $2x - 7 = 5x$

$$2x = 18 \qquad\qquad -7 = 3x$$

$$x = 9 \qquad\qquad -\tfrac{7}{3} = x$$

9.   **a.** $\dfrac{P}{100} = \dfrac{3}{20}$ **b.** $\dfrac{25}{W} = \dfrac{80}{12}$

$$P = 15 \qquad\qquad W = 3.75$$

10.
$$x^2 = 4x + 5$$
$$x^2 - 4x - 5 = 0$$
$$(x - 5)(x + 1) = 0$$
$$x = 5, \; -1$$

11.
$$4x^2 + 1 = 6x$$
$$4x^2 - 6x + 1 = 0$$
$$x = \frac{6 \pm \sqrt{36 - 4(4)(1)}}{2(4)}$$
$$x = \frac{6 \pm \sqrt{20}}{8} = \frac{6 \pm 2\sqrt{5}}{8} = \frac{3 \pm \sqrt{5}}{4}$$

12. $3 < -x$
   $x < -3$

13. $2 - x \geq 4$
   $-x \geq 2$
     $x \leq -2$

14. $3x + 2 \leq x + 6$
     $2x \leq 4$
     $x \leq 2$

15. . $14 > 5x - 1$
     $15 > 5x$
     $3 > x$
     $x < 3$

16.   **a.** $\dfrac{2}{3}$ —— $\dfrac{67}{100}$ **b.** $\dfrac{3}{4}$ —— $\dfrac{75}{100}$ **c.** $\dfrac{23}{27}$ —— $\dfrac{92}{107}$ **d.** 0.05 —— 0.1 **e.** 0.99 —— 0.909

   $2(100) \; ? \; 3(67)$ $3(100) \; ? \; 4(75)$ $23(107) \; ? \; 27(92)$ $0.05 \; ? \; 0.10$ $0.990 \; ? \; 0.909$

   $200 \quad < \quad 201$ $300 \; = \; 300$ $2{,}461 \; < \; 2{,}484$ $<$ $>$

17.   1 is to 5 as 2.5 is to how much? $\dfrac{1}{5} = \dfrac{2.5}{x}$; this means $x = 5(2.5) = 12.5$. $12\tfrac{1}{2}$ cups of flour.

18.   Let x, $x + 2$, $x + 4$, and $x + 6$ be the four consecutive even integers. Then

$$x + (x + 2) + (x + 4) + (x + 6) = 100$$
$$4x + 12 = 100$$
$$4x = 88$$
$$x = 22$$

The integers are 22, 24, 26, and 28.

19. $(T + t)^2 = T^2 + 2Tt + t^2$

Genotypes: 27% tall: $T^2 = (0.52)^2 = 0.2704$;

50% tall (recessive short): $2Tt = 2(0.52)(0.48) = 0.4992$;

23% short: $t^2 = (0.48)^2 = 0.2304$

Phenotypes: 77% tall: $0.2704 + 0.4992 = 0.7696$;

23% short: 0.2304

20. (NEW YORK TO CHICAGO) + (CHICAGO TO SF) + (SF TO HONOLULU) $= 4,980$

$\uparrow$ $\uparrow$

(CHICAGO TO SF $-$ 1,140)   (CHICAGO TO SF $+$ 540)

Let $x =$ DISTANCE FROM CHICAGO TO SF

$$(x - 1,140) + x + (x + 540) = 4,980$$
$$3x - 600 = 4,980$$
$$3x = 5,580$$
$$x = 1,860$$

The distance from New York to Chicago is 720 mi, Chicago to San Francisco is 1,860 mi, and from San Francisco to Honolulu is 2,400 mi.

Group Research Problems, page 293

G18. Suppose $\sqrt{2}$ is rational. Then $\sqrt{2} = \frac{p}{q}$ where $\frac{p}{q}$ is a reduced fraction. Then

$\sqrt{2} = \frac{p}{q}$ *Given*

$2 = \frac{p^2}{q^2}$ *Square both sides.*

$2q^2 = p^2$ *Multiply both sides by q^2.*

p^2 is even   *Since p^2 is 2 times a number, it is even.*

p is even   *If p were odd, then p^2 is odd, so p must be even.*

Let $p = 2k$ *Since p is even, it can be written in the form 2k.*

$2q^2 = (2k)^2$ *Substitution*

$2q^2 = 4k^2$

$q^2 = 2k^2$

q^2 is even, so q is even. This means that $\frac{p}{q}$ is not reduced.

Since a fraction (namely $\frac{p}{q}$) cannot be reduced and not reduced at the same time, we know that no such number exists which implies that $\sqrt{2}$ is not rational.

G19.
$$x^2 + px = q$$
$$x^2 + px - q = 0$$

From the quadratic formula,

$$x = \frac{-p \pm \sqrt{p^2 - 4(1)(-q)}}{2(1)}$$

$$= -\frac{p}{2} \pm \frac{1}{2}\sqrt{p^2 + 4q}$$

$$= \sqrt{\frac{p^2 + 4q}{4}} - \frac{p}{2}$$

$$= \sqrt{\frac{p^2}{4} + q} - \frac{p}{2}$$

$$= \sqrt{\left(\frac{p}{2}\right)^2 + q} - \frac{p}{2}$$

G20. The number 1,324 raised to any power ends in a 6 or a 4. Also, 731 and 1,961 raised to any power ends in a 1. But any number ending in 6 or 4 added to a number ending in 1 cannot produce a number ending in 1. Thus, this equation has no solution for *any n*.

G21. The list 1, 10, 100, $\cdots$ contains infinitely many such integers.

G22. To divide a cube into two other cubes, a fourth power or in general any power whatever into two powers of the same denomination above the second is impossible, and I have assuredly found an admirable proof of this but the margin is too narrow to contain it. *Pierre de Fermat*

Individual Research Problems, page 294

P5.1. All were mathematics teachers at one time in their career.

P5.2. Answers vary. There are many good articles on this subject in *The Mathematics Teacher.* Check the journal's index for some ideas.

P5.3.
$$ax^2 + bx + c = 0 \qquad a \neq 0 \qquad \text{*This is a process known as completing the square.*}$$

$$x^2 + \frac{b}{a}x = -\frac{c}{a}$$

$$x^2 + \frac{b}{a}x + \left(\frac{b}{2a}\right)^2 = -\frac{c}{a} + \frac{b^2}{4a^2}$$

$$\left(x + \frac{b}{2a}\right)^2 = \frac{b^2}{4a^2} + \frac{-4ac}{4a^2}$$

$$x + \frac{b}{2a} = \pm\sqrt{\frac{b^2 - 4ac}{4a^2}}$$

$$x = -\frac{b}{2a} \pm \frac{\sqrt{b^2 - 4ac}}{2a}$$

$$x = \frac{-b \pm \sqrt{b^2 - 4ac}}{2a}$$

P5.4. Answers vary, but in answering this question you should see the PBS video, *The Proof.*

CHAPTER 6

THE NATURE OF GEOMETRY

6.1 Geometry, page 296

Problem 1 (Transparency 16) makes a good introduction to geometry because it is an ambiguous drawing. Transparency 17 shows the old woman and Transparency 18 shows the young woman. This old/young woman can be used to motivate a discussion about the nature of geometry.

Level 1, page 301

1. Answers vary; both a young woman and an old woman.   **2.** Answers vary; you cannot give an argument based on what you think you see in a geometric figure.   **3.** In order to construct a line segment congruent to a given line segment, copy a segment AB on any line ℓ. First fix the compass so that the pointer is on point A and the pencil is on B, as shown in the figure at the left. Then, on line ℓ, choose a point C. Next, without changing the compass setting, place the pointer on C and strike an arc at D, as shown in the figure below. Segment $\overline{CD}$ has been constructed to be congruent to the given segment $\overline{AB}$.

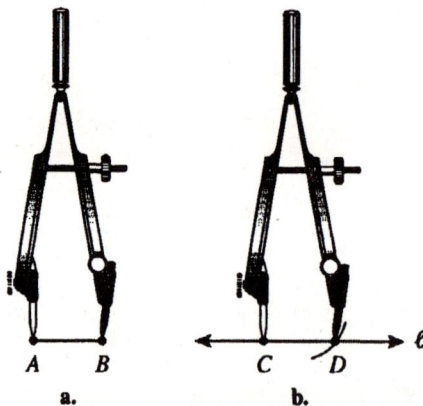

4.   Given a point and a radius of length $|AB|$, as shown in the figure at the left. Set the legs of the compass on the ends of radius $\overline{AB}$; move the pointer to point O without changing the setting, as shown in the middle figure. Hold the pointer at point O and move the pencil end of the compass to draw the circle, as shown below.

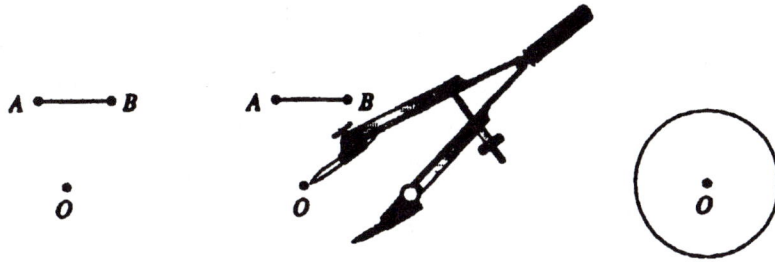

5. In order to define all words in a mathematical system, it is necessary to begin with certain words whose meaning is not given; it is agreed that these words are called undefined terms.   **6.** Line symmetry is best described by saying that a figure has line symmetry if the paper is folded along a line so that the left side folds onto the right side so that the figures on the left and on the right are identical.   **7.** An *axiom* is a statement accepted as fact without proof, and a theorem is a statement which is proved using undefined terms, axioms, other proved theorems, and laws of logic.

8. Geometry can be classified as traditional and transformational geometry.

9.

10.

11.

12.   **13.**   **14.**

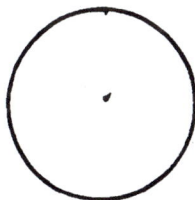

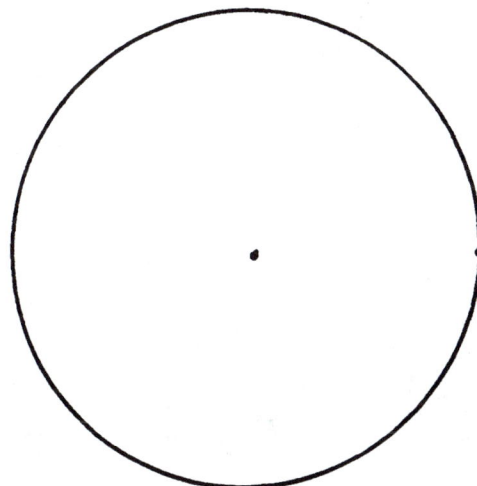

15.

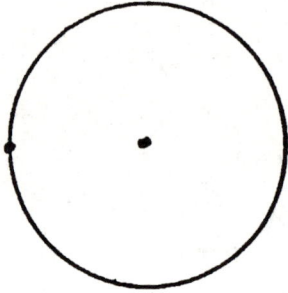

16.

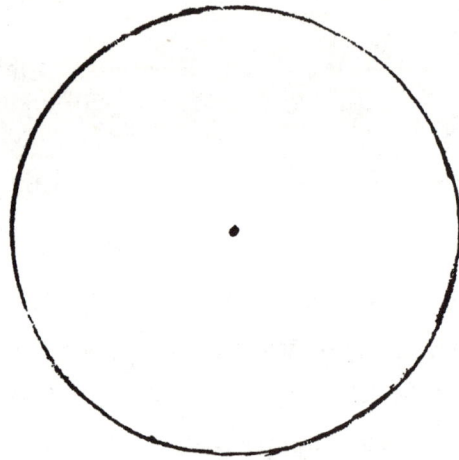

17.

18.

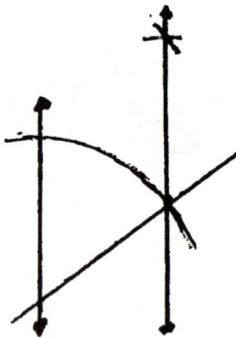

19.

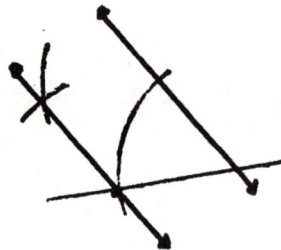

20.

21.

22.

23.

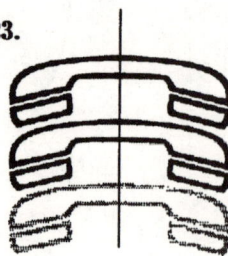

24.

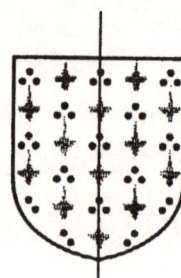

25. **26.** **27.** **28.**

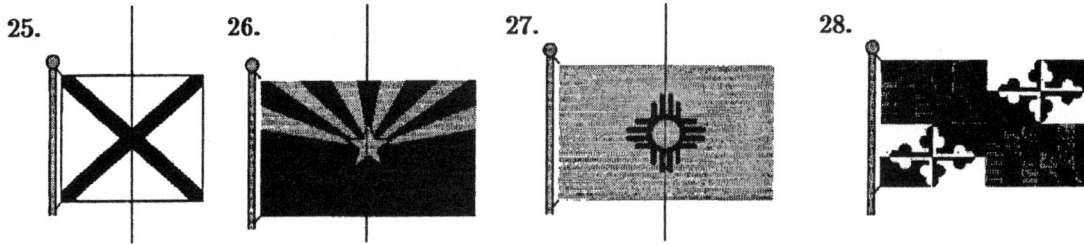

Note symmetric; it is symmetric with respect to the middle point, but it does not have line symmetry.

Level 2, page 302

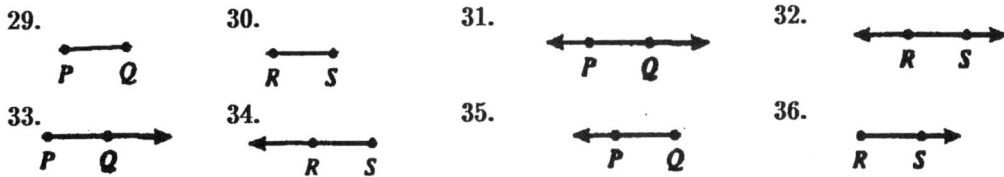

29.   **30.**   **31.**   **32.**

33.   **34.**   **35.**   **36.**

37. not symmetric   **38.** symmetric   **39.** symmetric   **40.** symmetric (except for words)

41. not symmetric   **42.** symmetric   **43.** not symmetric   **44.** symmetric   **45.** not symmetric

Level 3, page 304

46. reflection   **47.** rotation

Answers for Problems 48-53 may vary. If you have difficulty with this type of spatial visualization, try finding a die or other cube around your house and use post 'em on the faces.

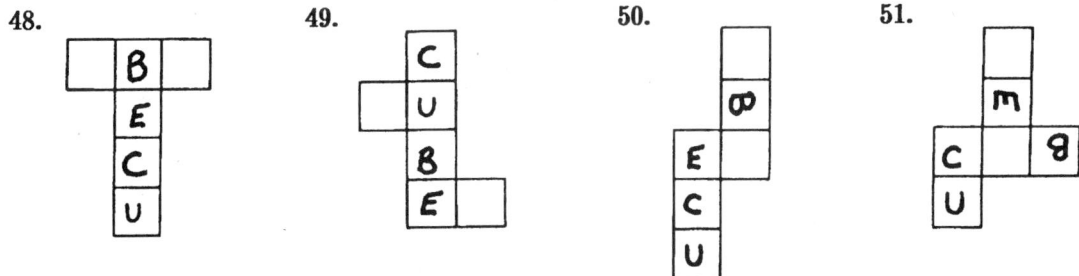

48.   **49.**   **50.**   **51.**

52.   **53.**   **54.** B   **55.** C   **56.** D   **57.** C   **58.** B

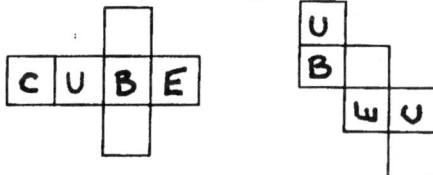

Level 3 Problem Solving, page 305

59. (1) letters with no symmetry;   (2) letters with horizontal line symmetry;   (3) letters with vertical line symmetry;   (4) letters with symmetry around both vertical and horizontal lines

60. The right-hand page offers from the left-hand page in the following ways: (1) Boy seated below lion is barefooted. (2) Girl has joined boy on rock. (3) Decoration on stone next to the boy is missing. (4) Dragon is missing horns. (5) Lady with cape in doorway has short hair. (6) Lion changes from male to female. (7) Man in front of unicorn isn't waving. (8) Birdman's tail loses some feathers. (9) Drummer has different drum. (10) Unicorn is missing horn. (11) Bird is in different window. (12) Triangular design on stone arch changes to oval. (13) Ball missing on top of flagpole. (14) Hornblower's sash changes sides. (15) Hornblower's clarions straighten. (16) Shingles on turret roofs change shape. (17) Queen tosses one less flower. (18) Little man in doorway behind unicorn has different hat. (19) Human face in archway under hornblower's changes to a pig's face. (20) Monster under lower bridge has eye closed. (21) No artist signature. (22) Dragon has gained some scales. (23) Cross removed from knothole in tree. (24) The word *substantial* changes to *substantive.* (25) Monster under lower bridge has eye closed.

6.2 Polygons and Angles, page 305
Level 1, page 311

1. An angle is two rays with a common endpoint.   **2.** Equal angles are angles that are the same whereas congruent angles are angles whose measures are the same.   **3.** A ray includes the endpoint and a half-line may or may not include the endpoint.   **4.** A polygon with four sides; trapezoid, parallelogram, rhombus, rectangle, and square.   **5.** In a diagram on a printed page, any line that is parallel to the top and bottom edge of the page is considered horizontal. Lines that are perpendicular to a horizontal line are considered to be vertical.   **6.** A right angle is an angles whose measure is $90°$; it is forms a square corner. An acute angles is one whose measure is between $0°$ and $90°$, and an obtuse angle is one whose measure is between $90°$ and $180°$.   **7.** Consider two intersecting lines forming four angles; adjacent angles are two angles which, when taken together, form a straight line and vertical angles are angles that are not adjacent. Now, consider two parallel lines cut by a transversal; corresponding angles are two nonadjacent angles whose interiors lie on the same side of the transversal such that one angle lies between the parallel lines and the other does not. Corresponding angles are congruent.   **8.** Two lines are parallel if they are in the same plane and do not intersect.

9. a. quadrilateral **b.** pentagon   **10. a.** octagon **b.** decagon   **11. a.** triangle **b.** hexagon

12. a. octagon **b.** dodecagon   **13. a.** quadrilateral **b.** heptagon   **14. a.** octagon **b.** nonagon

15. a. T **b.** T   **16. a.** T **b.** F   **17. a.** T **b.** F   **18. a.** T **b.** T   **19. a.** F **b.** F

20. a. yes **b.** yes **c.** yes **d.** yes **e.** yes   **21. a.** yes **b.** yes **c.** yes **d.** yes **e.** yes

22. a. yes **b.** yes **c.** yes **d.** no **e.** no   **23. a.** no **b.** no **c.** no **d.** no **e.** no

24. a. yes **b.** yes **c.** yes **d.** no **e.** no

Level 2, page 311

25. 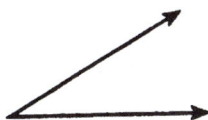   **26.** 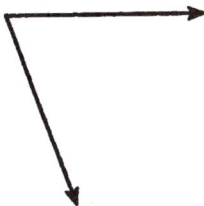   **27.**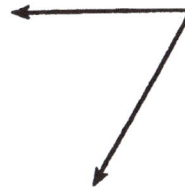

28. 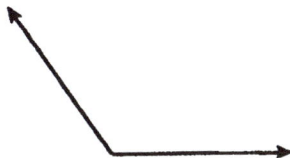   **29.**    **30.**

31. a. acute **b.** right   **32. a.** acute **b.** acute   **33. a.** right **b.** obtuse   **34. a.** obtuse

b. obtuse   **35. a.** acute **b.** acute   **36.** vertical   **37.** adjacent and supplementary

38. adjacent and supplementary   **39.** adjacent and supplementary   **40.** corresponding

41. vertical   **42.** alternate interior angles   **43.** alternate interior angles

44. F   **45.** F   **46.** F   **47.** T   **48.** T   **49.** F

50. $m\angle 1 = m\angle 3 = m\angle 5 = m\angle 7 = 110°;\ \ m\angle 2 = m\angle 4 = m\angle 6 = m\angle 8 = 70°$

51. $m\angle 1 = m\angle 3 = m\angle 5 = m\angle 7 = 115°;\ \ m\angle 2 = m\angle 4 = m\angle 6 = m\angle 8 = 65°$

52. $m\angle 1 = m\angle 3 = m\angle 5 = m\angle 7 = 161°;\ \ m\angle 2 = m\angle 4 = m\angle 6 = m\angle 8 = 19°$

53. $m\angle 1 = m\angle 3 = m\angle 5 = m\angle 7 = 153°;\ \ m\angle 2 = m\angle 4 = m\angle 6 = m\angle 8 = 27°$

54. $m\angle 1 = m\angle 3 = m\angle 5 = m\angle 7 = 120°;\ \ m\angle 2 = m\angle 4 = m\angle 6 = m\angle 8 = 60°$

55. $m\angle 1 = m\angle 3 = m\angle 5 = m\angle 7 = 163°;\ \ m\angle 2 = m\angle 4 = m\angle 6 = m\angle 8 = 17°$

Level 3, page 312

56. $x = 135°$; $y = 45°$ **57.** $x = 132.5°$; $y = 47.5°$ **58.** $x = 136°$; $y = 44°$

Level 3 Problem Solving, page 313

59.

60. Answers may vary; the number of sides of a golygon must be a multiple of 8.

6.3 Triangles, page 313

Level 1, page 317

1. A *triangle* is a polygon with three sides.   **2.** 180°   **3.** It means that $\triangle ABC$ and $\triangle DEF$ are congruent; also, it means that A corresponds to D, B corresponds to E, and C corresponds to F.

4. It has three sides but does not have three angles (it has one corner cut off).

5. $\overline{AB} \simeq \overline{ED}$; $\overline{AC} \simeq \overline{EF}$; $\overline{CB} \simeq \overline{FD}$; $\angle A \simeq \angle E$; $\angle B \simeq \angle D$; $\angle C \simeq \angle F$

6. $\overline{GH} \simeq \overline{G'H'}$; $\overline{GI} \simeq \overline{G'I'}$; $\overline{HI} \simeq \overline{H'I'}$; $\angle G \simeq \angle G'$; $\angle H \simeq \angle H'$; $\angle I \simeq \angle I'$

7. $\overline{RS} \simeq \overline{TU}$; $\overline{RT} \simeq \overline{RT}$; $\overline{ST} \simeq \overline{UR}$; $\angle SRT \simeq \angle UTR$; $\angle S \simeq \angle U$; $\angle STR \simeq \angle URT$

8. $\overline{WX} \simeq \overline{YZ}$; $\overline{WZ} \simeq \overline{YX}$; $\overline{XZ} \simeq \overline{ZX}$; $\angle W \simeq \angle Y$; $\angle WXZ \simeq \angle YZX$; $\angle WZX \simeq \angle YXZ$

9. $\overline{JL} \simeq \overline{PN}$; $\overline{LK} \simeq \overline{NM}$; $\overline{JK} \simeq \overline{PM}$; $\angle J \simeq \angle P$; $\angle L \simeq \angle N$; $\angle K \simeq \angle M$

10. $\overline{TP} \simeq \overline{XB}$; $\overline{BO} \simeq \overline{PO}$; $\overline{TO} \simeq \overline{XO}$; $\angle X \simeq \angle T$; $\angle B \simeq \angle P$; $\angle TOP \simeq \angle BOX$

11. 88°   **12.** 62°   **13.** 145°   **14.** 35°   **15.** 56°   **16.** 49°   **17.** 75°   **18.** 95°   **19.** 80°

20. 95°   **21.** 100°   **22.** 115°

23.

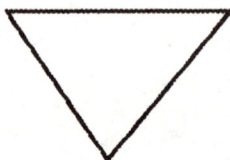

24.

25.

26.

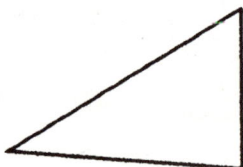

27.

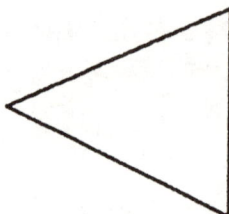

28.

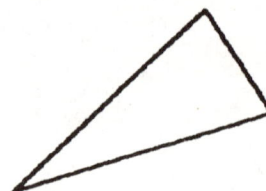

Level 2, page 318

29. An isosceles right triangle is a right triangle whose legs have the same length.   **30.** Answers vary; it rests on the fact that the shortest distance between two points is a straight line.

31. 20°   **32.** 10°   **33.** 21°   **34.** 12.5°   **35.** 6°   **36.** 5°   **37.** 60°; 60°; 60°   **38.** 45°; 45°; 90°

39. 50°; 60°; 70°   **40.** 60°; 50°; 70°   **41.** 13°; 53°; 114°   **42.** 25°; 65°; 90°

Level 3, page 319

Answers to Problems 43-50 may vary.

43.

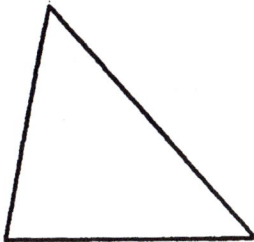

44.

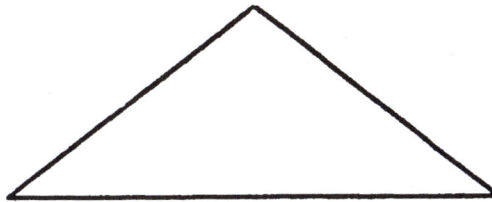

45.

46.

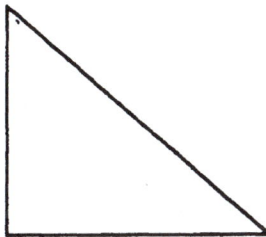

47.

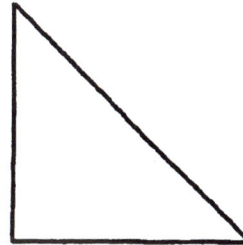

48. impossible   **49.**

50.

51.

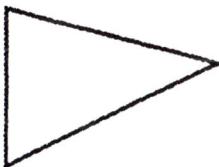

52.

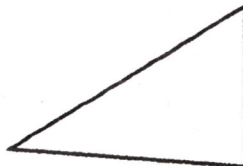

53.

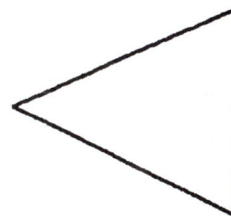

54.

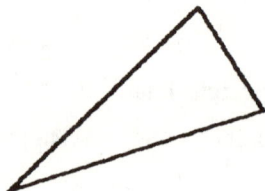

55. If $\angle x$ and $\angle y$ are complements, then

$$m\angle x + m\angle y = 90°$$

If $\angle z$ is also a complement of $\angle y$, then

$$m\angle z + m\angle y = 90°$$

Then

$$m\angle x + m\angle y = m\angle z + m\angle y$$

$$m\angle x = m\angle y \qquad \textit{Subtract } m\angle y \textit{ from both sides.}$$

56. If $\triangle ABC$ is equilateral, then it is isosceles so the the base angles are equal. Repeat for another two sides to conclude $\triangle ABC$ is equilateral.   **57.** Tape the results of your experiment onto your paper to show your work.   **58. a.** Follow the directions on your own paper.   **b.** Answers vary.   **c.** $180°$ **d.** $180°$ **e.** $360°$ **f.** Yes; formula is $S = 180(n - 2)$ where S is the sum of the measures of the angles of an n-sided polygon.

Level 3 Problem Solving, page 319

59. $540°$ **60.** $1{,}080°$

6.4 Similar Triangles, page 319

Level 1, page 322

1. Answers vary; congruent triangles have the same size and shape, whereas similar triangles assure only the same shape.   **2.** If the sides of two triangles are labeled a, b, c and d, e, f, respectively, then when we say the sides are proportional, we mean $\frac{a}{b} = \frac{d}{e}$; $\frac{a}{c} = \frac{d}{f}$; $\frac{b}{c} = \frac{e}{f}$; $\frac{b}{a} = \frac{e}{d}$; $\frac{c}{a} = \frac{f}{d}$; and $\frac{c}{b} = \frac{f}{e}$.

3. similar   **4.** not similar   **5.** not similar   **6.** similar   **7.** similar   **8.** similar

9. $m\angle A = m\angle A' = 25°$; $m\angle B = m\angle B' = 75°$; $m\angle C = m\angle C' = 80°$

10. $m\angle K = m\angle H = 38°$; $m\angle L = m\angle I = 110°$; $m\angle J = m\angle G = 32°$

11. $m\angle A = m\angle D = 38°$; $m\angle B = m\angle E = 68°$; $m\angle C = m\angle F = 74°$

12. $m\angle A = m\angle A' = 44°$; $m\angle B = m\angle B' = 46°$; $m\angle C = m\angle C' = 90°$

13. $m\angle A = m\angle A' = 54°$; $m\angle B = m\angle B' = 36°$; $m\angle C = m\angle C' = 90°$

14. $m\angle E = m\angle E' = 20°$; $m\angle F = m\angle F' = 100°$; $m\angle D = m\angle D' = 60°$

15. $|\overline{AC}| = |\overline{AB}| = 11$; $|\overline{A'C'}| = |\overline{A'B'}| = 22$; $|\overline{BC}| = 5$; $|\overline{B'C'}| = 10$

16. $\frac{8}{16} = \frac{|\overline{HI}|}{22}$, so $|\overline{HI}| = 11$; $\frac{|\overline{IG}|}{22} = \frac{12}{16}$, so $|\overline{IG}| = 16.5$; $|\overline{HG}| = 22$; $|\overline{DE}| = 16$; $|\overline{EF}| = 8$; $|\overline{DF}| = 12$

17. $|\overline{GH}| = 14$; $|\overline{HI}| = 12$; $|\overline{GI}| = 16$; $|\overline{DF}| = 20$; $\dfrac{|\overline{DE}|}{20} = \dfrac{14}{16}$, so $|\overline{DE}| = 17.5$; $\dfrac{|\overline{EF}|}{20} = \dfrac{12}{16}$, so $|\overline{EF}| = 15$ **18.** $\dfrac{|\overline{AB}|}{15} = \dfrac{5}{3}$, so $|\overline{AB}| = 25$; $\dfrac{|\overline{AC}|}{15} = \dfrac{4}{3}$, so $|\overline{AC}| = 20$; $|\overline{BC}| = 15$; $|\overline{A'B'}| = 5$; $|\overline{A'C'}| = 4$; $|\overline{B'C'}| = 3$ **19.** $|\overline{AB}| = 10$; $|\overline{AC}| = 6$; $|\overline{BC}| = 8$; $|\overline{B'C'}| = 5$; $\dfrac{|\overline{A'B'}|}{5} = \dfrac{10}{8}$, so $|\overline{A'B'}| = 6.25$; $\dfrac{|\overline{A'C'}|}{5} = \dfrac{6}{8}$, so $|\overline{A'C'}| = 3.75$ **20.** $|\overline{DF}| = 9$; $\dfrac{|\overline{DE}|}{9} = \dfrac{17}{6}$, so $|\overline{DE}| = 25.5$; $\dfrac{|\overline{EF}|}{9} = \dfrac{15}{6}$, so $|\overline{EF}| = 22.5$; $|\overline{DF}| = 9$; $|\overline{D'E'}| = 17$; $|\overline{E'F'}| = 15$; $|\overline{D'F'}| = 6$

21. $\sqrt{80}$ or $4\sqrt{5}$ **22.** $\sqrt{58}$ **23.** 4   **24.** 9   **25.** $\dfrac{16}{3}$ **26.** $\dfrac{15}{7}$ **27.** $\dfrac{8}{3}$ **28.** $\dfrac{10}{9}$

29. $\dfrac{h}{30+25} = \dfrac{10}{25}$, so $h = 22$ **30.** $\dfrac{h}{40+60} = \dfrac{20}{40}$, so $h = 50$ **31.** $\dfrac{x}{18} = \dfrac{10}{5+18}$, so $x \approx 7.8$

32. $\dfrac{x}{5} = \dfrac{3}{5+2}$, so $x \approx 2.1$ **33.** Let h = height of the larger triangle. Then,

$$h^2 + (20+30)^2 = 55^2$$

$$h^2 = 525 \qquad\qquad \frac{h}{20+30} = \frac{y}{30}$$

$$h = \sqrt{525} \qquad\qquad y \approx 13.7$$

34. $\dfrac{z}{14+20} = \dfrac{15}{14}$, so $z \approx 36.4$

Level 2, page 324

35. $m\angle BMA = m\angle BMC$ *Right angles because* $\overline{AC} \perp \overline{MB}$.

   $m\angle A = m\angle C$ *Isosceles triangle property because* $\triangle ABC$ *is equilateral*

   $\triangle ABM \sim \triangle CBM$ *Similar triangle theorem*

36. $m\angle D = m\angle E$ *Given*

   $m\angle DBA = m\angle EBC$ *Vertical angles*

   $\triangle ABD \sim \triangle CBE$ *Similar triangle theorem*

37. $m\angle ACT = m\angle TCO$ *Given that they are bisectors.*

   $m\angle ATC = m\angle OTC$ *Given*

   $\triangle CAT \sim \triangle COT$ *Similar triangle theorem*

38. $m\angle R_1 AO = m\angle R_2 AO$ *Bisector given*

   $m\angle R_1 OW = m\angle R_2 OW$ *Bisector given*

   $m\angle R_1 OA = 180° - m\angle R_1 OW$ *Straight angle*

   $m\angle R_2 OA = 180° - m\angle R_2 OW$ *Straight angle*

   $m\angle R_1 OA = m\angle R_2 OA$ *Substitution*

   $\triangle AOR_1 \sim \triangle AOR_2$ *Similar triangle theorem*

39. Let L = length of the lake;
$$\frac{L}{210 + 140} = \frac{50}{140}$$
$$L = 125$$

The lake is 125 ft long.

40. Let L = length of the lake;
$$\frac{L}{150 + 90} = \frac{50}{90}$$
$$L = 133\frac{1}{3}$$

The lake is about 133 ft long.

41. Let h = height of the house;
$$\frac{6}{4} = \frac{h}{16}$$
$$h = 24$$

The height of the building is 24 ft.

42. Let h = height of the house;
$$\frac{5 + \frac{8}{12}}{4} = \frac{h}{16}$$
$$h = 22\frac{2}{3}$$

The building is 22 ft 8 in.

43. Let h = height of the building;

Since a yardstick is 3 ft,
$$\frac{3}{5} = \frac{h}{75}$$
$$h = 45$$
The building is 45 ft tall.

44. Let h = height of the building;
$$\frac{3}{5} = \frac{h}{75 + \frac{3}{12}}$$
$$h = 45.15$$

The building is 45 ft 2 in. tall.

45. Let h = height of the tower;
Since a yardstick is 36 in.,
$$\frac{36 \text{ in.}}{23 \text{ in.}} = \frac{h \text{ ft}}{45 \text{ ft}}$$
$$h \approx 70.4348$$

The bell tower is 70 ft tall.

46. Let t = height of the tree;
$$\frac{t}{12} = \frac{6}{2.5}$$
$$t = 28.8$$

The tree is 29 ft tall.

47. Let t = height of the tree
$$\frac{t}{10} = \frac{5}{3}$$
$$t \approx 16.7$$

The tree is 17 ft tall.

48. Let t = height of the tree
$$\frac{t}{99} = \frac{70}{31} \quad \textit{Convert to inches.}$$
$$t \approx 223.55$$

The tree is 18 ft 8 in.

49. Let t = height of tree

$$\frac{t}{53} = \frac{69}{46} \quad \textit{Convert to inches.}$$
$$t = 79.5$$

The tree is 6 ft 8 in. tall.

50. *NYC* to Washington, D.C is 210 mi and *NYC* to Buffalo is 250 mi.

51. Denver to New Orleans is 1,080 mi and Chicago to Denver is 1,020 mi.

52. **a.** $\triangle HAP$ and $\triangle TBP$

b. Yes; $m\angle HPA = m\angle BPT$ (vertical angles) and $m\angle HAP = m\angle PBT$ (right angles)

    c. $\frac{6}{10} = \frac{|\overline{BT}|}{40}, |\overline{BT}| = 24$; it is 24 ft to the bottom of the canyon.

53. **a.** $\triangle HSP$ and $\triangle BTP$

    b. Yes; $m\angle HPS = m\angle TPB$ (vertical angles) and $m\angle HSP = m\angle PTB$ (right angles); thus

        $\triangle HSP \sim \triangle BTP$ (similar triangle theorem)

    c. $\frac{6}{10} = \frac{|\overline{BT}|}{35}, |\overline{BT}| = 21$; the footbridge is 21 ft above the base.

Level 3, page 325

54. $|\overline{AB}| = |\overline{BC}|$ *Given*

    $\triangle ABC$ is isosceles.   *Definition of isosceles triangle*

    $m\angle A = m\angle C$ *Isosceles triangle property*

    $\triangle BAM \simeq \triangle BCM$ *SAS (Figure 6.33)*

    $m\angle BMA = m\angle BMC$ *Corresponding parts of congruent triangles*

    $\triangle ABM \sim \triangle BMC$ *Similar triangle theorem*

55. Given $m\angle D = m\angle E$, show $\triangle ADC \sim \triangle CBA$

*	$m\angle D = m\angle E$	*Given*
	$m\angle DBA = m\angle EBC$	*Vertical angles*
$m\angle D + m\angle DBA + m\angle BAD = 180°$		*Sum of the measures of the angles of a triangle.*
$m\angle E + m\angle EBC + m\angle BCE = 180°$		*Sum of the measures of the angles of a triangle.*
Thus,	$m\angle BAD = m\angle BCE$	*Sum of the measures of the angles of a triangle.*
	$m\angle BAC = m\angle BCA$	*Isosceles triangle property*
	$m\angle DAC = m\angle BAD + m\angle BCA$	
	$m\angle ECA = m\angle BCE + m\angle BAC$	
*Therefore,	$m\angle DAC = m\angle ECA$	
so that	$\triangle ADC \sim \triangle CEA$	*Similar triangle property (two angles are equal from steps marked *)*

56. Let $\overline{CD}$ in Figure 6.48 be the given statement. Draw $\overline{AB}$ so that $|\overline{AF}| = 2$ and $|\overline{BF}| = 3$, so that

$$\frac{|\overline{AF}|}{|\overline{BF}|} = \frac{2}{3}$$

Draw line $\overleftrightarrow{PF}$ so that the intersection point E on the given segment is also in a 2-to-3 ratio.

57. Let $\overline{CD}$ in Figure 6.48 be the given segment. Draw $\overline{AB}$ so that $|\overline{AF}| = 3$ and $|\overline{BF}| = 7$. Repeat the steps outlined in Problem 18 to find E on the given segment in a 3-to-7 ratio.

58. An equilateral triangle is equiangular so all angles are 60°, since the sum of the angles of a triangle is 180°. This means that two angles of one are equal to two angles of the other, and are therefore similar.

59. $\angle C$ is a right angle, so $m\angle ACD + m\angle BCD = 90°$. Also, $m\angle A + m\angle ACD = 90°$ so $m\angle A = m\angle BCD$. Since $\overline{CD} \perp \overline{AB}$, $\angle ADC$ and $\angle BDC$ are right angles, so $m\angle ADC = m\angle BDC$. Thus, by the similar triangle property $\triangle ADC \sim \triangle BDC$.

60.   If $\triangle ABC \sim \triangle GHI$ and $\triangle DEF \sim \triangle GHI$, then $\triangle ABC \sim \triangle DEF$. By the definition of similar triangles, $m\angle A = m\angle G$, $m\angle B = m\angle H$ and $m\angle D = m\angle G$, $m\angle E = m\angle H$. Thus, $m\angle A = m\angle D$ and $m\angle B = m\angle E$. So by the similar triangle property, $\triangle ABC \sim \triangle DEF$.

6.5 Right-Triangle Trigonometry, page 326

Level 1, page 332

1.   For any right triangle with sides with lengths a and b and hypotenuse with length c,
$$a^2 + b^2 = c^2$$

Also, if a, b, and c are lengths of sides of a triangle so that
$$a^2 + b^2 = c^2,$$

then the triangle is a right triangle.

2.   In a right triangle ABC with right angle at C, $\sin A = \dfrac{\text{length of opposite side of } A}{\text{length of hypotenuse}}$.

3.   In a right triangle ABC with right angle at C, $\cos A = \dfrac{\text{length of adjacent side of } A}{\text{length of hypotenuse}}$.

4.   In a right triangle ABC with right angle at C, $\tan A = \dfrac{\text{length of opposite side of } A}{\text{length of adjacent side of } A}$.

5.   Rope A would form a right triangle.   **6.** a **7.** b **8.** b **9.** a **10.** c **11.** $\frac{a}{c}$ **12.** $\frac{b}{c}$

13. $\frac{b}{c}$ **14.** $\frac{a}{c}$ **15.** $\frac{a}{b}$ **16.** $\frac{b}{a}$ **17.** 0.8290   **18.** 0.2588   **19.** 0.8746   **20.** 0.3090

21. 0.5878   **22.** 0.9903   **23.** 0   **24.** 0.8290   **25.** 0.4452   **26.** 1.2799   **27.** 3.7321

28. 57.2900   **29.** 30°   **30.** 0°   **31.** 45°   **32.** 45°   **33.** 60°   **34.** 0°   **35.** 56°   **36.** 20°   **37.** 37°

Level 2, page 332

38. $\sin A = \frac{5}{13}$; $\cos A = \frac{12}{13}$; $\tan A = \frac{5}{12}$ **39.** $\sin A = \frac{12}{13}$; $\cos A = \frac{5}{13}$; $\tan A = \frac{12}{5}$

40. $\sin A = \frac{3}{5}$; $\cos A = \frac{4}{5}$; $\tan A = \frac{3}{4}$

41. $\sin A = \frac{\sqrt{35}}{6} \approx 0.9860$; $\cos A = \frac{1}{6} \approx 0.1667$; $\tan A = \frac{\sqrt{35}}{1} \approx 5.9161$

42. $\sin A = \frac{1}{6} \approx 0.1667$; $\cos A = \frac{\sqrt{35}}{6} \approx 0.9860$; $\tan A = \frac{1}{\sqrt{35}} \approx 0.1690$

43. $\sin A = \frac{1}{\sqrt{5}} \approx 0.4472$; $\cos A = \frac{2}{\sqrt{5}} \approx 0.8944$; $\tan A = \frac{1}{2} = 0.5000$

44. $\sin A = \frac{\sqrt{3}}{2} \approx 0.8660$; $\cos A = \frac{1}{2} = 0.5000$; $\tan A = \sqrt{3} \approx 1.7321$

45. $\sin A = \frac{1}{2} = 0.5000$; $\cos A = \frac{\sqrt{3}}{2} \approx 0.8660$; $\tan A = \frac{1}{\sqrt{3}} \approx 0.5774$

46. $\sin A = \frac{6}{\sqrt{37}} \approx 0.9864$; $\cos A = \frac{1}{\sqrt{37}} \approx 0.1644$; $\tan A = 6 = 6.000$

47. $\sin A = \frac{\sqrt{2}}{3} \approx 0.4714$; $\cos A = \frac{\sqrt{7}}{3} \approx 0.8819$; $\tan A = \frac{\sqrt{2}}{\sqrt{7}} \approx 0.5345$

48. $\tan 38° = \frac{h}{90}$, so $h \approx 70.3157$; the height is 70 ft.   **49.** $\tan 52° = \frac{h}{85}$, so $h \approx 108.795$; the height is 109 ft.   **50.** $\tan 35° = \frac{h}{30}$, so $h \approx 21.006$; the height of the building is 21 m.   **51.** $\tan 37° = \frac{150}{x}$, so $x \approx 199.0567$; the distance is 199 m.   **52.** $\tan 71° = \frac{1,000}{x}$, so $x \approx 344.3276$; the distance is 344 ft.

53. $\sin 52° = \frac{x}{16}$, so $x = 12.608$; the top of the ladder is 12 ft 7 in. from the ground.

54. $\tan 58.15° = \frac{h}{1,000}$, $h = 1,609.6966$; the Barrington Space Needle is 1,610 ft high.   **55.** $\tan 51.36° = \frac{h}{1,000}$, $h \approx 1,250.8863$; the chimney stack is 1,251 ft tall.   **56.** $\tan 52° = \frac{h}{351}$, $h \approx 449.2595$; the height of the Pyramid of Khufu is 449 ft.

Level 3 Problem Solving, page 333

57. The legs are x, $\sqrt{3}\,x$, so by the Pythagorean theorem, $x^2 + 3x^2 = c^2$, where c is the hypotenuse. Solving, $c = 2x$, so $\cos 30° = \frac{\sqrt{3}\,x}{2x} = \frac{\sqrt{3}}{2}$, $\sin 30° = \frac{x}{2x} = \frac{1}{2}$, and $\tan 30° = \frac{x}{\sqrt{3}\,x} = \frac{1}{\sqrt{3}} = \frac{\sqrt{3}}{3}$.

58. $r = 5.5$ in. (The diameter of the circle is the same as the width of the rectangle.)

59. $w = 12$ in. (Form a right triangle which is 1/4 of the entire rectangle. Then $x^2 + 8^2 = 10^2$, so $x = 6$, and then the width of the rectangle is 12.)   **60. a.** $\sin 47° = \frac{x}{92,900,000}$, so $x \approx 67,942,759$; the distance is 67,900,000 mi.   **b.** $\cos 47° = \frac{y}{92,900,000}$, so $y \approx 63,357,647.65$; the distance is 63,400,000 mi.

6.6 Golden Rectangles, page 334

I introduce this section by having a "beauty contest" of rectangles. I put many rectangles on the board and then the class votes on "the most beautiful rectangle." This leads us into a discussion of the golden ratio. The students also enjoy comparing their measurements with those shown by DaVinci and Dürer in Figure 6.60.

Level 1, page 339

1. $\tau = \dfrac{1+\sqrt{5}}{2} \approx 1.6180339888$　　2. A golden rectangle is a rectangle whose height h and width w

satisfies the proportion $\dfrac{h}{w} = \dfrac{w}{h+w}$.　　3. The divine proportion is the proportion $\dfrac{h}{w} = \dfrac{w}{h+w}$.

4. τ is the symbol used to denote the golden ratio; it is an irrational number

5. Answers vary. **a.** 6, 8, 14, 22, 36, $\cdots$　**b.** $\frac{8}{6} \approx 1.33$; $\frac{14}{8} = 1.75$; $\frac{22}{14} \approx 1.57$; $\frac{36}{22} \approx 1.63$

6. Answers vary.

7. 1, 3, 4, 7, 11, 18, 29, 47, 76, 123; the ratios are 3, 1.33, 1.75, 1.57, 1.64, 1.61, 1.62, 1.62, 1.62; it is τ

8. 1.62 (about the same)

9. Ratio of s to $\frac{1}{2}b$ is 1.62; b to h is 1.57; these are both about the same as τ.

10. Ratio of b to s is 1.7; $\frac{1}{2}b$ to h is 1.62; these are both about the same as τ.

11. Ratio is 1.875; this is close to τ.

12. Ratio is 2.17; this is not too close to τ.

Level 2, page 339

13. **a.** $5'' \times 3''$　　　　　　　　　**b.** 1.67 (about the same)

14. **a.** 8 in. by 4 in. (actually $3\frac{1}{2}$ in.)　**b.** 2.0 (not too close) to 2.28 (for the actual size)

15. **a.** 10 in. by 8 in.　　　　　　　　**b.** 1.25 (not close)

Problems 16-18, answers vary.

19. **a.** 2　**b.** 8　**c.** 144　**d.** 1, 2, 1.50, 1.67, 1.60, 1.63, 1.62, 1.62, 1.62, 1.62, 1.62; close to τ.

20. **a.** 3 reflections, 5 paths

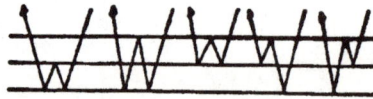

b. 4 reflections, 8 paths

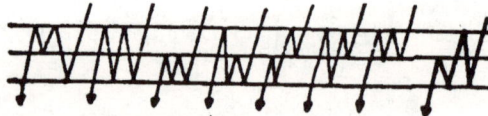

c. Look for a pattern for the number of paths: 1, 2, 3, 5, 8, 13, $\cdots$. Each term is found by adding the previous two terms.

21. $\frac{2}{1} = 2$; $\frac{3}{2} = 1.5$; $\frac{5}{3} \approx 1.67$; $\frac{8}{5} = 1.6$; $\frac{13}{8} \approx 1.63$; $\frac{21}{13} \approx 1.62$; close to τ.

22. $\dfrac{2}{w} = \dfrac{w}{2+w}$; $w = 1 + \sqrt{5}$　　23. Solve $\dfrac{5}{w} = \dfrac{w}{5+w}$ and $\dfrac{w}{5} = \dfrac{5}{w+5}$ and reject the negatives in both

cases to find $\dfrac{-5 + 5\sqrt{5}}{2} \approx 3.1$ ft and $\dfrac{5 + 5\sqrt{5}}{2} \approx 8.1$ ft

24. $\dfrac{\text{LENGTH}}{\text{WIDTH}} = \tau$, so $\begin{aligned} \text{LENGTH} &= 18\tau \\ &= 16\left(\dfrac{1 + \sqrt{5}}{2}\right) \\ &\approx 29.12 \end{aligned}$ The canvas should be 29 in. long.

25. $\dfrac{\text{HEIGHT}}{\text{WIDTH}} = \tau$, so $\begin{aligned} \text{HEIGHT} &= 9\tau \\ &= 9\left(\dfrac{1 + \sqrt{5}}{2}\right) \\ &\approx 14.56 \end{aligned}$ The photograph should be 15 cm high.

26. $\dfrac{\text{WIDTH}}{\text{HEIGHT}} = \tau$ $\begin{aligned} \text{WIDTH} &= 60\tau \\ &= 60\left(\dfrac{1 + \sqrt{5}}{2}\right) \\ &\approx 97.08 \end{aligned}$ The Parthenon should be 97 ft.

Level 3 Problem Solving, page 340

27. The length-to-width ratio will remain unchanged if

$$\frac{L}{W} = \frac{2W}{L}$$

$$\left(\frac{L}{W}\right)^2 = 2$$

$$\frac{L}{W} = \sqrt{2}$$

$$\approx 1.41421$$

28. $L = 9.25$, $W = 8$ so $\dfrac{L}{W} \approx 1.16$; It is closer to the librarian's ratio than it is to the golden ratio.

29. **a.** $x^2 - x - 1 = 0$

$$x = \frac{1 \pm \sqrt{1 - 4(-1)}}{2}$$

$$= \frac{1 \pm \sqrt{5}}{2}$$

   b. They are negative reciprocals.

30. Use a calculator to find the requested Fibonacci numbers **a.** 5 **b.** 55 **c.** 6,765

6.7 Projective and Non-Euclidean Geometries, page 340

Level 1, page 344

1. A non-Euclidean geometry is a geometry that does not assume Euclid's fifth postulate. **2.** See

Table 6.3. **Euclidean geometry:** Euclid (about 300 B.C.); Given a point not on a line, there is one and only one line through the point parallel to the given line. Geometry is on a plane, and the sum of the angles of a triangle is 180° and lines are infinitely long. **Hyperbolic geometry:** Gauss, Bolyai, Lobachevski (ca. 1830); Given a point not on a line, there are an infinite number of lines through the point that do not intersect the given line. Geometry is on a pseudosphere, and the sum of the angles of a triangle is less than 180°, and lines are infinitely long. **Elliptic geometry:** Riemann (ca. 1850); there are no parallels. Geometry is on a sphere: Geometry is on a pseudosphere, and the sum of the angles of a triangle is more than 180°, and lines are finite in length.   **3.** Answers vary; it is prevalent because it describes the real world around us.   **4. a.** hyperbolic **b.** elliptic **c.** Euclidean **d.** hyperbolic **e.** hyperbolic   **5.** Euclidean   **6.** elliptic   **7.** Euclidean   **8.** elliptic **9.** elliptic   **10.** Euclidean   **11.** hyperbolic   **12.** elliptic   **13.** It is a Saccheri quadrilateral. **14.** It is not a Saccheri quadrilateral.   **15.** It is not a Saccheri quadrilateral.   **16.** It is not a Saccheri quadrilateral.

Level 2, page 345

Answers to Problems 17-20 may vary. **17.** B   **18.** B   **19.** C   **20.** C   **21.** Their sum is greater than 180°.   **22.** Their sum is greater than 180°.

Level 3, page 345

23. a. A great circle is a circle on a sphere with a diameter equal to the diameter of the sphere.
b. The great circle ℓ is a line, but m is not. **c.** Yes   **24.** Two lines intersect in two points.
25. Lines are great circles, and the circle labeled m is not a great circle.   **26.** North Pole
27. True; answers vary.   **28.** Yes; answers vary.

Level 3 Problem Solving, page 346

29. Just down from the North Pole there is a parallel that has a circumference of exactly one mile. If you begin anywhere on the circle that is a parallel 300 ft north of this parallel, the conditions are satisfied.

30. Tessellations vary.

Chapter 6 Review Questions, page 345

1. **a.** 0

   b. 8 (corners of cube)

   c. 24 (2 such pieces on each of the 12 edges)

   d. 24 (4 on each face)

   e. 8 (interior unpainted pieces)

2. **a.** Flag is symmetric; picture is not (if you include the flagpole).   **b.** no   **c.** no   **d.** yes

3. **a.** corresponding angles   **b.** adjacent angles and supplementary angles   **c.** vertical angles

   d. alternate interior angles   **e.** alternate exterior angles   **f.** vertical angles

   g. ℓ_1 and ℓ_2 are parallel (given); ℓ_3 is horizontal.

4. **a.** 41°   **b.** 131°   **c.** 41°

5. $(3x + 20) + (2x - 40) + (x - 16) = 180$

$$6x - 36 = 180$$

$$6x = 216$$

$$x = 36$$

6. $$\frac{h}{10} = \frac{10}{h + 10}$$

   $$h(h + 10) = 100$$

   $h^2 + 10h - 100 = 0$

   $$h = \frac{-10 \pm \sqrt{10^2 - 4(1)(-100)}}{2}$$

   $$\approx 6.18, \; -16.18 \qquad \textit{Disregard negative value.}$$

   The width is about 6 in.

7. $$13^2 = x^2 + 12^2$$

   $$169 - 144 = x^2$$

   $$25 = x^2$$

   $$x = \pm 5 \qquad \textit{Disregard negative value.}$$

   The other leg is 5 in.

8. **a.** $\sin 59° \approx 0.8572$ **b.** $\tan 0° = 0$ **c.** $\cos 18° \approx 0.9511$ **d.** $\tan 82° \approx 7.1154$

9. Use the definition of the tangent ratio:

$$\tan 12° = \frac{160}{d}$$

Let d be the distance to the lighthouse.

$$d \tan 12° = 160$$

Multiply both sides by d.

$$d = \frac{160}{\tan 12°}$$

By calculator: $\boxed{160}\ \boxed{\div}\ \boxed{12}\ \boxed{\tan}\ \boxed{=}$
Display: 752.7408175

$$d \approx 753$$

By table: Use 0.2126 for tan $12°$.

10. A Saccheri quadrilateral is a quadrilateral $ABCD$ with base angles A and B right angles and with sides AC and BD with equal lengths. If the summit angles C and D are right angles, then the result is Euclidean geometry. If they are acute, then the result is hyperbolic geometry. If they are obtuse, the result is elliptic geometry.

Group Research Problems, page 349
G23.

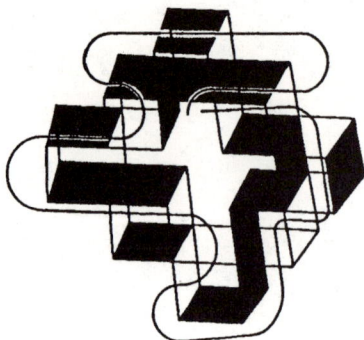

G24.

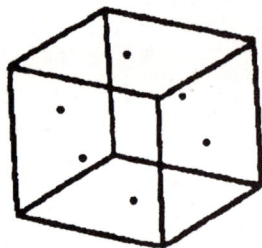

G25. Fold the bill in half lengthwise, then fold each half again, lengthwise again. It will now support the half dollar.

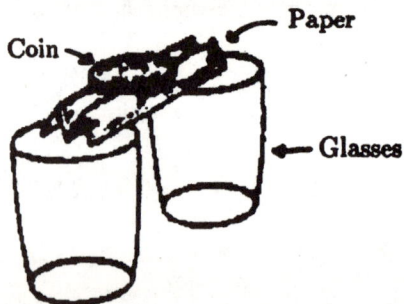

Individual Research Problems, page 350

P6.1 They all have Ph.D.'s in mathematics.

P6.2 to **P6.12** Answers vary.

CHAPTER 7

THE NATURE OF MEASUREMENT

7.1 Perimeter, page 352

Level 1, page 358

One the best zingers that I have found involves a problem that I call the Steel Band Problem (see Problem 60 of this section and Transparency 19). I use it as an extra credit problem to solve, but I also use it to talk about measurements, precision, accuracy, and estimation. I do not give the class the answer (Transparency 20) for at least a week.

1. Answers vary; precision refers to the unit of measurement used, whereas the accuracy refers to the answers. In this book, the accuracy of an answer should not exceed the precision of the measurement. For example, a measurement of the length of the side of a square to the nearest tenth, say 2.1 in. might be used in calculating the area. On a calculator, $2.1^2 = 4.41$. The precision of the measurement is 2.1 (one decimal place) and the accuracy 4.41 is two two decimal places. Since the accuracy should not exceed the accuracy of the measurement (one decimal places), we say the area of the square is 4.4 in.^{2} 2. All measurements are as precise as given in the text. If you are asked to make a measurement, the precision will be specified. Carry out all calculations without rounding. After you obtain a final answer, round this answer to be as accurate as *the least precise measurement*. Round money answers to the nearest cent.   3. perimeter of a square: $P = 4s$; perimeter of a rectangle: $P = 2(\ell + w)$; perimeter of an equilateral triangle: $P = 3s$; perimeter of a regular pentagon: $P = 5s$ 4. $C = \pi d$ or $C = 2\pi r$ 5. π is the ratio of the circumference to the diameter of a circle. It is an irrational number without an exact decimal or fractional representation.

6. a. ——————   b. ———   c. ———————————

7. a. ———————————————   b. ——   c. ———

8. a. ————————————————————

b. ————————————————————————

c. ——————————

9. A   10. B   11. B   12. B   13. A   14. C   15. C   16. B   17. B   18. A

19. B   **20.** C   **21.** C   **22.** A   **23.** C   **24.** A   **25.** C   **26.** B   **27.** A   **28.** B

29. C   **30.** B   **31.** A   **32.** C   **33.** A   **34. a.** 3 cm   **b.** 2.6 cm   **c.** 1 in.   **d.** 1 in.

35. a. 3 cm   **b.** 3.4 cm   **c.** 1 in.   **d.** $1\frac{3}{8}$ in.   **36. a.** 4 cm   **b.** 4.3 cm   **c.** 2 in.

d. $1\frac{6}{8}$ in. (or $1\frac{3}{4}$)

Level 2, page 359

37. 2(4 in. + 5 in.) = 18 in.   **38.** 2(12 ft + 3 ft) = 30 ft   **39.** 3(3 dm) = 9 dm

40. 3(6 yd) = 18 yd   **41.** $2\pi(2.40$ m$) \approx 15.08$ m   **42.** $\pi(50.00$ ft$) \approx 157.08$ ft

43. $\pi(50$ ft$) + 2(120$ ft$) = 397.08$ ft   **44.** 2(4 in. + 18 in. + 4 in.) = 52 in.

45. 2(14 ft + 9 ft) = 46 ft   **46.** 40 cm + 160 cm + 50 cm + 280 cm = 530 cm

47. $\frac{1}{2}\pi(100$ cm$) + 100$ cm ≈ 257.08 cm   **48.** 2(1.70 cm + 3.50 cm + 1.00 cm + 1.40 cm) = 15.20 cm

49. $\frac{1}{4}\pi(2)(10.00$ in.$) + 10.00$ in. + 10.00 in. ≈ 35.71 in.

50. 2.00 cm + 3.00 cm + 3.00 cm + $\frac{1}{2}\pi(2.00$ cm$) \approx 11.14$ cm

51. 2.00 cm + 3.00 cm + 3.00 cm + $\frac{1}{2}\pi(2.00$ cm$) \approx 11.14$ cm

52. $P = 2(\ell + w)$

    410 ft = 2(140 ft + w)

    205 ft = 140 ft + w

    65 ft = w The width is 65 ft.

53. $P = 2(\ell + w)$ *Perimeter formula*

    750 m = 2(ℓ + 75 m)   *Substitute known values.*

    375 m = ℓ + 75 m   *Divide both sides by 2.*

    300 m = ℓ The length is 300 m.

54. $P = 2(\ell + w)$ *Perimeter formula*

    54 cm = 2[(3w − 5 cm) + w]   *Since $\ell = 3w - 5$ cm*

    27 cm = 3w − 5 + w *Divide both sides by 2.*

    32 cm = 4w *Add 5 to both sides; add similar terms.*

    8 cm = w *Divide both sides by 4.*

   The length is 3(8 cm) − 5 cm = 19 cm, so the dimensions of the rectangle are 8 cm by 19 cm.

Level 3, page 359

55. $P = a + b + c$

117 in. $= 2x + 3x + 4x$

117 in. $= 9x$

 13 in. $= x$ The sides are $2x = 26$ in., $3x = 39$ in., and $3x = 52$ in.

56. $P = |\overline{AB}| + |\overline{BC}| + |\overline{CD}| + |\overline{DE}| + |\overline{EA}|$

280 cm $= 2x + 3x + 3x + 2x + 4x$

280 cm $= 14x$

 20 cm $= x$ The sides are 40 cm, 60 cm, 60 cm, 40 cm, and 80 cm.

57. Answers vary; 48 cm on my body is smaller than the 52.5 cm in l'Louvre.

58. Answers vary; you should remember these sizes for comparisons and reference.

Level 3 Problem Solving, page 360

59. From Problem 57, 1 cubit ≈ 0.525 m; thus,

300 cubits ≈ 157.50 m,   50 cubits ≈ 26.25 m,   and 30 cubits ≈ 15.75 m

The approximate measurements of the ark are 158 m long, 26 m wide, and 16 m high.

60.

Equator

a. Answers vary

b. Let r and c be the radius and circumference of the earth and R and C be the radius and circumference of the band, respectively.
Then, $R = r + 6$; and

AMOUNT OF EXTRA MATERIAL $= C - c$

$$= 2\pi R - 2\pi r$$

$$= 2\pi(R - r)$$

Since $R = r + 6$, then $R - r = 6$ and we have

AMOUNT OF EXTRA MATERIAL $= 2\pi(6)$

$$= 12\pi$$

$$\approx 37.7$$

Notice that the answer is independent of the radius! It would take about 38 ft.

7.2 Area, page 360

Another excellent zinger is the Extra Square Inch Problem (see Transparency 21). I have the students actually construct an 8 in. by 8 in. square, cut into four pieces as shown. Next, I have them actually rearrange the pieces to form a 5 in. by 13 in. rectangle. I then ask them to answer the question, "Where did the extra square inch come from?" I do not give the answer when I pose the question, but give them at least a week to work on this problem before I give the answer: the extra square inch is distributed along the diagonal of the 5 in. by 13 in. rectangle.

Level 1, page 366

1. The number representing its linear dimension, as measured from end to end, is called its length.

2. The area of a plane figure is the number of square units it contains.   **3.** Answers vary; the appropriate formulas are: $A = \pi r^2$; $C = 2\pi r$ **4. a.** 3 cm^2 **b.** 3 cm^2 **c.** 8 cm^2

5. **a.** 5 cm^2 **b.** 7 cm^2 **c.** 12 cm^2 **6.** 10 cm^2 **7.** 6 cm^2 **8.** C   **9.** C   **10.** A   **11.** B

12. C   **13.** A   **14.** A   **15.** B   **16.** C   **17.** C   **18.** B

Level 2, page 367

19. 15 in.^{2} **20.** 12 ft^2 **21.** 1,196 m^2 **22.** 36 mi^2 **23.** 100 mm^2 **24.** 250 in.^{2}

25. 7,560 ft^2 **26.** 13.5 m^2 **27.** 136.5 dm^2 **28.** 40 ft^2 **29.** 6,600 cm^2 **30.** 55 ft^2

31. 28 in.^{2} **32.** 145 mm^2 **33.** 314.2 in.^{2} **34.** 706.9 in.^{2} **35.** 78.5 in.^{2} **36.** 19.6 dm^2

37. 307.9 in.^{2} **38.** 353.4 cm^2 **39.** 7.6 cm^2 **40.** 4.4 cm^2 **41.** A; \$2.86/ft^2 for Lot A and \$3.19/ft^2 for Lot B.   **42.** C; \$5.59/ft^2 for Lot C and \$6.14/ft^2 for Lot D.

43. $\dfrac{(750 \text{ ft})(1,290 \text{ ft})}{43,560 \text{ ft}^2} \approx 22.2$ acres   **44.** 196,020 ft^2 **45.** 216 in.^{2} **46.** 13,875 ft^2

47. $93\frac{1}{2}$ in.^{2} **48.** 8 ft   **49.** 20 pounds are necessary; \$117   **50. a.** 28 in.^{2} **b.** 79 in.^{2} **c.** 113 in.^{2} **d.** 154 in.^{2}

Level 3, page 369

51. $s = 11.5$; $A \approx 20$ ft^2 **52.** $s = 365$; $A \approx 12,928$ ft^2

53. $s = 230$; area of triangle $\approx 9,592$; area of figure $\approx 52,792$ ft^2

54. The area is s^2; from the Pythagorean theorem

$$s^2 + s^2 = 7^2$$
$$2s^2 = 49$$
$$s^2 = \frac{49}{2}$$

Thus, the area is 24.5 in.^{2}.

55. Since a pentagon has five sides, there are 10 right triangles with $b = 5$ and $h \approx 6.88$.

$$A = 10(\tfrac{1}{2}bh)$$
$$\approx 5(5)(6.88)$$
$$= 172$$

The area of the pentagon is 172 in.^{2}.

56. Since a hexagon has six sides, there are 12 right triangles with $b = 5$ and $h \approx 8.66$.

$$A = 12(\tfrac{1}{2}bh)$$
$$\approx 6(5)(866)$$
$$= 259.8$$

The area of the hexagon is 260 in.^{2}.

57. Since an octagon has eight sides, there are 16 right triangles with $b = 5$ and $h \approx 12.07$.

$$A = 16(\tfrac{1}{2}bh)$$
$$\approx 8(5)(12.07)$$
$$= 482.8$$

The area of the octagon is 483 in.^{2}.

Level 3 Problem Solving, page 369

58. Notice that the unshaded portion is a circle that has been "turned inside out."

AREA OF SQUARE $-$ AREA OF CIRCLE $= 10^2 - 5^2\pi \approx 21$ in.^{2}

59. Notice that the shaded portion is a circle that has been "turned inside out."

AREA OF CIRCLE $= 5^2\pi \approx 79$ in.^{2}

60. The edges of the two triangles and two trapezoids do not really form a diagonal in the new rectangle.

7.3 Surface Area, Volume, and Capacity, page 370

Level 1, page 377

1. Answers vary; length is one dimensional (in., ft, cm, etc.), area is two dimensional (in.^{2}, ft^2, cm^2, etc.), and volume is three dimensional (in.^{3}, ft^3, cm^3, etc.).   **2.** The surface area of a three dimensional object is the sum of the area of its faces or sides.   **3.** Answers vary; volume is the number of cubic units a solid object contains, whereas capacity is the amount of liquid a container will hold. **4.** A square inch is: ⬜   and a square centimeter is ⬜

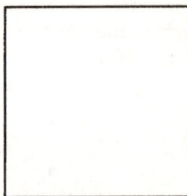

5. Answers vary; a liter is a bit larger than a quart.   **6.** Answers vary; a meter is a bit larger than a yard.   **7.** 3,125 cm^2 **8.** 600 cm^2 **9.** 780 cm^2 **10.** 11 m^2 **11.** 35,000 cm^2 **12.** 262 in.^{2}

13. 32π cm^2 + 4π cm^2 = 36π cm^2 $\approx$ 113.1 cm^2 **14.** 16π in.^{2} + 4π in.^{2} = 20π in.^{2} $\approx$ 62.8 in.^{2}

15. π ft + $\frac{\pi}{4}$ ft = 1.25π ft^2 $\approx$ 3.9 ft^2 **16.** 2(45 cm^2) + 78 cm^2 + 150 cm^2 + 117 cm^2 = 435 cm^2

17. 2(1.25 in.^{2}) + 1.35 in.^{2} + 1.95 in.^{2} + 2 in.^{2} = 7.80 in.^{2} **18.** 25 in.^{2} + 4(15 in.^{2}) = 85 in.^{2}

= 85 in.^{2} **19.** 60 cm^3 **20.** 80 cm^3 **21.** 125 ft^3 **22.** 1,728 in.^{3} **23.** 8,000 cm^3

24. 1 yd^3 **25.** 24 ft^3 **26.** 400 mm^3 **27.** 96,000 cm^3 **28.** 3,600 in.^{3}

29. a. 2 c **b.** 16 oz   **30. a.** 11 oz **b.** 320 ml   **31. a.** 13 oz **b.** 380 ml   **32. a.** 1$\frac{3}{4}$ c

b. 14 oz **c.** 420 ml   **33. a.** 25 ml **b.** 75 ml **c.** 70 ml   **34.** A   **35.** B   **36.** C   **37.** A

38. A   **39.** C   **40.** B   **41.** B   **42.** C   **43.** B   **44.** A   **45.** 4.3 gal   **46.** 15.6 L

47. 2.5 L   **48.** 1 L   **49.** 9 L   **50.** 17.5 L   **51.** 500 L   **52.** 3.7 gal

Level 2, page 380

53. a. 45.375 ft^3 **b.** 26.375 ft^3 **54.** No; exterior dimensions are 24 ft^3, so it cannot be 27 ft^3

inside.   **55.** 2.5 yd^3 **56. a.** 10 yd^3 **b.** 6 yd^3 **c.** 1 yd^3 **d.** 4 yd^3 **57.** 112 kl

58. 15,080 gal

Level 3 Problem Solving, page 381

59. a. $\dfrac{2.8 \times 10^7 \text{ ft}^2}{1 \text{ mi}^2} \times \dfrac{1 \text{ person}}{50 \text{ ft}^2} = 560{,}000$ people/mi^2

b. $\dfrac{1 \text{ mi}^2}{560{,}000 \text{ people}} \times (6.2 \times 10^9 \text{ people}) \approx 11{,}071$ mi^2

This is about the size of the state of Maryland.

c. $(5.2 \times 10^7) \times 640$ acres; divide this by the population 6.2×10^9 and the result is about 5.4 acres per person!

60. a. Answers vary

b. $\frac{1}{2}$ mi = 2640 ft; volume of box is 1.8×10^{10} ft^3.

If each person takes 2 ft^3, then this box could contain about 9.2×10^9 people. This box would contain 1$\frac{1}{2}$ times the present population.

7.4 Miscellaneous Measurements, page 381

Level 1, page 388

1. Answers vary.   **2.** Answers vary.   **3.** Answers vary.   **4. a.** kilometer **b.** centimeter

5. a. centimeter **b.** meter   **6. a.** milliliter **b.** milliliter   **7. a.** liter **b.** kiloliter

8. a. gram **b.** kilogram   **9. a.** Celsius **b.** Celsius   **10.** A   **11.** C   **12.** B   **13.** C

14. B   **15.** B   **16.** B   **17.** C   **18.** C   **19.** A   **20.** C   **21.** B   **22.** A   **23.** C

24. A   **25.** B   **26.** D   **27.** B   **28.** A   **29.** A

30. 0.009 kilometer; 0.09 hectometer; 0.9 dekameter; **9 meter**; 90 decimeter; 900 centimeter; 9,000 millimeter

31. 0.000063 kiloliter; 0.00063 hectoliter; 0.0063 dekaliter; 0.063 liter; 0.63 deciliter; 6.3 centiliter; **63 milliliter**

32. 0.00006 kilometer; 0.0006 hectometer; 0.006 dekameter; 0.06 meter; 0.6 decimeter; **6 centimeter**; 60 millimeter

33. 0.0035 kiloliter; 0.035 hectoliter; 0.35 dekaliter; **3.5 liter**; 35 deciliter; 350 centiliter; 3,500 milliliter

34. **4 kilometer**; 40 hectometer; 400 dekameter; 4,000 meter; 40,000 decimeter; 400,000 centimeter; 4,000,000 millimeter

35. 0.08 kiloliter; 0.8 hectoliter; **8 dekaliter**; 80 liter; 800 deciliter; 8,000 centiliter; 80,000 milliliter

36. 0.0015 kilometer; 0.015 hectometer; 0.15 dekameter; 1.5 meter; 15 decimeter; **150 centimeter**; 1,500 millimeter

37. 0.31 kiloliter; **3.1 hectoliter**; 31 dekaliter; 310 liter; 3,100 deciliter; 31,000 centiliter; 310,000 milliliter

Level 2, page 389

38. a. 10 **b.** 0.01 **c.** 0.00001 **d.** 0.1   **39. a.** 0.001 **b.** 0.000001 **c.** 1,000 **d.** 100

40. a. 0.001 **b.** 0.01 **c.** 0.001 **d.** 0.000001   **41. a.** 1,000 **b.** 10 **c.** 0.001 **d.** 1,000,000

42. a. 10,000 **b.** 1,000 **c.** 1,000 **d.** 100   **43.** $S = (3.5 \text{ cm})^2 + 4(3.5 \text{ cm})(4.2 \text{ cm}) \approx 71 \text{ cm}^2$

44. $S = \pi(1.6 \text{ cm})^2 + (2.4 \text{ cm})[2\pi(1.6 \text{ cm})] = 10.24\pi \text{ cm}^2 \approx 32 \text{ cm}^2$

45. $S = 4\pi(15 \text{ cm})^2 = 900\pi \text{ cm}^2 \approx 2,827 \text{ cm}^2$

46.
$$S = \pi r \sqrt{r^2 + h^2} \quad \text{(no top on the solid)}$$
$$= \pi(3 \text{ in.})\sqrt{(3 \text{ in.})^2 + (5 \text{ in.})^2}$$
$$= \pi(3 \text{ in.})\sqrt{34 \text{ in.}^2}$$
$$\approx 54 \text{ in.}^2$$

47. $V = (5 \text{ ft})(3 \text{ ft})(2 \text{ ft}) = 30 \text{ ft}^3$ **48.** $V = \frac{1}{3}Bh = \frac{1}{3}\left[\frac{1}{2}(5 \text{ in.})(12 \text{ in.})\right](3 \text{ in.}) = 30 \text{ in.}^3$

49. $V = Bh = \left[\frac{1}{2}(10 \text{ cm})(6 \text{ cm})\right](20 \text{ cm}) = 600 \text{ cm}^3$ **50.** $V = \frac{1}{3}Bh = \frac{1}{3}(4 \text{ cm})(8 \text{ cm})(6 \text{ cm}) = 64 \text{ cm}^3$

51. $S = 4\pi r^2 = 4\pi(6 \text{ in.})^2 = 144\pi \text{ in.}^2 \approx 452 \text{ in.}^2$; $V = \frac{4}{3}\pi r^3 = \frac{4}{3}\pi(6 \text{ in.})^3 = 288\pi \text{ in.}^3 \approx 905 \text{ in.}^3$

52. $S = \pi r\sqrt{r^2 + h^2} + \pi r^2 = \pi(3 \text{ in.})\sqrt{(3 \text{ in.})^2 + (4 \text{ in.})^2} + \pi(3 \text{ in.})^2 = 15\pi \text{ in.}^2 \approx 47 \text{ in.}^2$;

$V = \frac{1}{3}Bh = \frac{1}{3}[\pi(3 \text{ in.})^2](4 \text{ in.}) = 12\pi \text{ in.}^3 \approx 38 \text{ in.}^3$ **53.** $A = \ell w = (2\ell)(3w) = 6\,\ell w$; The area is

increased six-fold.   **54.** $V = \ell wh = (2\ell)(3w)(2h) = 12\,\ell wh$; The volume is increased twelve-fold.

55. $S = 4\pi r^2 = 4\pi(3r)^2 = 36\pi r^2$; The area is increased nine-fold.   **56.** $V = \frac{1}{3}Bh = \frac{1}{3}B(2h) = \frac{2}{3}Bh$;

The volume is doubled.

Level 3, page 389

57. $A = \pi r^2$, or (in terms of the diameter), $A = \pi\left(\frac{d}{2}\right)^2 = \frac{\pi d^2}{4}$. Now, if we double the diameter, we

have $A = \frac{\pi(2d)^2}{4} = \frac{4\pi d^2}{4}$, so we see that the area is increased four-fold.   **58.** $V = \ell wh = (2\ell)(2w)(2h)$

$= 8\ell wh$; The volume is increased eight-fold, so it will take 80 hours to fill the similarly-shaped pool.

Level 3 Problem Solving, page 390

59.

60. a. given   **b.** 5   **c.** 6   **d.** 6   **e.** 6, 8   **f.** 8, 12   **g.** 12, 20   **h.** 20, 12

Pattern: NUMBER OF SIDES + NUMBER OF VERTICES = NUMBER OF EDGES + 2

Chapter 7 Review Questions, page 383

1. a. _____   **b.** 1.1 cm   **c.** Answers vary; $\frac{1}{2}$ in.; 1.5 cm

2. **a.** First find the distance around the semicircle: $C = \frac{1}{2}\pi d = 4\pi \approx 12.56637061$.

    The distance around is $10 + 7 + 4\pi + 7 + 6$. To the nearest inch, the perimeter is 43 in.

    b. First find the area of the semicircle: $A = \frac{1}{2}\pi(4)^2 = 8\pi \approx 25.13274123$.

    The area of the trapezoid is $A = \frac{1}{2}(8)(7 + 13) = 80$. The area of the entire figure is

    $8\pi + 80 \approx 105.1327412$.   To the nearest square inch, the area is 105 in.^{2}.

3. **a.** $V = \ell w h$ **b.** Since 1 L = 1 dm^3, we see from part **a** that the box holds 30

    $ = 2(3)(5)$ liters.

    $ = 30$ **c.**   BASE + 2(SIDE) + 2(FRONT) $= 2 \cdot 3 + 2(2 \cdot 5) + 2(3 \cdot 5)$

    The volume is 30 dm^3. $= 56$ dm^2

4. **a.** $S = 4\pi r^2 = 4\pi(1 \text{ ft})^2 = 4\pi \text{ ft}^2$

    b. $4\pi \text{ ft}^2 \approx 12.6 \text{ ft}^2$

    c. $V = \frac{4}{3}\pi r^3 = \frac{4}{3}\pi(1 \text{ ft})^3 = \frac{4}{3}\pi \text{ ft}^3$

    d. $\frac{4}{3}\pi \approx 4.188790205$; the volume is 4.2 ft^3.

5. **a.** feet and meters   **b.** Answers vary; 40°C; 100°F   **c.** one-thousandth   **d.** capacity

    e. 10 km = 1,000,000 cm (Move the decimal point 5 places to the right.)

6. **a.** $S = (80 \text{ ft})(30 \text{ ft}) = 2{,}400 \text{ ft}^2$ **b.** $V = (80 \text{ ft})(30 \text{ ft})(3 \text{ ft}) = 7{,}200 \text{ ft}^3$

    c. Since 1 ft$^2 \approx 7.48$ gal, and the volume is 7,200 ft^3, the total capacity of the swimming pool is

    $(7.48 \text{ gal})(7{,}200) \approx 53{,}856$ gal

7. $11 \text{ ft} \times 16 \text{ ft} = 3\frac{2}{3} \text{ yd} \times 5\frac{1}{3} \text{ yd}$

    $= \frac{11}{3} \times \frac{16}{3} \text{ yd}^2$

    $= \frac{176}{9} \text{ yd}^2$

    $\approx 19.6 \text{ yd}^2$; You need to purchase 20 square yards.

8. **a.** The 6-in. pizza is about \$0.13/in.^{2}; the 10-in. and 12-in. pizzas are about \$0.10/in.^{2}, and the

    14-in. pizza is about \$0.09/in.^{2}. The best value is the large size.

    b. Answers vary; the size is 201 in.^{2}; price for a large should be about \$0.09 per square inch; I

    would price it at \$17.95 (\$16.95 to \$18.25 is acceptable; calculate $201 \times 0.09 \approx 18.09$).

9. **a.** $5^2 + 12^2 = 25 + 144 = 169 = 13^2$

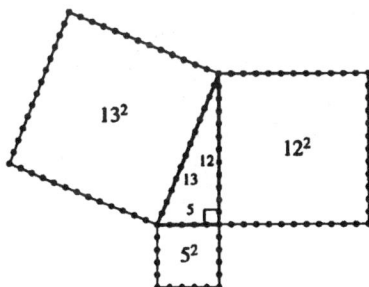

b. $25^2 + 312^2 = 313^2$, so the answer is yes,

it is possible to form such a triangle.

10.   Use the diagram to give a geometric justification of the distributive law.

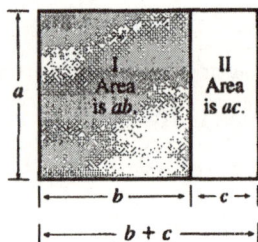

Area of large rectangle is
$a(b + c)$ and is the same as
the areas I and II combined.

Area of rectangle I: ab
Area of rectangle II: ac
Area of large rectangle is $ab + ac$.

Group Research Problems, page 395

G26. The volume of the house is $34 \times 26.75 \times 8 + 22 \times 13 \times 2 = 7{,}848$ ft^3. We need to approximate the number of marbles in a cubic foot. Begin by calculating that 1 ft^3 = (12 in.)^3 = 1,728 in.^{3} Next, estimate (conservatively) that there are 8 marbles per cubic inch, so that there are $1{,}728 \times 8 = 13{,}824$ marbles per cubic ft. Thus, the house would contain about 108,490,752 marbles. (Note: We will be able to use this information for some estimations in the next chapter.)

G27. Answers vary.

   a.   The spa contains $7^2\pi(4) \approx 615.75$ ft^3; since 1 ft$^3 \approx 7.48$ gal we estimate the spa contains about 4,606 gallons. A little research is necessary to find that water weighs about 52.4 lb/ft^3, so the water in the spa weights about 32,265 lb. If we add the weight of the spa (550 lb) and the number of people the spa could contain (at maximum) 10 people (5,000 lb), you should build the deck strong enough to support 38,000 lb.

   b.   The plans for the deck may vary.

   c.   The list of materials will depend on the plans.

   d.   The cost will depend on the list of materials.

G28.   Answers vary; see the computer output.

	A	B	C	D	E	F	G	H
1	Pythagorean Triplets							
2								
3	Number, m a	b		c				
4	1 +2*(A4+1)	+A4*(A4+2)	+SQRT(B4^2+C4^2)					
5	+A4+1 replicate	replicate	replicate					
6	replicate							
7								
8								
9								
10								

Here is a sample output:

Pythagorean Triplets

Number, m	a	b	c	Number, m	a	b	c
1	4	3	5	26	54	728	730
2	6	8	10	27	56	783	785
3	8	15	17	28	58	840	842
4	10	24	26	29	60	899	901
5	12	35	37	30	62	960	962
6	14	48	50	31	64	1023	1025
7	16	63	65	32	66	1088	1090
8	18	80	82	33	68	1155	1157
9	20	99	101	34	70	1224	1226
10	22	120	122	35	72	1295	1297
11	24	143	145	36	74	1368	1370
12	26	168	170	37	76	1443	1445
13	28	195	197	38	78	1520	1522
14	30	224	226	39	80	1599	1601
15	32	255	257	40	82	1680	1682
16	34	288	290	41	84	1763	1765
17	36	323	325	42	86	1848	1850
18	38	360	362	43	88	1935	1937
19	40	399	401	44	90	2024	2026
20	42	440	442	45	92	2115	2117
21	44	483	485	46	94	2208	2210
22	46	528	530	47	96	2303	2305
23	48	575	577	48	98	2400	2402
24	50	624	626	49	100	2499	2501
25	52	675	677	50	102	2600	2602

Individual Research Problems, page 396

The projects of this section are all research papers, so the answers will, of course, vary. Be sure to check out the web resources at **www.mathnature.com**.

CHAPTER 8

THE NATURE OF GROWTH

8.1 Exponential Equations, page 398

Level 1, page 404

1. Short answer: a *logarithm* is an exponent. Complete answer: $\log_b x$ is the exponent on a base b that gives the answer x.   **2.** A *common logarithm*, written $\log N$, is the exponent on a base 10 that gives N.   **3.** A *natural logarithm*, written $\ln N$, is the exponent on a base e that gives N.

4. a. $\log N = \log_{10} N$ **b.** $\ln N = \log_e N$ **c.** $\log_b N$ is the exponent on a base b that gives N.

5. Answers vary depending on calculator model. **a.** $\log N$; $\boxed{\log}\ \boxed{N}$ **b.** $\ln N$; $\boxed{\ln}\ \boxed{N}$

c. $\log_b N$; $\boxed{\log}\ \boxed{N}\ \boxed{\div}\ \boxed{\log}\ \boxed{b}$ **6.** An *exponential equation* is an equation in which a variable appears in an exponent.   **7.** $10^x = N$; $e^x = N$; and $b^x = N$ **8.** Solve for the exponential (10^x, e^x, or b^x) by using opposite operations and then solve for the exponent using the definition of logarithm.

9. a. $6 = \log_2 64$ **b.** $2 = \log 100$ **c.** $p = \log_n m$ **10. a.** $3 = \log 1{,}000$ **b.** $2 = \log_9 81$

c. $-1 = \ln \frac{1}{e}$ **11. a.** $-1 = \log \frac{1}{10}$ **b.** $2 = \log_6 36$ **c.** $n = \log_t s$ **12. a.** $3 = \log_5 125$

b. $-2 = \log_{1/3} 9$ **c.** $c = \log_b a$ **13. a.** 1 **b.** 3 **c.** -5 **14. a.** 2 **b.** -1 **c.** -3

15. a. 1 **b.** 2 **c.** -3 **16. a.** 1 **b.** 2 **c.** -4 **17. a.** 1 **b.** 3 **c.** -6 **18. a.** n **b.** n

c. n **19. a.** $\log_4 5$ **b.** $\log_5 8$ **c.** $\log_6 4.5$ **20. a.** $\log_\pi 10$ **b.** $\log_{\sqrt{2}} 5$ **c.** $\ln 9$

21. a. $\log 15$ **b.** $\log 2.5$ **c.** $\log 45$ **22. a.** $\ln 6$ **b.** $\ln 1.8$ **c.** $\ln 34.2$ **23. a.** 0.63 **b.** 2.00

c. 3.00   **24. a.** 0.03 **b.** 4.00 **c.** 0.50   **25. a.** 4.85 **b.** 2.00 **c.** 0.00   **26. a.** 1.28 **b.** 5.00

c. 0.00   **27. a.** -1.38 **b.** -0.49 **c.** -1.27 **28. a.** 0.82 **b.** 2.82 **c.** 1.99   **29. a.** 2.30

b. 4.61 **c.** 6.91   **30. a.** 0.43 **b.** 0.87 **c.** 1.30   **31.** $\log_3 45 = \dfrac{\log 45}{\log 3} \approx 3.4650$

32. $\log_5 91 = \dfrac{\log 91}{\log 5} \approx 2.8028$ **33.** $\log_6 10 = \dfrac{1}{\log 6} \approx 1.2851$ **34.** $\log_5 304 = \dfrac{\log 304}{\log 5} \approx 3.5522$

35. $\log_4 3.05 = \dfrac{\log 3.05}{\log 4} \approx 0.8044$ **36.** $\log_2 1{,}513 = \dfrac{\log 1{,}513}{\log 2} \approx 10.5632$

37. $\log_2 0.0056 = \dfrac{\ln 0.0056}{\ln 2} \approx -7.4804$ **38.** $\log_{8.3} 105 = \dfrac{\ln 105}{\ln 8.3} \approx 2.1991$

39. $\log_8 10 = \dfrac{\ln 10}{\ln 8} \approx 1.1073$ **40.** $\log_\pi e^2 = \dfrac{2}{\ln \pi} \approx 1.7471$ **41.** $\log_8 e = \dfrac{1}{\ln 8} \approx 0.4809$

42. $\log_{1.08} 5{,}450 \approx \dfrac{\ln 5{,}450}{\ln 1.08} \approx 111.7886$

Level 2, page 405

43. a. $x = \log_2 128 = 7$ **b.** $x = \log_3 243 = 5$

44. a. $x = \log_{125} 25 = \frac{2}{3} \approx 0.666666667$ **b.** $x = \log_{216} 36 = \frac{2}{3} \approx 0.666666667$

45. a. $x = \log_4 \frac{1}{16} = -2$ **b.** $x = \log_{27} \frac{1}{81} = -\frac{4}{3} \approx -1.333333333$

46. a. $x = \log_8 3 \approx 0.528320834$ **b.** $x = \log_{64} 5 \approx 0.386988016$

47. a. $x = \ln 4 \approx 1.386294361$ **b.** $x = \ln 25 \approx 3.218875825$

48. a. $x = \log 42 \approx 1.623249290$ **b.** $\log 0.0234 \approx -1.630784143$

49. a. $x = \frac{1}{5}\log 5 \approx 0.139794001$ **b.** $x = \frac{1}{3}\log 0.45 \approx -0.115595829$

50. a. $5 - 3x = \log 0.041;\ x = \frac{1}{3}(5 - \log 0.041) \approx 2.129072048$ **b.** $2x - 1 = \log 515;$
$x = \frac{1}{2}(\log 515 + 1) \approx 1.855903615$ **51. a.** $1 - 2x = \ln 3;\ x = \frac{1}{2}(1 - \ln 3) \approx -0.049306144$
b. $1 - 5x = \ln 25;\ x = \frac{1}{5}(1 - \ln 25) \approx -0.443775165$ **52. a.** $x = -\ln 8 \approx -2.079441542$
b. $x = -\log 125 \approx -2.096910013$

Level 3, page 405

53.
$$2 \cdot 3^x + 7 = 61$$
$$2 \cdot 3^x = 54$$
$$3^x = 27$$
$$x = 3$$

54.
$$3 \cdot 5^x + 30 = 105$$
$$3 \cdot 5^x = 75$$
$$5^x = 25$$
$$x = 2$$

55. $\left(1 + \dfrac{0.08}{360}\right)^{360x} = 2$
$$360x = \log_b 2 \text{ where } b = 1 + \frac{0.08}{360}$$
$$x = \frac{\log_b 2}{360}$$
$$= \frac{1}{360}\left[\frac{\log 2}{\log\left(1 + \frac{0.08}{360}\right)}\right]$$
$$\approx 8.665302427$$

56. $\left(1 + \dfrac{0.055}{12}\right)^{12x} = 2$
$$12x = \log_b 2 \text{ where } b = 1 + \frac{0.055}{12}$$
$$x = \frac{\log_b 2}{12}$$
$$= \frac{1}{12}\left[\frac{\log 2}{\log\left(1 + \frac{0.055}{12}\right)}\right]$$
$$\approx 12.63153513$$

57.
$$P = P_0 e^{rt}$$
$$\frac{P}{P_0} = e^{rt}$$
$$rt = \ln \frac{P}{P_0}$$
$$t = \frac{1}{r}\ln \frac{P}{P_0}$$

58.
$$I = I_0 e^{-rt}$$
$$\frac{I}{I_0} = e^{-rt}$$
$$-rt = \ln \frac{I}{I_0}$$
$$r = -\frac{1}{t}\ln \frac{I}{I_0}$$

59. $V = \frac{4}{3}\pi r^3 = \frac{4}{3}\pi (3{,}963 \text{ mi})^3 \approx 2.60711883 \times 10^{11} \text{ mi}^3$

Change this to in.^{3} using $(1 \text{ mi})^3 = (5{,}280 \text{ ft})^3 = [5{,}280(12 \text{ in.})]^3$ and

$(1 \text{ ft})^3 = (12 \text{ in.})^3$; $V \approx 6.631416905 \times 10^{25} \text{ in.}^3$

Since 1 gallon $= 231 \text{ in.}^3$, $V = 2.870743249 \times 10^{23} \text{ gal}$

Solve the exponential equation

$$2^x = V$$
$$x = \log_2 V = \frac{\log V}{\log 2} \approx 77.92577049$$

It would take about 78 doublings, the machine would reach the required capacity in the next minute; that is in about 1 hr 18 minutes.

60. Weight is about 5.9×10^{21} metric tons $= 5.9 \times 10^{27}$ g; solve the exponential equation

$$\left(\frac{1}{2}\right)^x = 5.9 \times 10^{27}$$
$$x = \log_{1/2}(5.9 \times 10^{27}) = \frac{\log(5.9 \times 10^{27})}{\log 0.5} = -92.25277352$$

The negative is because we are starting with the large number and counting down.

This means it will be cut in half about 93 times to reach one gram. This is about one hour 33 minutes.

8.2 Logarithmic Equations, page 406

Level 1, page 412

1. A *logarithmic equation* is an equation in which there is a logarithm on one or both sides.

2. Type I: The unknown is the logarithm.   Type II: The unknown is the base.   Type III: The logarithm of an unknown is equal to a number.   Type IV: The logarithm of an unknown is equal to the logarithm of a number.   **3.** Solve for the logarithm; this will lead to one of the forms:

(1) $\log_b M = \log_b N$ which implies $M = N$; (2) $\log_b M = N$ which implies $b^N = M$

Reasons in Problems 4-20 may vary.   **4.** T   **5.** F; a common logarithm is a logarithm in which the base is 10.   **6.** F; a natural logarithm is a logarithm in which the base is e.   **7.** F; the exponent is $\log_b N$.   **8.** F; divide $\log N$ by $\log 5$.   **9.** T; note that $81^2 = 6{,}561$.   **10.** T; divide both sides by 2.   **11.** F; x is the exponent on 1.5.   **12.** F; try $x = 10$ for a counterexample.   **13.** F; try $A = 10$ and $B = 20$ for a counterexample.   **14.** F; try $A = 10$ and $B = 20$ for a counterexample.   **15.** F; try $A = 10$ and $B = 20$ for a counterexample.   **16.** T   **17.** F; try $A = 10$ and $B = 20$ for a counterexample.   **18.** F; try $A = 10$ and $B = 20$ for a counterexample.   **19.** F; $\log_b N$ is not defined when N is negative.   **20.** F; $\log N$ is negative when $0 < N < 1$.   **21. a.** 23 **b.** 3.4 **c.** x **22. a.** 4.2 **b.** 3   **c.** x **23. a.** x **b.** x **24. a.** 2 **b.** 7   **25. a.** 4 **b.** 3   **26. a.** -1

b. 4   27. a. 3   b. -3 28. a. 10^5 b. e 29. a. e^2 b. e^3 30. a. e^4 b. 14

31. a. 9.3   b. 109   32. a. $\log 24$ b. $\log 2$ c. $\ln(x^2 y^3 / z^4)$ 33. a. $\ln 6$ b. $\ln 4$ c. $\ln \frac{2}{243}$

34. a. $\ln \frac{3}{2}$ b. $\ln \frac{1}{48}$ c. $\ln \frac{3}{1,024}$ 35. a. $\log(x-3)$ b. $\log(x-3)$ c. $\ln(x-2)$

Level 2, page 412

*The **a** and **b** parts of Problems 36-43 are designed to show the students that logarithmic equations are essentially the same as first degree equations.*

36. a.
$$\frac{1}{2}x - 2 = 2$$
$$\frac{1}{2}x = 4$$
$$x = 8$$

   b.
$$\frac{1}{2}\log x - \log 100 = 2$$
$$\frac{1}{2}\log x = 4 \quad \textit{Note: } \log 100 = 2$$
$$\log x = 8$$
$$x = 10^8$$

37. a.
$$3 + 2x = 11$$
$$2x = 8$$
$$x = 4$$

   b.
$$\ln e^3 + 2\log x = 11$$
$$2\log x = 8 \quad \textit{Note: } \ln e^3 = 3$$
$$\log x = 4$$
$$x = 10^4$$

38. a.
$$\frac{1}{2}x = 3 - x$$
$$\frac{3}{2}x = 3$$
$$x = 2$$

   b.
$$\frac{1}{2}\log_b x = 3\log_b 5 - \log_b x$$
$$\frac{3}{2}\log_b x = 3\log_b 5 \quad \textit{Add } \log_b x \textit{ to both sides.}$$
$$\log_b x = 2\log_b 5$$
$$\log_b x = \log_b 25 \quad \textit{Note: } 2\log_b 5 = \log_b 5^2$$
$$x = 25$$

39. a.
$$x - 2 = 2$$
$$x = 4$$

   b.
$$\log 10^x - 2 = \log 100$$
$$x - 2 = 2 \quad \textit{Note: } \log 10^x = x;\ \log 100 = 2$$
$$x = 4$$

40. a.
$$1 = x - 1$$
$$2 = x$$

   b.
$$\ln e = \ln \frac{\sqrt{2}}{x} - \ln e$$
$$1 = \ln \frac{\sqrt{2}}{x} - 1 \quad \textit{Note: } \ln e = 1$$
$$2 = \ln \frac{\sqrt{2}}{x}$$
$$e^2 = \frac{\sqrt{2}}{x}$$
$$x = \frac{\sqrt{2}}{e^2}$$

41. a. $1 = \frac{3}{2} - x$

$\quad x = \frac{1}{2}$

b. $\log 10 = \log \sqrt{1{,}000} - \log x$

$\quad\quad 1 = \frac{3}{2} - \log x \quad\quad$ *Note:* $\log 10 = 1; \log \sqrt{1{,}000} = \frac{3}{2}$

$\quad \log x = \frac{1}{2}$

$\quad \sqrt{10} = x \quad\quad\quad\quad$ *Definition of logarithm*

42. a. $3 - x = 1$

$\quad\quad 3 = x + 1$

$\quad\quad 2 = x$

b. $\ln e^3 - \ln x = 1$

$\quad\quad 3 - \ln x = 1 \quad$ *Note:* $\ln e^3 = 3$

$\quad\quad\quad 2 = \ln x$

$\quad\quad\quad x = e^2$

43. $0 + x = 2$

$\quad\quad x = 2$

b. $\ln 1 + \ln e^x = 2$

$\quad\quad 0 + x = 2 \quad$ *Note:* $\ln 1 = 0; \ln e^x = x$

$\quad\quad\quad x = 2$

44. $\log(\log x) = 1$

$\quad \log x = 10^1$

$\quad\quad x = 10^{10}$

45. $\ln[\log(\ln x)] = 0$

$\quad \log(\ln x) = e^0$

$\quad\quad \ln x = 10^1 \quad\quad$ *Note:* $e^0 = 1$

$\quad\quad\quad x = e^{10}$

46. $x^2 5^x = 5^x$

$\quad\quad x^2 = 1$

$\quad x^2 - 1 = 0$

$(x - 1)(x + 1) = 0$

$\quad\quad\quad x = 1, -1$

47. $x^2 3^x = 9(3^x)$

$\quad\quad x^2 = 9$

$\quad x^2 - 9 = 0$

$(x - 3)(x + 3) = 0$

$\quad\quad\quad x = 3, -3$

48. $\log x = 1.8 + \log 4.8$

$\quad \log x - \log 4.8 = 1.8$

$\quad\quad \log \frac{x}{4.8} = 1.8$

$\quad\quad 10^{1.8} = \frac{x}{4.8}$

$\quad\quad\quad x = 4.8 \cdot 10^{1.8}$

49. $\ln x = 1.8 - \ln 4.8$

$\quad \ln x + \ln 4.8 = 1.8$

$\quad\quad \ln 4.8x = 1.8$

$\quad\quad 4.8x = e^{1.8}$

$\quad\quad\quad x = \frac{e^{1.8}}{4.8}$

50. $\ln x - \ln 8 = 12$

$\quad\quad \ln \frac{x}{8} = 12$

$\quad\quad \frac{x}{8} = e^{12}$

$\quad\quad x = 8e^{12}$

51. $\log x + \log 8 = 12$

$\quad\quad \log 8x = 12$

$\quad\quad 8x = 10^{12}$

$\quad\quad x = \frac{10^{12}}{8}$

52. $\log 2 = \frac{1}{4}\log 16 - x$

$\quad\quad x = \log(2^4)^{1/4} - \log 2$

$\quad\quad\quad = \log 2 - \log 2$

$\quad\quad\quad = 0$

53. $\log_8 5 + \frac{1}{2}\log_8 9 = \log_8 x$

$\quad\quad \log_8 5 + \log_8 9^{1/2} = \log_8 x$

$\quad\quad\quad \log_8 5(3) = \log_8 x$

$\quad\quad\quad\quad x = 15$

54. $\log x + \log(x - 3) = 2$

$\quad\quad\quad \log x(x - 3) = 2$

$\quad\quad\quad\quad x(x - 3) = 10^2$

$\quad x^2 - 3x - 100 = 0$

$$x = \frac{3 \pm \sqrt{9 - 4(1)(-100)}}{2}$$

$$= \frac{3 \pm \sqrt{409}}{2}$$

Level 3, page 412

55. **a.** $a = 1,000;\ N = 1,500 + 300\ln 1,000 \approx 3,572.33$; Could expect to sell 3,572 items.

b. $a = 50,000;\ N = 1,500 + 300\ln 50,000 \approx 4,745.93$; Could expect to sell 4,746 items.

c. $N = 4,000$

$$4,000 = 1,500 + 300\ln a$$

$$2,500 = 300\ln a$$

$$\frac{25}{3} = \ln a$$

$$a = e^{25/3} \approx 4,160.26$$

Need to spend about $4,200.

56. **a.** $\text{pH} = -\log[2.86 \times 10^{-4}] \approx 3.5$

b. $\text{pH} = -\log[6.31 \times 10^{-7}] \approx 6.2$

c. $\quad 8.1 = -\log[\text{H}^+]$

$\quad -8.1 = \log[\text{H}^+]$

$10^{-8.1} = [\text{H}^+]$

57. **a.** $M = \dfrac{\log 15^{15} - 11.8}{1.5} \approx 3.89$

b. $M = \dfrac{\log 10^{25} - 11.8}{1.5} = 8.8$

c. $8.0 = \dfrac{\log E - 11.8}{1.5}$

$\quad 12 = \log E - 11.8$

$\quad 23.8 = \log E$

$\quad E = 10^{23.8}$

The energy is about $10^{23.8}$ ergs.

d. $M = \dfrac{\log E - 11.8}{1.5}$

$\quad 1.5M = \log E - 11.8$

$\quad 1.5M + 11.8 = \log E$

$\quad E = 10^{1.5M+11.8}$

58. a.

$$10 = 80 - 27 \ln t$$
$$27 \ln t = 70$$
$$\ln t = \frac{70}{27}$$
$$t = e^{70/27} \approx 13.364$$

After about 13 seconds.

b.

$$R = 80 - 27 \ln t$$
$$27 \ln t = 80 - R$$
$$\ln t = \frac{80 - R}{27}$$
$$t = e^{(80 - R)/27}$$

Level 3 Problem Solving, page 413

59. a. $1 - N/80$ must be positive, so the maximum number of words is 80. If we assume that N is a counting number, then the largest value for N is 79 (since $\ln 0$ is not defined).

b.

$$t = -62.5 \ln\left(1 - \frac{N}{80}\right)$$
$$\frac{t}{-62.5} = \ln\left(1 - \frac{N}{80}\right)$$
$$1 - \frac{N}{80} = e^{-t/62.5}$$
$$1 - e^{-t/62.5} = \frac{N}{80}$$
$$N = 80(1 - e^{-t/62.5})$$

60. Let $m = \log_b A$; then

$$b^m = A \qquad \textit{Definition of logarithm}$$
$$(b^m)^p = A^p \qquad \textit{Raise both sides to the pth power.}$$
$$b^{pm} = A^p \qquad \textit{Property of exponents}$$
$$pm = \log_b A^p \qquad \textit{Definition of logarithm}$$
$$p\log_b A = \log_b A^p \qquad \textit{Substitute } m = \log_b A.$$

8.3 Applications of Growth and Decay, page 413
Level 1, page 420

1. The *growth/decay* formula is $A = A_0 e^{rt}$ where A_0 is the initial amount, A is the future amount, t is the time (in years), and r is the annual growth rate (if it is positive) and the annual decay rate (if it is negative).   **2.** The *half-life* is the length of time it takes for a given quantity to decay to half its original amount.   **3.** A *logarithmic scale* is a scale on a graph in which logarithms are used to make data more manageable by expanding small variations and compressing large ones. (See Problem 58, for example.)   **4.** A *micrometer* (denoted by μm) is one millionth of a meter.

5. $5,996,215,340 = 5,918,624,368 e^r$; $r = \ln(5,996,215,340/5,918,624,368) \approx 0.0130244419$, which means the growth rate is about 1.3%. The answers for the day's population depends on the date.

6. $6 = 5.3 e^{0.0098t}$; $0.0098t = \ln(6/5.3)$ $t \approx 12.65843354$, so the projected date is 12 years, 8 months, or in other words February, 2003.

For Problems 7-14,

     a. *The growth rate is $r = 0.1\ln(P/P_0)$ where P is the population in 1990 and P_0 is the population in 1980.*

     b. *Use the number r from part **a** (not the rounded value) with $P = P_0 e^{4r}$, where P_0 is the population in 1990.*

     c. *Answers vary.*

     d. *Use r from part **a** (not the rounded value) with $P = P_0 e^{15r}$, where P_0 is the population in 1990.*

7. a. 1.6% **b.** 678,662 **d.** 811,078   **8. a.** 2.9% **b.** 415,074 **d.** 572.093   **9. a.** 0.7%

b. 502,112 **d.** 541,917   **10. a.** 1.9% **b.** 556,509 **d.** 687,478   **11. a.** 0.006% **b.** 365,362

d. 365,608   **12. a.** 0.20% **b.** 578,862 **d.** 591,642   **13. a.** 0.07% **b.** 273,041 **d.** 275,270

14. a. -0.42% **b.** 195,227 **d.** 186,457   **15.** $r \approx 0.1614549977$ (from Example 3), so $A \approx 209,693 e^{9r} \approx 896,716$ **16.** $r = \frac{1}{4}\ln(450,000/209,693) \approx 0.1909007566$; $A = 209,693 e^{10r} \approx 1,414,671$ **17.** $r = \frac{1}{30}\ln 0.5 \approx -0.023104906$

18. $r = \frac{1}{86}\ln 0.5 \approx -0.0080598509$ **19.** $t = \ln 0.5/(-0.0246) \approx 28.18$; The half-life is about 28 years.   **20.** $t = \ln 0.5/(-0.0641) \approx 10.81$; The half-life is about 11 years.

21. $M = \log(5.1 \times 10^2) + 4.0 \approx 6.7$ **22.** $M = \log(8.2 \times 10^5) + 1.7 \approx 7.6$ **23.** $6 - 4 = \log \frac{A_1}{A_2}$, so $10^2 = \frac{A_1}{A_2}$ or $A_1 = 100 A_2$. This means that the earthquake with magnitude 6 is 100 times stronger than an earthquake of magnitude 4.   **24.** $8 - 7 = \log \frac{A_1}{A_2}$, so $A_1 = 10 A_2$. This means the earthquake with magnitude 8 is 10 times stronger than that of an earthquake of magnitude 7.

25. $D = 10 \log(10^{-13}/10^{-16}) = 30$ dB   **26.** $D = 10 \log(10^{-8}/10^{-16}) = 80$ dB

27. $D = 10 \log(5.23 \times 10^{-6}/10^{-16}) \approx 107$ dB   **28.** $D = 10 \log(3.16 \times 10^{-10}/10^{-16}) \approx 65$ dB

29. $D = 10 \log(2.53 \times 10^{-5}/10^{-16}) \approx 114$ dB   **30.** $D = 10 \log(10^{-5}/10^{-16}) \approx 110$ dB

31. $\log E = 11.8 + 1.5(8.3)$, so $E = 10^{24.25}$ ergs

Level 2, page 420

32. a. Solve $18 = 10 e^{4r}$, so $r \approx 0.1469466662$; at 6:00 P.M., $P = 18 e^{4r} \approx 32.4$

      The population at 6:00 P.M. is about 32.4 million.

  b. Solve $2 = e^{rt}$ where r is the rate from part **a** (not the rounded value); $t = \ln 2/r \approx 4.7$

The population will be double its 10 A.M. population at 2:43 P.M.

33. *Note:* Do not use the rounded growth rates shown here, but rather the growth rates shown on

your calculator. Use the formula $A = A_0 e^{rt}$; 1980 population, 2,968,528; 1990 population,

3,485,557, and 1994 population, 3,448,613.

    a. 1980 population is 2,968,528; 1990 population is 3,485,557. Use

      $r = \frac{1}{10}\ln(3,485,557/2,968,528)$ as the growth rate (it is about 1.6%);

      Predict the 2004 population by $A = 3,485,557 e^{14r} \approx 4,364,110 \approx 4,364,000$

    b. Use $r = \frac{1}{4}\ln(3,448,613/3,485,557)$ as the growth rate (it is about -0.27%)

      Predict the 2004 population by $A = 3,448,613 e^{10r} \approx 3,357,957 \approx 3,358,000$

    c. Use $r = \frac{1}{14}\ln(3,448,613/2,968,528)$ as the growth rate (it is about 1.07%);

      Predict the 2004 population by $A = 2,968,528 e^{24r} \approx 3,838,370 \approx 3,838,000$

34. From Example 5, $r \approx -1.209680943\mathrm{E}-4$, so

$$\text{PERCENT PRESENT} = e^{2047r} \approx 0.7807; \text{ this is about } 78.1\%$$

35. $0.28 = e^{rt}$ where $r \approx -1.209680943\mathrm{E}-4$ (from Example 5), so $t = \ln 0.28/r \approx 10,523$

The artifact is about 10,500 years old.

36. $0.5 = e^{rt}$ where $t = 2.52 \times 10^5$; we find $r = \ln 0.5/(2.52 \times 10^5) \approx -2.75058405\mathrm{E}-6$

$0.973 = e^{rt}$ so $t = \ln 0.973/r \approx 9951.04905$

The length of time is about 10,000 years.

37. $0.5 = e^{rt}$ where $t = 2.6$; we find $r = \ln 0.5/2.6 \approx -0.2665950694$

$15.5 = 100 e^{rt}$ so that $t = \ln(15.5/100)/r \approx 6.993115686$

The elapsed time is about 7 years.

38. From Example 9, $\dfrac{E_1}{E_2} = \dfrac{10^{1.5(6.6)+11.8}}{10^{1.5(7.1)+11.8}} = 10^{-0.75}$, so $10^{0.75}E_1 = E_2$

The earthquake with a magnitude of 7.1 releases $10^{0.75} \approx 5.6$ times more energy than an
earthquake with magnitude 6.6.

39. From Example 9, $\dfrac{E_1}{E_2} = \dfrac{10^{1.5(8.3)+11.8}}{10^{1.5(9.5)+11.8}} = 10^{-1.8}$, so $10^{1.8}E_1 = E_2$

The earthquake with a magnitude of 9.5 releases $10^{1.8} \approx 63$ times more energy than an
earthquake with magnitude 8.3.

40. $0.12 = e^{rt}$ where $r \approx -1.209680943\mathrm{E}-4$, so $t = \ln 0.12/r \approx 17,527$

The artifact is about 17,500 years old.

41. $0.85 = e^{rt}$ where $r \approx -1.209688094\mathrm{E}-4$, so $t = \ln 0.85/r \approx 1,343$

The artifact is about 1,300 years old.

42.　$0.7336 = e^{rt}$ when $t = 24$, so $r = \ln 0.7336/24 \approx -0.0129079733$; then

$0.43 = e^{rt}$, so $t = \ln 0.43/r \approx 65.38362405$

The time is about 65 hours. The half life is $t = \ln 0.5/r \approx 53.69914913$, or about 54 hours.

43.　From Example 9, $\dfrac{E_1}{E_2} = \dfrac{10^{1.5(7.0)+11.8}}{10^{1.5(7.1)+11.8}} = 10^{-0.15}$, so $10^{0.15}E_1 = E_2$

The earthquake with a magnitude of 7.1 released $10^{0.15} \approx 1.41$ times more energy than an earthquake with magnitude 7.0.

44.　From Example 9, $\dfrac{E_1}{E_2} = \dfrac{10^{1.5(6)+11.8}}{10^{1.5(3)+11.8}} = 10^{4.5}$, so $10^{4.5}E_2 = E_1$

The earthquake with a magnitude of 6 releases $10^{4.5} \approx 31{,}623$ times more energy than an earthquake with magnitude 3.

45.　From Example 9, $\dfrac{E_1}{E_2} = \dfrac{10^{1.5(8)+11.8}}{10^{1.5(7)+11.8}} = 10^{1.5}$, so $10^{1.5}E_2 = E_1$

The earthquake with a magnitude of 8 releases $10^{1.5} \approx 31.6$ times more energy than an earthquake with magnitude 3.

Level 3, page 421

46.　$0.73 = e^{r_1 t_1}$ where $r_1 \approx -1.209680943E-4$, so $t_1 = \ln 0.73/r \approx 2601.601245$

For radium, $0.32 = e^{r_2 t_1}$, so $r_2 = \ln 0.32/t_1 \approx -4.37974227E-4$

Finally, the half life is $0.50 = e^{r_2 h}$, so $h = \ln 0.5/r_2 \approx 1582.62095$

The half-life of radium is about 1,583 yr.

47.　$50 = 100e^{rt}$, where $t = 46.5$, so $r = \ln 0.5/46.5 \approx -0.014906391$

Amount present in 35 hours: $A = 100e^{35r} \approx 59.35$ mg

Amount present in 30 hours: $A = 100e^{30r} \approx 63.94$ mg

The amount lost is 4.59 mg. The percent lost is $4.59/63.94 \approx 0.072$ or 7.2% lost.

The percent is the same for any five hour period. We can show this as follows:

$100e^{30r} - 100e^{35r} = 100e^{30r}(1 - e^{5r})$, so the percent lost during a five-hour period is

$\dfrac{100e^{30r}(1 - e^{5r})}{100e^{30r}} = 1 - e^{5r}$, which is independent of the initial time.

48.　$T = 72 + (100 - 72)10^{-0.01(20)} \approx 89.67$; the water is about 90°F.

49.　$T = 74 + (250 - 74)10^{-0.075(10)} \approx 105.30$; the cookie is about 105°F.

50.　$k = \dfrac{1}{60} \log \dfrac{120 - 20}{70 - 20} \approx 0.0050171666$; $T = 20 + (120 - 20)10^{-k(30)} \approx 90.71$; The temperature in 30 minutes is about 91°C.

51. $k = \frac{1}{30} \log \frac{100 - 22}{45 - 22} \approx 0.0176788922$; $T = 22 + (100 - 22)10^{-k(40)} \approx 37.31$; The

temperature in 40 minutes is about 37°C.

52. $13.23 = 14.7e^{-0.21a}$, so $a = \dfrac{\ln(13.23/14.7)}{-0.21} \approx 0.5017167412$ miles;

In feet, the altitude is $5,280a \approx 2,649$ ft.

53. $11.9 = 14.7e^{-0.21a}$, so $a = \dfrac{\ln(11.9/14.7)}{-0.21} \approx 1.006$ miles;

In feet, the altitude is $5,280a \approx 5,313$ ft.

54. $10.2 = 14.7e^{-0.21a}$, so $a = \dfrac{\ln(10.2/14.7)}{-0.21} \approx 1.740284636$ miles;

In feet, the altitude is $5,280a \approx 9,188$ ft.

55.
$$P = 50e^{-t/250}$$

$$e^{-t/250} = \frac{P}{50}$$

$$-\frac{t}{250} = \ln \frac{P}{50}$$

$$t = -250 \ln \frac{P}{50}$$

If $P = 30$ W, then $t = -250 \ln(30/50) \approx 127.7$; The satellite will operate for about 128 days.

56. From Problem 56, $t = -250 \ln(10/50) \approx 402.36$; The satellite will operate for about 402 days.

Level 3 Problem Solving, page 422

57. From Example 5, $r \approx -1.209680943E - 4$, and from $P = P_0 e^{rt}$ where $P/P_0 = 0.923$,

$t = \ln 0.923 / r \approx 662.37337$

Since the sample was measured in 1988, this dates the artifact at $1988 - t \approx 1326$. This means the shroud was probably new when d'Arcis wrote the memo to the Pope, and that it is probably not the burial robe of Christ.

58. Answers vary.

59. When $x = 75$, $y \approx 0.16$ or no O-rings failures; when $x = 32$, $y \approx 4.80$ or 5 O-ring failures.

60.
$$\frac{1}{\eta} = Ae^{-E/(RT)}$$

$$\frac{1}{\eta A} = e^{-E/(RT)}$$

$$-\frac{E}{RT} = \ln\left(\frac{1}{\eta A}\right)$$

$$-\frac{E}{R\ln\left(\frac{1}{\eta A}\right)} = T$$

Chapter 8 Review Questions, page 423

1. $\log 100 + \log\sqrt{10} = 2 + \frac{1}{2}$
$$= 2\frac{1}{2}$$

2. $\ln e + \ln 1 + \ln e^{542} = 1 + 0 + 542$
$$= 543$$

3. $\log_8 4 + \log_8 16 + \log_8 8^{2.3} = \log_8 64 + 2.3$
$$= 2 + 2.3$$
$$= 4.3$$

4. $10^{\log 0.5} = 0.5$

5. $\ln e^{\log 1,000} = \ln e^3$
$$= 3$$

6. **a.** $\log 8.43 \approx 0.93$ **b.** $\log 9,760 \approx 3.99$

7. **a.** $\ln 2 \approx 0.69$ **b.** $\ln 0.125 \approx -2.08$

8. **a.** $\log_2 10 \approx 3.32$ **b.** $\log_\pi \frac{1}{\pi} = \log_\pi(\pi)^{-1} = -1$

9. $10^x = 85$
$$x = \log 85$$
$$\approx 1.929418926$$

10. $e^x = 500$
$$x = \ln 500$$
$$\approx 6.214608098$$

11. $435^x = 890$
$$x = \log_{435} 890$$
$$\approx 1.117832865$$

12. $e^{3x+1} = 45$
$$3x + 1 = \ln 45$$
$$x = \frac{\ln 45 - 1}{3}$$
$$\approx 0.935554163$$

13. $\log_6 x = 4$
$$x = 6^4$$

14. $2^{3x-1} = 6$
$$3x - 1 = \log_2 6$$
$$x = \frac{\log_2 6 + 1}{3}$$

15. $10^{2x} = 5$
$$2x = \log 5$$
$$x = \frac{\log 5}{2}$$

16. $\log(x+1) = 2 + \log(x-1)$
$$\log(x+1) - \log(x-1) = 2$$
$$\log\frac{x+1}{x-1} = 2$$
$$\frac{x+1}{x-1} = 10^2$$
$$x + 1 = 100x - 100$$
$$101 = 99x$$
$$x = \frac{101}{99}$$

17. $3\ln\frac{e}{\sqrt[3]{5}} = 3 - \ln x$
$$\ln\frac{e^3}{5} + \ln x = 3$$
$$\ln\frac{e^3 x}{5} = 3$$
$$\frac{e^3 x}{5} = e^3$$
$$x = 5$$

18. $A = P(1 + i)^x$

$\frac{A}{P} = (1 + i)^x$

$x = \log_{(1+i)} \frac{A}{P}$

19. $A = A_0 e^{-0.1t}$

$\frac{A}{A_0} = e^{-0.1t}$

$-0.1t = \ln \frac{A}{A_0}$

$t = -10 \ln \frac{A}{A_0}$

$= -10 \ln 0.5$

≈ 6.93 days

It would take about 7 days.

20. $A = A_0 e^{r(14/12)}$

$\frac{A}{A_0} = e^{r(14/12)}$

$r = \frac{12}{14} \ln 2$

≈ 0.59

Equation is $A = A_0 e^{0.59t}$, where A_0 is the number of teen infected and t is the number of years after 1992.

Group Research Problems, page 424

G29. Answers vary.

G30. a. $A = \frac{A_0}{k}(e^{rT} - 1)$

$Ak = A_0(e^{rT} - 1)$

$\frac{Ak}{A_0} = e^{rT} - 1$

$\frac{Ak}{A_0} + 1 = e^{rT}$

$rT = \ln\left(\frac{Ak}{A_0} + 1\right)$

$T = \frac{1}{r}\ln\left(\frac{Ak}{A_0} + 1\right)$

   b.   Answers vary;   One possibilities is $y = 29.6 e^{0.03x}$; another possibility is $y = -39 e^{-0.11x} + 61$.

   c.   For the first possibility, world reserves will last about 8 years; for the second possibility, the world reserves should last about 15 years.

Individual Research Problems, page 425

The projects of this section are all research papers, so the answers will, of course, vary. Be sure to check out the web resources at **www.mathnature.com**.

CHAPTER 9
THE NATURE OF
FINANCIAL MANAGEMENT

9.1 Interest, page 428

I introduce this section by asking some questions about interest: what it is, when you pay interest, etc. I usually ask someone in class if they have recently given up smoking. Someone usually says yes so I ask what he/she saves per day. I use this as a basis for introducing future value. This section usually takes me two days to cover.

Throughout this chapter I use the same variables in all of the formulas (see Transparency 22):

P for present value (or principal)
A for future value
I for interest
r for annual percentage rate
t for the time (in years)

In addition, we later define:

m as the amount of periodic payment
n for the number of times the interest is compounded per year; in this book we assume that the compounding period is the same period as the periodic payment (usually monthly).

Level 1, page 438

1. *Interest* is a rental fee paid for the use of another's money.   2. Answers vary. The *amount of interest I* is the dollar amount paid for the use of another's money, whereas the *interest rate* refers to a percent r in the formula $I = Prt$.   3. Answers vary. *Simple interest* is calculated according to the formula $I = Prt$, whereas *compound interest* pays interest on the accrued interest after a given length of time (called the compounding period).   4. Answers vary. The *present value* is the worth of a lump sum today, whereas the *future value* is the value of a lump sum of money at some time in the future. Present value and future value are related by the formula $A = P + I$ where I is the amount of interest.   5. It is about opening a savings account for a child to be used at the child's retirement. It shows how much can be accumulated if you plan ahead. Today, 12% is not an unrealistic rate of return, and the example shows that a *one-time* deposit of just over \$400 would result in a million dollars at retirement. If you want \$10 million you need a one-time deposit of \$4,102.70.   6. A;

interest rate is not stated, but you should still recognize a reasonable answer.   **7.** C; interest rate is not stated, but you should still recognize a reasonable answer.   **8.** C; price is not stated, but you should still recognize a reasonable answer.   **9.** B is the most reasonable.   **10.** C is the most reasonable.   **11.** A   **12.** D   **13.** D   **14.** C   **15.** B   **16.** $400   **17.** $1,500   **18.** $1,200   **19.** $4,200   **20.** $383.25   **21.** $1,028.25   **22.** $755.59   **23.** $16,536.79   **24.** $50.73   **25.** $1,132,835.66   **26.** $4,524.89   **27.** $1,661.44   **28.** $69,737.61   **29.** $165,298.89   **30.** $109,276.64   **31. a.** $28,940.57 **b.** $29,222.27 **c.** $29,367.30 **d.** $29,465.60 **e.** $29,513.58 **f.** $29,515.22 **g.** $29,515.24 **h.** $22,800.00   **32. a.** $255,358.46 **b.** $264,027.28 **c.** $268,622.83 **d.** $271,790.29 **e.** $273,352.17 **f.** $273,406.19 **g.** $273,406.40 **h.** $105,915.00

Level 2, page 439

33. $\$1,500\left(1 + 0.21 \cdot \frac{55}{360}\right) = \$1,548.13$ **34.** $\$8,553\left[1 + 0.165 \cdot \left(3 + \frac{125}{360}\right)\right] = \$13,276.75$

35. a. $1.18 **b.** $1.96 **c.** $3.06 **d.** $2.43 **e.** $940.99 **f.** $21,956.37 **g.** $43,912.74 **h.** $25,092.99

36. a. $8.24 **b.** $24.65 **c.** $98.92 **d.** $107.17 **e.** $74.19 **f.** $140.14 **g.** $610.03 **h.** $989.23

37. $1,000 = 750(1 + 0.08t)$

$\frac{4}{3} = 1 + 0.08t$

$0.08t = \frac{1}{3}$

$t = 4.1\overline{6}$

4 years, 61 days

38. $3,500 = 3,000(1 + 0.06t)$

$\frac{1}{6} = 0.06t$

$t = 2.7\overline{7}$

2 years, 284 days

39. $5,000 = 3,500\left(1 + \frac{0.05}{365}\right)^{365t}$

$\frac{5,000}{3,500} = \left(1 + \frac{0.05}{365}\right)^{365t}$

$365t = \log_{(1+0.05/365)}\frac{5,000}{3,500}$

$t \approx 7.133987462$

7 years, 49 days

40. $5,000 = 3,500 e^{0.045t}$

$0.045t = \ln\frac{5,000}{3,500}$

$t \approx 7.926109865$

7 years, 338 days

41. $A = 1,000\left(1 + \frac{0.04}{4}\right)^{5 \cdot 4}$

$\approx \$1,220.19$

42. $A = 2,400 e^{(0.12)5}$

$= \$4,373.09$

$I = A - P = \$1,973.09$

43. $A = \$1,000 e^{(0.16)15}$

$= \$11,023.18$

44. $I = A - P$

$= \$480 \cdot 12 \cdot 5 - \$18,490$

$= \$10,310$

45. $I = A - P$

$= \$410.83 \cdot 12 \cdot 4 - \$14,500$

$= \$5,219.84$

46. $I = A - P$

$= \$2,293.17 \cdot 12 \cdot 30 - \$285,000$

$= \$540,541.20$

47. $I = A - P$

$= \$1,247.40 \cdot 12 \cdot 30 - \$170,000$

$= \$279,064$

48. $A = \$125,000 e^{0.10(30)}$

$\approx \$2,510,692.12$

49. $A = \$650 e^{0.10(30)}$

$\approx \$13,055.60$

50. $P = \$50,000(1 + 0.06)^{-30}$

$\approx \$8,705.51$

51. $I = Prt$

$\$5,075 = P(0.05) \cdot 1$

$P = \$101,500$

52. $I = Prt$

$\$45.33 = P(0.03)\left(\frac{240}{360}\right)$

$P = \$2,266.50$

53. **a.** $\dfrac{5.7 \times 10^{12}}{3 \times 10^8} = \$19,000$

The debt is about \$19,000 per person.

b. Let $s = 365 \cdot 24 \cdot 60 \cdot 60$, which is the number of seconds in a year.

$I = Prt$

$= (5.7 \times 10^{12})(0.06)\left(\frac{1}{s}\right)$

$= \$10,844.75$

This is about 10,845 dollars/sec interest.

In Problems 54-56, the amount needed in the retirement account is simple interest because the interest is removed from the account each month and consequently does not accrue any interest.

54. $I = Prt$

$\$10,000 = P(0.15)\left(\frac{1}{12}\right)$

$P = \$800,000$

55. $I = Prt$

$\$50,000 = P(0.12)\left(\frac{1}{12}\right)$

$P = \$5,000,000$

56. $I = Prt$

$\$1,000 = P(0.06)\left(\frac{1}{12}\right)$

$P = \$200,000$

Level 3, page 440

57. a. $\quad 1,250 = 1,000\left(1 + \dfrac{0.07}{12}\right)^{12t}$

$\left(1 + \dfrac{0.07}{12}\right)^{12t} = \dfrac{1,250}{1,000}$

$12t = \log_{(1+0.07/12)}\left(\dfrac{1,250}{1,000}\right)$

$t \approx 3.197053654$

or 3 years, 2.4 months; this means that the necessary time to wait is 3 years, 3 months

b. $2{,}000 = 1{,}000\left(1 + \frac{0.07}{12}\right)^{12t}$

$12t = \log_{(1+0.07/12)} 2$

$t \approx 9.930955715$

or 9 years, 11.2 months; this means

you must wait 10 years.

58. a. $1{,}250 = 1{,}000e^{0.05t}$

$0.05t = \ln \frac{1{,}250}{1{,}000}$

$t \approx 4.462871026$

or 4 years, 169 days

c. $2{,}000 = 1{,}000e^{5r}$

$5r = \ln 2$

$r \approx 0.1386294361$ The annual rate is about 13.9%.

c. $2 = \left(1 + \frac{r}{12}\right)^{12(5)}$

$2^{1/60} = 1 + \frac{r}{12}$ *Raise both sides to the 1/60 power.*

$r = 12(2^{1/60} - 1)$

≈ 0.1394332836

The interest rate is about 14%.

b. $2{,}000 = 1{,}000e^{0.05t}$

$0.05t = \ln 2$

$t \approx 13.86294361$

or 13 years, 315 days

Level 3 Problem Solving, page 440

59. We want to find Y so that the compounded amount is equal to the amount with the simple interest formula; that is:

$$P\left(1 + \frac{r}{n}\right)^{nt} = P(1 + Yt)$$
$$P\left(1 + \frac{r}{n}\right)^{n} = P(1 + Y) \quad t = 1 \text{ (one year)}$$
$$\left(1 + \frac{r}{n}\right)^{n} = 1 + Y \quad \text{Divide both sides by } P.$$
$$\left(1 + \frac{r}{n}\right)^{n} - 1 = Y \quad \text{Subtract 1 from both sides.}$$

60. a. 6.14% **b.** 6.17% **c.** 4.04% **d.** 4.08%

9.2 Installment Buying, page 441

Level 1, page 447

1. Answers vary. *Add-on interest* is a method of calculating interest which uses the simple interest formula to calculate the amount of interest and then adds this total onto the original principal in order to find the amount that needs to be repaid.

2. Answers vary. *APR* refers to annual percentage rate; it is the actual interest rate paid when the rate is based on the actual amount owed for the length of time that it is owed. It is calculated according to the formula APR $= 2Nr/(N + 1)$ where N is the number of payments.

3. *Open-ended credit* authorizes a pre-approved line of credit; it usually takes the form of a credit card loan. *Closed-ended credit* refers to a fixed amount that is borrowed with a promise to repay the loan in a specified manner.

4. Answers vary.   For credit card interest, use the simple interest formula, $I = Prt$. **Previous balance method:** Interest is calculated on the previous month's balance. With this method, $P =$ previous balance, $r =$ annual rate, and $t = \frac{1}{12}$. **Adjusted balance method:** Interest is calculated on the previous month's balance *less* credits and payments. With this method, $P =$ adjusted balance, $r =$ annual rate, and $t = \frac{1}{12}$. **Average daily balance method:** Add the outstanding balances for *each day* in the billing period, and then divide by the number of days in the billing period to find what is called the *average daily balance*. With this method, $P =$ average daily balance, $r =$ annual rate, and $t =$ number of days in the billing period divided by 365.

5. Answers vary.   **6.** Answers vary; find the dealer's cost and then make an offer that is no more than 1.05% of the dealers cost.   **7.** B (Don't forget the interest must also be paid.)   **8.** B (Don't forget the interest must also be paid.)   **9.** B   **10.** A   **11.** B   **12.** C   **13.** B   **14.** A; rate is not given but you can still choose a reasonable answer.   **15.** B   **16.** A   **17.** B   **18.** C   **19.** C   **20.** B   **21.** A   **22.** A; (it should be the most expensive)   **23.** B; (it should be the least expensive)   **24.** C; (it should be the middle amount)   **25.** 15%   **26.** 16%   **27.** 18%   **28.** 10%   **29.** 8%   **30.** 11%   **31. a.** $4.50 **b.** $3.75 **c.** $3.95   **32. a.** $4.50 **b.** $0.75 **c.** $1.97   **33. a.** $37.50 **b.** $36.88 **c.** $36.58   **34. a.** $37.50 **b.** $6.25 **c.** $16.44

Level 2, page 448

35. $13,378(1 + 0.06) \approx \$14,181$ **36.** $12,412.70(1 + 0.05) \approx \$13,033$

37. $17,250(1 + 0.10) = \$18,975$ **38.** $28,916(1 + 0.10) \approx \$31,808$

39. $\text{APR} = \dfrac{2(0.12)36}{37} \approx 23.4\%$ **40.** $\text{APR} = \dfrac{2(0.12)24}{25} \approx 25.0\%$

41. $\text{APR} = \dfrac{2(0.11)24}{25} \approx 21.1\%$ **42.** $\text{APR} = \dfrac{2(0.14)36}{37} \approx 27.2\%$

43. **a.** $m = \dfrac{\$20,650}{60} = \$344.17$

  b. $P = \$20,650 - \$2,000 = \$18,650$;

$$I = Prt = \$18,650(0.025)(5) = \$2,331.25;$$

$$m = \frac{\$18,650 + \$2,331.25}{60} = \$349.69$$

   c. $\text{APR} = \dfrac{2(0.025)(60)}{61} \approx 4.9\%$ **d.** The 0% financing is better.

44. **a.** $m = \dfrac{\$62{,}650}{60} = \$1{,}044.17$

    b. $P = \$62{,}650 - \$6{,}000$ $I = Prt = \$56{,}650(0.025)(5) = \$7{,}081.25;$

        $= \$56{,}650$ $m = \dfrac{\$56{,}650 + \$7{,}081.25}{60} = \$1{,}062.19$

    c. $\text{APR} = \dfrac{2(0.025)(60)}{61} \approx 4.9\%$ **d.** The 0% financing is better.

45. **a.** $m = \dfrac{\$42{,}700}{60} = \$711.67$

    b. $P = \$42{,}700 - \$5{,}100 = \$37{,}600;$

       $I = Prt = \$37{,}600(0.025)(5) = \$4{,}687.50;$

       $m = \dfrac{\$37{,}600 + \$4{,}687.50}{60} = \$704.79$

    c. $\text{APR} = \dfrac{2(0.025)(60)}{61} \approx 4.9\%$ **d.** The 2.5% financing is better.

46. **a.** $m = \dfrac{\$36{,}500}{60} = \$608.33$

    b. $P = \$36{,}500 - \$4{,}200 = \$32{,}300;$

       $I = Prt = \$32{,}300(0.025)(5) = \$4{,}037.50;$

       $m = \dfrac{\$32{,}300 + \$4{,}037.50}{60} = \$605.63$

    c. $\text{APR} = \dfrac{2(0.025)(60)}{61} \approx 4.9\%$ **d.** The 2.5% financing is better.

Level 3 page 448

47. **a.** $A = \$317.50(36)$ **48.** **a.** $A = \$141.62(36)$

         $= \$11{,}430$ $= \$5{,}098.32$

    b. $I = A - P$ **b.** $I = A - P$

         $= \$2{,}430$ $= \$1{,}098.32$

    c. $I = Prt$ **c.** $I = Prt$

     $2{,}430 = 9{,}000r(3)$ $1{,}098.32 = 4{,}000r(3)$

         $r = 0.09$ $r \approx 0.0915266667$

    d. $\text{APR} = \dfrac{2r(36)}{37}$ **d.** $\text{APR} = \dfrac{2r(36)}{37}$

       $\approx 17.5\%$ $\approx 17.8\%$

49. a. $A = \$488.40(48)$

$= \$23,443.20$

b. $I = A - P$

$= \$9,093.20$

c. $I = Prt$

$9,093.20 = 14,350r(4)$

$r \approx 0.1584181185$

d. APR $= \dfrac{2r(48)}{49}$

$\approx 31.0\%$

50. a. $A = \$339.97(48) + \$3,290$

$= \$19,608.56$

b. $I = A - P$

$= \$3,158.56$

c. $I = Prt$

$3,158.56 = 13,160r(4)$

$r \approx 0.0600030395$

d. APR $= \dfrac{2r(48)}{49}$

$\approx 11.8\%$

51. a. $A = \$168.51(48) + \$798$

$= \$8,886.48$

b. $I = A - P$

$= \$2,088.48$

c. $I = Prt$

$2,088.48 = 6,000r(4)$

$r \approx 0.08702$

d. APR $= \dfrac{2r(48)}{49}$

$\approx 17.0\%$

52. a. $A = \$662.06(60) + \$6,000$

$= \$45,723.41$

b. $I = A - P$

$= \$15,273.41$

c. $I = Prt$

$15,273.41 = 24,450r(5)$

$r \approx 0.1249358691$

d. APR $= \dfrac{2r(60)}{61}$

$\approx 24.6\%$

53. 8% add-on rate; APR $= \dfrac{2(0.08)48}{49} \approx 15.7\%$ **54.** 5% add-on rate; APR $= \dfrac{2(0.05)48}{49} \approx 9.8\%$

55. 11% add-on rate; APR $= \dfrac{2(0.11)48}{49} \approx 21.6\%$

Level 3 Problem Solving, page 449

56. Over \$28.50 the interest would be more than \$0.50, but because of rounding, any amount over \$28.50 would incur at least a \$0.50 service charge.

57. Previous balance method, \$45; Adjusted balance method, \$40.50; Average daily balance method, \$43.35

58. Sears' add-on rate is 29.2% APR; bank is better.

59. $P = \$1,598(1.05) = \$1,677.90$

$I = 1,677.90(0.15)(3) = 755.055$

$m = \dfrac{A}{36} = \dfrac{\$1,677.90 + \$755.055}{36} \approx \$67.58$

60. **a.** \$126.00 **b.** \$64.17 **c.** finance charge is \$4,488; \$2,541.67 **d.** finance charge is \$25,500; \$2,382.79

9.3 Sequences, page 450

I introduce this section by using the "Are You A Genius?" test (see Transparency 23). Most students enjoy doing that sort of thing, and it leads nicely to a discussion on number sequences.

Level 1, page 459

1. A *sequence* is a list of numbers having a first term, a second term, a third term, and so on. **2.** A general term is a formula which gives a specific term of a sequence. **3.** An *arithmetic sequence* is a sequence whose consecutive terms differ by the same real number, called the common difference.

4. A *geometric sequence* is a sequence whose consecutive terms have the same quotient, called the common ratio. **5.** A *Fibonacci-type sequence* is a sequence for which a first and second number are given, and then consecutive numbers are found by adding the two previous terms. **The Fibonacci** sequence is $1, 1, 2, 3, 5, 8, 13, 21, \cdots$. **6. a.** arithmetic **b.** $d = 2$ **c.** 10 **7. a.** geometric **b.** $r = 2$ **c.** 32 **8. a.** Fibonacci-type **b.** $s_1 = 2, s_2 = 4$ **c.** 16 **9. a.** arithmetic **b.** $d = 10$ **c.** 35 **10. a.** geometric **b.** $r = 3$ **c.** 135 **11. a.** Fibonacci-type **b.** $s_1 = 5, s_2 = 15$ **c.** 35 **12. a.** geometric **b.** $r = 5$ **c.** 125 **13. a.** geometric **b.** $r = \frac{1}{5}$ **c.** $\frac{1}{5}$ **14. a.** geometric **b.** $r = \frac{1}{3}$ **c.** $\frac{1}{3}$ **15. a.** geometric **b.** $r = 3$ **c.** 27 **16. a.** none of the classified types **b.** one more subtracted than on previous subtraction; subtract 1, subtract 2, $\cdots$ **c.** 6 **17. a.** none of the classified types. **b.** differences are $-2, 1, -2, 1, -2, \cdots$ **c.** 5 **18. a.** arithmetic **b.** $d = 3$ **c.** 17 **19. a.** geometric **b.** $r = 2$ **c.** 96 **20. a.** geometric **b.** $r = -3$ **c.** $-1,215$ **21. a.** both arithmetic and geometric **b.** $d = 0$ or $r = 1$ **c.** 10 **22. a.** Fibonacci-type **b.** $s_1 = 2, s_2 = 5$ **c.** 19 **23. a.** Fibonacci-type **b.** $s_1 = 3, s_2 = 6$ **c.** 24 **24. a.** none of the classified types **b.** perfect cubes of consecutive counting numbers **c.** $6^3 = 216$ **25. a.** geometric **b.** $r = \frac{3}{2}$ **c.** $\frac{81}{2}$ **26. a.** geometric **b.** $r = 3^3$ or 27 **c.** 3^{14} **27. a.** geometric **b.** $r = 4^{-1}$ or $\frac{1}{4}$ **c.** 4 **28. a.** none of the classified types **b.** $\frac{1}{2}$ followed by all not previously listed reduced common fractions **c.** $\frac{5}{6}$ **29. a.** arithmetic **b.** $d = \frac{1}{10}$ **c.** $\frac{6}{10} = \frac{3}{5}$ **30. a.** geometric **b.** $r = \frac{3}{2}$ **c.** $\frac{27}{4}$ **31. a.** arithmetic **b.** $d = \frac{1}{12}$ **c.** $\frac{11}{12}$

Level 2, page 459

32. a. 1, 5, 9   **b.** arithmetic; $d = 4$ **33. a.** 0, 3, 6   **b.** arithmetic; $d = 3$ **34. a.** 10, 20, 30

b. arithmetic; $d = 10$ **35. a.** 1, 0, -1 **b.** arithmetic; $d = -1$ **36. a.** 4, 1, -2

b. arithmetic; $d = -3$ **37. a.** 0, -10, -20 **b.** arithmetic; $d = -10$ **38. a.** 2, 1, $\frac{2}{3}$

b. neither   **39. a.** 0, $\frac{1}{2}$, $\frac{2}{3}$ **b.** neither   **40. a.** 0, $\frac{1}{3}$, $\frac{1}{2}$ **b.** neither   **41. a.** 1, 3, 6   **b.** neither

42. a. 1, 9, 36   **b.** neither   **43. a.** -1, 1, -1 **b.** geometric; $r = -1$ **44. a.** -5, -5, -5

b. both; $d = 0$ or $r = 1$ **45. a.** $\frac{2}{3}$, $\frac{2}{3}$, $\frac{2}{3}$ **b.** both; $d = 0$ or $r = 1$ **46. a.** 1, -1, 1

b. geometric, $r = -1$ **47. a.** -2, 3, -4 **b.** neither   **48.** 57   **49.** -200 **50.** 21

51. 625   **52.** 2, 6, 18, 54, 162   **53.** 3, 1, $\frac{1}{3}$, $\frac{1}{9}$, $\frac{1}{27}$ **54.** 1, 1, 2, 3, 5   **55.** 1, 2, 3, 5, 8

Level 3 Problem Solving, page 459

56. Consider 1, 1, 2, 3, 5, 8, $\cdots$ which is the Fibonacci sequence. The sequence formed by adding one to each of these terms is 2, 2, 3, 4, 6, 8, $\cdots$ which is not a Fibonacci-type sequence. Thus, the answer is no.

57. It is Fibonacci. Reference: the pattern follows Binet's formula:

$$f_n = \frac{1}{\sqrt{5}}\left[\left(\frac{1+\sqrt{5}}{2}\right)^n - \left(\frac{1-\sqrt{5}}{2}\right)^n\right]$$

58. It is Fibonacci. Consider, for example a five-floor building. The fifth floor can be either yellow or blue. If it is yellow, the rest of the building can be painted in any possible way for a four-floor building. If it is blue, then the fourth floor must be yellow, and the rest of the building can be painted in any possible way for a three-floor building. So $a_5 = a_4 + a_3$, as it is for Fibonacci numbers.

59.

1st number:	x
2nd number:	y
3rd number:	$x + y$
4th number:	$x + 2y$
5th number:	$2x + 3y$
6th number:	$3x + 5y$
7th number:	$5x + 8y$
8th number:	$8x + 13y$
9th number:	$13x + 21y$
10th number:	$21x + 34y$
SUM:	$55x + 88y = 11(5x + 8y)$

See commentary for Section 9.4 for more information on this problem and solution.

60. a. $8 - d, 8, 8 + d, 8 + 2d, 8 + 3d = 27, \cdots$

Thus, $3d = 19$

$$d = \frac{19}{3}$$

$1\frac{2}{3}, 8, 14\frac{1}{3}, 20\frac{2}{3}, 27, 33\frac{1}{3}, \cdots$

b. $\frac{8}{r}, 8, 8r, 8r^2, 8r^3 = 27, \cdots$

Thus, $r^3 = \frac{27}{8}$

$$r = \frac{3}{2}$$

$\frac{16}{3}, 8, 12, 18, 27, \frac{81}{2}, \cdots$

c. Answers vary; $s_n = s_{n-1} + s_{n-2}$, where $n \geq 3$; $s_1 = \frac{3}{2}$; $s_2 = 8$; the sequence is
$1.5, 8, 9.5, 17.5, 27, 44.5$

9.4 Series, page 460

To introduce this section, I use a zinger based on Problem 59 of Section 9.3, but will be new to most students. It is a pattern based on multiplication by 11:

$$11 \times 10 = 110$$
$$11 \times 11 = 121$$
$$11 \times 12 = 132$$
$$11 \times 13 = 1\ \ 3$$

original two digits

$$= 143$$

sum of two digits
$$\vdots$$
$$11 \times 52 = 572$$

sum of the two digits

$11 \times 74 = 7 \boxed{11} 4$ or 814 doing the usual carry.

I ask the students to pick any two numbers (a_1 and a_2) on Transparency 24. I then ask students to add those numbers (I do it on Transparency 24, but I have covered up the solution shown in the last column. The class and I do it together down to the 7th term (which is the term I need). I then ask them to complete the list and after they do that to sum the ten terms. In the meantime, I write the answer on the board. I can find the sum in my head by simply multiplying the 7th term and 11 — which is easy to do mentally. After I do this with the class, I ask them to explain why. This leads us to an introduction of series.

Level 1, page 468

1. A *sequence* is a list of numbers having a first term, a second term, and so on. A *series* is the indicated sum of the terms of a sequence.   2. Answers vary; the example given in the text is:

This is the last natural number in the domain. It is called the *upper limit.*

$$\sum_{k=1}^{10} 2k\} \leftarrow \text{This is the function being evaluated. It is called the } \textit{general term.}$$

This is the first natural number in the domain. It is called the *lower limit.*

Thus, $\displaystyle\sum_{k=1}^{10} 2k = 2(1)+2(2)+2(3)+2(4)+2(5)+2(6)+2(7)+2(8)+2(9)+2(10) = 110$

3. A *partial sum* of an infinite series is $G_1 = g_1$, $G_2 = g_1 + g_2$; $G_3 = g_1 + g_2 + g_3$; $\cdots$

4. A *geometric series* has n terms and an *infinite geometric series* has infinitely many terms.

5. Arithmetic sequence with common difference $d = -3$, $s_1 = 12$, $s_5 = 0$
$$S_5 = A_5 = 5\left(\frac{12 + 0}{2}\right) = 30$$

6. $s_1 = 5$, $s_8 = 40$; this is an arithmetic sequence with $d = 5$, so
$$S_8 = A_8 = 8\left(\frac{5 + 40}{2}\right) = 180$$

7. $s_1 = 10$; this is a geometric sequence with $r = 2$, so
$$S_4 = G_4 = \frac{10(1 - 2^4)}{1 - 2} = 150$$

8. $s_1 = -1$; this is a geometric sequence with $r = -1$, so
$$S_6 = G_6 = \frac{-1[1 - (-1)^6]}{1 - (-1)} = 0$$

9. $s_1 = -1$; this is a geometric sequence with $r = -1$, so
$$S_7 = G_7 = \frac{-1[1 - (-1)^7]}{1 - (-1)} = -1$$

10. $s_1 = 40$; this is a geometric sequence with $r = 5$, so
$$S_3 = G_3 = \frac{40(1 - 5^3)}{1 - 5} = 1,240$$

11. $\displaystyle\sum_{k=3}^{5} k = 3 + 4 + 5$
$$= 12$$

12. $\displaystyle\sum_{k=1}^{4} k^2 = 1^2 + 2^2 + 3^2 + 4^2$
$$= 30$$

13. $\displaystyle\sum_{k=2}^{6} k^2 = 2^2 + 3^2 + 4^2 + 5^2 + 6^2$

$\qquad\qquad = 90$

14. $\displaystyle\sum_{k=2}^{5} (100 - 5k) = (100 - 10) + (100 - 15) + (100 - 20) + (100 - 25)$

$\qquad\qquad\qquad = 400 - 70$

$\qquad\qquad\qquad = 330$

15. $\displaystyle\sum_{k=1}^{10} [1^k + (-1)^k] = 0 + 2 + 0 + 2 + 0 + 2 + 0 + 2 + 0 + 2$

$\qquad\qquad\qquad = 10$

16. $\displaystyle\sum_{k=1}^{5} (-2)^{k-1} = (-2)^0 + (-2)^1 + (-2)^2 + (-2)^3 + (-2)^4$

$\qquad\qquad\qquad = 1 - 2 + 4 - 8 + 16$

$\qquad\qquad\qquad = 11$

17. $\displaystyle\sum_{k=0}^{4} 3(-2)^k = 3(-2)^0 + 3(-2)^1 + 3(-2)^2 + 3(-2)^3 + 3(-2)^4$

$\qquad\qquad\qquad = 3[1 - 2 + 4 - 8 + 16]$

$\qquad\qquad\qquad = 33$

18. $\displaystyle\sum_{k=1}^{3} (-1)^k(k^2 + 1) = (-1)^1(1^2 + 1) + (-1)^2(2^2 + 1) + (-1)^3(3^2 + 1)$

$\qquad\qquad\qquad = -2 + 5 - 10$

$\qquad\qquad\qquad = -7$

Level 2, page 468

19. 2   **20.** no sum; $r > 1$ **21.** $\frac{3}{2}$ **22.** 200   **23.** $-\frac{40}{3}$ **24.** $-\frac{200}{3}$ **25.** 25   **26.** 30

27. 15   **28.** 100   **29.** 110   **30.** 55   **31.** 10,000   **32.** 10,100   **33.** 5,050   **34.** n^2

35. $n(n + 1)$ **36.** $\frac{n}{2}(n + 1)$ **37.** 11,500   **38.** 5,375   **39.** 2,030   **40.** 4,048   **41.** 120

42. 36; 84   **43.** 56; 64   **44.** 465   **45.** 9,330   **46.** 406 blocks   **47.** 3,828 blocks   **48.** 5,050

blocks   **49.** 1,024,000,000 present after 10 days

50. $P = 1,000,000 \cdot 2^d$ **51.** $G_6 = \dfrac{16(1 - 0.5^5)}{1 - 0.5} = 31$ games

52. $G_3 = \dfrac{9\left(1 - \left(\frac{1}{3}\right)^3\right)}{1 - \frac{1}{3}} = 13$ games

53. $729/3 = 243$ which is the number playing the first round.
$$G_6 = \dfrac{243\left(1 - \left(\frac{1}{3}\right)^6\right)}{1 - \frac{1}{3}} = 364 \text{ games}$$

54. $G = \dfrac{20}{1 - 0.90} = 200$ cm

55. $G_6 = \dfrac{25}{1 - 0.75} = 100$ cm

56. $G = \dfrac{375}{1 - 0.75} = 1{,}500$ revolutions

57. $G = \dfrac{500}{1 - \frac{2}{3}} = 1{,}500$ revolutions

58.
$$10 + 2 \cdot 9 + 2 \cdot 8.1 + \cdots = 10 + 2 \cdot 10(0.9) + 2 \cdot 10(0.9)^2 + \cdots$$
$$= 10 + 2 \cdot 10(0.9)[1 + 0.9 + 0.9^2 + \cdots]$$
$$= 10 + 18\left(\frac{1}{1 - 0.9}\right)$$
$$= 10 + 18(10)$$
$$= 190 \qquad \text{The distance is 190 ft.}$$

59.
$$10 + 2 \cdot 10 \cdot \left(\tfrac{2}{3}\right) + 20\left(\tfrac{2}{3}\right)^2 + \cdots = 10 + 20\left(\tfrac{2}{3}\right) + 20\left(\tfrac{2}{3}\right)^2 + \cdots$$
$$= 10 + \tfrac{40}{3}\left[1 + \tfrac{2}{3} + \left(\tfrac{2}{3}\right)^2 + \cdots\right]$$
$$= 10 + \tfrac{40}{3}\left(\frac{1}{1 - \frac{2}{3}}\right)$$
$$= 10 + \tfrac{40}{3}(3)$$
$$= 50 \qquad \text{The distance is 50 ft.}$$

Level 3 Problem Solving, page 470

60. **a.** $1^2 + 2^2 + 3^2 + 4^2 + 5^2 = \dfrac{5(5 + 1)(2 \cdot 5 + 1)}{6} = 55$

b. $1^2 + 2^2 + \cdots + n^2 = \dfrac{n(n + 1)(2n + 1)}{6}$; for $n = 50$,

$$\dfrac{50(50 + 1)(2 \cdot 50 + 1)}{6} = 42{,}925 \text{ blocks}$$

9.5 Annuities, page 470

Level 1, page 476

1. Answers vary; a *lump sum problem* is a financial problem which deals with a single amount of money.   **2.** Answers vary; an *annuity* is the future value of periodic payments to an interest bearing account. Since an annuity is an accumulation of payments and interest, it is a periodic payment problem.   **3.** An *annuity* is a financial problem which seeks the future value of periodic payments to an interest bearing account.   **4.** A *sinking fund* is a financial problem which seeks the amount of monthly payment necessary to accumulate (including added interest) to a known future value.

5. Answers vary; an annuity seeks the future value, given a monthly payment, whereas a sinking fund seeks the monthly payment, given the future value.   **6.** Answers vary; it shows the benefit of *early* savings for retirement.   **7.** $1,937.67   **8.** $1,966.81   **9.** $2,026.78   **10.** $2,153.84

11. $41,612.93   **12.** $50,225.75   **13.** $74,517.97   **14.** $174,748.21   **15.** $15,528.23

16. $16,387.93   **17.** $18,294.60   **18.** $23,003.87   **19.** $170,413.86   **20.** $213,706.54

21. $540,968.11   **22.** $652,934.78   **23.** $56,641.61   **24.** $39,529.09   **25.** $59,497.67

26. $13,211.40   **27.** $99,386.46   **28.** $6,629.90   **29.** $38,754.39   **30.** $112,885.15

31. $2,204.31   **32.** $60,030.54   **33.** $252,057.07   **34.** $136,340.66   **35.** $1,193.20

36. $4,014.26   **37.** $1,896.70   **38.** $1,219.65   ▲**39.** $7,493.53   **40.** $55,281.31

41. $8,277.87   **42.** $2,153.09   **43.** $2,261.06   **44.** $441.06   **45.** $489.00   **46.** $98.36

Level 2, page 476

47. annuity; $m = \$20,000$, $n = 1$, $t = 20$, $r = 0.08$; $A = \$915,239.29$

48. annuity; $m = \$650$, $n = 4$, $r = 0.08$, $t = 5$; $A = \$15,793.29$

49. annuity; $m = \$1,000$, $n = 4$, $r = 0.08$, $t = 5.5$; $A = \$27,298.98$

50. sinking fund; $A = \$50,000$, $t = 4$, $r = 0.08$, $n = 2$; $m = \$5,426.39$

51. sinking fund; $A = \$70,000$, $t = 5$, $n = 4$, $r = 0.08$; $m = \$2,880.97$

52. annuity; $m = \$50,000$, $t = 20$, $n = 1$, $r = 0.05$; $A = \$1,653,297.71$

Level 3 Problem Solving, page 476

53. annuity; $m = \$20,000$, $t = 29$, $n = 1$, $r = 0.05$; $A = \$1,246,454.24$

     TOTAL = ANNUITY + FINAL PAYMENT = $1,666,454.24

54. sinking fund (principal); $A = \$4,000,000$, $t = 30$, $n = 1$, $r = 0.08$; $A = \$35,309.73$;

     interest: $I = Prt$

                   $= \$4,000,000(0.055)(1)$

                    $= \$220,000$

     TOTAL = PRINCIPAL + INTEREST = $255,309.73

55. sinking fund (principal); $A = \$200,000$, $t = 10$, $n = 2$, $r = 0.08$; $A = \$6,716.35$

     interest: $I = Prt$

                   $= \$200,000(0.06)(\tfrac{1}{2})$

                    $= \$6,000$

     TOTAL = PRINCIPAL + INTEREST = $12,716.35

56. annuity; $m = \$150$, $t = 5$, $n = 4$, $r = 0.0672$; $A = \$3,530.70$

57. annuity; $m = \$1,200$, $t = 40$, $n = 1$, $r = 0.13$; $A = \$1,216,445.09$

58. annuity; $m = \$5,263.80$, $t = 40$, $n = 1$, $r = 0.04$; $A = \$500,195.00$

59. Answers vary.　　60. Answers vary.

9.6 Amortization, page 477

After introducing the notion of present value of an annuity, I work Problems 4 and 5 and then ask the students to try to explain why a lower interest rate in Problem 4 gives a larger present value when compared to Problem 5. This idea is the key to understanding the relationship between interest rates and fluctuations of bond prices. I do this to reinforce what we are finding with the present value of an annuity. It is the amount that we can borrow with a fixed monthly payment. This means, if I am paying $500 per month and the interest rates *go down*, then I should be able to *borrow more*. Once the students understand this concept, they will understand what present value of an annuity means.

Level 1, page 483

1. Answers vary; *amortization* is the process of paying off a debt by systematically making partial payments until the debt and accrued interest are repaid. 　　2. Answers vary; the financial formula called *present value of an annuity* is used when the monthly payment is known and the amount of loan is desired. It is a periodic payment formula for which the monthly payment is known and the present value of future payments is known. 　　3. m is the amount of a periodic payment (usually a monthly payment); n is the number of payments made each year; t is the number of years; r is the annual interest rate; A is the future value; and P is the present value

4. $m = 50$, $r = 0.05$, $t = 5$; $P = \$2,649.54$　　5. $m = 50$, $r = 0.06$, $t = 5$; $P = \$2,586.28$

6. $m = 50$, $r = 0.08$, $t = 5$; $P = \$2,465.92$　　7. $m = 150$, $r = 0.05$, $t = 30$; $P = \$27,942.24$

8. $m = 150$, $r = 0.06$, $t = 30$; $P = \$25,018.74$　　9. $m = 150$, $r = 0.08$, $t = 30$; $P = \$20,442.52$

10. $m = 1,050$, $r = 0.05$, $t = 30$; $P = \$195,595.70$　　11. $m = 1,050$, $r = 0.06$, $t = 30$; $P = \$175,131.20$

12. $P = 14,000$, $r = 0.05$, $t = 5$; $m = \$264.20$　　13. $P = 14,000$, $r = 0.10$, $t = 5$; $m = \$297.46$

14. $P = 14,000$, $r = 0.19$; $t = 5$; $m = \$363.17$　　15. $P = 150,000$, $r = 0.08$, $t = 30$; $m = \$1,100.65$

16. $P = 150,000$, $r = 0.09$, $t = 30$; $m = \$1,206.93$　　17. $P = 150,000$, $r = 0.10$, $t = 30$; $m = \$1,316.36$

18. $P = 260,000$, $r = 0.12$, $t = 30$, $m = \$2,674.39$　　19. $P = 260,000$, $r = 0.09$, $t = 30$; $m = \$2,092.02$

20. $5,628.89$　　21. $6,882.42$　　22. $5,655.87$　　23. $10,827.33$　　24. $16,649.54$

25. $5,429.91$　　26. $7,948.90$　　27. $12,885.18$　　28. $1,479.55$　　29. $7,407.76$　　30. $5,761.54$

31. $9,299.39$

Level 2, page 484

32. amortization; $P = 500$, $n = 12$, $t = 1$, $r = 0.12$; $m = \$44.42$

33. amortization; $P = 100$, $n = 12$, $t = 1.5$, $r = 0.18$; $m = \$6.38$

34. amortization; $P = 4{,}560$, $n = 12$, $t = \frac{20}{12}$, $r = 0.12$; $m = \$272.19$

35. amortization; $P = 3{,}520$, $n = 12$, $t = \frac{30}{12}$, $r = 0.19$; $m = \$148.31$

36. amortization; $P = 2{,}300$, $n = 12$, $t = 2$, $r = 0.15$; $m = \$111.52$

37. amortization; $P = 12{,}450$, $n = 12$, $t = \frac{30}{12}$, $r = 0.029$, $m = \$430.73$

38. amortization; $P = 3{,}456$, $n = 12$, $t = 3$, $r = 0.23$; $m = \$133.78$

39. amortization; $P = 985$, $n = 12$, $t = \frac{15}{12}$, $r = 0.17$; $m = \$73.35$

40. amortization; $P = 112{,}000(0.80) = 89{,}600$, $n = 12$, $t = 30$, $r = 0.115$; $m = \$887.30$

41. amortization; $P = 108{,}000(0.70) = 75{,}600$, $n = 12$, $t = 30$, $r = 0.1205$; $m = \$780.54$

42. amortization; $P = 450{,}000$, $n = 12$, $t = 30$, $r = 0.125$; $m = \$4{,}802.66$

43. amortization; $P = 859{,}000$, $n = 12$, $t = 20$, $r = 0.132$; $m = \$10{,}186.47$

44. $P = \$112{,}000(0.80) = \$89{,}600$

Total interest (monthly payment from Problem 40) is

$$I = A - P$$
$$= \$887.30(12)(30) - \$89{,}600$$
$$= \$229{,}828$$

Monthly payment for $t = 15$, $n = 12$, $r = 0.115$ is $m = \$1{,}046.70$

Total interest is

$$I = A - P$$
$$= \$1{,}046.70(12)(15) - \$89{,}600$$
$$= \$98{,}806$$

Savings: $131,022 (to the nearest dollar).

45. $P = \$108{,}000(0.70) = \$75{,}600$

Total interest (monthly payment from Problem 41) is

$$I = A - P$$
$$= \$780.54(12)(30) - \$75{,}600$$
$$= \$205{,}394.40$$

Monthly payment for $t = 15$, $n = 12$, $r = 0.1205$ is $m = \$909.76$

Total interest is

$$I = A - P$$
$$= \$909.76(12)(15) - \$75,600$$
$$= \$88,156.80$$

Savings: $117,238 (to the nearest dollar).

46. Present value of an annuity; $m = 500$, $n = 1$, $t = 5$, $r = 0.125$; $P = \$1,780.28$

47. Present value of an annuity; $m = 50,000$, $n = 1$, $t = 20$, $r = 0.1225$; $P = \$367,695.71$

48. Present value of an annuity; $m = 250$, $n = 12$, $t = 20$, $r = 0.10$; $P = \$25,906.15$
Since this is greater than the lump-sum payment, the annuity is the better choice.

49. Present value of an annuity; $m = 750$, $n = 12$, $t = 20$, $r = 0.09$; $P = \$83,358.72$
Since this is considerably greater than the lump-sum payment, the annuity is the better choice.

50. Present value of an annuity; $m = 250$; $n = 12$; $t = 4$; $r = 0.13$; $P = \$9,318.80$. This is the maximum loan you can afford. Thus, if you add to this the amount of the down payment, plus the value of the trade-in, you will know the amount you can pay for the car.

Level 3 Problem Solving, page 484

51. Present value of an annuity ($m = \$1,000,000$, $r = 0.05$, $n = 1$, and $t = 20$) is $12,462,210.34. This would be a fair price to receive for the $20,000,000 lottery prize.

52. Simple interest;
$$I = Prt$$
$$= 6000(0.20)\left(\frac{90}{360}\right)$$
$$= 300$$

$$A = P + I$$
$$= 6,000 + 300$$
$$= 6,300 \qquad\qquad \text{The future value is } \$6,300.$$

53. $5,500 - \$625 = \$4,875$
$m = 0.36(\$4,875) = \$1,755$
present value of an annuity; $m = 1,755$, $n = 12$, $t = 30$, $r = 0.0965$; $P = \$206,029.43$

54. $4,550 + \$3,980 - \$1,235 = \$7,295$
$m = 0.36(\$7,295) = \$2,626.20$
present value of an annuity; $n = 12$, $t = 30$, $r = 0.1185$; $P = \$258,209.54$
The down payment is $355,000 - \$258,209.54 = \$96,790$; to the nearest hundred dollars, the down payment is $96,800. The percent is
$$\frac{\$96,790.46}{\$355,000} = 27.3\%$$

55. amortization; $P = \$225,000(0.75) = \$168,750$, $t = 30$, $n = 12$, $r = 0.1024$; $m = \$1,510.92$

Total monthly payments are $\$1,510.92(12)(30) = \$543,931.20$.

For $t = 15$, the amortization gives $m = \$1,838.25$ for total payments of

$\$1,838.25(12)(15) = \$330,885.00$

Savings: $\$543,931.20 - \$330,885.00 = \$213,046.20$.

56. Price range of $\$187,509.52$ to $\$219,011.42$.

57. Price range of $\$175,322.31$ to $\$204,386.77$.

58. Price range of $\$198,088.59$ to $\$231,706.31$.

59. **a.** increases **b.** decreases **c.** increases

60. **a.** decreases **b.** increases **c.** decreases

9.7 Summary of Financial Formulas, page 485

In some respects, this is the most important section in the book. I tell my classes that if I can put some extra dollars in their pockets at some time in their life as a direct result of studying this chapter, then this class has been successful *regardless of their grade in the class*. I relate to them all kinds of stories about how a knowledge of these formulas has saved me (literally) tens of thousands of dollars in my life.

At the time of this writing (September 1, 2002), current passbook savings rates are 3%, Treasury bill rates (10-yr) are 4.16%, the prime rate is 4.75%, credit card interest is 9%-22% (with teaser rates as low as 5.9%), and stock market rates of over 20%. Car loans are 8.4%-21% (with teaser rates as low as 0%), and personal loans 18%-26%. The interest rates used in the book are consistent with these rates. See **http://www.bankrate.com/brm/ratehm.asp** for a listing of current rates. By the way, on March 1, 2003, the national debt was approximately **$6,425,000,000**.

The important thing to emphasize with this section is the proper classification of financial formulas. I expect my classes to know the formulas for lump sums (present value and future value). I also expect them to know what each variable stands for. Finally, I provide them with the formulas, but expect them to match the appropriate formula with the financial type.

I have found the following spreadsheet program for comparison rates very useful since in practice it is necessary to compare not one or two financial institutions, but many. If you check your local newspaper or banks and savings and loans you will see that rates for home loans are divided into three parts: APR, points, and fees. Points refer to a percentage point fee that you must pay when the loan is obtained. A formula I use for comparing the rates is:

$$\text{COMPARISON RATE} = \text{APR} + 0.125\left(\text{POINTS} + \frac{\text{FEES}}{\text{AMOUNT OF THE LOAN}}\right)$$

You will notice that the fees and points charged are *one-time* fees, whereas the APR represents an annual amount. If you keep the loan for 30 years, we would divide the one-time fees by 30 to obtain the *annual* rate. On the other hand, if you keep the loan for 2 years, we would divide the one-time fees by 2 to obtain the annual rate. Since you do not know know long you will keep the loan, we use the average length of home ownership to divide the one-time fees by 8. (Note, dividing by 8 is the same as multiplying by 0.125.)

Here is a spreadsheet program for finding comparison rates.

	A	B	C	D	E	F
1	Comparison Rates for Home Loans					
2	What is the amount of the loan?		[enter amount]			
3						
4	Bank	APR	POINTS	FEES	COMPARISON RATE	
5					+B5+0.125*(+C5/100+D5/D2)	
6					replicate	
7						
8						
9						
10						

	A	B	C	D	E	F
1	Comparison Rates for Home Loans					
2	What is the amount of the loan?		$100,000.00			
3						
4	Bank	APR	POINTS	FEES	COMPARISON RATE	
5	Bank of America	6.23%	2	$400	6.53%	
6	Central Bank	6.80%	0	$200	6.83%	
7	River City	6.30%	1.5	$0	6.49%	
8	City Bank	5.50%	4	$300	6.04%	
9	First Interstate	6.28%	1	$150	6.42%	
10						

Level 1, page 486

1. $P = m\left[\dfrac{1 - (1 + \frac{r}{n})^{-nt}}{\frac{r}{n}}\right]$; the unknown is the present value   2. $m = \dfrac{P(\frac{r}{n})}{1 - (1 + \frac{r}{n})^{-nt}}$; the

unknown is the periodic payment   **3.** Answers vary; typical is 30%, although many first buyer's programs offer 5% down.   **4.** Answers vary; it is not unusual to obtain a car loan that is 100% of value (that is, 0% down). Car payments vary greatly depending on the value of the car, but typically are $250-$450 per month.   **5.** Answers vary; a good procedure is to ask a series of questions. **Is it a lump sum problem?** If it is, then what is the unknown? If FUTURE VALUE is the unknown, then it is a *future value* problem. If PRESENT VALUE is the unknown, then it is a *present value* problem. **Is it a periodic payment problem?** If it is, then is the periodic payment known? If the PERIODIC PAYMENT IS

KNOWN and you want to find the future value then it is an *ordinary annuity* problem.   If the PERIODIC PAYMENT IS KNOWN and you want to find the present value then it is a *present value of an annuity* problem.   If the PERIODIC PAYMENT IS UNKNOWN and you know the future value then it is a *sinking fund* problem.   If the PERIODIC PAYMENT IS UNKNOWN and you know the present value then it is an *amortization* problem.   **6.** future value   **7.** present value   **8.** present value of an annuity   **9.** annuity   **10.** sinking fund   **11.** amortization

12. $A = m\left[\dfrac{\left(1 + \frac{r}{n}\right)^{nt} - 1}{\frac{n}{r}}\right]$; Annuity

$P = m\left[\dfrac{1 - \left(1 + \frac{r}{n}\right)^{-nt}}{\frac{n}{r}}\right]$; Present value of an annuity

$m = \dfrac{A\left(\frac{r}{n}\right)}{\left(1 + \frac{r}{n}\right)^{nt} - 1}$; Sinking fund

$m = \dfrac{P\left(\frac{r}{n}\right)}{1 - \left(1 + \frac{r}{n}\right)^{-nt}}$; Amortization

13. future value; $P = \$1,000$, $n = 1$, $t = 3$, $r = 0.12$; $A = \$1,404.93$

14. annuity; $m = \$300$, $n = 1$, $t = 10$, $r = 0.12$; $A = \$5,264.62$

15. present value; $A = \$10,000$; $n = 1$, $t = 5$, $r = 0.12$; $P = \$5,674.27$

16. sinking fund; $A = \$10,000$; $n = 1$, $t = 5$, $r = 0.12$; $m = \$1,574.10$

Level 2, page 487

17. a. future value   **b.** $P = \$1,000$, $t = 2.5$, $r = 0.12$, $n = 12$; $A = \$1,347.85$

18. a. annuity   **b.** $m = \$300$, $r = 0.12$, $n = 1$, $t = 10$; $A = \$5,264.62$

19. a. present value   **b.** $A = \$10,000$, $t = 5$, $r = 0.12$, $n = 4$; $P = \$5,536.76$

20. a. present value of an annuity   **b.** $t = 5$, $n = 1$, $m = \$300$, $r = 0.10$, $P = \$1,137.24$

21. a. sinking fund   **b.** $A = \$10,000$, $t = 5$, $r = 0.09$, $n = 1$; $m = \$1,670.92$

22. a. amortization   **b.** $P = \$5,000,000$, $t = 10$, $n = 1$, $r = 0.14$; $m = \$958,567.70$

23. a. future value   **b.** $r = 0.0625$, $P = \$20,000$, $t = 5$, $n = 1$; $A = \$27,081.62$

24. a. amortization   **b.** $P = \$9,500$, $r = 0.07$, $t = 4$, $n = 12$; $m = \$227.49$

25. a. present value of an annuity   **b.** $t = 33$, $n = 1$, $m = \$500$, $r = 0.08$; $P = \$5,756.94$

26. a. present value of an annuity   **b.** $m = \$875$, $r = 0.065$, $n = 12$, $t = 30$; $P = \$138,434.47$

27. **a.** amortization **b.** $n = 12$, $P = \$125,000(0.80) = \$100,000$, $t = 30$, $r = 0.12$; $m = \$1,028.61$

28. **a.** present value **b.** $A = \$80,000$, $t = 4$, $r = 0.09$, $n = 2$; $P = \$56,254.81$

29. **a.** annuity **b.** $m = \$730$, $n = 1$, $r = 0.10$, $t = 15$; $A = \$23,193.91$

30. **a.** future value **b.** $P = \$12,500$, $t = 20$, $r = 0.11$, $n = 365$; $A = \$112,775.28$

31. **a.** present value **b.** $A = \$1,000,000$, $r = 0.09$, $n = 12$, $t = 55$; $P = \$7,215.46$

32. **a.** present value **b.** $A = \$150,000$, $t = \frac{270}{360}$, $r = 0.18$, $n = 4$; $P = \$131,444.49$

33. **a.** annuity **b.** $t = 5$, $m = \$900$, $n = 4$, $r = 0.08$; $A = \$21,867.63$

34. **a.** sinking fund **b.** $n = 12$, $r = 0.18$, $A = \$100,000$, $t = \frac{25}{3}$; $m = \$437.06$

35. **a.** present value **b.** $t = 5$, $A = \$300,000$, $r = 0.12$, $n = 12$; $P = \$165,134.88$

36. **a.** present value **b.** $t = \frac{7}{2}$, $A = \$45,000$, $r = 0.12$, $n = 2$; $P = \$29,927.57$

37. **a.** amortization **b.** $P = \$45,000$, $r = 0.12$, $n = 12$, $t = \frac{7}{2}$; $m = \$1,317.40$

38. **a.** annuity **b.** $m = \$4,000$, $n = 4$, $r = 0.10$, $t = \frac{15}{2}$; $A = \$175,610.81$

39. **a.** annuity **b.** $m = \$1,000$, $n = 12$, $r = 0.18$, $t = \frac{25}{3}$; $A = \$228,803.04$

40. **a.** future value **b.** $P = \$112,000$, $t = 5$, $r = 0.14$, $n = 12$; $A = \$224,628.30$

41. **a.** future value **b.** $P = \$800$, $t = 1$, $r = 0.10$, $n = 360$; $A = \$884.12$

42. **a.** future value **b.** $P = \$9,000$, $t = 4$, $r = 0.20$, $n = 12$; $A = \$19,898.24$

43. **a.** future value **b.** $P = \$5,000$, $n = 1$, $r = 0.055$, $t = 12$; $A = \$9,506.04$

44. **a.** future value **b.** $P = \$10,000$, $n = 1$, $r = 0.08$, $t = 18$, $A = \$39,960.19$

45. **a.** present value **b.** $A = \$5,000$, $t = 3$, $n = 1$, $r = 0.11$; $P = \$3,655.96$

46. **a.** present value **b.** $A = \$20,000$, $t = 5$, $r = 0.09$, $n = 2$; $P = \$12,878.55$

47. **a.** present value **b.** $t = 3$, $A = \$200,000$, $r = 0.10$, $n = 12$; $P = \$148,348$

48. **a.** present value **b.** $t = \frac{5}{2}$, $A = \$2,900$, $r = 0.11$, $n = 2$; $P = \$2,219$

49. **a.** present value **b.** $t = \frac{18}{12}$, $A = \$560,000$, $r = 0.075$, $n = 360$; $P = \$500,420$

50. **a.** present value of an annuity **b.** $m = \$1,000,000$, $t = 5$, $n = 1$, $r = 0.14$; $P = \$3,433,081$

Level 3 Problem Solving, page 488

51. We can compare present values or future values; we choose to compare present values.

    $A = \$45,000$, $t = 1$, $r = 0.10$, $n = 12$; $P = \$40,734.56$; add the present payment of \$10,000 to obtain \$50,734.56 which is better than the \$50,000 payment. Note: if you compare future values the same choice is better and the future values are \$56,047.13 vs \$55,235.65.

52. amortization; $t = 30$, $P = \$185,500$, $r = 0.0775$, $n = 12$; $m = \$1,328.94$

53. The total amount paid is $\$1,328.94(12)(30) = \$478,418.40$, so the amount of interest is

    $I = A - P = \$478,418.40 - \$185,500 = \$292,918.40$

54. Amortization schedule is required (not shown here).

55. A 20-yr loan increases the monthly payment to \$1,522.86. This loan has total payments of \$365,486.40 with total interest of \$179,986.40. The interest savings is \$112,932.

56. amortization; $t = 30$, $P = \$418{,}500$, $r = 0.08375$, $n = 12$; $m = \$3{,}180.90$

57. The total amount paid is $\$3{,}180.90(12)(30) = \$1{,}145{,}124$, so the amount of interest is

$$I = A - P = \$1{,}145{,}124 - \$418{,}500 = \$726{,}624$$

58. Amortization schedule is required (not shown here).

59. A 22-yr loan increases the monthly payment to \$3,474.80. This loan has total payments of \$917,347.20 with total interest of \$498,847.20. The interest savings is \$227,776.80.

60. This is a future value problem. Let $r = 0.16$; $t = 20$, $P = \$15$, $n = 1$; $A = \$291.91$;

For the actual inflation rate, solve

$$80 = 15(1 + r)^{20}$$
$$\frac{80}{15} = (1 + r)^{20}$$
$$r = \left(\frac{80}{15}\right)^{1/20} - 1 \approx 0.0873$$

The actual inflation rate is about 8.7%.

Chapter 9 Review Questions, page 491

1. A sequence is a list of numbers having a first term, a second term, and so on; a series is the indicated sum of the terms of a sequence. An arithmetic sequence is one that has a common difference $[a_n = a_1 + (n - 1)d]$, a geometric sequence is one that has a common ratio $[r_n = g_1 r^{n-1}]$, and a Fibonacci-type sequence is one that, given the first two terms, the next is found by adding the previous two terms $[s_n = s_{n-1} + s_{n-2}]$. The sum of an arithmetic sequence is $A_n = n\left(\dfrac{a_1 + a_n}{2}\right)$ or $A_n = \frac{n}{2}[2a_1 + (n - 1)d]$, and the sum of a geometric sequence is $G_n = \dfrac{g_1(1 - r^n)}{1 - r}$. **2.** Answers vary; a good procedure is to ask a series of questions. **Is it a lump-sum problem?** If it is, then what is the unknown? If FUTURE VALUE is the unknown, then it is a *future value* problem. If PRESENT VALUE is the unknown, then it is a *present value* problem. **Is it a periodic payment problem?** If it is, then is the periodic payment known? If the PERIODIC PAYMENT IS KNOWN and you want to find the future value, then it is an *ordinary annuity* problem. If the PERIODIC PAYMENT IS KNOWN and you want to find the present value, then it is a *present value of an annuity* problem. If the PERIODIC PAYMENT IS UNKNOWN and you know the future value, then it is a *sinking fund* problem. If the PERIODIC PAYMENT IS UNKNOWN and you know the present value, then it is an *amortization* problem.

3. a. arithmetic; $a_n = 5n$　**b.** geometric; $g_n = 5 \cdot 2^{n-1}$　**c.** Fibonacci-type; $s_1 = 5$, $s_2 = 10$,

$s_n = s_{n-1} + s_{n-2}, n \geq 3$ **d.** none of these (add $5, 10, 15, 20, \cdots$); $55, 80$ **e.** geometric;
$g_n = 5 \cdot 10^{n-1}$ **f.** none of these (alternate terms); $5, 50$

4. **a.** $\displaystyle\sum_{k=1}^{3} (k^2 - 2k + 1) = (1^2 - 2(1) + 1) + (2^2 - 2(2) + 1) + (3^2 - 2(3) + 1) = 5$

   b. $\displaystyle\sum_{k=1}^{4} \frac{k-1}{k+1} = \frac{0}{2} + \frac{1}{3} + \frac{2}{4} + \frac{3}{5} = \frac{43}{30}$ or $1.4\overline{3}$

5.   parents, grandparents, great-grandparents, $\cdots$; that is, $2, 4, 8, 16, \cdots$. Find G_{10} where $g_1 = 2$

   and $r = 2$; $G_{10} = \dfrac{2(1 - 2^{10})}{1 - 2} = 2{,}046$. Thus, there are a minimum of $2{,}046$ people.

6.   There will be 72 divisions in 24 hours; $g_{73} = 2^{10} \cdot 2^{72} = 2^{82}$

7. $18{,}579(1 + 0.05) = \$19{,}507.95$. You should offer $\$19{,}500$ for the car.

8.   **a.** $I = Prt = 13{,}500(0.029)(2) = 783$;

   $A = P + I = 13{,}500 + 783 = 14{,}283$; monthly payment is $\$14{,}283 \div 24 = \$595.13$.

   The total interest is $\$783$, and the monthly payment is $\$595.13$.

   b. $\text{APR} = \dfrac{2Nr}{N+1}$

   $\qquad = \dfrac{2(24)(0.029)}{25}$

   $\qquad = 0.05568$ The APR is 5.568%.

9. $A = 48(353.04) = 16{,}945.92$; $I = A - P = 16{,}945.92 - 11{,}450.00 = 5{,}495.92$; also,

$$I = Prt$$
$$5{,}495.92 = 11{,}450(r)(4) \qquad \textit{48 months is 4 years, so } t = 4.$$
$$1{,}373.98 = 11{,}450r \qquad \textit{Divide both sides by 4.}$$
$$0.12 \approx r \qquad \textit{Divide both sides by 11,450.}$$

Finally, $\text{APR} = \dfrac{2Nr}{N+1} = \dfrac{2(48)r}{49} \approx 0.235$. The APR is about 23.5%.

10. The adjusted daily balance method is most advantageous to the consumer.

PREVIOUS BALANCE METHOD	ADJUSTED BALANCE METHOD	AVERAGE DAILY BALANCE
$I = Prt$	$I = Prt$	$I = Prt$
$\quad = 525(0.09)(\frac{1}{12})$	$\quad = (525 - 100)(0.09)(\frac{1}{12})$	$\quad = \dfrac{525 \times 7 + 425 \times 24}{31}(0.09)(\frac{31}{365})$
$\quad = 3.9375$	$\quad = 3.1875$	$\quad = 3.421232877$
The finance charge is $\$3.94$.	The finance charge is $\$3.19$.	The finance charge is $\$3.42$.

11. $A = \$1,000,000$; $t = 50$; $r = 0.09$; $n = 12$; Thus,

$$P = A\left(1 + \frac{r}{n}\right)^{-nt} = 1,000,000\left(1 + \frac{0.09}{12}\right)^{-(50)12} \approx 11,297.10$$

Deposit \$11,297.10 to have a million dollars in 50 years.

12. **a.** Amortization; where $P = 154,000$, $n = 12$, $r = 0.08$, and $t = 20$. Thus,

$$m = \frac{P\left(\frac{n}{r}\right)}{1 - \left(1 + \frac{r}{n}\right)^{-nt}} = \frac{154,000\left(\frac{0.08}{12}\right)}{1 - \left(1 + \frac{0.08}{12}\right)^{-(12)(20)}} \approx 1,288.117706 \quad \text{(by calculator)}$$

The monthly payments are \$1,288.12.

b. 80%(PRICE OF HOME) = \$154,000

PRICE OF HOME = \$192,500

c. Future value (continuous compounding): $A = Pe^{rt} = \$192,500e^{0.04(20)} = \$428,416.63$

13. Present value; where $A = \$100,000$; $r = 0.064$; $t = 3\frac{4}{12} = \frac{40}{12}$; $n = 12$; thus,

$$P = A\left(1 + \frac{r}{n}\right)^{-nt} = \$100,000\left(1 + \frac{0.064}{12}\right)^{-(12)(40/12)} \approx \$80,834.49$$

14. present value; $A = \$420,000$, $t = 30$, $n = 1$, $r = 0.05$; $P = \$97,178.53$

15. annuity; $m = \$20,000$, $t = 29$, $n = 1$, $r = 0.05$; $A = \$1,246,454.24$

16. present value of an annuity; $m = \$20,000$, $t = 29$, $n = 1$, $r = 0.05$; $P = \$302,821.47$

17. present value; $A = \$420,000$, $t = 30$, $n = 1$, $r = 0.05$; $P = \$97,178.53$;

You need to set aside \$97,178.53.

18. $I = Prt$

$= \$200,000(0.05)(1)$

$= \$10,000$

19. You can compare present values of each or the future values of each. It is easier to compare present values.

Cash value: \$300,000

Annuity: \$302,821.47 (see Problem 16)
Preset value of cash: \$97,178.53
Total: \$400,000

Since \$400,000 is worth more than \$300,000, take the installments.

20. Compare present values.

Cash value: \$300,000

Present value of annuity;
$m = \$20,000$, $t = 29$, $n = 1$, $r = 0.10$
\$187,392.12
Present value of cash:
$A = \$420,000$, $t = 30$, $n = 1$, $r = 0.10$
\$24,069.59
Total: \$211,461.71

Since \$300,000 is worth more than \$211,461.71, take the one-time payment.

Group Research Problems, pages 492-494

G31. Answers vary.   **G32.** Answers vary.

G33. Consider the total amount of compensation associated with each offer.

Option A

1st year: 21,000

2nd year: $21,000 + (21,000 + 1,200) = 43,200$

3rd year: $21,000 + (21,000 + 1,200) + (21,000 + 1,200 + 1,200) = 66,600$

$$\vdots$$

nth year: $21,000 + (21,000 + 1,200) + (21,000 + 2 \cdot 1,200) + \cdots + [21,000 + (n - 1) \cdot 1,200]$

$$= 21,000n + 1,200[1 + 2 + 3 + \cdots + (n - 1)]$$
$$= 21,000n + 1,200(\frac{n-1}{2})[1 + (n - 1)]$$
$$= 21,000n + 600n^2 - 600n$$

So, for $n = 10$, Option A provides $21,000(10) + 600(10)^2 - 600(10) = 264,000$

Option B

1st year: $10,500 + (10,500 + 300) = 21,300$

2nd year: $10,500 + (10,500 + 300) + (10,500 + 2 \cdot 300) + (10,500 + 3 \cdot 300)$

$$= 43,800$$

3rd year: $10,500 + (10,500 + 300) + (10,500 + 2 \cdot 300) + (10,500 + 3 \cdot 300)$

$$+ (10,500 + 4 \cdot 300) + (10,500 + 5 \cdot 300) = 67,500$$

$$\vdots$$

nth year: $10,500 + (10,500 + 300) + (10,500 + 2 \cdot 300) + \cdots + [10,500 + (2n - 1) \cdot 300]$

$$= 10,500(2n) + 300[1 + 2 + 3 + \cdots + (2n - 1)]$$
$$= 21,000n + 300(\frac{2n-1}{2})[1 + (2n - 1)]$$
$$= 21,000n + 600n^2 - 300n$$

Option C

1st year: $5,250 + (5,250 + 75) + (5,250 + 2 \cdot 75) + (5,250 + 3 \cdot 75) = 21,450$

2nd year: $5,250 + (5,250 + 75) + (5,250 + 2 \cdot 75) + (5,250 + 3 \cdot 75)$

$$+ (5,250 + 4 \cdot 75) + (5,250 + 5 \cdot 75) + (5,250 + 6 \cdot 75) + (5,250 + 7 \cdot 75)$$
$$= 44,100$$

3rd year: $12(5,250) + 75(1 + 2 + \cdots + 11) = 67,950$

$$\vdots$$

nth year: $4n(5,250) + 75\left[\frac{(4n - 1)(4n)}{2}\right] = 21,000n + 600n^2 - 150n$

Option D

nth year: $12n(1,750) + 10\left[\dfrac{(12n-1)(12n)}{2}\right] = 21,000n + 720n^2 - 60n$

So, for $n = 10$,

Option A provides $21,000(10) + 600(10)^2 - 600(10) = 264,000$
Option B provides $21,000(10) + 600(10)^2 - 300(10) = 267,000$
Option C provides $21,000(10) + 600(10)^2 - 150(10) = 268,500$
Option D provides $21,000(10) + 720(10)^2 - 60(10) = 281,400$

Option D is best (not only for 10 years) but for any number of years.

G34. **a.** The letter is an attempt to get the homeowner to change from a monthly payment to a biweekly payment.

b. There are no advantages for accepting this offer. It *is* better to make biweekly payments, but it is not necessary to "sign up" to do this; the homeowner can just do it. The company making this offer might say that it will force the homeowner to make the biweekly payments, rather than leaving it voluntary.

c. 52-weeks a year is 26 biweekly payments; this is equivalent to 13 four week months.

d. The income is $93,750,000$

e. $I = Prt = \$1,000(250,000)(0.05)(2/52) = \$480,769.23$ every two weeks. This is an annual income of $12,500,000$.

Individual Research Problems, page 495

P9.1 Answers vary. **P9.2** $\$1,000,000 = \$39,000e^{0.06t}$, or $0.06t = \ln(\$1,000,000/\$39,000)$ which means that $t \approx 54$ years. **P9.3** Answers vary. **P9.4** Answers vary.

P9.5 Answers vary; continuous interest would use the formula $A = Pe^{rt}$. **P9.6** Answers vary.

P9.7 Answers vary.

CHAPTER 10

THE NATURE OF SETS AND COUNTING

10.1 Sets, Subsets, and Venn Diagrams, page 498

The ideas of this section are sometimes difficult for the students, particularly ideas involving the empty set. I have found that students can understand the material more easily if I use { } instead of ∅. The idea that the empty set is a subset of every set as well as the difference between ∅ and { ∅ } need some very careful development. A device that I have used to make it easy for the students to see the difference between an element of a set and a subset of a set is to bring a bunch of little boxes of cutout cardboard pieces representing elements; the boxes represent sets and the pieces of cardboard represent elements. We then discuss the ideas of this section using these boxes as models. This might seem like a juvenile demonstration, but the students need something concrete to look at, and I've found that they understand the material much better when I use this model.

Level 1, page 504

1. Answers vary. *Union:* $A \cup B$ is the set of all elements in either A or B (or possibly in both). *Intersection:* $A \cap B$ is the set of all elements in both A and B. *Complementation:* $\overline{A}$ is the set of all elements not in A.   2. Answers vary; the set of people over 20 ft tall.   3. Answers vary; the number of people living today.   4. Answers vary; Let $U =$ students accepted at the public university; $C =$ those accepted at the local private college. If James adds the number of elements in U

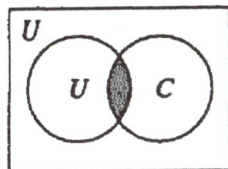

and the number of elements in C, he will have counted the number of elements in the intersection (shaded portion) twice. For example, suppose 100 are accepted at the public university, and 100 are accepted at the local private college. If 30 people are accepted at both schools, then we see that a total of 170 have been accepted at one of these schools, leaving 30 people not accepted at all.

5. $\{2, 6, 8, 10\}$ 6. $\{6, 8\}$ 7. $\{2, 3, 5, 6, 8, 9\}$ 8. $\emptyset$ 9. $\{3, 4, 5\}$ 10. $\{1, 3, 4, 5, 6, 7, 10\}$

11. $\{1, 2, 3, 4, 5, 6, 7\}$ 12. $\{3, 4, 6, 8, 10\}$ 13. $\emptyset$ 14. U 15. $\{1, 2, 3, 4, 5, 6\}$ 16. $\{1, 2\}$

17. $\{1, 2, 3, 4, 5, 7\}$ 18. $\{3\}$ 19. $\{5\}$ 20. $\{1, 2, 3, 5, 6, 7\}$ 21. $\{5, 6, 7\}$ 22. $\{3, 4, 7\}$

23. $\{1, 2, 4, 6\}$ 24. $\emptyset$ 25. A 26. $\emptyset$ 27. $A \cap B$ 28. $A \cup B$

29.   30.   31.

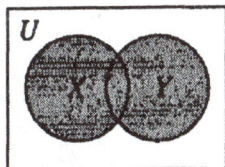   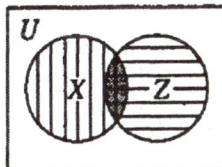   

32.

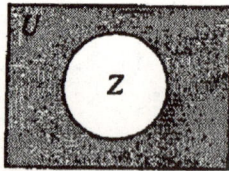

Level 2, page 504

33.

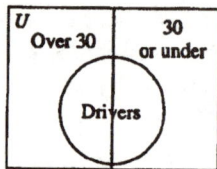

34.

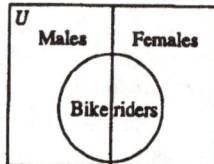

35.

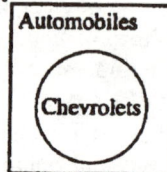

36.

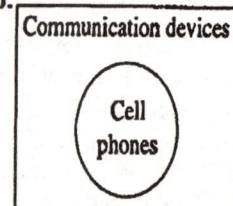

37.   True; every element of the first set is also in the second.

38.   True; m is an element of $\{m, a, t, h\}$.

39.   False; m is an element of the second set, but $\{m\}$ is not. The correct statement would be $\{m\} \subset \{m, a, t, h\}$.

40.   True; the sets are equal, so one is a proper subset of the other.

41.   False; the two sets are equal, so one cannot be a proper subset of the other. The correct statement would be $\{m, a, t, h\} \subseteq \{h, t, a, m\}$.

42.   False; the word "math" should not be considered to be the same as the individual letters.

43.   True; math and history are certainly high school subjects, and there is at least one high school subject, say English, that is not in the first set.

44.   True; the empty set is a subset of every set.

45.   False; blue is a color of the rainbow but $\{\text{blue}\} \neq$ blue. A correct statement would be blue $\in$ {colors of the rainbow} or {blue} $\subseteq$ {colors of the rainbow}.

46.   True; 1 is an element of the second set.

47.   False; correct statements would be $1 \in \{1, 2, 3, 4, 5\}$ or $\{1\} \subseteq \{1, 2, 3, 4, 5\}$.

48.   False; 1 is not an element of the second set.

49.   True; $\{1\}$ is an element of the second set.

50.   False; $\{1\}$ is an element, but not a subset.

51.   False; the number zero and the empty set are very different.

52.   True; both are representations for the empty set.

53. False; the set containing the empty set is not the same as the empty set.

54. Let $U = \{\text{SRJC students}\}$, $F = \{\text{females}\}$, and

$A = \{\text{students over 25}\}$;

$|U| = 29{,}000$;

$|F| = 0.58(29{,}000) = 16{,}820$;

$|\overline{F}| = 29{,}000 - 16{,}820 = 12{,}180$;

$|A| = 0.62(29{,}000) = 17{,}980$;

$|\overline{F} \cap A| = 0.40(17{,}980) = 7{,}192$

$|F \cap A| = 0.60(17{,}980) = 10{,}788$.

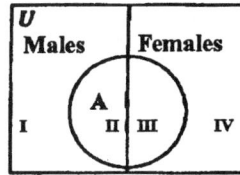

Thus, I: $12{,}180 - 7{,}192 = 4{,}988$ II: $7{,}192$ III: $10{,}788$ IV: $16{,}820 - 10{,}788 = 6{,}032$

55. We work this problem without Venn diagrams to show an alternate method of solution.

$|B \cup S| = |B| + |S| - |B \cap S|$

$\qquad = 50 + 36 - 14$

$\qquad = 72$

Yes, they can get by with two buses.

56.

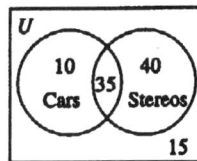

a. $10 + 40 = 50$

b. 15

57.

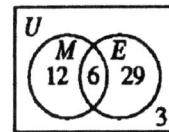

Three people are not taking either.

58. $|F \cup B| = |F| + |B| - |B \cap F|$

$\qquad = 25 + 16 - 7$

$\qquad = 34$

There are 34 people playing.

59. 51,000 booklets are needed.

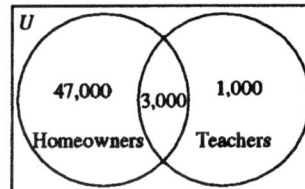

Level 3 Problem Solving, page 505

60. $A = 3$, $B = 6$, $C = 1$, $D = 5$, $E = 8$, $F = 4$, $G = 9$, $H = 2$, $I = 7$
or
$A = 3$, $B = 6$, $C = 1$, $D = 2$, $E = 8$, $F = 7$, $G = 9$, $H = 5$, $I = 4$

10.2 Combined Operations with Sets, page 505

Level 1, page 511

1. *De Morgan's laws* relate the operations of complements, unions, and intersections. There are two statements: $\overline{X \cup Y} = \overline{X} \cap \overline{Y}$ and $\overline{X \cap Y} = \overline{X} \cup \overline{Y}$.

2. Draw a Venn diagram, and then fill in the number of elements in the various regions. Fill in the innermost region first, and then work your way outward (using subtraction) until the number of elements of the eight regions formed by the three sets is known.

3.
$$(A \cup B) \cap C = (\{1, 2, 3, 4\} \cup \{1, 2, 5, 6\}) \cap \{3, 5, 7\}$$
$$= \{1, 2, 3, 4, 5, 6\} \cap \{3, 5, 7\}$$
$$= \{3, 5\}$$

4.
$$A \cup (B \cap C) = \{1, 2, 3, 4\} \cup (\{1, 2, 5, 6\} \cap \{3, 5, 7\})$$
$$= \{1, 2, 3, 4\} \cup \{5\}$$
$$= \{1, 2, 3, 4, 5\}$$

5.
$$\overline{A \cup B} \cap C = \overline{\{1, 2, 3, 4, 5, 6\}} \cap \{3, 5, 7\} \qquad \text{(A $\cup$ B) from Problem 3.}$$
$$= \{7\} \cap \{3, 5, 7\}$$
$$= \{7\}$$

6.
$$A \cup \overline{B \cap C} = \{1, 2, 3, 4\} \cup \overline{\{5\}} \qquad \text{(B $\cap$ C) from Problem 4.}$$
$$= \{1, 2, 3, 4\} \cup \{1, 2, 3, 4, 6, 7\}$$
$$= \{1, 2, 3, 4, 6, 7\}$$

7.
$$\overline{A} \cup (B \cap C) = \overline{\{1, 2, 3, 4\}} \cup \{5\} \qquad \text{(B $\cap$ C) from Problem 4.}$$
$$= \{5, 6, 7\} \cup \{5\}$$
$$= \{5, 6, 7\}$$

8.
$$(A \cup B) \cap \overline{C} = \{1, 2, 3, 4, 5, 6\} \cap \overline{\{3, 5, 7\}}$$
$$= \{1, 2, 3, 4, 5, 6\} \cap \{1, 2, 4, 6\}$$
$$= \{1, 2, 4, 6\}$$

9.
$$\overline{(A \cup B) \cap C} = \overline{\{3, 5\}} \qquad \text{From Problem 3.}$$
$$= \{1, 2, 4, 6, 7\}$$

10.
$$\overline{A} \cup (\overline{B} \cap \overline{C}) = \{5, 6, 7\} \cup (\overline{\{1, 2, 5, 6\}} \cap \{1, 2, 4, 6\}) \qquad \text{Complements from Problems 7 and 8.}$$
$$= \{5, 6, 7\} \cup (\{3, 4, 7, 8\} \cap \{1, 2, 4, 6\})$$
$$= \{5, 6, 7\} \cup \{4\}$$
$$= \{4, 5, 6, 7\}$$

11. $\overline{X} \cup \overline{Y}$ **12.** $\overline{X \cup Y}$ **13.** $\overline{X \cap Y}$ **14.** $\overline{X} \cap \overline{Y}$ **15.** $\overline{X} \cup Y$ **16.** $X \cup \overline{Y}$

17. $\overline{X} \cap \overline{Y}$ **18.** $\overline{X \cap Y}$

19. 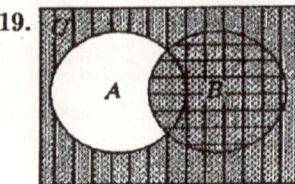   **20.** 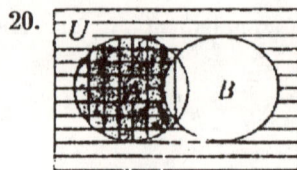   **21.**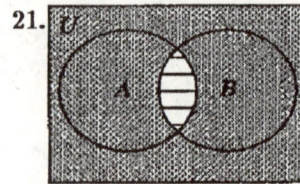

22.
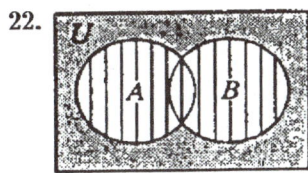

23.
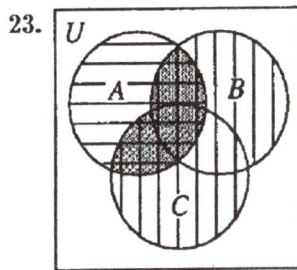

24.
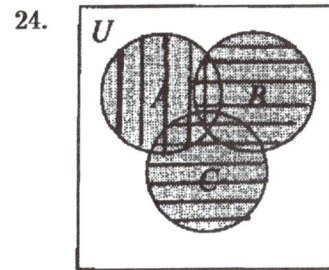

25.
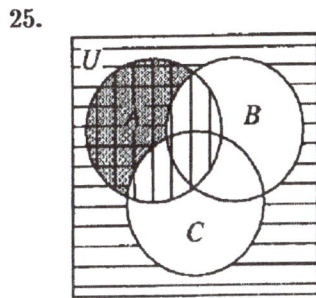

26.
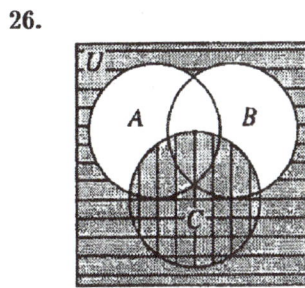

27.
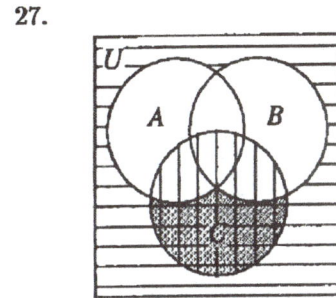

28.
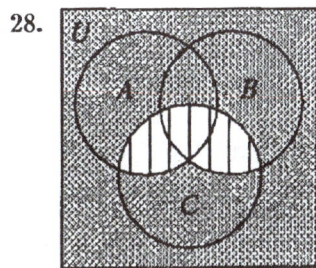

29.
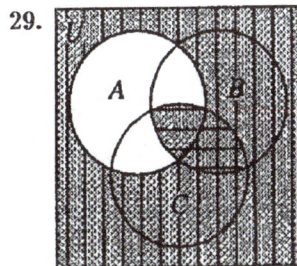

30.
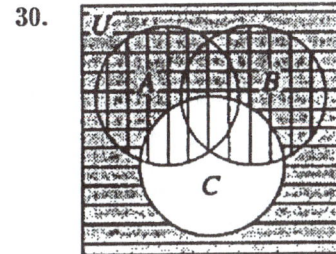

31.
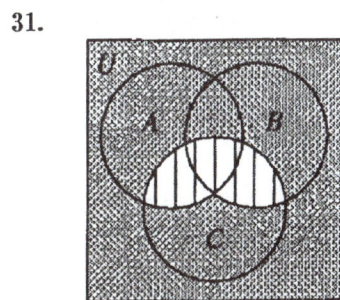

32.
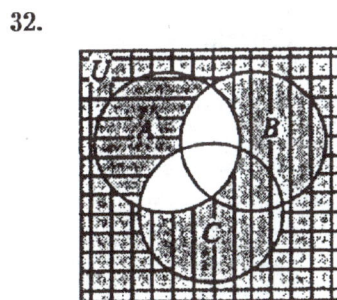

33.
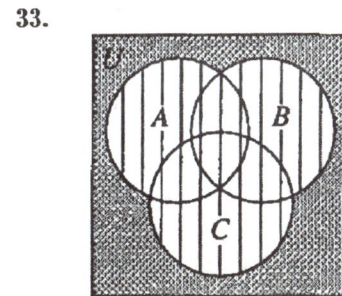

34.
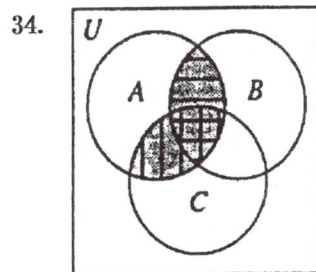

35.

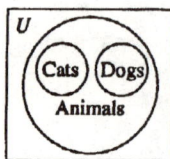

36.

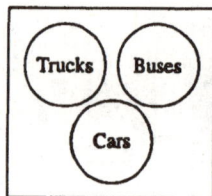

37.

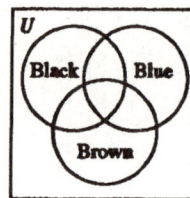

38.

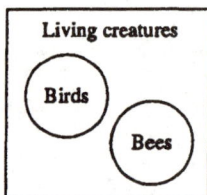

Level 2, page 511

39. *Note:* We interpret white, black, or Hispanic to mean both parents are of that race.

Let $U = \{$U. S. population$\}$ where $|U| = 263$
(assume all numbers are in millions)
As we can see from the Venn diagram, there
are 8 regions. Let V be the region representing
persons with one white and one black parent:

$$0.005(263) = 1.315;$$

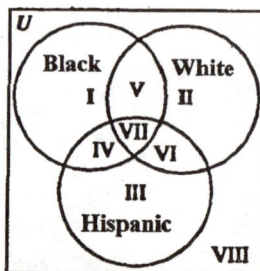

IV the region representing persons with one
black and one Hispanic parent:

$$0.02(263) = 5.26$$

VI the region representing persons with one white
and one Hispanic parent: $0.01(263) = 2.63$.
Let I be the region representing persons with 2 black parents: $0.12(263) = 31.56$
Region II with 2 white parents: $0.80(263) = 210.4$
Region III with 2 Hispanic parents. $0.09(263) = 23.67$
Region VIII includes everyone not otherwise classified: $263 - 274.835 = 11.835$
Region VII is empty since everyone has two parents and are accounted for elsewhere.
Here are the answers:

I: 31,560,000   II: 210,400,000   III: 23,670,000   IV: 5,260,000   V: 1,315,000   VI: 2,630,000

VII: 0   VIII: 11,835,000

40. Clinton, VII; Kennedy, VII; Bayh, II; Inouye, V; Campbell, I; Helms, I; Thurmond, I; Lott, I

41. $\overline{A \cup B}$ **42.** $\overline{A} \cap B$ **43.** $A \cap (B \cup C)$ **44.** $[\overline{(A \cup B)} \cap C] \cup [(A \cap B) \cap C]$

45. Steffi Graf, II; Michael Stich, V; Martina Navratilova, III; Stefan Edberg, V; Chris Evert Lloyd,

VI; Boris Becker, V; Evonne Goolagong, II; Pat Cash, V; Virginia Wade, II; John McEnroe, IV

46.   false

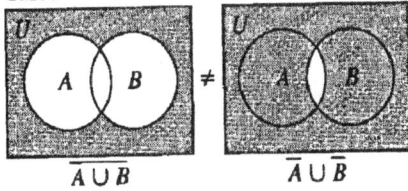

$$\overline{A \cup B} \neq \overline{A} \cup \overline{B}$$

47.   true

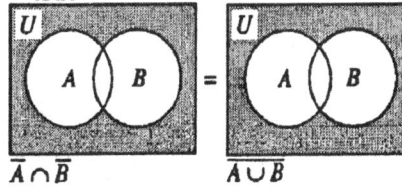

$$\overline{A \cap B} = \overline{A} \cup \overline{B}$$

48.   true

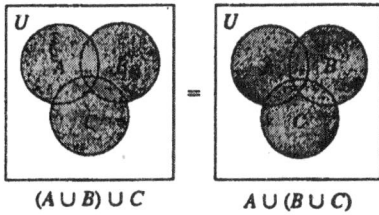

$$(A \cup B) \cup C = A \cup (B \cup C)$$

49.   true

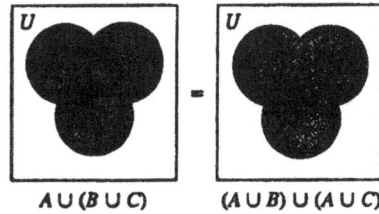

$$A \cup (B \cup C) = (A \cup B) \cup (A \cup C)$$

50.   false

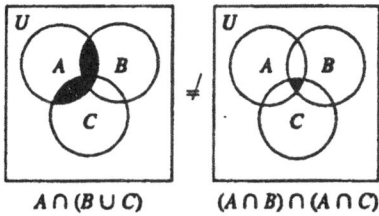

$$A \cap (B \cup C) \neq (A \cap B) \cap (A \cap C)$$

51.   true

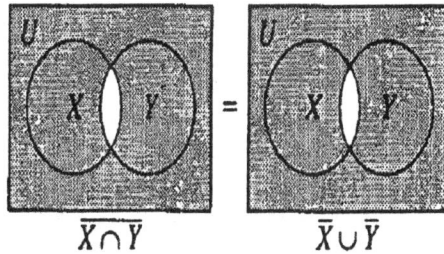

$$\overline{X \cap Y} = \overline{X} \cup \overline{Y}$$

52.

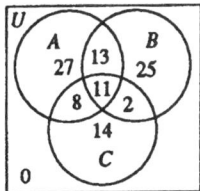

$A = \{$use shampoo $A\}$
$B = \{$use shampoo $B\}$
$C = \{$use shampoo $C\}$

53.

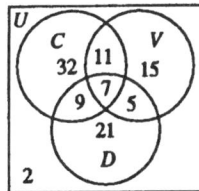

$C = \{$like comedies$\}$
$V = \{$like variety$\}$
$D = \{$like drama$\}$
No, there were 102 persons polled, not 100.

54.

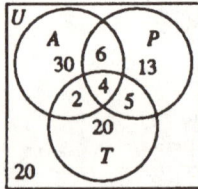

A = {drive alone}
P = {carpool}
T = {public transportation}

20 used none of the above
means of transportation

55. a.

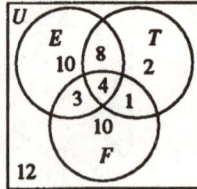

E = {favor Prop. 8}
T = {favor Prop. 13}
F = {favor Prop. 5}
b. 12
c. 7

Level 3 Problem Solving, page 513

56. a.   Let U = {people who live in the United States} and C = {people who use computers}

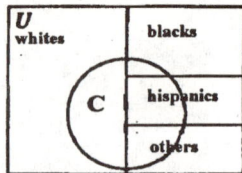

b.   Cannot add percentages. From Problem 34 (or an almanac), $|W| = 210.4$, $|B| = 31.56$, $|H| = 23.67$ so $0.27|W| \approx 57$, $0.14|B| \approx 4.4$, and $0.13|H| \approx 3$. Thus, among the blacks and Hispanics, 7.4 million who use computers does not even come close to the 57 million white people who use the computer.

57.

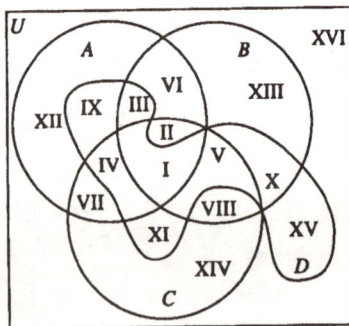

58. Answers vary.

a. Inside E but outside A, B, C, and D;

$\overline{(A \cup B \cup C \cup D)} \cap E$

b. Inside A and E but outside B, C, and D;

$(A \cap E) \cap \overline{(B \cup C \cup D)}$

c. Inside B, E, and D but outside A and C;

$(B \cap E \cap D) \cap \overline{(A \cup C)}$

d. Inside all sets; $A \cap B \cap C \cap D \cap E$

e. Inside A, B and E but outside C and D;

$(A \cap B \cap E) \cap \overline{(C \cup D)}$

59. a. region 32　**b.** region 2　**60.**

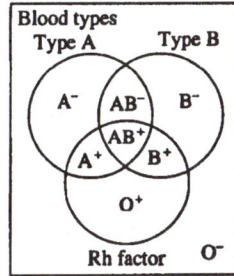

10.3 Permutations, page 513

Level 1, page 513

1. A permutation of r elements selected from a set S with n elements is an ordered arrangement of those r elements selected without repetitions. The formula is $_nP_r = \dfrac{n!}{(n-r)!}$.

2. $_9P_1 = 9$　**3.** $_9P_2 = 9 \cdot 8 = 72$　**4.** $_9P_3 = 9 \cdot 8 \cdot 7 = 504$　**5.** $_9P_4 = 9 \cdot 8 \cdot 7 \cdot 6 = 3{,}024$

6. $_9P_0 = 1$　**7.** $_5P_4 = 5 \cdot 4 \cdot 3 \cdot 2 \cdot 1 = 120$　**8.** $_{52}P_3 = 52 \cdot 51 \cdot 50 = 132{,}600$　**9.** $_7P_2 = 7 \cdot 6 = 42$

10. $_4P_4 = 4! = 24$　**11.** $_{100}P_1 = 100$　**12.** $_{12}P_5 = 12 \cdot 11 \cdot 10 \cdot 9 \cdot 8 = 95{,}040$

13. $_5P_3 = 5 \cdot 4 \cdot 3 = 60$　**14.** $_8P_4 = 8 \cdot 7 \cdot 6 \cdot 5 = 1{,}680$　**15.** $_8P_0 = 1$　**16.** $_gP_h = \dfrac{g!}{(g-h)!}$

17. $_{92}P_0 = 1$　**18.** $_{52}P_1 = 52$　**19.** $_7P_5 = 7 \cdot 6 \cdot 5 \cdot 4 \cdot 3 = 2{,}520$　**20.** $_{16}P_3 = 16 \cdot 15 \cdot 14 = 3{,}360$

21. $_nP_4 = \dfrac{n!}{(n-4)!}$　**22.** $_7P_3 = 7 \cdot 6 \cdot 5 = 210$　**23.** $_5P_5 = 5! = 120$

24. $_{50}P_4 = 50 \cdot 49 \cdot 48 \cdot 47 = 5{,}527{,}200$　**25.** $_{25}P_1 = 25$　**26.** $_mP_3 = \dfrac{m!}{(m-3)!}$

27. $_8P_3 = 8 \cdot 7 \cdot 6 = 336$　**28.** $_{12}P_0 = 1$　**29.** $_{10}P_2 = 10 \cdot 9 = 90$

30. $_{11}P_4 = 11 \cdot 10 \cdot 9 \cdot 8 = 7{,}920$　**31.** $_nP_5 = \dfrac{n!}{(n-5)!}$　**32.** $_5P_r = \dfrac{5!}{(5-r)!}$

33. $_xP_y = \dfrac{x!}{(x-y)!}$　**34.** $\dbinom{7}{1,\,1,\,1,\,1,\,1,\,1,\,1} = 7! = 5{,}040$　**35.** $\dbinom{5}{1,\,2,\,1,\,1} = 60$

36. $\dbinom{6}{2,\,1,\,1,\,1,\,1} = 360$　**37.** $\dbinom{11}{2,\,1,\,1,\,1,\,1,\,1,\,1,\,1,\,1,\,1} = 19{,}958{,}400$　**38.** $\dbinom{11}{1,\,4,\,4,\,2} = 34{,}650$

39. $\dbinom{13}{2,\,2,\,3,\,1,\,2,\,1,\,1,\,1} = 129{,}729{,}600$　**40.** $\dbinom{11}{1,\,2,\,2,\,2,\,1,\,1,\,1,\,1} = 4{,}989{,}600$

41. $\dbinom{11}{1,\,1,\,3,\,2,\,1,\,1,\,1,\,1} = 3{,}326{,}400$　**42.** $\dbinom{11}{1,\,2,\,2,\,3,\,2,\,1} = 831{,}600$

43. $\dbinom{13}{2,\,2,\,1,\,1,\,2,\,1,\,1,\,1,\,1,\,1} = 778{,}377{,}600$

Level 2, page 520

44. $_5P_4 = 120$ **45.** $_{15}P_2 = 210$ **46.** $_{10}P_3 = 720$ **47.** $_{25}P_2 = 25 \cdot 24 = 600$ **48.** $2 \cdot 8 = 16$

49. $_8P_8 = 40{,}320$ **50.** $_8P_3 = 336$ **51.** $_{362}P_3 = 47{,}045{,}520$

52. $8 \cdot 8 \cdot 10 \cdot 10 \cdot 10 \cdot 10 \cdot 10 = 6{,}400{,}000$ **53.** $5 \cdot 3 \cdot 4 = 60$ **54.** $10 \cdot 10 \cdot 10 \cdot 10 = 10{,}000$

55. $5 \cdot 5 \cdot 5 \cdot 5 \cdot 5 = 3{,}125$ **56.** $_8P_8 = 8! = 40{,}320$ **57.** $_9P_9 = 9! = 362{,}880$ **58.** $9 \cdot 8 \cdot 10 = 720$

59. $_7P_3 = 210$

Level 3 Problem Solving, page 521

60. $2^6 - 1 = 63$

10.4 Combinations, page 521

Level 1, page 525

1. $\binom{9}{1} = 9$ **2.** $\binom{9}{2} = 36$ **3.** $\binom{9}{3} = 84$ **4.** $\binom{9}{4} = 126$ **5.** $\binom{9}{0} = 1$ **6.** $\binom{5}{4} = 5$

7. $\binom{52}{3} = 22{,}100$ **8.** $\binom{7}{2} = 21$ **9.** $\binom{4}{4} = 1$ **10.** $\binom{100}{1} = 100$ **11.** $\binom{7}{3} = 35$ **12.** $\binom{5}{5} = 1$

13. $\binom{50}{48} = 1{,}225$ **14.** $\binom{25}{1} = 25$ **15.** $\binom{g}{h} = \dfrac{g!}{h!(g-h)!}$ **16.** $\binom{n}{4} = \dfrac{n!}{4!(n-4)!}$

17. $_kP_4 = \dfrac{k!}{(k-4)!}$ **18.** $_nC_5 = \dfrac{n!}{5!(n-5)!}$ **19.** $_mC_n = \dfrac{m!}{n!(m-n)!}$ **20.** $_mP_n = \dfrac{m!}{(m-n)!}$

21. $\binom{10}{4,\,3,\,3} = \dfrac{10!}{4!3!3!} = 4{,}200$ **22.** $\binom{7}{2,\,2,\,3} = \dfrac{7!}{2!2!3!} = 210$ **23.** $\binom{9}{2,\,3,\,4} = \dfrac{9!}{2!3!4!} = 1{,}260$

24. $\binom{9}{1}\binom{8}{4}\binom{4}{4} = \dfrac{9!}{1!4!4!} = 630$ **25.** $\binom{7}{1}\binom{6}{4}\binom{2}{2} = \dfrac{7!}{1!4!2!} = 105$ **26.** $\binom{10}{6}\binom{4}{2}\binom{2}{1} = \dfrac{10!}{6!2!1!} = 2{,}520$

27. $_{12}C_5 = 792$ **28.** $_{100}C_5 = 75{,}287{,}520$ **29.** $_4C_3 = 4$ **30.** $_4C_2 = 6$ **31.** $_4C_4 = 1$

32. $_{13}C_5 = 1{,}287$ **33.** $_{13}C_5 = 1{,}287$ **34.** $_4C_3 \cdot {_4C_2} = 4 \cdot 6 = 24$ **35.** $_4C_3 \cdot {_4C_2} = 24$

36. $_{15}C_5 = 3{,}003$

Level 2, page 525

37. combination; $_{30}C_5 = 142{,}506$ **38.** permutation; $_{30}P_5 = 17{,}100{,}720$

39. combination; $_{20}C_2 = 190$ **40.** combination; $_5C_3 = 10$ **41.** combination; $_{31}C_3 = 4{,}495$

42. FCP (fundamental counting principle) and combinations; $_{31}C_3 \cdot _8C_3 = 251{,}720$ **43.** neither;

distinguishable permutation; 1,260   **44.** neither; $6 \cdot 3 \cdot 2 \cdot 2 \cdot 1 \cdot 1$; 72   **45.** permutation; $_{10}P_2$; 90

46. combination; $_{52}C_2$; 1,326   **47.** permutation; $_6P_5$; 720   **48.** permutation; $_6P_6$; 720

49. combination; $_{12}C_{10}$; 66   **50.** combination; $_7C_4$; 35   **51.** permutation; $_7P_7$; 5,040

52. permutation; $_7P_2$; 42   **53.** neither; $2 \cdot 2 \cdot 2 \cdot \cdots \cdot 2 = 2^{10}$; 1,024

54. neither; $3 \cdot 3 \cdot 3 \cdot \cdots \cdot 3 = 3^{10}$; 59,049   **55.** neither; $5 \cdot 5 \cdot 5 \cdot \cdots \cdot 5 = 5^{10}$; 9,765,625

Level 3, page 526

56. a. $_3C_3 = 1$ **b.** $_4C_3 = 4$ **c.** $_6C_3 = 20$ **d.** $_nC_3$ **57.** $_nC_5$

58. $_{35}P_4 \cdot _{31}C_6$; approximately $9.252401558 \times 10^{11}$

59. $_{16}P_2 \cdot _{19}P_2 \cdot _{31}C_6$; approximately $6.043394448 \times 10^{10}$

60.
$$
\begin{aligned}
_nC_r &= \frac{n!}{r!(n-r)!} \\
&= \frac{n!}{(n-r)!\,r!} \\
&= \frac{n!}{(n-r)![n-n+r]!} \\
&= \frac{n!}{(n-r)![n-(n-r)]!} \\
&= _nC_{n-r}
\end{aligned}
$$

10.5 Counting without Counting, page 526

Level 1, page 533

1. The *fundamental counting principle* gives the number of ways of performing two or more tasks. If task A can be performed in m ways, and if, after task A is performed, a second task B, can be performed in n ways, then task A followed by task B can be performed in $m \cdot n$ ways.   **2.** A *permutation* of a set of objects is an arrangement of certain of these objects in a specified order. A *combination* of a set of objects is an arrangement of certain of these objects without regard to their order.

3. FCP; $3 \cdot 5 = 15$ **4.** FCP; $26 \cdot 10 \cdot 10 \cdot 10 \cdot 10 \cdot 10 = 2{,}600{,}000$

5. FCP; $26 \cdot 10 \cdot 9 \cdot 8 \cdot 7 \cdot 6 = 786{,}240$ **6.** $_{15}P_3 = 2{,}730$ **7.** $_7C_2 = 21$

8. FCP; $26 \cdot 26 \cdot 26 \cdot 10 \cdot 10 \cdot 10 - 245 = 17{,}575{,}755$ **9.** $_{30}P_5 = 17{,}100{,}720$

10. a. $|A| = 50$ **b.** $|B| = 33$ **c.** $|A \cap B| = 16$ (These are the multiples of 6.)

d. $|A \cup B| = |A| + |B| - |A \cap B| = 50 + 33 - 16 = 67$ **11.** FCP; $6^4 = 1{,}296$ **12.** Four flips has $2^4 = 16$ possibilities, so at least one head is $2^4 - 1 = 15$ ways.   **13.** Five flips has $2^5 = 32$ possibilities, so at least one head is $2^5 - 1 = 31$ ways.   **14.** Six flips has $2^6 = 64$ possibilities, so at least one head is $2^6 - 1 = 63$ ways.   **15.** Seven flips has $2^7 = 128$ possibilities, so at least one head is $2^7 - 1 = 127$ ways.   **16.** $2^n - 1$ **17.** $_{15}P_3 = 2{,}730$ **18.** $_{30}P_3 = 24{,}360$ **19.** 26 letters, 10 numerals, 1 space, and 4 symbols is a total of 41 possibilities. Since there are seven spaces, the total number of plates is $41^7 \approx 1.95 \times 10^{11}$.   **20.** FCP; $2^5 = 32$ **21.** FCP; $5 \cdot 3 \cdot 3 \cdot 3 = 135$

22. $_{15}C_4 = 1{,}365$ **23.** $_{30}C_4 = 27{,}405$ **24.** permutation; $_8P_5 = 6{,}720$

25. permutation; $_6P_3 = 120$ **26.** permutation; $_5P_5 = 120$ **27.** combination; $_{13}C_3 = 286$

28. combination; $_{100}C_6 = 1{,}192{,}052{,}400$ **29.** permutation; $_{125}P_{125} = 125!$

30. permutation; $_6P_6 = 720$ **31.** permutation; $_6P_4 = 360$ **32.** neither (it is distinguishable permutation); $\left(\begin{matrix} & 4 & \\ 1, & 1, 1, & 1 \end{matrix} \right) = 4! = 24$ **33.** neither (FCP); $1 \cdot {_5P_5} \cdot {_5P_5} = 14{,}400$

34. permutation; $_{10}P_2 = 90$ **35.** permutation; $_4P_3 = 24$ **36.** combination; $_{52}C_2 = 1{,}326$

37. permutation; $_6P_5 = 720$ **38.** neither (it is a distinguishable permutation); $\left(\begin{matrix} & 6 & \\ 1, & 1, 1, 1, & 1, 1 \end{matrix} \right) = 6! = 720$ **39.** $_{12}C_{10} = 66$

Level 2, page 534

40. Assume that an outfit will consist of choosing one of the blouses, one of the slacks or skirts, 1 sweater or jacket, and each of these may or may not have a scarf. This means that

$_4C_1 \cdot {_4C_1} \cdot {_3C_1} \cdot 2 = 96$; This does not seem correct, so let's say there are 4 tops and 4 bottoms which give 16 basic outfits (FCP); next, choose the sweater, or one of the jackets, or no wrap for 4 additional possibilities with each of the 16 basic outfits for 64 possibilities. Finally, each of the 64 outfits could appear with or without a scarf for a grand total of 128 outfits. Even if we allow the possibility with wearing a skirt over the slacks, we do not come up with the claimed 122 different outfits.

41. Subsets: $\{a, b\}$; 2 arrangements: (a, b), (b, a) **42.** Subsets: $\{c, d, e\}$; 6 arrangements: (c, d, e), (c, e, d), (d, c, e), (d, e, c), (e, c, d), (e, d, c) **43.** Subsets: $\{a, b, c, d\}$; 24 arrangements:
(a, b, c, d), (a, b, d, c), (a, c, b, d), (a, c, d, b), (a, d, b, c), (a, d, c, b), (b, a, c, d), (b, a, d, c),
(b, c, a, d), (b, c, d, a), (b, d, a, c), (b, d, c, a), (c, a, b, d), (c, a, d, b), (c, b, a, d), (c, b, d, a),
(c, d, a, b), (c, d, b, a), (d, a, b, c), (d, a, c, b), (d, c, a, b), (d, c, b, a), (d, b, a, c), (d, b, c, a).

44. Subsets: $\{a, b, c, d, e\}$; there are $5! = 120$ arrangements.

45. $9 \cdot 9 \cdot 10 \cdot 10 \cdot 10 \cdot 10 \cdot 10 \cdot 10 \cdot 10 = 810{,}000{,}000$ **46.** $_3P_1 \cdot {_4}P_2 = 36$

47. There are three possibilities for each question, so the total is $3^{20} = 3{,}486{,}784{,}401$

48. $13^5 = 371{,}293$; the claim is correct.   **49.** $_{26}P_{26} = 26!$

Level 3 Problem Solving, page 535

50. No; all numbers but 13 are possible.   **51.** Count the number of kidney beans that will fit into one cubic inch. Suppose this number is 15. Then, calculate the volume of the jar; the cylinder is about $1{,}176$ in.^{3} Finally, multiply to estimate the number of beans to be 17,671.   **52.** 1 min = 50 jumps; 1 hr = 3,000 jumps; 1 day = 24,000 jumps; 1 year = 6×10^6 jumps; 1 century = 6×10^8 jumps; thus, 5,480,523,297,162 times would take about 9,134.2 centuries!   **53.** About a half a year; more precisely, 30 weeks working five days per week, 8 hours per day.   **54.** Answers vary; in Pascal's triangle, two numbers are added to obtain the number below; in this arrangement the triangle is inverted and the numbers are subtracted to find the numbers below.   **55.** There must be at least one bald person with 0 hairs and any number times 0 is 0; the answer is 0.   **56.** $1{,}048{,}576 \div 365 \approx 2{,}872.81$; about 2,873 years   **57.** The number of possibilities for one pizza is:

$$_{11}C_5 + {_{11}}C_4 + {_{11}}C_3 + {_{11}}C_2 + {_{11}}C_1 + {_{11}}C_0 = 1{,}024;$$

The number of possible (different) pizzas are $_{1{,}024}C_2 + 1{,}024 = 524{,}800$.

58. $\left[_{11}C_5 + {_{11}}C_4 + {_{11}}C_3 + {_{11}}C_2 + {_{11}}C_1 + {_{11}}C_0 \right]^2 = 1{,}048{,}576$; the advertisement is correct.

59. *Number of 0-topping pizzas:* $_{11}C_0 = 1$

Number of 1-topping pizzas: $_{11}C_1 = 11$

Number of 2-topping pizzas: $_{11}C_2 + {_{11}}C_1 = 55 + 11 = 66$

Note: $_{11}C_2$ is the number of pizzas with two different toppings;

$_{11}C_1$ is the number of pizza with one topping taken twice.

Number of 3-topping pizzas: $_{11}C_3 + {_{11}}C_1 \cdot {_{10}}C_1 = 165 + 110 = 275$

Note: $_{11}C_1 \cdot {_{10}}C_1$ is the number of pizzas where 1 topping is double and the other topping is single.

Number of 4-topping pizzas: $_{11}C_4 + {_{11}C_2} + {_{11}C_1} \cdot {_{10}C_2} = 330 + 55 + 495 = 880$

Note: $_{11}C_2$ is the number of pizzas where 2 toppings are doubled;

$_{11}C_1 \cdot {_{10}C_2}$ is the number of pizzas where 1 topping is double and the other two toppings are single.

Number of 5-topping pizzas: $_{11}C_5 + {_{11}C_1} \cdot {_{10}C_3} + {_{11}C_2} \cdot {_9C_1} = 462 + 1{,}320 + 495 = 2{,}277$

Thus, the total is $1 + 11 + 66 + 275 + 880 + 2{,}277 = 3{,}510$. Finally, for two pizzas, there are

$_{3510}C_2 + 3{,}510 = 6{,}161{,}805$ pizzas.

60.
$$|A \cup B \cup C| = |(A \cup B) \cup C|$$
$$= |A \cup B| + |C| - |(A \cup B) \cap C|$$
$$= |A| + |B| - |A \cap B| + |C| - |(A \cap C) \cup (B \cap C)|$$
$$= |A| + |B| + |C| - |A \cap B| - \big[|A \cap C| + |B \cap C) - [(A \cap C) \cap (B \cap C)]|\big]$$
$$= |A| + |B| + |C| - |A \cap B| - |A \cap C| - |B \cap C| + |(A \cap C) \cap (B \cap C)|$$
$$= |A| + |B| + |C| - |A \cap B| - |A \cap C| - |B \cap C| + |(A \cap B \cap C)|$$

10.6 Binomial Theorem, page 536

Level 1, page 540

1. For any positive integer n,

$$(a+b)^n = \binom{n}{0}a^n + \binom{n}{1}a^{n-1}b + \binom{n}{2}a^{n-2}b^2 + \cdots + \binom{n}{r}a^{n-r}b^r$$
$$+ \cdots + \binom{n}{n-2}a^2b^{n-2} + \binom{n}{n-1}ab^{n-1} + \binom{n}{n}b^n$$

2. 2^t **3.** Look at the 8th row, 1st diagonal of Pascal's triangle: 8; $_8C_1 = 8$ **4.** Look at the 5th row, fourth diagonal: 5; $_5C_4 = 5$ **5.** Look at the 8th row, second diagonal: 28; $_8C_2 = 28$ **6.** Look at the 7th row, fifth diagonal: 21; $_7C_5 = 21$ **7.** Look at the 8th row, third diagonal: 56; $_8C_3 = 56$
8. Look at the 9th row, fifth diagonal: 126; $_9C_5 = 126$ **9.** Look at the 12th row, first diagonal; 12; $_{12}C_1 = 12$ **10.** The 0th entry in any row is 1; $_{15}C_0 = 1$ **11.** The last entry in any row is 1; $_{20}C_{20} = 1$ **12.** the next-to-the-last element in the nth row is n, so in the 32 row it 32; $_{32}C_{31} = 32$
13. The second entry in the nth row is $n(n-1)/2$: $\frac{18(17)}{2} = 153$; $_{18}C_2 = 153$ **14.** The second entry in the nth row is $n(n-1)/2$: $\frac{46(45)}{2} = 1{,}035$; $_{46}C_2 = 1{,}035$
15. $(x+1)^4 = x^4 + 4x^3 + 6x^2 + 4x + 1$
16. $(x+1)^8 = x^8 + 8x^7 + 28x^6 + 56x^5 + 70x^4 + 56x^3 + 28x^2 + 8x + 1$
17. $(x+4)^4 = x^4 + 16x^3 + 96x^2 + 256x + 256$

18. $(a + b)^6 = a^6 + 6a^5b + 15a^4b^2 + 20a^3b^3 + 15a^2b^4 + 6ab^5 + b^6$

19. $(a + b)^7 = a^7 + 7a^6b + 21a^5b^2 + 35a^4b^3 + 35a^3b^4 + 21a^2b^5 + 7ab^6 + b^7$

20. $(x - 1)^5 = x^5 - 5x^4 + 10x^3 - 10x^2 + 5x - 1$

21. $(x - 1)^9 = x^9 - 9x^8 + 36x^7 - 84x^6 + 126x^5 - 126x^4 + 84x^3 - 36x^2 + 9x - 1$

22. $(x - y)^6 = x^6 - 6x^5y + 15x^4y^2 - 20x^3y^3 + 15x^2y^4 - 6xy^5 + y^6$

23. $(x - y)^5 = x^5 - 5x^4y + 10x^3y^2 - 10x^2y^3 + 5xy^4 - y^5$

24. $(x + 2)^5 = x^5 + 10x^4 + 40x^3 + 80x^2 + 80x + 32$

25. $(x - 2)^6 = x^6 - 12x^5 + 60x^4 - 160x^3 + 240x^2 - 192x + 64$

26. $(x - 3)^5 = x^5 - 15x^4 + 90x^3 - 270x^2 + 405x - 243$

27. $(2x + 3y)^4 = 16x^4 + 96x^3y + 216x^2y^2 + 216xy^3 + 81y^4$

28. $(x - 2y)^8 = x^8 - 16x^7y + 112x^6y^2 - 448x^5y^3 + 1,120x^4y^4 - 1,792x^3y^5 + 1,792x^2y^6 - 1,024xy^7 + 256y^8$

Level 2, page 540

29. $\binom{11}{6}a^5(-b)^6 = 462a^5b^6$; the coefficient is 462   **30.** $\binom{11}{7}a^4b^7 = 330a^4b^7$; the coefficient is 330

31. $\binom{14}{4}x^{10}y^4 = 1,001x^{10}y^4$; the coefficient is 1,001   **32.** $\binom{15}{5}x^{10}(-y)^5 = -3,003x^{10}y^5$; the

coefficient is $-3,003$ **33.** $\binom{16}{4}x^{12}(-1)^4 = 1,820x^{12}$; the coefficient is 1,820

34. $\binom{12}{4}y^8(1)^4 = 495y^8$; the coefficient is 495   **35.** $\binom{9}{4}r^6(2)^4 = 126r^6(16) = 2,016r^6$; the coefficient

is 2,016   **36.** $\binom{10}{5}s^5(-2)^5 = 252s^5(-32) = -8,064s^5$; the coefficient is $-8,064$

37. $\binom{8}{1}a^7(-2b) = 8a^7(-2b) = -16a^7b$; the coefficient is -16

38. $\binom{8}{4}a^4(2b)^4 = 70a^4(16b^4) = 1,120a^4b^4$; the coefficient is 1,120

39. False; $(a + b)^3 = a^3 + 3a^2b + 3ab^2 + b^3$ **40.** True; the *n*th degree has $n + 1$ terms.

41. False; look at the coefficient of H^3T^2 **42.** False; it is 2^5.

43. $(x - y)^{15} = x^{15} - 15x^{14}y + 105x^{13}y^2 - 455x^{12}y^3 + \ldots$

44. $(x + 2y)^{16} = x^{16} + 32x^{15}y + 480x^{14}y^2 + 4,480x^{13}y^3 + \ldots$

45. $(x - 2y)^{12} = x^{12} - 24x^{11}y + 264x^{10}y^2 - 1,760x^9y^3 + \ldots$

46. $(x - 3y)^{10} = x^{10} - 30x^9y + 405x^8y^2 - 3,240x^7y^3 + \ldots$

47. $(ab - 2b)^{15} = a^{15}b^{15} - 30a^{14}b^{15} + 420a^{13}b^{15} - 3,640a^{12}b^{15} + \ldots$

48. $(rs - 3t)^{13} = r^{13}s^{13} - 39r^{12}s^{12}t + 702r^{11}s^{11}t^2 - 7,722r^{10}s^{10}t^3 + \ldots$

49. $(z^2 + 5k)^{11} = z^{22} + 55z^{20}k + 1,375z^{18}k^2 + 20,625z^{16}k^3 + \ldots$

50. $(z^3 - k^2)^7 = z^{21} - 7z^{18}k^2 + 21z^{15}k^4 - 35z^{12}k^6 + \ldots$

51. $2^7 = 128$ **52.** 2^{100} **53.** Look at the coefficient of B^2G^3 in the expansion of $(B + G)^5$;

$\binom{5}{3} = 10$ **54.** Look at the coefficient of B^3G^3 in the expansion of $(B + G)^6$; $\binom{6}{3} = 20$

55. Look at the coefficient of B^4G^3 in the expansion of $(B + G)^7$; $\binom{7}{3} = 35$

56. Look at the coefficient of H^5T^3 in the expansion of $(H + T)^8$; $\binom{8}{3} = 56$

57. Look at the coefficient of H^4T^6 in the expansion of $(H + T)^{10}$; $\binom{10}{6} = 210$

58. Look at the coefficient of H^2T^7 in the expansion of $(H + T)^9$; $\binom{9}{7} = 36$

Level 3 Problem Solving, page 540

59. $(1 + 1)^n = \binom{n}{0}1^n + \binom{n}{1}1^{n-1}1^1 + \binom{n}{2}1^{n-2}1^2 + \cdots + \binom{n}{n-1}1^1 1^{n-1} + \binom{n}{n}1^n$

$2^n = \binom{n}{0} + \binom{n}{1} + \binom{n}{2} + \cdots + \binom{n}{n-1} + \binom{n}{n}$

Thus, the sum of the entries of the nth row of Pascal's triangle is 2^n.

60.

$$\binom{n-1}{r-1} + \binom{n-1}{r} = \frac{(n-1)!}{(r-1)![(n-1)-(r-1)]!} + \frac{(n-1)!}{r![(n-1)-r]!}$$

$$= \frac{(n-1)!}{(r-1)!(n-r)!} + \frac{(n-1)!}{r!(n-r-1)!}$$

$$= \frac{(n-1)!}{(r-1)!(n-r)!} \cdot \frac{r}{r} + \frac{(n-1)!}{r!(n-r-1)!} \cdot \frac{n-r}{n-r}$$

$$= \frac{r(n-1)!}{r!(n-r)!} + \frac{(n-r)(n-1)!}{r!(n-r)!}$$

$$= \frac{r(n-1)! + (n-r)(n-1)!}{r!(n-r)!}$$

$$= \frac{[r + (n-r)](n-1)!}{r!(n-r)!}$$

$$= \frac{n(n-1)!}{r!(n-r)!}$$

$$= \frac{n!}{r!(n-r)!}$$

$$= \binom{n}{r}$$

10.7 Rubik's Cube and Instant Insanity, page 541

Level 1, page 543

1. 331,776 sec = 5,529.6 min = 92.16 hr, or about 92 hr, 10 min

2. 331,776 possibilities = 331 sec = 5.5 min, or about 5 min, 30 sec

3. a.   b.   c.   d.

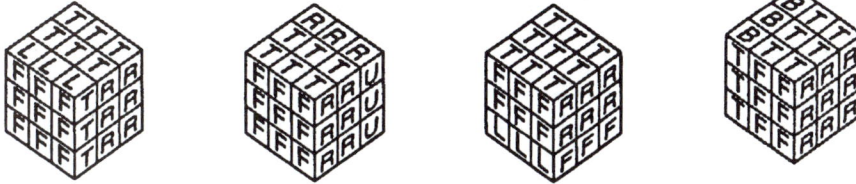

4. a.   b.   c.   d.

5. a.   b.   c.   d.

6. a.   b.   c.   d.

7.   8.   9.

10.    **11.**    **12.**

13.    **14.**    **15.**

Level 2, page 543

16. F^{-1} **17.** T^2 or $(T^{-1})^2$ **18.** F or $(F^{-1})^3$ **19.** $U^{-1}T^{-1}$ **20.** BF^{-1} **21.** L

22. TF^{-1} **23.** BF

Level 3, page 543

24. no   **25.** no   **26.** no; answers vary

Level 3 Problem Solving, page 543

27. To find the number of different possibilities we will not count the order of the cubes (certainly once we have found a solution, it doesn't matter how we switch the four cubes around — as long as we don't rotate any of them). Also, we will not count as a different solution the rotation of the final $1 \times 1 \times 4$ block around its long axis. Thus, if we do not count rotations, we see that the first cube can be placed in any one of three different ways. There is only one way of placing this correctly, but there are three ways of arranging this cube. The key here is to get the "proper faces" along the axis that is never shown. This is, if this cube looks like the one shown below, it will have an R-R that "doesn't come up."

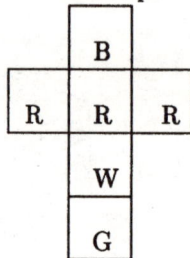

The other two possibilities have W-B and R-G along the axis as shown below:

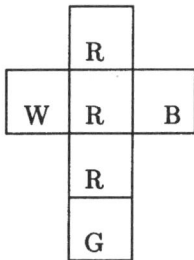
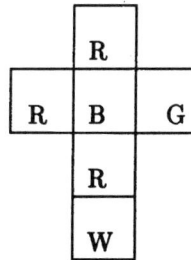

Thus, the first cube can come up three ways. Now, the other cubes do not have this much freedom of choice, since the first cube has been "fixed" in position. That is, each of the remaining cubes cannot be rotated independently, but must be rotated in conjunction with the first cube. Thus, each of the remaining cubes has 24 possibilities. If we use the Fundamental Counting Principle, we see that there are

$$3 \cdot 24 \cdot 24 \cdot 24 = 41{,}472$$

possibilities, of which there is only one solution.

28.    **29.**    **30.**

Chapter 10 Review Questions, page 545

1. **a.** $A \cup B = \{1, 3, 5, 7, 9\} \cup \{2, 4, 6, 9, 10\}$
$= \{1, 2, 3, 4, 5, 6, 7, 9, 10\}$

   b. $A \cap B = \{1, 3, 5, 7, 9\} \cap \{2, 4, 6, 9, 10\}$
$= \{9\}$

2. **a.** $\overline{B} = \overline{\{2, 4, 6, 9, 10\}}$
$= \{1, 3, 5, 7, 8\}$ *You need to look at the universal set, U, to obtain this result.*

   b. $|\emptyset| = 0$

3. **a.** $|U| = 10$ **b.** $|A| = 5$

4. **a.** $\overline{A \cap B} = \overline{\{1, 3, 5, 7, 9\} \cap \{2, 4, 6, 9, 10\}}$
$= \overline{\{9\}}$ *Intersection first, then complement.*
$= \{1, 2, 3, 4, 5, 6, 7, 8, 10\}$

   b. $\overline{A} \cup \overline{B} = \overline{\{1, 3, 5, 7, 9\}} \cup \overline{\{2, 4, 6, 9, 10\}}$
$= \{2, 4, 6, 8, 10\} \cup \{1, 3, 5, 7, 8\}$ *Complements first, then union.*
$= \{1, 2, 3, 4, 5, 6, 7, 8, 10\}$

5. $\overline{A} \cap (B \cup A) = \{2, 4, 6, 8, 10\} \cap \{1, 2, 3, 4, 5, 6, 7, 9, 10\}$
$= \{2, 4, 6, 10\}$

6. $\overline{(A \cup B)} \cap A = \overline{\{1, 2, 3, 4, 5, 6, 7, 9, 10\}} \cap \{1, 3, 5, 7, 9\}$

$\qquad\qquad\quad = \overline{\{1, 3, 5, 7, 9\}}$

$\qquad\qquad\quad = \{2, 4, 6, 8, 10\}$

7. **a.** $N = \{1, 3, 5, \cdots, 47, 49\}; |N| = 25$

    b. $P = \{2, 3, 5, 7, 11, 13, 17, 19, 23, 29, 31, 37, 41, 43, 47\}; |P| = 15$

    c. $N \cap P = \{3, 5, 7, \cdots 43, 47\}; |N \cap P| = 14$

    d. $|N \cup P| = |N| + |P| - |N \cap P|$

$\qquad\qquad\quad = 25 + 15 - 14$

$\qquad\qquad\quad = 26$

8. **a.**   **b.**   **c.**

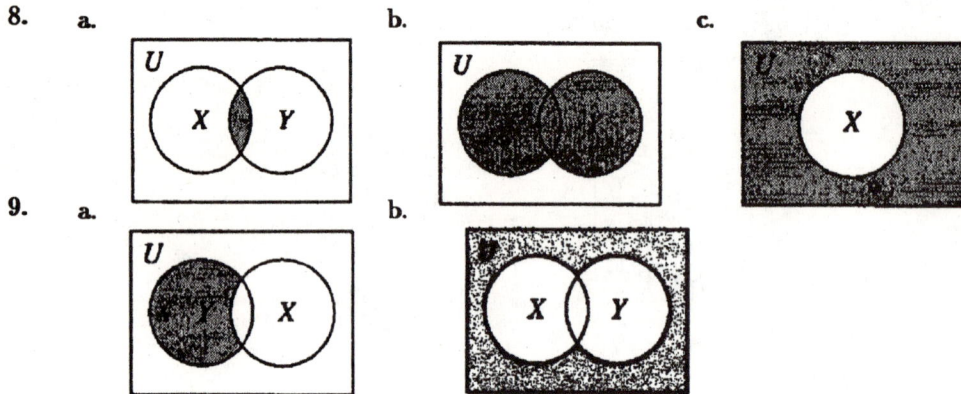

9. **a.**   **b.**

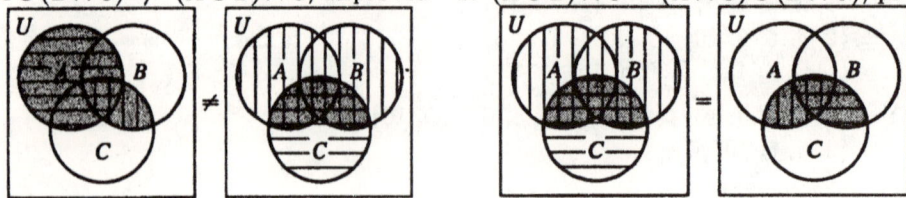

10. **a.** $8! - 3! = 40,320 - 6 = 40,314$ **b.** $8 - 3! = 8 - 6 = 2$ **c.** $(8 - 3)! = 5! = 120$

d. $\left(\dfrac{8}{2}\right)! = 4! = 24$ **d.** $\dbinom{8}{2} = 28$ **11.** Answers for **a-d** are found using Pascal's triangle:

a. $_5C_3 = 10$ **b.** $_8P_3 = 3! \cdot {}_8C_3 = 6 \cdot 56 = 336$ **c.** $_{12}P_0 = 1 \cdot 1 = 1$ **d.** $_{14}C_4 = 1,001$

e. $_{100}P_3 = \dfrac{100!}{97!} = 100 \cdot 99 \cdot 98 = 970,200$ **12.** $_{12}C_3 = 220$; they can form 220 different committees.

13. $_5P_5 = 5! = 120$; they can form 120 different lineups at the teller's window.

14. happy: $\dbinom{5}{1, 1, 2, 1} = \dfrac{5!}{1!1!2!1!} = 60$ college: $\dbinom{7}{1, 1, 2, 2, 1} = \dfrac{7!}{2!2!} = 1,260$

15. If you draw three balls, then to have at least one red, you'd need to have:

1 red and 2 whites or 2 red and 1 white or 3 red and 0 white $= \dbinom{4}{1}\dbinom{6}{2} + \dbinom{4}{2}\dbinom{6}{1} + \dbinom{4}{3}$

$\qquad\qquad = 4 \cdot 15 + 6 \cdot 6 + 4 = 100$

16. **a.** $A \cup (B \cap C) \ne (A \cup B) \cap C$; disproved **b.** $(A \cup B) \cap C = (A \cap C) \cup (B \cap C)$; proved

17. Draw a Venn diagram; begin with the innermost part first:
15 have all three.

    Next, use the information for two overlapping sets
to fill in 2, 20, and 10.

    Fill in the regions in each circle not yet completed
to fill in 5, 5, and 3.

    Finally, total all the numbers to see that 10 of the students
have none of these items.

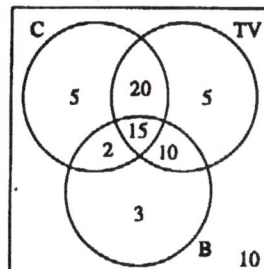

18. Fundamental counting principle; each item can be included or not included, and there are 8 items, so the number of possibilities is $2^8 = 256$. The claim is correct.

19. The number of arrangements is $_{10}P_{10} = 10! = 3,628,800$. This is almost 10,000 years, so that day will never come for the members of this club.

20. **a.** $24^5 = 7,962,624$ **b.** Divide by 60 to change to minutes; divide again by 60 to change to hours; then divide by 24 to change to days. The result is more than 92 days (nonstop).

Group Research Problems, pages 546-548

G35. Use the formula

$$|A \cup B \cup C| = |A| + |B| + |C| - |A \cap B| - |A \cap C| - |B \cap C| + |A \cap B \cap C|$$

Let $U = \{1, 2, 3, \cdots\}$, $S = \{\text{perfect squares}\}$, $|S| = 1,000$; $C = \{\text{perfect cubes}\} = 100$;

$F = \{\text{perfect fifth powers}\}$, $|F| = 15$ since $16^5 > 1,000,000$.

From the commentary, $|S \cap C| = 10$.

$S \cap F = \{\text{perfect tenth powers}\} = \{1^{10}, 2^{10}, 3^{10}\}$, so $|S \cap F| = 3$

$C \cap F = \{\text{perfect fifteenth powers}\} = \{1^{15}, 2^{15}\}$, so $|C \cap F| = 2$

$S \cap C \cap F = \{\text{perfect thirtieth powers}\} = \{1^{30}\}$, so $|S \cap C \cap F| = 1$

$$|S \cup C \cup F| = |S| + |C| + |F| - |S \cap C| - |S \cap F| - |C \cap F| + |S \cap C \cap F|$$

$$= 1,000 + 100 + 15 - 10 - 3 - 2 + 1$$

$$= 1,101$$

Thus, the millionth positive integer that is *not* a perfect square, cube, or fifth power is 1,001,101.

G36. 17% (Solution by using a Venn diagram.)

G37. A *paradox* is a statement that is inescapable, but yet impossible. Consider the *barber's rule*. If he does shave himself, then according to the barber's rule, he does not shave himself. On the other hand, if he does not shave himself, then, according to the barber's rule, he shaves himself. We can only conclude that there can be no such barber's rule. But why not? The paper you should write will address this question. The *barber's rule* is better known as the famous Russell paradox.

G38. Try to build a hexahexaflexagon... it is worth the effort!

G39. Answers vary.   **G40.** Answers vary.

Individual Research Problems, page 549

P10.1. The Tower of Hanoi consists of 64 discs of decreasing size. The problem is to move the discs from one tower to another but never put a bigger disc on top of a smaller one using no more than three center stakes. To solve this Tower of Hanoi problem, consider simpler cases and look for a pattern. For four discs, number them 1, 2, 3, 4 (in order of smallest to largest):

```
Start  1st   2nd    3rd   4th   5th   6th   7th   8th   9th   10th
1
2      2                                     1     1
3      3     3      3 1         1 1   1 2   2     2     2 1    1
4      4 1   4 1 2  4 2   4 3 2 4 3 2 4 3   4 3   3 4   3 4    2 3 4
```

```
11th   12th  13th   14th  15th        For 5 discs, 2^5 − 1 = 31 moves
```

$$\text{For 5 discs, } 2^5 - 1 = 31 \text{ moves}$$

```
                          1
                    2     2
1      1    3       3     3
2 3 4  2 4  2 1 4   1 4   4
```

For 64 discs, it would take $2^{64} - 1$ moves. The legend connected with this problem is that when the priests finish this task of moving the rings, the world will end.

P10.2. Answers vary.   **P10.3.** Answers vary.   **P10.4** Answers vary.

P10.5

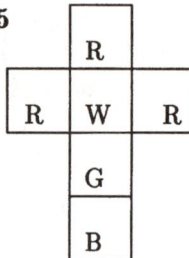
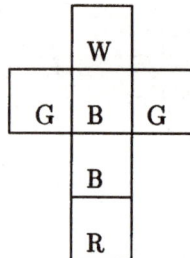
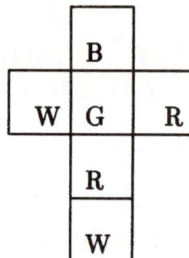
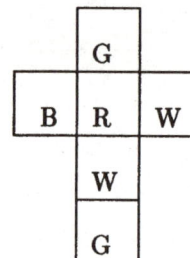

CHAPTER 11

THE NATURE OF PROBABILITY

11.1 Introduction to Probability, page 552

I often introduce the concepts of probability by experimentation, allowing the students to make conjectures and discoveries about properties of probabilities. If time permits, I give the students a handout with four experiments. I have included these experiments on pages 234-237 so you can duplicate them for use with your class, if you wish. These should be considered group projects in which the class works in small groups. As the class is doing these experiments, I introduce them to some ideas, as indicated below.

EXPERIMENT 1: ROLLING ONE DIE (Problem 48)

The possible outcomes are 1, 2, 3, 4, 5, and 6. This list is called the *sample space* of the experiment. This experiment introduces the students to the procedure of gathering *data*. When a theoretical definition of probability is made, the class can compare the definition with their empirical results. This sometimes leads to a discussion of *model building*. A mathematical model is good only insofar as the empirical probabilities (our data) and the theoretical probabilities are "about the same." Notice also, if we specify the probability of rolling a one as a percentage of occurrence, we see that the percentage must be between 0% and 100%. Therefore, if we specify the probability as a fraction, the fraction must be between 0 and 1. The closer to 0 the less likely the event is to occur; the closer to 1, the more likely the event is to occur. This will lead to the properties of probability that will be discussed later: $0 \leq P(E) \leq 1$ for some event E, $P(S) = 1$ for the sample space S, and $P(\emptyset) = 0$.

EXPERIMENT 2: ROLLING A PAIR OF DICE (Problem 49)

This experiment introduces events that are not equally likely (compare to Experiment 1). Have your students make a conjecture about the probability of each. In this experiment, it is clear that there is no tendency for each to be equally likely, as there was in Experiment 1. In fact, it seems as if there is a strong tendency for the middle outcomes to be more probable. Have the students try to explain why, but don't force them to make any "right" conjectures.

EXPERIMENT 3: TOSSING A COIN AND ROLLING A DIE (Problem 51)

After the students have made comments or conjectures about this experiment, you might ask the following questions:

1. What if the experiment is performed with a die and coin separately; will there be any differences? Notice that the events of rolling a 3 and obtaining a head are *independent*.
3. Suppose we call "success" obtaining an even number or a three. Retabulate the data. Notice that the events of rolling a 3 and obtaining an even number are *mutually exclusive*.

EXPERIMENT 4: THREE CARD PROBLEM (Problem 54)

This is Problem 54 in the text.

EXPERIMENT 1
ROLLING ONE DIE

Name of Team _____

_____

Directions: One team member rolls a single die 50 times. The other person records the outcomes using tally marks on the table below. For Trial 2, switch places and repeat the experiment.

OUTCOME	TRIAL 1 FREQUENCY	TRIAL 2 FREQUENCY	TOTAL	PERCENT
1				
2				
3				
4				
5				
6				

Total the tally marks for Trials 1 and 2. Next, calculate the percentage of occurrence (divide the total number of times for Outcome 1 by 100 to find the percentage of occurrence for Outcome 1.

EXPERIMENT 2: ROLLING TWO DICE

Directions: One team member rolls a pair of dice 50 times. The other person records the outcomes of the sum of the top faces using tally marks on the table below. For Trial 2, switch places and repeat the experiment.

OUTCOME	TRIAL 1 FREQUENCY	TRIAL 2 FREQUENCY	TOTAL	PERCENT
1				
2				
3				
4				
5				
6				
7				
8				
9				
10				
11				
12				

Total the tally marks for Trials 1 and 2 and calculate the percentage of occurrence.

EXPERIMENT 3: A COIN AND A DIE

Directions: One team member simultaneously tosses a coin and rolls a single die 50 times. The other person records the outcomes using tally marks on the table below. For Trial 2, switch places and repeat the experiment.

OUTCOME	TRIAL 1 FREQUENCY	TRIAL 2 FREQUENCY	TOTAL	PERCENT
1H				
1T				
2H				
2T				
3H				
3T				
4H				
4T				
5H				
5T				
6H				
6T				

Total the tally marks for Trials 1 and 2 and calculate the percentage of occurrence.
How does $P(3H)$ compare with $P(3)$ and $P(H)$?
How does $P(\text{even number or } 3)$ compare with $P(\text{even})$ and $P(3)$?
How does $P(\text{even number or head})$ compare with $P(\text{even})$ and $P(\text{head})$?
Make a conjecture about $P(E \text{ or } F)$.

EXPERIMENT 4: THREE CARD PROBLEM

Directions: You need to prepare three 3×5 cards so that they are indistinguishable except for the color. One card is white on both sides; a second is black on both sides, and the third is white on one side and black on the other. One person shuffles the three cards under the table so that the cards cannot be seen. Be sure to flip some cards over and back and forth so you don't know which side is "up" or which card is on top. Now select one card at random and place it flat on the table — be careful not to look at the bottom of this card. You will see either a black or white card. Record the color in Column A. This is not the probability with which we are concerned. Rather, we are interested in predicting the probability of the *other* side being black or white. Record the color of the second side in Column B. Repeat the experiment 100 times.

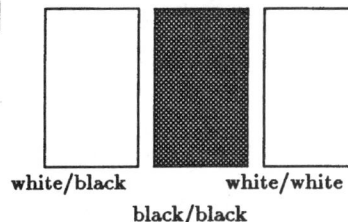

white/black   white/white

black/black

Top	**COLUMN A**	OUTCOME Bottom	**COLUMN B**	**COLUMN C**
WHITE		WHITE		
		BLACK		
BLACK		WHITE		
		BLACK		

The total number of tally marks in Column *A* should be 100. Find the percentage of occurrence in Column *C*. To find the percentage of white/white, divide your entry in Column *B* (white) by the entry in Column *A* (white). To find the percentage of white/black, divide your entry in Column C (black) by the entry in Column A (white). These two percentages should add up to 100%. Now, do the same to find the percentage of black/white and black/black. Are these the results you expected? Why or why not?

Transparencies 25 and 26 can be used throughout this chapter whenever using dice and cards. Do not assume that all of your students know these sample spaces.

Level 1, page 560

1. Empirical probability is done by experimentation and gives the percent of occurrence, whereas a theoretical probability is defined as a ratio found without performing an experiment.

2. If an experiment can result in any of n mutually exclusive and equally likely outcomes, and if s of these outcomes are considered favorable, then the **probability** of an event E, denoted by $P(E)$, is
$$P(E) = \frac{s}{n} = \frac{\text{NUMBER OF OUTCOMES FAVORABLE TO } E}{\text{NUMBER OF ALL POSSIBLE OUTCOMES}}$$

3. Answers vary; since E is a subset of S the number s must be between 0 and n

$$\frac{0}{n} \leq \frac{s}{n} \leq \frac{n}{n}$$

$$0 \leq \frac{s}{n} \leq 1$$

$$0 \leq P(E) \leq 1 \qquad \text{Since } P(E) = \frac{s}{n}.$$

4. a. $\frac{1}{3}$ b. $\frac{1}{3}$ c. $\frac{1}{3}$ 5. a. $\frac{1}{4}$ b. $\frac{1}{4}$ c. $\frac{1}{2}$ 6. a. $\frac{1}{6}$ b. $\frac{1}{6}$ c. $\frac{1}{6}$ 7. a. $\frac{1}{2}$ b. $\frac{1}{3}$ c. $\frac{2}{3}$

8. about 0.02   9. about 0.05   10. about 0.21   11. 0.19   12. a. $\frac{5}{18}$ b. $\frac{4}{18} = \frac{2}{9}$

c. $\frac{9}{18} = \frac{1}{2}$ 13. $P(\text{royal flush}) \approx 0.000001539077169$ 14. $P(\text{straight flush}) \approx 0.00001385169452$

15. $P(\text{four of a kind}) \approx 0.0002400960384$ 16. $P(\text{full house}) \approx 0.00144057623$

17. $P(\text{flush}) \approx 0.00196540155$ 18. $P(\text{straight}) \approx 0.003924646782$

19. $P(\text{three of a kind}) \approx 0.02112845138$ 20. $P(\text{two pair}) \approx 0.04753901561$

21. $P(\text{one pair}) \approx 0.42256902761$ 22. $P(\text{no pair or better}) \approx 0.501177394$

Level 2, page 561

23. a. $P(\text{five of clubs}) = \frac{1}{52}$ b. $P(\text{five}) = \frac{1}{13}$ c. $P(\text{club}) = \frac{1}{4}$

24. a. $P(\text{jack}) = \frac{1}{13}$ b. $P(\text{spade}) = \frac{1}{4}$ c. $P(\text{jack of spades}) = \frac{1}{52}$

25. a. $P(\text{five and a jack}) = 0$ b. $P(\text{five or a jack}) = \frac{2}{13}$

26. a. $P(\text{heart and a jack}) = \frac{1}{52}$ b. $P(\text{heart or a jack}) = \frac{4}{13}$

27. a. $\frac{1}{12}$ b. $\frac{7}{12}$ c. $\frac{1}{12}$ 28. a. $\frac{1}{2}$ b. $\frac{7}{12}$ c. $\frac{2}{3}$ 29. a. $\frac{1}{4}$ b. $\frac{3}{4}$ 30. a. $\frac{1}{12}$ b. $\frac{7}{12}$

31. $P(\text{five}) = \frac{1}{9}$ 32. $P(\text{six}) = \frac{5}{36}$ 33. $P(\text{seven}) = \frac{1}{6}$ 34. $P(\text{eight}) = \frac{5}{36}$ 35. $P(\text{nine}) = \frac{1}{9}$

36. $P(\text{two}) = \frac{1}{36}$ 37. $P(\text{four } or \text{ five}) = \frac{7}{36}$ 38. $P(\text{even number}) = \frac{1}{2}$ 39. $P(\text{eight } or \text{ ten}) = \frac{2}{9}$

40. Player A can spin a 5 or 3 and B can spin a 6 or 1.

$\backslash B$	6	1
A		
5	B wins	A wins
3	B wins	A wins

$P(A) = P(B) = \frac{1}{2}$; it is a fair game so pick either.

41. Player A can spin a 5 or 3 and C can spin 2 or 4; look at the sample space:

$A\backslash C$	4	2
5	A wins	A wins
3	C wins	A wins

Pick A: $P(A \text{ winning}) = \frac{3}{4}$

42. Player B can spin a 1 or 6 and C can spin a 2 or 4.

$B\backslash C$	4	2
6	B wins	B wins
1	C wins	C wins

$P(B) = P(C) = \frac{1}{2}$; pick either.

43. Player D can spin 1, 7 or 8 and E can spin a 4, 5, or 6.

$D\backslash E$	4	5	6
1	E	E	E
7	D	D	D
8	D	D	D

$P(D) = \frac{2}{3}$; $P(E) = \frac{1}{3}$; so pick D.

44.   Player E can spin a 4, 5 or 6 and F can spin a 2, 3, or 9.

E＼F	2	3	9
4	E	E	F
5	E	E	F
6	E	E	F

$P(E) = \frac{2}{3}$; $P(F) = \frac{1}{3}$, so pick E.

45.   Player D can spin a 1, 7 or 8 and F can spin a 2, 3, or 9.

D＼F	2	3	9
1	F	F	F
7	D	D	F
8	D	D	F

$P(F) = \frac{5}{9}$; $P(D) = \frac{4}{9}$, so pick F.

46.   Player B can spin a 1 or 6, and D can spin a 1, 7, or 8. If both players spin a 1, play again.

B＼D	1	7	8
1	Tie	D	D
6	B	D	D

$P(B) = \frac{1}{5}$; $P(D) = \frac{4}{5}$; pick D.

47.   Player C can spin a 2 or 4, and F can spin a 2, 3, or 9. If both players spin a 2, play again.

C＼F	2	3	9
2	Tie	F	F
4	C	C	F

$P(C) = \frac{2}{5}$; $P(F) = \frac{3}{5}$; pick F.

Level 3, page 562

48. Answers vary; $P(H) = \frac{1}{2}$ **49.** Answers vary; the answers here are the theoretical probabilities, but the problem requests empirical probabilities.   **a.** $P(\text{two}) = 0.0278$ **b.** $P(\text{three}) = 0.0556$

c. $P(\text{four}) = 0.0833$ **d.** $P(\text{five}) = 0.1111$ **e.** $P(\text{six}) = 0.1389$ **f.** $P(\text{seven}) = 0.1667$

g. $P(\text{eight}) = 0.1389$ **h.** $P(\text{nine}) = 0.1111$ **i.** $P(\text{ten}) = 0.0833$ **j.** $P(\text{eleven}) = 0.0556$

k. $P(\text{twelve}) = 0.0278$ **50.** Answers vary; results are not equally likely. **a.** $P(\text{three heads}) = 0.125$

b. $P(\text{two heads and one tail}) = 0.375$ **c.** $P(\text{two tails and one head}) = 0.375$ **d.** $P(\text{three tails}) = 0.125$

51. yes   **52.** Assume that a year is 365.25 days or 8,766 hr and it is given that you are at work 1,920 hours per year. This means that $P(\text{earthquake}) = \dfrac{1,920}{8,766} \approx 22\%$ **53.** $P(\text{earthquake}) = \dfrac{5(174)}{8,766} \approx 10\%$

54. Answers vary; here are the theoretical probabilities: Frequency of white or black, $\frac{1}{2}$ for each;

$P(\text{white on the bottom, given that there is a white on top}) = \frac{2}{3}$

$P(\text{black on the bottom, given that there is a white on top}) = \frac{1}{3}$

$P(\text{white on the bottom, given that there is a black on top}) = \frac{1}{3}$

$P(\text{black on the bottom, given that there is a black on top}) = \frac{2}{3}$

55. a. $\frac{8}{36} = \frac{2}{9}$ **b.** $\frac{4}{36} = \frac{1}{9}$ **56.** $\frac{1}{36}$

57. a.

	1	2	3	4
1	(1, 1)	(1, 2)	(1, 3)	(1, 4)
2	(2, 1)	(2, 2)	(2, 3)	(2, 4)
3	(3, 1)	(3, 2)	(3, 3)	(3, 4)
4	(4, 1)	(4, 2)	(4, 3)	(4, 4)

b. $P(\text{two}) = \frac{1}{16}$ **c.** $P(\text{three}) = \frac{1}{8}$ **d.** $P(\text{four}) = \frac{3}{16}$ **e.** $P(\text{five}) = \frac{1}{4}$ **f.** $P(\text{six}) = \frac{3}{16}$

g. $P(\text{seven}) = \frac{1}{8}$

58.

	1	2	3	4	5	6	7	8
1	(1, 1)	(1, 2)	(1, 3)	(1, 4)	(1, 5)	(1, 6)	(1, 7)	(1, 8)
2	(2, 1)	(2, 2)	(2, 3)	(2, 4)	(2, 5)	(2, 6)	(2, 7)	(2, 8)
3	(3, 1)	(3, 2)	(3, 3)	(3, 4)	(3, 5)	(3, 6)	(3, 7)	(3, 8)
4	(4, 1)	(4, 2)	(4, 3)	(4, 4)	(4, 5)	(4, 6)	(4, 7)	(4, 8)
5	(5, 1)	(5, 2)	(5, 3)	(5, 4)	(5, 5)	(5, 6)	(5, 7)	(5, 8)
6	(6, 1)	(6, 2)	(6, 3)	(6, 4)	(6, 5)	(6, 6)	(6, 7)	(6, 8)
7	(7, 1)	(7, 2)	(7, 3)	(7, 4)	(7, 5)	(7, 6)	(7, 7)	(7, 8)
8	(8, 1)	(8, 2)	(8, 3)	(8, 4)	(8, 5)	(8, 6)	(8, 7)	(8, 8)

Level 3 Problem Solving, page 563

59. There are three possibilities. The jar has either two brains (B_1 and B_2) or a brain and a kidney (B and K).

Outside jar:	B_1	B_2	B
Inside jar:	B_2	B_1	K

In 2 out of 3 of these possibilities there is a brain in the jar, so $P(\text{brain}) = \frac{2}{3}$.

60. Answers vary.

11.2 Mathematical Expectation, page 563

The expectation of various contests and games are computed in this section, which leads us to a definition of mathematical expectation. In the example given in the text we assume that we draw with replacement (which is probably unrealistic for an actual contest). It might be pointed out what the arithmetic would look like if it were drawn without replacement, and at the same time, it might be pointed out that, for a large sample space, the difference in the final results between performing the experiment with or without replacement is *very* small (it is not necessary to justify this result formally; it is intuitively appealing to the students). That is, point out that, for large sample spaces, we can perform the experiment with or without replacement, whichever is most convenient, and the difference in the final result is negligible.

St. Petersburg's Paradox (see Problems 53-56) is a good way to motivate this material. Let's consider a particular example. Flip a coin continuously until a head appears for the first time. If a head does not appear by the 20th flip, you will be paid a billion dollars. Would you pay $100 to play this game? Most people would not accept this game. However, suppose we calculate the mathematical expectation:

$$E = \text{AMOUNT TO WIN(PROB OF WINNING)} + \text{AMOUNT TO LOSE (PROB OF LOSING)}$$

$$= \$1,000,000,000\left(\frac{1}{524,288}\right) + (-\$100)\left(\frac{524,287}{524,288}\right)$$

$$\approx \$1,807.35$$

You should be willing to pay up to the amount of mathematical expectation to play this game.

Here is another game which I found in John Paulos' book *Innumeracy*. Which would you choose: a sure $30,000 or an 80% chance of winning $40,000 and a 20% chance of winning nothing? Most of your students will choose the $30,000, but consider the mathematical expectation:

$$E = \text{AMOUNT TO WIN(PROB OF WINNING)} + \text{AMOUNT TO LOSE (PROB OF LOSING)}$$

$$= \$40,000(0.8) + \$0(0.2)$$

$$= \$32,000$$

The expectation of this choice is better than that of selecting the certain $30,000.

Now ask your class the exact opposite: Which would you choose, a certain loss of $30,000 or an 80% chance of losing $40,000 and a 20% chance of losing nothing? It seems that people tend to avoid risk when seeking gains, but choose risk to avoid loss.

Level 1, page 568

1. False; there are also green spots. $P(\text{black}) = \frac{18}{38} = \frac{9}{19}$.

2. False; if you play the game a number of times, the *average* winnings per game should be $5.

3. True; play with positive expectation, do not play with negative expectation; zero expectation means that it is a fair game.

4. True; in the book we answer by assuming that we wish to maximize winning.

5. False; there is a cost of playing: $E = \$50,000\left(\frac{1}{1,000,000}\right) + -\$1 = -\$0.95$.

6. A   7. B   8. C   9. B   10. C   11. B

12. You should be willing to pay any amount up to the mathematical expectation; $E = \frac{1}{6}(5) \approx \$0.83$

13. $E = \frac{1}{1,000,000}(80,000) = \$0.08$

14. $E = \$1(\frac{1}{5}) + \$5(\frac{1}{5}) + \$10(\frac{1}{5}) + \$20(\frac{1}{5}) + \$100(\frac{1}{5}) = \$27.20$

15. $E = \$1(\frac{1}{5}) + \$5(\frac{1}{5}) + \$10(\frac{1}{5}) + \$20(\frac{1}{5}) + \$100(\frac{1}{5}) - \$20 = \$7.20$

16. $E = \frac{5}{25}(0.25) + \frac{5}{25}(0.10) + \frac{10}{25}(0.05) + \frac{5}{25}(0.01) = 0.092$; The barber's expectation is about \$0.09,

    but he is most likely to receive a nickel.

17. $E = \frac{1}{4}(0.50) = 0.125$; A fair price would be to pay \$0.25 for two plays of the game.

18. $E = 70,000(\frac{1}{10,000,000}) + 7,000(\frac{7}{10,000,000}) + 700(\frac{77}{10,000,000}) + 70(\frac{777}{10,000,000}) = 0.022729$;

    About \$0.02

19. $E = 100(\frac{1}{100}) + 10(\frac{5}{100}) = \$1.50$

Level 2, page 569

20. $E = \$500(0.001) - \$1 = -\$0.50$

21. $E = \frac{1}{20}(10,000) + \frac{2}{20}(5,000) + \frac{1}{20}(1,000) = 1,050$.

    Since this is more than \$1,000, the contestant should play.

22. $E = 0.4(0.05 \cdot 350,000) + 0.3(0.025 \cdot 350,000) + 0.30(0) - 2,400 = 7,225$

23. $E = 0.5(0.06 \cdot 185,000) + 0.3(0.03 \cdot 185,000) + 0.20(0) - 800 = 6,415$

24. $E = 425,000(\frac{1}{40}) + 125,000(\frac{1}{20}) - 25,000 = -8,125$

    They should not dig since the expectation is negative.

25. $E = 825,000(\frac{1}{40}) + 225,000(\frac{1}{20}) - 25,000 = 6,875$;

    They should dig the well because the expectation is positive.

26. $E = 2(\frac{3}{6}) + 1(\frac{2}{6}) + (-11)(\frac{1}{6}) = -\frac{1}{2}$; the expectation is a loss of \$0.50 per play. It is not a good

    deal.

27. $E = 0.50(\frac{4}{52}) + 0.25(\frac{12}{52}) + 1.00(\frac{1}{52}) - 0.10 \approx 0.02$; yes, you should play the game.

28. $E = 0.000005(10,000) + 0.00002(1,000) + 0.0002(100) + 0.010886(10) = 0.19886$; the expectation

    is about \$0.20.

29. $E = 25\left(\frac{1}{1,635}\right) + 3\left(\frac{1}{163}\right) + 1\left(\frac{1}{68}\right) = 0.0484013102$ the expectation is about \$0.05.

30. $E = 55(0.001) + 60(0.022) + 65(0.136) + 70(0.341) + 75(0.341) + 80(0.136) + 85(0.022)$

    $+ 90(0.001) = 72.5$

    The expected height is 72.5 in.

31. $E = 0(0.15) + 1(0.25) + 2(0.31) + 3(0.21) + 4(0.08) = 1.82$

The expected number of tardies is 1.82.

32. $E = 5,000,000\left(\frac{1}{197,000,000}\right) + 100,000\left(\frac{1}{197,000,000}\right) + 50,000\left(\frac{2}{197,000,000}\right) + 20,000\left(\frac{3}{197,000,000}\right)$

$+ 5,000\left(\frac{4}{197,000,000}\right) + 500\left(\frac{10}{197,000,000}\right) + 100\left(\frac{400}{197,000,000}\right) + 89\left(\frac{58,147}{197,000,000}\right) = \frac{10,500,083}{197,000,000}$

≈ 0.0532999137; the expectation is about $0.05.

33. $E = \$1,000(0.12) + \$800(0.38) + \$500(0.45) + \$200(0.05) = \$659$

If the cost is $500, then the expected profit is $159. This is not at least the needed $200, so they should not purchase them.

34. $E = 1\left(\frac{18}{38}\right) + (-1)\left(\frac{20}{38}\right) \approx -0.05$ **35.** $E = 1\left(\frac{18}{38}\right) + (-1)\left(\frac{20}{38}\right) \approx -0.05$

36. $E = 35\left(\frac{1}{38}\right) + (-1)\left(\frac{37}{38}\right) \approx -0.05$ **37.** $E = 17\left(\frac{2}{38}\right) + (-1)\left(\frac{36}{38}\right) \approx -0.05$

38. $E = 11\left(\frac{3}{38}\right) + (-1)\left(\frac{35}{38}\right) \approx -0.05$ **39.** $E = 8\left(\frac{4}{38}\right) + (-1)\left(\frac{34}{38}\right) \approx -0.05$

40. $E = 6\left(\frac{5}{38}\right) + (-1)\left(\frac{33}{38}\right) \approx -0.08$ **41.** $E = 5\left(\frac{6}{38}\right) + (-1)\left(\frac{32}{38}\right) \approx -0.05$

42. $E = 2\left(\frac{12}{38}\right) + (-1)\left(\frac{26}{38}\right) \approx -0.05$ **43.** $E = 2\left(\frac{12}{38}\right) + (-1)\left(\frac{26}{38}\right) \approx -0.05$

Level 3, page 571

44. $E = 8\left(\frac{1}{2}\right) + 2\left(\frac{1}{2}\right) - 5 = 0$; fair game

45. $E = 5\left(\frac{1}{2}\right) + 0\left(\frac{1}{4}\right) + 10\left(\frac{1}{4}\right) - 5 = 0$; fair game

46. $E = 2\left(\frac{1}{2}\right) + 8\left(\frac{2}{3} \cdot \frac{1}{2}\right) + 5\left(\frac{1}{3} \cdot \frac{1}{2}\right) - 5 = -0.5$; not a fair game; don't play

47. $E = 14\left(\frac{1}{8}\right) + 8\left(\frac{1}{8}\right) + 4\left(\frac{1}{4}\right) + 2\left(\frac{1}{2}\right) - 5 = -\frac{1}{4}$; not a fair game; don't play

48. $E = \frac{1}{2}(1) + \frac{1}{4}(6) + \frac{1}{8}(10) + \frac{1}{16}(8) + \frac{1}{16}(4) = 4$; pay up to $4.00 to play this game

49. $E = \frac{1}{2}(-5) + \frac{1}{4}(16) + \frac{1}{8}(8) + \frac{1}{16}(4) + \frac{1}{32}(2) + \frac{1}{32}(1) \approx 2.84$

You should be willing to pay $2.84 to play the game.

Level 3 Problem Solving, page 571

50. $E = \$150,000\left(\frac{1}{4}\right) + \$200,000\left(\frac{1}{4}\right) + \$1,000,000\left(\frac{1}{4}\right) + \$0 = \$337,500$

No, the player should not stop for $100,000 since the mathematical expectation is $337,500.

51. Since the contestant has already won $350,000, this is the amount at risk.

$E = \$1,000,000\left(\frac{1}{2}\right) + \$0 = \$500,000$. Since this is more than the amount at risk, the player should continue.

52. Answers vary; this problem is mathematically equivalent to Problem 5. In this book, we are answering by maximizing the gain, whereas in real life we would often want to minimize our loses. Very few reasonable people would risk the security of their home on the chance of winning a million dollars even though the mathematical expectation directs us to do so:

$$E = \tfrac{1}{2}(\$1,000,000) - \tfrac{1}{2}(\$350,000) = \$325,000$$

53. $E = \$1(\tfrac{1}{2}) + \$2(\tfrac{1}{4}) + \$4(\tfrac{1}{8}) = \$1.50$ **54.** $E = \$1(\tfrac{1}{2}) + \$2(\tfrac{1}{4}) + \$4(\tfrac{1}{8}) + \cdots + \$2^9(\tfrac{1}{2^{10}}) = \$5.00$

55. $E = \$1(\tfrac{1}{2}) + \$2(\tfrac{1}{4}) + \$4(\tfrac{1}{8}) + \cdots + \$2^{999}(\tfrac{1}{2^{1,000}}) = \$500.00$

56. $E = \$1(\tfrac{1}{2}) + \$2(\tfrac{1}{4}) + \$4(\tfrac{1}{8}) + \$8(\tfrac{1}{16}) + \cdots = \$0.50 + \$0.50 + \$0.50 + \cdots$

Pay *any* amount to play the game since the sum is infinite. This result is unacceptable to many mathematicians and is consequently called a paradox. See Martin Gardner's *Mathematical Puzzles and Diversions* (New York: Simon and Schuster, 1959, p. 145) for a discussion of the problem.

57. answers vary   **58.** answers vary   **59.** answers vary   **60.** answers vary

11.3 Probability Models, page 572

This section presents conditional probability. The students are expected to compute conditional probability by considering the altered sample space rather than by using Bayes' theorem. However, for a more capable class, you could go on and *define* independence as follows: *Two events E and F are independent if and only if $P(E|F) = P(E)$.* That is to say, *E* and *F* are independent if the occurrence of event *F* has no effect on the probability of event *E*.

Here is a little quiz you can give your students to motivate the material in this section (Problem 6, Section 11.3):

1. Of these three means of travel, which is the safest?

    A.   car
    B.   train
    C.   plane

Answer: B; The odds of dying when traveling by train are 1 to 6.6 million; by plane, 1 to 3.3 million, and by car 1 to 215,000.

2. Which of the following is the most probable (Problem 7, Section 11.3)?
    A.   Winning the grand prize in a state lottery
    B.   Being struck by lightning
    C.   Appearing on the *Tonight Show*

Answer: C; The odds of appearing on the *Tonight Show* are 1 to 490,000; being struck by lightning 1 to 600,000, and winning the grand prize about 1 to 5,000,000.
Here is an interesting example that I bring to class (it is based on Problem 54, Section 11.1).

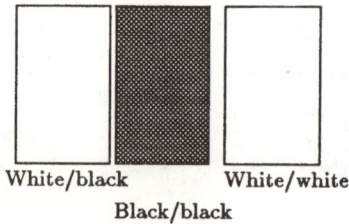

Suppose your friend George shows you three cards. One is white on both sides, one is black on both sides, and the last is black on one side and white on the other. He mixes the cards and lets you select one at random and place it on the table. George then calls out the color on the underside. If he selects the right color, you lose \$1; if he does not select the correct color on the underside, you win \$1. Should you play?

White/black   White/white
       Black/black

I use Polya's problem-solving guidelines to solve this example.

Understand the Problem. You might experiment with three cards to "get a feel" for this game. First, decide whether there is an "entry fee to play." Since you and George are each putting up \$1, there is no entry fee. However, suppose that you are not sure you want to play, so George explains: "Suppose you select a card, and we see that it's black on top. We know it's not the white-white card, so it must be either the black-black card or the black-white card. This gives you a 50-50 chance of winning, so it's a fair game. Come on, let's play!"

Devise a Plan. We will look at the sample space and calculate the mathematical expectation. If the expectation is positive, then we'll play, and if it is negative we will not. If the expectation is 0, then it is a fair game and we can play "for fun."

Carry Out the Plan. Before you agree to play, consider the sample space. Let's start by distinguishing between the front (side 1) and the back (side 2) of each card. The sample space of *equally likely* events is as follows:

Result	Card 1		Card 2		Card 3	
Side showing (side 1)	B_1	B_2	B	W	W_1	W_2
Side not showing (side 2)	B_2	B_1	W	B	W_2	W_1

Let's also assume that, after we select the card we see that a black side is face up. (If it is a white side that we see, we can repeat the same argument, with colors reversed.) We also see that

$$P(\text{black is face down}) = \frac{3}{6} = \frac{1}{2} \quad \textit{(This is George's incorrect argument.)}$$

But we have additional information. We wish to compute the *conditional probability* of black face down, given that a black card is face up. Alter the sample space to take into account the additional information:

Result	Card 1		Card 2		Card 3	
Side showing (side 1)	B_1	B_2	B	W	W_1	W_2
Side not showing (side 2)	B_2	B_1	W	~~B~~	~~W_2~~	~~W_1~~

                                                    ↑   ↑   ↑
                                          Cross this out (black on top)

$$P(\text{black is face down} \mid \text{black on top}) = \frac{2}{3}$$

This means that, if George picks the color on the bottom to match the color on top he will have a probability of winning of 2/3, so the probability that we will win is 1/3. We now calculate the expectation:

$$\text{EXPECTATION} = (\$1)\left(\tfrac{1}{3}\right) + (-\$1)\left(\tfrac{2}{3}\right) \approx -\$.33$$

Look Back. Since the expectation is negative, you should not play.

Here is a related problem you can distribute to class (Transparency 27):

Suppose that your friend George shows you three cards. One card is white on both sides, one is black on both sides, and the last is black on one side and white on the other. He mixes the cards and tells you to select one at random and place it on the table. Suppose the upper side turns out to be black. It is not the white-white card; it must be either the black-black or black-white card. "Thus," says George, "I'll bet you $1 that the other side is black." Would you play? Perhaps you hesitate. Now George says he feels generous. You need to pay him only 75¢ if you lose, and he will still pay you $1 if he loses. Would you play now?

The solution is No, $E = -\$.17$.

Level 1, page 580

1. $P(E) = 1 - P(\overline{E})$ 2. Answers vary; $P(E) = \frac{s}{n}$ whereas the odds in favor of E are $\frac{s}{f}$ where

$s + f = n$ 3. Answers vary; $P(E \mid F)$ is the probability of the event E, given that the event F has

occurred. That is, reevaluate E by looking at a sample space altered by considering the information

that F has occurred.   4. The *fundamental counting principle* gives the number of ways of performing

two or more tasks. If task A can be performed in m ways, and if, after task A is performed, a second

task B, can be performed in n ways, then task A followed by task B can be performed in $m \cdot n$ ways.

5. Since there are 9 possible squares in which to place the X (first move), the total number of

possibilities is $9 \cdot 8 \cdot 7 \cdot 6 \cdot 5 = 15,120$

Problems 6-11 are to be done by estimation. These are really exercises in subjective probabilities in the sense they are asking for an opinion of which is more probable. As directed by the directions, do NOT calculate these probabilities.

6. B; odds of dying in a car are 1 to 215,000, train 1 to 6.6 million, and a plane 1 to 3.3 million

7. C; odds of winning the lottery about 1 to 5,000,000, being struck by lightning about 1 to 600,000,

and of appearing on the *Tonight Show* about 1 to 490,000   8. B   9. B   10. They are the same

11. B; choice A the odds are about 1 to 10,000,000, and choice B the odds are about 1 to 2,000,000

12. $1 - 0.6 = 0.4$ 13. $1 - \frac{4}{5} = \frac{1}{5}$ 14. $1 - \frac{9}{13} = \frac{4}{13}$ 15. $1 - 0.005 = 0.995$

16. $P(\text{not a face card}) = 1 - P(\text{face card}) = 1 - \frac{12}{52} = \frac{10}{13}$

17. $P(\text{not a mult. of 5}) = 1 - P(\text{mult. of 5}) = 1 - \frac{1}{5} = \frac{4}{5}$

18. $P(\text{at least one head}) = 1 - P(\text{no head}) = 1 - \frac{1}{8} = \frac{7}{8}$

19. $P(\text{at least one head}) = 1 - P(\text{no head}) = 1 - \frac{1}{16} = \frac{15}{16}$ 20. $\frac{4}{48} = \frac{1}{12}$; odds of 1 to 12

21. $\frac{15}{1}$; odds of 15 to 1 **22.** $\frac{13}{39} = \frac{1}{3}$; odds of 1 to 3 **23.** odd in favor $= \frac{s}{f} = \frac{0.82}{0.18} = \frac{41}{9}$; odds in favor are about 5 to 1. **24.** $P(\text{event}) = \frac{1}{11}$ **25.** $P(\#1) = \frac{1}{3}$; $P(\#2) = \frac{1}{16}$; $P(\#3) = \frac{2}{5}$; $P(\#4) = \frac{5}{12}$; $P(\#5) = \frac{1}{2}$; Note: at a horse race odds are determined by track betting and the sum of the probabilities is not necessarily 1. **26.** $\frac{9}{1}$; $P(\text{bald}) = \frac{9}{9+1} = \frac{9}{10}$ **27.** $\frac{33}{1}$; $P(\text{lie}) = \frac{33}{33+1} = \frac{33}{34}$

28. **a.** BBBB; BBBG; BBGB; BBGG; BGBB; BGBG; BGGB; BGGG; GBBB; GBBG; GBGB; GBGG; GGBB; GGBG; GGGB; GGGG

b. $P(4 \text{ girls}) = P(4 \text{ boys}) = \frac{1}{16}$

c. $P(1 \text{ girl and } 3 \text{ boys}) = P(3 \text{ girls and } 1 \text{ boy}) = \frac{1}{4}$

d. $P(2 \text{ boys and } 2 \text{ girls}) = \frac{6}{16} = \frac{3}{8}$

e. 1

29. Reduced sample space: HHHH; HHHT; HHTH; HHTT; HTHH; HTHT; HTTH; ~~HTTT~~; THHH; THHT; THTH; ~~THTT~~; TTHH; ~~TTHT~~; ~~TTTH~~; ~~TTTT~~;

$P(\text{exactly 3H}|\text{at least 2H}) = \frac{4}{11}$

Level 2, page 581

30. **a.** $P(13 \text{ boys}) = \left(\frac{1}{2}\right)^{13} = 0.000122$ or about 1 in 10,000 **b.** $P(\text{next child a boy}) = \frac{1}{2}$

31. 1 **32.** $\frac{3}{12} = \frac{1}{3}$ **33.** $\frac{13}{39} = \frac{1}{3}$ **34.** $\frac{4}{40} = \frac{1}{10}$ **35.** $\frac{2}{4} = \frac{1}{2}$ **36.** $\frac{2}{26} = \frac{1}{13}$ **37.** $\frac{4}{51}$ **38.** $\frac{3}{51} = \frac{1}{17}$

39. $\frac{12}{51} = \frac{4}{17}$ **40.** $\frac{13}{51}$ **41.** $\frac{26}{51}$ **42.** $\frac{25}{51}$

43.
$$P(R) = 1 - P(\overline{R})$$
$$= 1 - \frac{26 \times 25 \times 10 \times 9 \times 8 \times 7}{26^2 \times 10^4}$$
$$\approx 0.5153$$

About 51.5% of the plates have repetitions.

44.
$$P(R) = 1 - P(\overline{R})$$
$$= 1 - \frac{10 \times 26 \times 25 \times 24 \times 9 \times 8 \times 7}{10 \times 26 \times 26 \times 26 \times 10 \times 10 \times 10}$$
$$\approx 0.5526$$

About 55.3% of the plates have repetitions.

45.
$$P(R) = 1 - P(\overline{R})$$
$$= 1 - \frac{26 \times 25 \times 24 \times 10 \times 9 \times 8}{26^3 \times 10^3}$$
$$\approx 0.3609$$

About 36.1% of the plates have repetitions.

46. $P(\text{eight girls in a row}) = \frac{1}{2^8} = \frac{1}{256} \approx 0.004$

We see $s = 1$, $n = 256$, so $f = n - s = 255$. The odds against are 255 to 1.

47. **a.** $P(N) = \frac{25}{80} = 0.3125$

    b. $P(S) = \frac{54}{80} = 0.675$

    c. $P(C \mid S) = \frac{21}{54} \approx 0.389$

    d. $P(S \mid C) = \frac{21}{30} = 0.70$

48. **a.** $P(6 \mid \text{odd}) = 0$ **b.** $P(5 \mid \text{odd}) = \frac{1}{3}$

Level 3, page 582

49. **a.** $P(\text{odd} \mid 6) = 0$ **b.** $P(\text{odd} \mid 5) = 1$

50. Look at the sample space (Figure 11.3 in the text).

    a. $P(7 \mid \text{odd}) = \frac{6}{18} = \frac{1}{3}$ **b.** $P(\text{odd} \mid 7) = 1$ **c.** $P(7 \mid \text{one die is a } 2) = \frac{2}{11}$

51. **a.** $P(5 \mid \text{one die is a } 2) = \frac{2}{10} = \frac{1}{5}$ **b.** $P(3 \mid \text{one die is a } 2) = \frac{2}{10} = \frac{1}{5}$ **c.** $P(2 \mid \text{one die is a } 2) = 0$

52. **a.** $P(8 \mid \text{double}) = \frac{1}{6}$ **b.** $P(\text{double} \mid 8) = \frac{1}{5}$

53. $\dfrac{P(\overline{E})}{P(E)} = \dfrac{\frac{f}{n}}{\frac{s}{n}} = \dfrac{f}{n} \cdot \dfrac{n}{s} = \dfrac{f}{s} = \text{odds against}$

54. **a.** $P(C_1) = \frac{1}{4}$ **b.** $P(\overline{C_1}) = \frac{3}{4}$ **c.** $P(C_2 \mid C_1) = \frac{12}{51}$ **d.** $P(\overline{C_2} \mid C_1) = \frac{39}{51}$ **e.** $P(C_2 \mid \overline{C_1}) = \frac{13}{51}$

    f. $P(\overline{C_2} \mid \overline{C_1}) = \frac{38}{51}$ **g.** $P(C_2) = \frac{13}{52} \cdot \frac{12}{51} + \frac{39}{52} \cdot \frac{13}{51} = \frac{1}{4}$

    h.

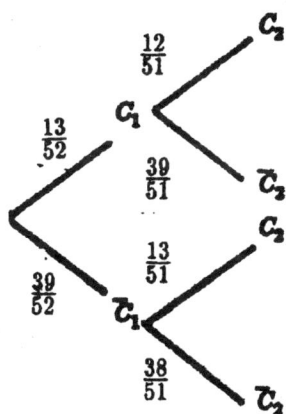

55. **a.** $P(L_1) = \frac{10}{35} = \frac{2}{7}$ **b.** $P(\overline{L}_1) = \frac{25}{35} = \frac{5}{7}$ **c.** $P(L_2 \mid L_1) = \frac{9}{34}$ **d.** $P(\overline{L}_2 \mid L_1) = \frac{25}{34}$

    e. $P(L_2 \mid \overline{L}_1) = \frac{10}{34} = \frac{5}{17}$ **f.** $P(\overline{L}_2 \mid \overline{L}_1) = \frac{24}{34} = \frac{12}{17}$ **g.** $P(L_2) = \frac{10}{35} \cdot \frac{9}{34} + \frac{25}{35} \cdot \frac{10}{34} = \frac{2}{7}$

h.

First selection   *Second selection*

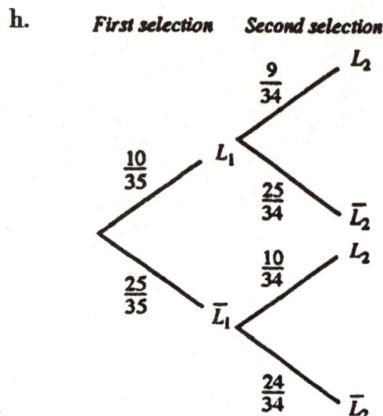

56. Answers vary. A jail cell measuring 8 ft $\times$ 10 ft $\times$ 8 ft = 640 ft^3 = 1,105,920. If a ping pong ball takes up 1 in.^{3}, then the odds against winning is about the same as reaching into cell filled with ping pong balls and drawing out a red ping pong ball if the room contains only one red ping pong ball painted red with all the others being the usual white ping pong ball.

57. Answers vary. A typical classroom is 20 ft $\times$ 30 ft $\times$ 10 ft = 10,368,000 in.^{3} If a ping pong ball takes up 1 in.^{3}, then, the odds against winning is about the same as reaching into a classroom filled with ping pong balls and drawing out a red ball if the room contains only one ping pong ball painted red.

58. Answers vary. See the answer to Problem 57 for ideas.

Level 3 Problem Solving, page 583

59. Answers vary; your limited resources and the betting limit imposed on the game.

60. Pat is correct.

11.4 Calculated Probabilities, page 583

After you have introduced the idea of independence, you might wish to use the following example from John Paulos' book *Innumeracy*.

> A more contemporary instance of the same sort of calculation involves the likelihood of acquiring AIDS heterosexually. It's estimated that the chance of contracting AIDS in a single unprotected heterosexual episode from a partner known to have the disease is about one in five hundred (the average of the figures from a number of studies). Thus, the probability of not getting it from a single such encounter is 499/500. If these risks are independent, as many assume them to be, then the chances

of not falling victim after two such encounters is $(499/500)^2$, and after N such encounters $(499/500)^N$. Since $(499/500)^{346}$ is $\frac{1}{2}$, one runs about a 50 percent chance of not contracting AIDS by having unsafe heterosexual intercourse every day for a year with someone who has the disease (and thus, equivalently, a 50 percent chance of contracting it).

With a condom the risk of being infected from a single unsafe heterosexual episode with someone known to have the disease falls to one in five thousand, and safe sex every day for ten years with such a person (assuming the victim's survival) would lead to a 50 percent chance of getting the disease yourself. If your partner's disease status is not known, but he or she is not a member of any known risk group, the chance per episode of contracting the infection is one in five million unprotected, one in fifty million with a condom. You're more likely to die in a car crash on the way home from such a tryst.

Another example (and perhaps less controversial) from the same book involves predictive dreams:

Assume the probability to be one out of 10,000 that a particular dream matches in a few vivid details some sequence of events in real life. This is a pretty unlikely occurrence, and means that the chances of a nonpredictive dream are an overwhelming 9,999 out of 10,000. Also assume that whether or not a dream matches experience one day is independent of whether or not some other dream matches experience some other day. Thus, the probability of having two successive nonmatching dreams is, by the multiplication principle for probability, the product of 9,999/10,000 and 9,999/10,000. Likewise, the probability of having N straight nights of nonmatching dreams is $(9,999/10,000)^N$; for a year's worth of nonmatching or nonpredictive dreams, the probability is $(9,999/10,000)^{365}$.

Since $(9,999/10,000)^{365}$ is about 0.964, we can conclude that about 96.4 percent of the people who dream every night will have only nonmatching dreams during a one-year span. But that means that about 3.6 percent of the people who dream every night will have a predictive dream. 3.6 percent is not such a small fraction; it translates into millions of apparently precognitive dreams every year. Even if we change the probability to one in a million for such a predictive dream, we'll still get huge numbers of them by chance alone in a country the size of the United States. There's no need to invoke any special parapsychological abilities; the ordinariness of apparently predictive dreams does not need any explaining. What would need explaining would be the nonoccurrence of such dreams.

Level 1, page 592

1. Events E and F are *independent* if the occurrence of one of these in no way influences the probability of the other. In terms of conditional probabilities, events E and F are independent if $P(E \mid F) = P(E)$.   **2.** If E and F are independent, then $P(E \cap F) = P(E \text{ and } F) = P(E) \cdot P(F)$ **3.** $P(E \cup F) = P(E \text{ or } F) = P(E) + P(F) - P(E \cap F)$ **4.** The *birthday problem* is the probability that there is at least one birthday (not year) match in a group of people.   **5.** Answers vary.   **6.** Answers vary; the betting system called "double on each loss" has the disadvantage of possibly betting a large amount of money to win $1. One the other hand, the betting system called "double on each win" risks a small amount of money ($1) to win a larger amount of money ($16), but is also not a good betting scheme because it carries a negative expectation. You win only if black occurs five times in a row. The probability of this occurring is $\left(\frac{18}{38}\right)^5 \approx 0.02$, or about 2% of the time.

7. $1 - \frac{1}{2} = \frac{1}{2}$ **8.** $1 - \frac{1}{3} = \frac{2}{3}$ **9.** $1 - \frac{1}{6} = \frac{5}{6}$

10. $\frac{1}{2} \cdot \frac{1}{3} = \frac{1}{6}$ **11.** $\frac{1}{2} \cdot \frac{1}{6} = \frac{1}{12}$ **12.** $\frac{1}{3} \cdot \frac{1}{6} = \frac{1}{18}$

13. $\frac{1}{2} + \frac{1}{3} - \frac{1}{2} \cdot \frac{1}{3} = \frac{2}{3}$ **14.** $\frac{1}{2} + \frac{1}{6} - \frac{1}{2} \cdot \frac{1}{6} = \frac{7}{12}$ **15.** $\frac{1}{3} + \frac{1}{6} - \frac{1}{3} \cdot \frac{1}{6} = \frac{4}{9}$

16. $1 - P(A \cap B) = 1 - \frac{1}{6} = \frac{5}{6}$ **17.** $1 - P(A \cap C) = 1 - \frac{1}{12} = \frac{11}{12}$

18. $1 - P(B \cap C) = 1 - \frac{1}{18} = \frac{17}{18}$ **19.** $1 - P(A \cup B) = 1 - \frac{2}{3} = \frac{1}{3}$

20. $1 - P(A \cup C) = 1 - \frac{7}{12} = \frac{5}{12}$ **21.** $1 - P(B \cup C) = 1 - \frac{4}{9} = \frac{5}{9}$

22. $\frac{1}{2} \cdot \frac{1}{3} \cdot \frac{1}{6} = \frac{1}{36}$ **23.** $1 - \frac{1}{36} = \frac{35}{36}$

24. $P[(A \cup B) \cap C] = P(A \cup B) \cdot P(C) = \frac{2}{3} \cdot \frac{1}{6} = \frac{1}{9}$

For Problems 25-36, note that $P(A) = \frac{1}{2}$, $P(B) = \frac{1}{6}$, $P(C) = \frac{1}{6}$, *and* $P(D) = \frac{1}{6}$.

25. no   **26.** yes   **27.** yes   **28.** yes

29. yes   **30.** no

31. not independent; $A \cap B = \{$first toss is a 3$\}$ so $P(A \cap B) = \frac{1}{6}$

32. independent; $P(A \cap C) = P(A) \cdot P(C) = \frac{1}{2} \cdot \frac{1}{6} = \frac{1}{12}$

33. $P(A \cup B) = P(A) + P(B) - P(A \cap B) = \frac{1}{2} + \frac{1}{6} - \frac{1}{6} = \frac{1}{2}$

34. $P(A \cup C) = P(A) + P(C) - P(A \cap C) = \frac{1}{2} + \frac{1}{6} - \frac{1}{12} = \frac{7}{12}$

35. $P(B \cup D) = P(B) + P(D) - P(B \cap D) = \frac{1}{6} + \frac{1}{6} - \frac{1}{36} = \frac{11}{36}$

36. $P(C \cup D) = P(C) + P(D) - P(C \cap D) = \frac{1}{6} + \frac{1}{6} - 0 = \frac{1}{3}$

Level 2, page 592

37. **a.** $P(3 \text{ bars}) = P(\text{bar}) \cdot P(\text{bar}) \cdot P(\text{bar}) = \left(\frac{1}{13}\right)^3 = \frac{1}{2,197} \approx 0.000455$

    b. $P(3 \text{ oranges}) = \frac{27}{2,197} \approx 0.01229$

    c. $P(3 \text{ plums}) = \frac{27}{2,197} \approx 0.01229$

    d. $P(1\text{st cherry}) = \frac{2}{13} \approx 0.1538$

    e. P(cherries on first two wheels) $= \left(\frac{2}{13}\right)^2 = \frac{4}{169} \approx 0.0237$

38. **a.** $P(3 \text{ bars}) = P(\text{bar}) \cdot P(\text{bar}) \cdot P(\text{bar}) = \frac{2}{20} \cdot \frac{4}{20} \cdot \frac{2}{20} = 0.002$

    b. $P(3 \text{ bells}) = \frac{1}{20} \cdot \frac{6}{20} \cdot \frac{1}{20} = 0.00075$

    c. $P(3 \text{ cherries}) = \frac{2}{20} \cdot \frac{5}{20} \cdot \frac{8}{20} = 0.01$

    d. $P(1\text{st cherry}) = \frac{2}{20} = 0.10$

e. $P(\text{cherries on first two wheels}) = \frac{2}{20} \cdot \frac{5}{20} = 0.025$

39. $E = 2(\frac{2}{20}) + 5(\frac{2}{20} \cdot \frac{5}{20}) + 10(\frac{2}{20} \cdot \frac{5}{20} \cdot \frac{2}{20}) + 10(\frac{2}{20} \cdot \frac{5}{20} \cdot \frac{8}{20}) + 14(\frac{6}{20} \cdot \frac{2}{20} \cdot \frac{3}{20})$

$\qquad + 14(\frac{3}{20} \cdot \frac{2}{20} \cdot \frac{6}{20}) + 20(\frac{1}{20} \cdot \frac{6}{20} \cdot \frac{1}{20}) + 50(\frac{2}{20} \cdot \frac{4}{20} \cdot \frac{2}{20}) - 1$

$\qquad = -0.309 \text{ coins}$

For playing 0.25 the mathematical expectation is

$\qquad \$0.25(-0.309) \approx -\$0.08$

40. $E = \$2{,}247\,P(\text{pick 1}) - \$749 = -\$187.25$

Another possible method of solution is to subtract the cost of paying as you go:

$E = (\$2{,}247 - \$749)P(\text{pick 1}) + (\$0 - \$749)P(\text{pick 0})$

$\qquad = \$1{,}498(\frac{1}{4}) + (-\$749)(\frac{3}{4}) = -\$187.25$

41. $E = 27{,}777\,P(\text{pick 6}) - 5 = 27{,}777\left(\frac{_{20}C_6}{_{80}C_6}\right) - 5 \approx \$1.42$

Here is another method:

$E = (27{,}777 - 5)P(\text{pick 6}) + (-5)P(\text{otherwise})$

$\qquad = 27{,}772\left(\frac{_{20}C_6}{_{80}C_6}\right) + (-5)\left(1 - \frac{_{20}C_6}{_{80}C_6}\right) \approx 27{,}772(1.289849391 \times 10^{-4}) - 5(0.99988710151)$

$\qquad \approx -\$1.42$

42. $E = \$12 \cdot P(\text{pick 2}) - \$1$

$\qquad = \$12\left(\frac{_{20}C_2}{_{80}C_2}\right) - \$1$

$\qquad = \$12(0.0601266) - \$1 \approx -\$0.28$

Here is another method:

$E = (12 - 1)P(\text{pick 2}) + (0 - 1)P(\text{pick 1}) + (0 - 1)P(\text{pick 0})$

$\qquad = 11\left(\frac{_{20}C_2}{_{80}C_2}\right) + (-1)\left(\frac{_{20}C_1 \cdot _{60}C_1}{_{80}C_2}\right) + (-1)\left(\frac{_{60}C_2}{_{80}C_2}\right)$

$\qquad \approx 11(0.0601266) - (0.3797468354) - (0.5601265823)$

$\qquad \approx -\$0.28$

43. $E = \$120 \cdot P(\text{pick 3}) + \$3 \cdot P(\text{pick 2}) - \$3$

$\qquad = \$120\left(\frac{_{20}C_3}{_{80}C_3}\right) + \$3\left(\frac{_{20}C_2 \cdot _{60}C_1}{_{80}C_3}\right) - \$3 \approx -\$0.92$

Here is another method:

$E = (120 - 3)P(\text{pick 3}) + (3 - 3)P(\text{pick 2}) + (0 - 3)P(\text{pick 1}) + (0 - 3)P(\text{pick 0})$

$\qquad \approx 117(0.0138753651) + 0 - 3(0.4308666018) - 3(0.4165043817)$

$\qquad \approx -\$0.92$

44. $E = \$565 \cdot P(\text{pick } 4) + \$15 \cdot P(\text{pick } 3) + \$5 \cdot P(\text{pick } 2) - \$5$

$$= \$565\left(\frac{_{20}C_4}{_{80}C_4}\right) + \$15\left(\frac{_{20}C_3 \cdot _{60}C_1}{_{80}C_4}\right) + \$5\left(\frac{_{20}C_2 \cdot _{60}C_2}{_{80}C_4}\right) - \$5 \approx -\$1.56$$

Here is another method:

$$E = (565-5)P(\text{pick } 4) + (15-5)P(\text{pick } 3) + (5-5)P(\text{pick } 2) + (-5)P(\text{pick } 1) + (-5)P(\text{pick } 0)$$

$$= 560\left(\frac{_{20}C_4}{_{80}C_4}\right) + 10\left(\frac{_{20}C_3 \cdot _{60}C_1}{_{80}C_4}\right) + 0\left(\frac{_{20}C_2 \cdot _{60}C_2}{_{80}C_4}\right) + (-5)\left(\frac{_{20}C_1 \cdot _{60}C_3}{_{80}C_4}\right) + (-5)\left(\frac{_{60}C_4}{_{80}C_4}\right)$$

$$\approx -1.56$$

The expectation is about $-\$1.56$.

45. $1 - \frac{_{365}P_5}{365^5} \approx 0.027$; the probability is about 2.7%.

46. $1 - \frac{_{365}P_{30}}{365^{30}} \approx 0.706$; the probability is about 70.6%.

47. $P(\text{five tails}) = \left(\frac{1}{2}\right)^5 = \frac{1}{32}$

48. $P(\text{at least one tail}) = 1 - P(\text{no tails}) = 1 - \frac{1}{32} = \frac{31}{32}$

49.   **a.** $\frac{5}{8}\cdot\frac{5}{8} = \frac{25}{64}$; $\frac{_5C_2}{_8C_2} = \frac{10}{28} = \frac{5}{14}$ **b.** $\frac{3}{8}\cdot\frac{3}{8} = \frac{9}{64}$; $\frac{_3C_2}{_8C_2} = \frac{3}{28}$ **c.** $\frac{5}{8}\cdot\frac{3}{8} + \frac{3}{8}\cdot\frac{5}{8} = \frac{15}{32}$; $\frac{_5C_1 \cdot _3C_1}{_8C_2} = \frac{15}{28}$

      d. $\frac{5}{8}\cdot\frac{3}{8} = \frac{15}{64}$; $\frac{_5P_1 \cdot _3P_1}{_8P_2} = \frac{5\cdot 3}{56} = \frac{15}{56}$

50.   **a.** $\frac{1}{4}, \frac{1}{4}$ **b.** $\left(\frac{1}{4}\right)^2 = \frac{1}{16}$; $\frac{_5C_2}{_{20}C_2} = \frac{1}{19} \approx 0.053$ **c.** $\frac{1}{64}$; $\frac{_5C_3}{_{20}C_3} = \frac{1}{114} \approx 0.009$

51.   **a.** $E = \frac{1}{4}(1) + \left(\frac{3}{4}\right)\left(\frac{1}{4}\right)(1) + \left(\frac{3}{4}\right)^2\left(\frac{1}{4}\right)(1) + \frac{27}{64}(-1) = 0.15625$; yes, play

      b. $E = \frac{1}{4}(1) + \left(\frac{3}{4}\right)\left(\frac{13}{51}\right)(1) + \left(\frac{3}{4}\right)\left(\frac{38}{51}\right)\left(\frac{13}{50}\right)(1) + \left(\frac{3}{4}\right)\left(\frac{38}{51}\right)\left(\frac{37}{50}\right)(-1) \approx 0.17294117647$;

      yes, play

52.   **a.** $E = \frac{12}{52}(1) + \left(\frac{40}{52}\right)\left(\frac{12}{51}\right)(1) + \left(\frac{40}{52}\right)\left(\frac{39}{51}\right)\left(\frac{12}{50}\right)(1) + \left(\frac{40}{52}\right)\left(\frac{39}{51}\right)\left(\frac{38}{50}\right)\left(\frac{12}{49}\right)(1) + \left(\frac{40}{52}\right)\left(\frac{39}{51}\right)\left(\frac{38}{50}\right)\left(\frac{37}{49}\right)(-1)$

        $= 0.32484994$; yes, play

      b. $E = \left(\frac{3}{13}\right)(1) + \left(\frac{10}{13}\right)\left(\frac{3}{13}\right)(1) + \left(\frac{10}{13}\right)^2\left(\frac{3}{13}\right)(1) + \left(\frac{10}{13}\right)^3\left(\frac{3}{13}\right)(1) + \left(\frac{10}{13}\right)^4(-1)$

        $= 0.2997444067$; yes, play

Level 3 Problem Solving, page 593

53.   This is equivalent to two independent rolls of a usual roulette game; thus, the probability is 1/38.

Look at it this way: the first ball specifies the slot, so the second one has one out of 38 chance of dropping into that slot.

54. First bet: $P(\text{at least one 6 in 4 rolls}) = 1 - P(\text{no 6 in 4 rolls})$

$$= 1 - \left(\tfrac{5}{6}\right)^4$$

$$\approx 0.5177$$

Second bet: $P(\text{at least one 12 in 24 rolls}) = 1 - P(\text{no 12 in 24 rolls})$

$$= 1 - \left(\tfrac{35}{36}\right)^{24}$$

$$\approx 0.4914$$

55. $P(3 \text{ black}) = \dfrac{{}_5C_3}{{}_7C_3} = \dfrac{10}{35} \approx 0.29;\; P(\text{winning}) = 1 - 0.29 = 0.71$

56. $E = (20 - 1)\dfrac{{}_{20}C_3}{{}_{80}C_3} + (2 - 1)\dfrac{{}_{20}C_2 \cdot {}_{60}C_1}{{}_{80}C_3} + (-1)\dfrac{{}_{20}C_1 \cdot {}_{60}C_2}{{}_{80}C_3} + (-1)\dfrac{{}_{60}C_3}{{}_{80}C_3} \approx -\$0.44$

57. $E = (50 - 1)\dfrac{{}_{20}C_4}{{}_{80}C_4} + (4 - 1)\dfrac{{}_{20}C_3 \cdot {}_{60}C_1}{{}_{80}C_4} + (1 - 1)\dfrac{{}_{20}C_2 \cdot {}_{60}C_2}{{}_{80}C_4} + (-1)\dfrac{{}_{20}C_1 \cdot {}_{60}C_3}{{}_{80}C_4}$

$$+ (-1)\dfrac{{}_{60}C_4}{{}_{80}C_4} \approx -\$0.46$$

58. $E = (250 - 1)\dfrac{{}_{20}C_5}{{}_{80}C_5} + (10 - 1)\dfrac{{}_{20}C_4 \cdot {}_{60}C_1}{{}_{80}C_5} + (2 - 1)\dfrac{{}_{20}C_3 \cdot {}_{60}C_2}{{}_{80}C_5} + (-1)(0.903327685)$

$$\approx -\$0.55$$

Note: $P(\text{lose } \$1) = 1 - \left[\dfrac{{}_{20}C_5}{{}_{80}C_5} + \dfrac{{}_{20}C_4 \cdot {}_{60}C_1}{{}_{80}C_5} + \dfrac{{}_{20}C_3 \cdot {}_{60}C_2}{{}_{80}C_5}\right] \approx 0.903327685$

59. $E = (1{,}000 - 1)\dfrac{{}_{20}C_6}{{}_{80}C_6} + (25 - 1)\dfrac{{}_{20}C_5 \cdot {}_{60}C_1}{{}_{80}C_6} + (4 - 1)\dfrac{{}_{20}C_4 \cdot {}_{60}C_2}{{}_{80}C_6} + (1 - 1)\dfrac{{}_{20}C_3 \cdot {}_{60}C_3}{{}_{80}C_6}$

$$+ (-1)(0.8384179112) \approx -\$0.55$$

Note: $P(\text{lose } \$1) = 1 - \left[\dfrac{{}_{20}C_6}{{}_{80}C_6} + \dfrac{{}_{20}C_5 \cdot {}_{60}C_1}{{}_{80}C_6} + \dfrac{{}_{20}C_4 \cdot {}_{60}C_2}{{}_{80}C_6} + \dfrac{{}_{20}C_3 \cdot {}_{60}C_3}{{}_{80}C_6}\right] \approx 0.8384179112$

60. The one spot is the best ticket.

11.5 The Binomial Distribution, page 594

Level 1, page 599

1. Consider an experiment with only two outcomes, A and $\overline{A}$. Let X represent the number of times that event A has occurred, with n repetitions of the experiment. The function X is called a binomial random variable.   **2.** A *Bernoulli trial* is an experiment that meets four conditions: 1. There must be a fixed number of trials. Denote this number by n. 2. There must be two possible mutually exclusive outcomes for each trial. Call them *success* and *failure*. 3. Each trial must be independent.

That is, the outcome of a particular trial is not affected by the outcome of any other trial. 4. The probability of success and failure must remain constant for each trial.

3. $P(X = 3) = \binom{5}{3}(0.3)^3(0.7)^2 \approx 0.132$ **4.** $P(X = 3) = \binom{4}{3}(0.25)^3(0.75)^1 \approx 0.047$

5. $P(X = 6) = \binom{12}{6}(0.65)^6(0.35)^6 \approx 0.128$ **6.** $P(X = 4) = \binom{10}{4}(0.8)^4(0.2)^6 \approx 0.006$

7. $P(X = 6) = \binom{6}{6}(0.5)^6(0.5)^0 \approx 0.016$ **8.** $P(X = 8) = \binom{8}{8}(0.75)^8(0.25)^0 \approx 0.100$

9. $P(X = 5) = \binom{7}{5}(0.1)^5(0.9)^2 \approx 1.701 \times 10^{-4}$

10. $P(X = 13) = \binom{15}{13}(0.4)^{13}(0.6)^2 \approx 2.537 \times 10^{-4}$

11. $P(X = 3) = \binom{5}{3}(0.5)^3(0.5)^2 = 0.3125$ **12.** $P(X = 4) = \binom{5}{4}(0.5)^4(0.5)^1 = 0.15625$

13. $P(X = 3) = \binom{6}{3}(0.5)^3(0.5)^3 = 0.3125$ **14.** $P(X = 5) = \binom{6}{5}(0.5)^5(0.5)^1 = 0.09375$

15. $P(X = 2) = \binom{5}{2}(\frac{1}{6})^2(\frac{5}{6})^3 \approx 0.161$ **16.** $P(X = 3) = \binom{5}{3}(\frac{1}{6})^3(\frac{5}{6})^2 \approx 0.032$

17. $P(X = 3) + P(X = 4) = \binom{4}{3}(0.5)^3(0.5)^1 + \binom{4}{4}(0.5)^4(0.5)^0 = 0.3125$

18. $P(X = 4) + P(X = 5) = \binom{5}{4}(0.5)^4(0.5)^1 + \binom{5}{5}(0.5)^5(0.5)^0 = 0.1875$

19. $P(X = 8) = \binom{10}{8}(0.5)^8(0.5)^2 \approx 0.044$ **20.** $P(X = 9) = \binom{10}{9}(0.5)^9(0.5)^1 = 0.010$

21. $P(X = 0) + P(X = 1) = \binom{10}{0}(0.5)^0(0.5)^{10} + \binom{10}{1}(0.5)^1(0.5)^9 \approx 0.0107$

22. $P(X = 9) + P(X = 10) = \binom{10}{9}(0.5)^9(0.5)^1 + \binom{10}{10}(0.5)^{10}(0.5)^0 \approx 0.0107$

23. $P(X = 0) = \binom{3}{0}(\frac{1}{3})^0(\frac{2}{3})^3 \approx 0.296$ **24.** $P(X = 1) = \binom{3}{1}(\frac{1}{3})^1(\frac{2}{3})^2 \approx 0.444$

25. $P(X = 2) = \binom{3}{2}(\frac{1}{3})^2(\frac{2}{3})^1 \approx 0.222$ **26.** $P(X = 3) = \binom{3}{3}(\frac{1}{3})^3(\frac{2}{3})^0 \approx 0.037$

27. $P(X = 4) = \binom{4}{4}(0.9)^4(0.1)^0 \approx 0.656$ **28.** $P(X = 3) = \binom{4}{3}(0.9)^3(0.1)^1 = 0.2916$

29. $P(X = 0) = \binom{4}{0}(0.9)^0(0.1)^4 = 0.0001$ **30.** $P(X = 1) = \binom{4}{1}(0.9)^1(0.1)^3 = 0.0036$

31. $P(X = 3) + P(X = 4) = \binom{4}{3}(0.9)^3(0.1)^1 + \binom{4}{4}(0.9)^4(0.1)^0 = 0.9477$

32.
$$P(X \le 2) = 1 - P(X > 2)$$
$$= 1 - [P(X = 3) + P(X = 4)]$$
$$= 1 - 0.9477 \qquad \textit{Use the answer from Problem 31.}$$
$$= 0.0523$$

You could also do this directly:

$$\binom{4}{0}(0.9)^0(0.1)^4 + \binom{4}{1}(0.9)^1(0.1)^3 + \binom{4}{2}(0.9)^2(0.1)^2 = 0.0523$$

33. $P(X = 0) = \binom{3}{0}(0.2)^0(0.8)^3 = 0.512$ **34.** $P(X = 3) = \binom{3}{3}(0.2)^3(0.8)^0 \approx 0.008$

35. $P(X = 2) = \binom{3}{2}(0.2)^2(0.8)^1 = 0.096$ **36.** $P(X = 1) = \binom{3}{1}(0.2)^1(0.8)^2 \approx 0.384$

37. $P(X \geq 1) = 1 - P(X < 1) = 1 - 0.512 = 0.488$

38. The sum of the results for Problems 33-36 is 1. Problem 37 is related to these parts since

$P(X \geq 1) = P(X = 1) + P(X = 2) + P(X = 3)$.

Level 2, page 600

39. $P(X = 2) = \binom{4}{2}(\tfrac{3}{4})^2(\tfrac{1}{4})^2 \approx 0.2109$

40. $P(X \geq 2) = \binom{4}{2}(\tfrac{3}{8})^2(\tfrac{5}{8})^2 + \binom{4}{3}(\tfrac{3}{8})^3(\tfrac{5}{8})^1 + \binom{4}{4}(\tfrac{3}{8})^4(\tfrac{5}{8})^0 \approx 0.4812$

41. $P(X = 1) = \binom{6}{1}(\tfrac{1}{6})^1(\tfrac{5}{6})^5 \approx 0.4019$ **42.** $P(X = 2) = \binom{12}{2}(\tfrac{1}{6})^2(\tfrac{5}{6})^{10} \approx 0.2961$

43. $P(X = 5) = \binom{6}{5}(0.7)^5(0.3)^1 \approx 0.3025$ **44.** $P(X = 1) = \binom{3}{1}(0.4)^1(0.6)^2 \approx 0.432$

45. $P(X = 4) = \binom{5}{4}(0.96)^4(0.04)^1 \approx 0.1699$ **46.** $P(X = 14) = \binom{15}{14}(0.96)^{14}(0.04)^1 \approx 0.3388$

47. $P(X = 15) = \binom{20}{15}(0.85)^{15}(0.15)^5 \approx 0.1028$

48. $P(X \geq 9) = P(X = 9) + P(X = 10) = \binom{10}{9}(0.85)^9(0.15)^1 + \binom{10}{10}(0.85)^{10}(0.15)^0 \approx 0.5443$

49. $P(X = 4) = \binom{7}{4}(\tfrac{3}{5})^4(\tfrac{2}{5})^3 \approx 0.2903$

50. To play 7 games, each team must win 3 games.

$P(\text{playing the 7th game}) = \binom{6}{3}(\tfrac{3}{5})^3(\tfrac{2}{5})^3 \approx 0.2765$

51. $P(X \geq 3) = P(X = 3) + P(X = 4) = \binom{4}{3}(0.3)^3(0.7)^1 + \binom{4}{4}(0.3)^4(0.7)^0 = 0.0837$

52. $P(X \geq 3) = 1 - P(X < 3) = 1 - [P(X = 0) + P(X = 1) + P(X = 2)]$

$= 1 - \left[\binom{6}{0}(0.2)^0(0.8)^6 + \binom{6}{1}(0.2)^1(0.8)^5 + \binom{6}{2}(0.2)^2(0.8)^4\right] \approx 0.0989$

Level 3, page 600

53. Find n such that $1 - \binom{n}{0}(0.1)^0(0.9)^n \geq 0.80$.

$$1 - \binom{n}{0}(0.1)^0(0.9)^n \geq 0.80$$
$$\binom{n}{0}(0.1)^0(0.9)^n < 0.20$$
$$(0.9)^n < 0.20$$
$$n \geq 15.3$$

The smallest number of missiles is 16.

54. **a.** If she is guessing then $p = 0.5$ and $q = 0.5$.

$$P(X \geq 5) = P(X = 5) + P(X = 6)$$
$$= \binom{6}{5}(0.5)^5(0.5)^1 + \binom{6}{6}(0.5)^6(0.5)^0$$
$$= 0.109375$$

   b. If her claim is correct, then $p = 0.75$ and $q = 0.25$.

$$P(X \geq 5) = P(X = 5) + P(X = 6)$$
$$= \binom{6}{5}(0.75)^5(0.25)^1 + \binom{6}{6}(0.75)^6(0.25)^0$$
$$\approx 0.5339$$

   Thus, the probability of denying her claim is $1 - P(X \geq 5) \approx 0.4661$

55. Part a is the same as the one shown in Problem 54. If the claim of 90% is correct, then $p = 0.9$ and $q = 0.1$.

$$P(X \geq 5) = P(X = 5) + P(X = 6)$$
$$= \binom{6}{5}(0.9)^5(0.1)^1 + \binom{6}{6}(0.9)^6(0.1)^0$$
$$= 0.885735$$

   Thus, the probability of denying the claim is $1 - P(X \geq 5) = 0.114265$.

56. $P(X \geq 1) = 1 - P(X = 0) = 1 - \binom{100}{0}(0.1)^0(0.9)^{100} \approx 0.99997$

Level 3 Problem Solving, page 601

57. $P(X = 1) + P(X = 2) = \binom{3}{1}(0.5)^1(0.5)^2 + \binom{3}{2}(0.5)^2(0.5)^1 = 0.75$

58. $P(X = 1) + P(X = 3) = \binom{4}{1}(0.5)^1(0.5)^3 + \binom{4}{3}(0.5)^3(0.5)^1 = 0.5$

59. $P(X = 1) + P(X = 4) = \binom{5}{1}(0.5)^1(0.5)^4 + \binom{5}{4}(0.5)^4(0.5)^1 = 0.3125$

59. $P(X = 1) + P(X = n - 1) = \binom{n}{1}(0.5)^1(0.5)^{n-1} + \binom{n}{n-1}(0.5)^{n-1}(0.5)^1 = 2 \cdot 2^{-n}n = n2^{1-n}$

Chapter 11 Review Questions, page 602

1. $P(\text{defective}) = \frac{4}{1,000} = 0.004$

2. The possible primes are 2, 3, and 5, so $P(\text{prime}) = \frac{3}{6} = \frac{1}{2}$.

3. Look at Figure 11.3; $P(\text{eight}) = \frac{5}{36}$

4. $P(\text{jack or better}) = \frac{16}{52} = \frac{4}{13}$

5. $P(\text{ace}) = \frac{3}{51} = \frac{1}{17}$

6. Look at Figure 11.3.

    a. $P(5 \text{ on at least one of the dice}) = \frac{11}{36}$

    b. $P(5 \text{ on one die or 4 on the other}) = \frac{20}{36} = \frac{5}{9}$

    c. $P(5 \text{ on one die and 4 on the other}) = \frac{2}{36} = \frac{1}{18}$

7. $P(E) = 0.01; \ P(\overline{E}) = 1 - P(E) = 1 - 0.01 = 0.99$

8. $P(E) = \frac{9}{10}; \ P(\overline{E}) = 1 - \frac{9}{10} = \frac{1}{10}$

    $$\text{Odds in favor} = \frac{P(E)}{P(\overline{E})} = \frac{\frac{9}{10}}{\frac{1}{10}} = \frac{9}{10} \cdot \frac{10}{1} = \frac{9}{1} \text{ or 9 to 1}$$

9. Let E be the event. Given $f = 1{,}000$ and $s = 1$,

    $$P(E) = \frac{s}{s + f} = \frac{1}{1 + 1{,}000} = \frac{1}{1{,}001}$$

10. $E = P(\text{one}) \cdot 12 = \frac{1}{6}(12) = 2;$ the expected value is \$2.

11. a. $P(\text{orange} \mid \text{orange}) = \frac{3}{5}$ b. $P(\text{orange} \mid \text{orange}) = \frac{2}{4} = \frac{1}{2}$

12.
$$\begin{aligned}
\overline{P(A \cup B)} &= 1 - P(A \cup B) \\
&= 1 - [P(A) + P(B) - P(A \cap B)] \\
&= 1 - P(A) - P(B) + P(A \cap B) \\
&= 1 - P(A) - P(B) + P(A)P(B) \\
&= 1 - \tfrac{2}{3} - \tfrac{3}{5} + \tfrac{2}{3} \cdot \tfrac{3}{5} \\
&= \tfrac{2}{15}
\end{aligned}$$

13. $P(X = x) = \binom{n}{x}(0.001)^x(0.999)^{n-x}$

14. $P(X = 3) = \binom{5}{3}(0.85)^3(0.15)^2 \approx 0.138$

15. $P(X = 4) = \binom{4}{4}(0.85)^4(0.15)^0 \approx 0.522$

16. List the sample space:

    $$\left.\begin{matrix} \text{MM} \\ \text{MF} \\ \text{FM} \end{matrix}\right\} \ \text{Two success out of 3 possibilities; } P(\text{male}) = \tfrac{2}{3}$$

    $$\text{FF} \ \left.\vphantom{\begin{matrix}a\end{matrix}}\right\} \ \text{Vet rules this out, so delete from the sample space.}$$

    $$F_2 F_1$$

    Your selection ↑ ↑ Cat left behind

17.
$$P(X \geq 1) = 1 - P(X = 0)$$
$$= 1 - \binom{4}{0}(0.35)^0(0.65)^4$$
$$\approx 0.821$$

18. Write out the possible sample spaces. The best possibility is with die C:

	D							
	5	5	5	5	1	1	1	1
2	5	5	5	5	2	2	2	2
2	5	5	5	5	2	2	2	2
2	5	5	5	5	2	2	2	2
C 2	5	5	5	5	2	2	2	2
2	5	5	5	5	2	2	2	2
2	5	5	5	5	2	2	2	2
6	6	6	6	6	6	6	6	6
6	6	6	6	6	6	6	6	6

We see that there are 64 possibilities and that die C wins 40 of those times; thus,

$$P(C \text{ wins}) = \frac{40}{64} = \frac{5}{8}$$

19. $P(\text{winning}) = \frac{1}{5}$. It is tempting to say 1 out of 25, but after you pick a key, one of five cars will fit that key.

20. $P(\text{winning a particular car}) = \frac{1}{5} \cdot \frac{1}{5} = \frac{1}{25}$

Group Research Problems, pages 603-604

G41. a. GO TO JAIL location is 20 spaces from the START. You can get there in one turn by rolling $\{12, 12, 6\}$, $\{11, 11, 8\}$, or $\{9, 9, 12\}$. You would "go to jail" by rolling any number 3 times in a row. You can also receive a "get out of jail free" card.

b. $P(7; \text{advance to Ill. Ave}; 7; 6; 2) = \frac{6}{36} \cdot \frac{1}{35} \cdot \frac{6}{36} \cdot \frac{5}{36} \cdot \frac{1}{36} = \frac{180}{36^5} = 2.976870904 \times 10^{-6}$; this is about one chance in 335,923.

G42. a. Answers vary.   **b.** $1 - \frac{365}{365} \times \frac{364}{365} \times \frac{363}{365} = 1 - \frac{_{365}P_3}{365^3} \approx 0.0054794521$

c. $1 - \frac{_{365}P_{23}}{365^{23}} \approx 0.5072972343$; verifications vary

d.

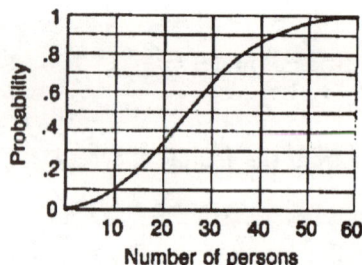

The number of people *necessary* to have a probability of one is 367 (don't forget possible leap years).

e. $1 - \dfrac{_{365}P_5}{365^5} \approx 0.0271355737$

$_{365}P_5 = 175{,}793{,}709{,}365$; this is the number of possibilities we need to account for:

exactly two: $_5C_2 = 10$ and for each of these the number of ways to select the four days is

$_{365}P_4$ so the number of possibilities is $10 \cdot 365 \cdot 364 \cdot 363 \cdot 362 = 174{,}586{,}011{,}600$

exactly three: $_5C_3 = 10$, with three days: $10 \cdot 365 \cdot 364 \cdot 363 = 482{,}281{,}800$

exactly four: $_5C_4 = 5$ with two days: $5 \cdot 365 \cdot 364 = 664{,}300$

all five: $_5C_5 = 1$ with one day: 365

G43. Answers vary; **a.** 0.85   **b.** 0.82   **c.** 0.31   **d.** 73 mm or 2.9 in. This problem is adapted from an article, "Coin Tossing," by Mako E. Haruta, Mark Flaherty, Jean McGivney, and Raymond J. McGivney in *The Mathematics Teacher*, November 1996, pp. 642-645.

Individual Research Problems, pages 605-606

P11.1   They were all practicing mathematicians at one time in their careers.

P11.2   Michael Rossides has come up with a scheme for eliminating coins. This scheme involves probability and the fact that most cash registers today are computers. Suppose that every cash register could be programmed with a random number generator; that is, suppose that it were possible to pick a random number from 1 to 99. Rossides' system works as follows. Suppose you purchase items totaling $15.89. The computer would choose a number from 1 to 99 and then compare it with the cents portion of the purchase. In this case, if the random number is between 1 and 89 the price would be rounded up to $16; if it is between 90 and 99 the price would be rounded down to $15. For example, if you purchase a cup of coffee for $1.20, and the random number generator produces a random number from 1 to 20, the price is $2, but if it produces a number from 21 to 99 the price is $1. Write a paper commenting on this scheme.

P11.3   It is not correct to say that the probability is $\frac{1}{7}$ because there are seven possible days in the week. In fact, it turns out that the 13th day of a month is more likely a Friday than any other day of the week. Answers vary.

P11.4   **a.** Answers vary.

   b. From smallest to largest, the order is (1) $5.00   (5) $5.00312   (4) $5.003235   (3) $5.003888 and (2) $5.088884

   c. answers vary

   d. Answers vary, but human preferences do not mirror mathematical preferences.

P11.5   **a.** Answers vary.

   b. (1) B   (2) C   (3) They are the same.

P11.6   Answers vary.

CHAPTER 12

THE NATURE OF STATISTICS

12.1 Frequency Distributions and Graphs, page 608

The SAAB advertisement (see Problem 49 and Transparency 28) ties together the ideas of problem solving and statistics. I ask the students to look at the advertisement and see if they can find the mistake. It is in second column of the advertisement: Outside, the car is 57 in. wide, but inside it is a "a full five feet across." Since 5 ft = 60 in., we see that it has an outside measurement of 57 in. but an inside measurement of 60 in. ... hmmmmm.

Level 1, page 616

1. Answers vary; a frequency distribution tabulates the number of outcomes in each class, whereas a stem and leaf plot is a systematic listing of each outcome.   **2.** Answers vary; it does not mean anything at all without some further information.   **3.** Answers vary; it does not mean very much without some further information.

4. a.

Height	Tally	Frequency
72			1
71				2
70					3
69						4
68					3
67	⊞	5
66					3
65					3
64						4
63				2

b.

```
6 | 3 3 4 4 4 5 5 5 6 6 6 7 7 7 7 8 8 8 9 9 9
7 | 0 0 0 1 1 2
```

c.

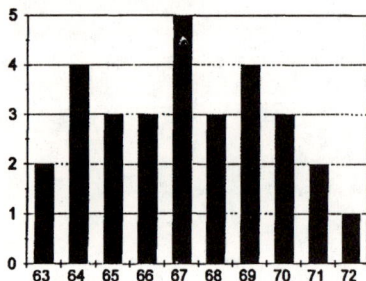

d.

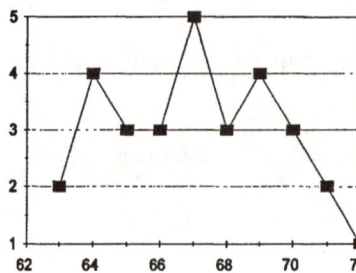

5. a.

Salary	Tally	Frequency
$60,000	|	1
$50,000	|	1
$35,000	|	1
$30,000	||	2
$25,000	|||	3
$20,000	|	1
$18,000	||	2
$16,000	|	1
$14,000	|	1

b. (in thousands)

1	4 6 8 8
2	0 5 5 5
3	0 0 5
4	
5	0
6	0

c.

d.

6.

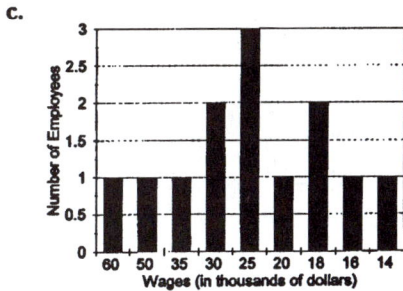

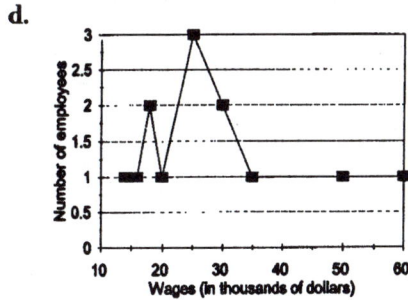

0	₩₩₩ ₩₩₩ ₩₩₩ ₩₩₩ ₩₩₩	25
1	|	1
2	|	1
3	₩₩₩ ₩₩₩ ₩₩₩ ||||	19
4	|	1
5	|||	3

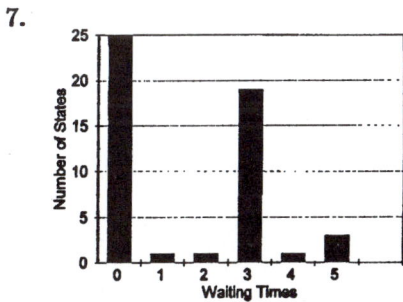

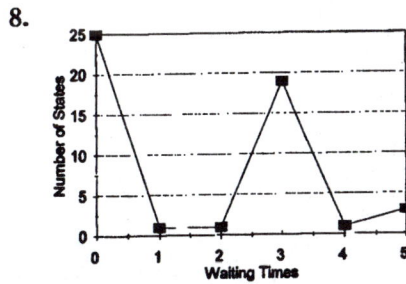

7.

8.

9.

Purchasing Power
1983 = $1.00

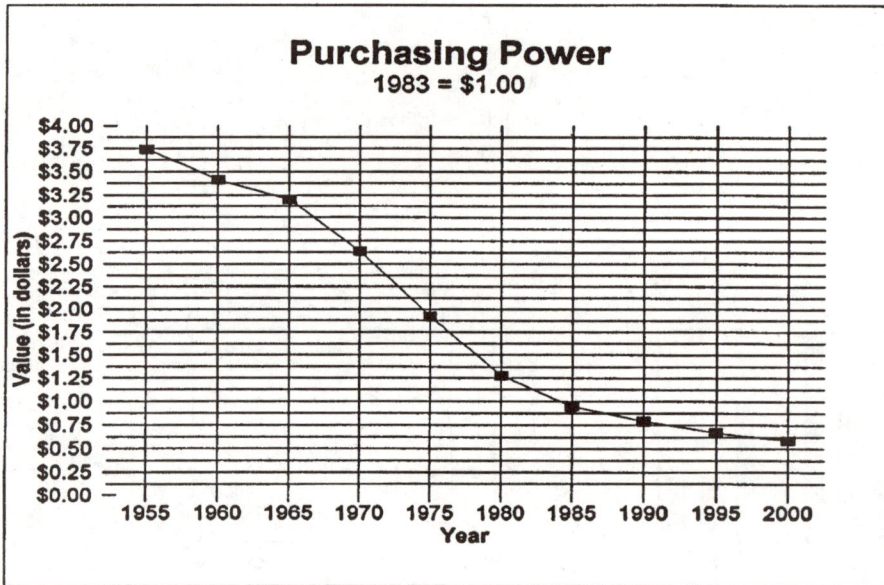

10.

Purchasing Power
1983 = $1.00

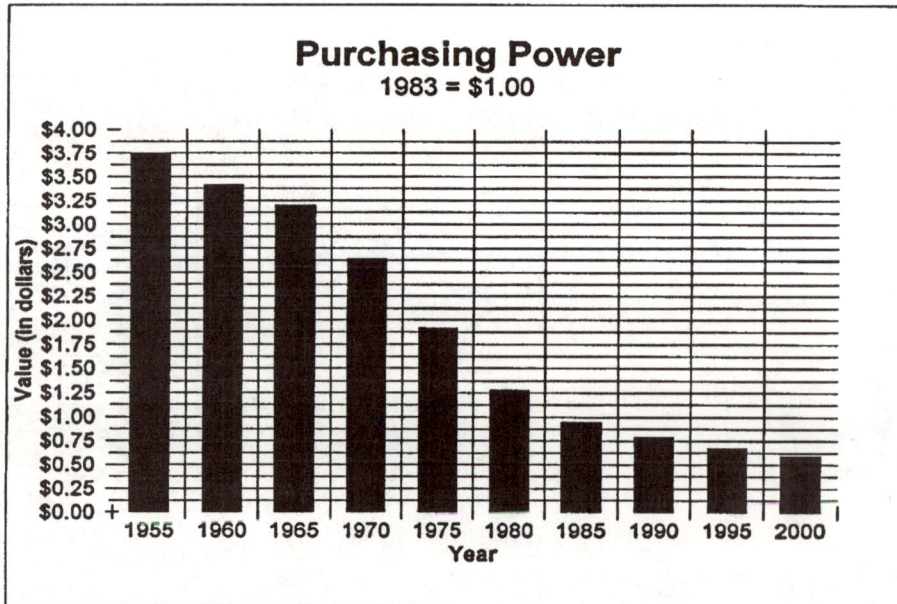

11.

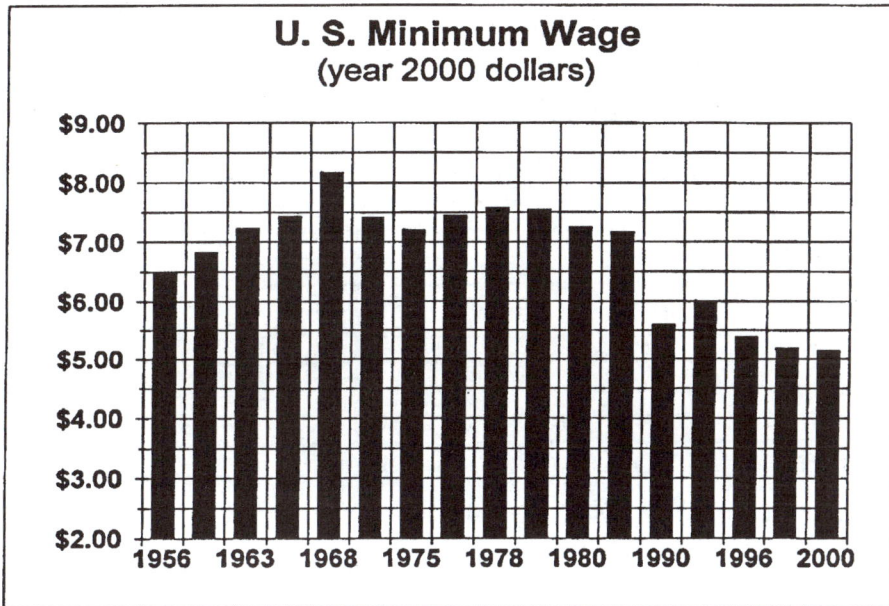

U. S. Minimum Wage
(year 2000 dollars)

12.

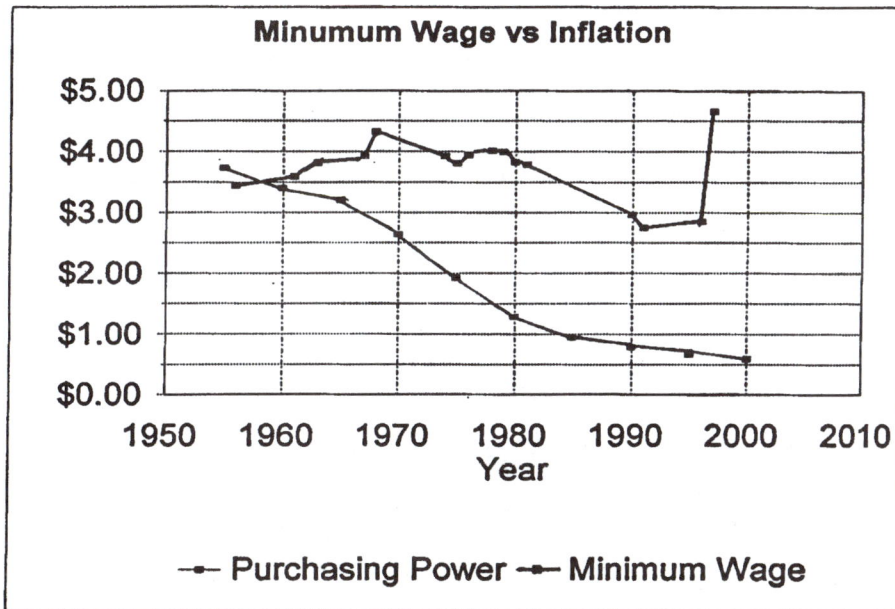

Minumum Wage vs Inflation

13.   They seem to cross in 1958.

14.

U.S. Minimum Wage
(Year 2000 dollars)

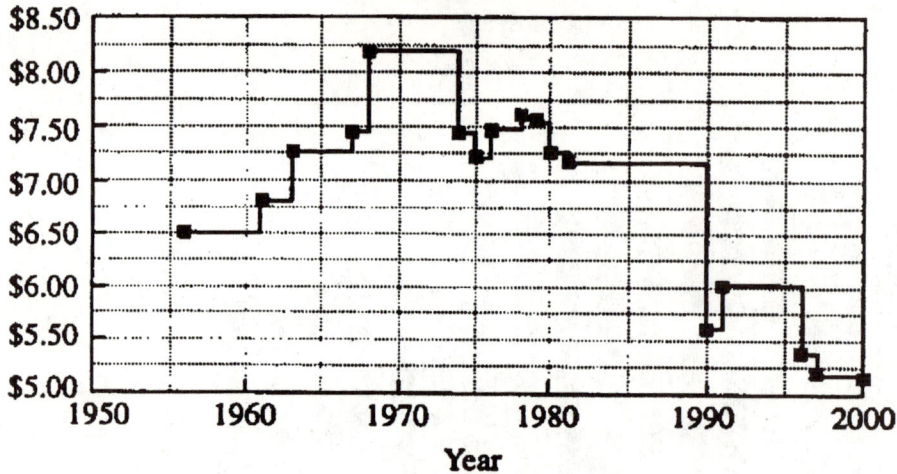

15.

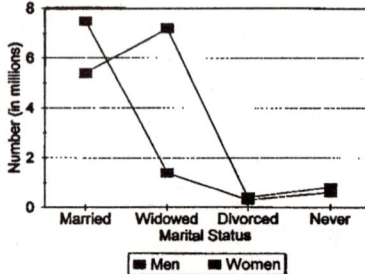

16. a. October   **b.** August

17. a. Gonorrhea cases by race and sex are graphed. The categories are black males, black females, white males, and white females.
    b. Black females and white males are almost the same in 1981 (acceptable answer), but the only one that shows the same on the graph is white males and white females in 1986.
    c. The number of cases in black females was increasing between 1984 and 1986.
    d. 100,000 cases

18. About 60%; C   **19.** Note $1,000 million is a billion; B   **20.** 3 networks; yes, FOX, UPN, WB, etc. Answers vary.   **21.** 1965 and 1990   **22.** ABC   **23.** 30%; NBC   **24.** 10%; ABC

25. Answers vary; the ratings for all three networks has generally declined from 1955 to 1997.

26. Answers vary; U.S. trade with Canada for the years 1988-1996   **27.** more imports

28. −$8.6 billion   **29.** $90 billion   **30.** Answers vary; U.S. trade with Mexico for the years 1988-1996   **31.** more imports   **32.** $1.4 billion   **33.** $56.8 billion

Level 2, page 619

34. 5 times   **35.** 25 times   **36.** maybe DUI (illegal if under 18 yr old)   **37.** maybe DUI (illegal if under 18 yr old)   **38.** 5   **39.** no

40.

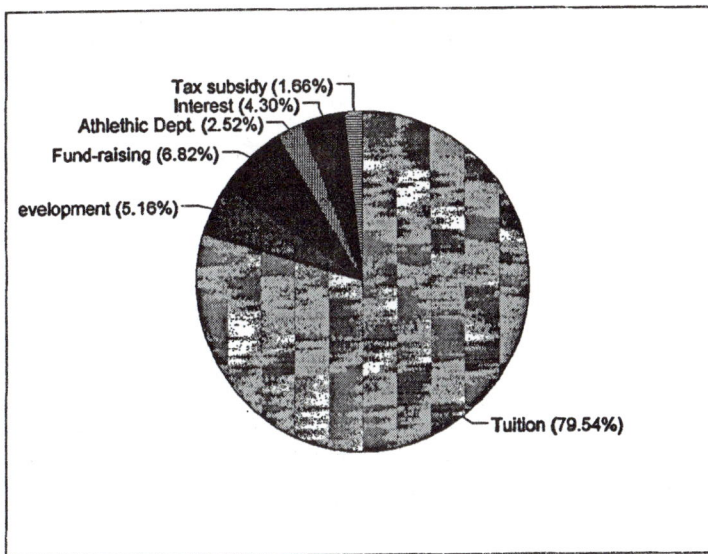

Pie chart labels: Tax subsidy (1.66%), Interest (4.30%), Athlethic Dept. (2.52%), Fund-raising (6.82%), evelopment (5.16%), Tuition (79.54%)

41.

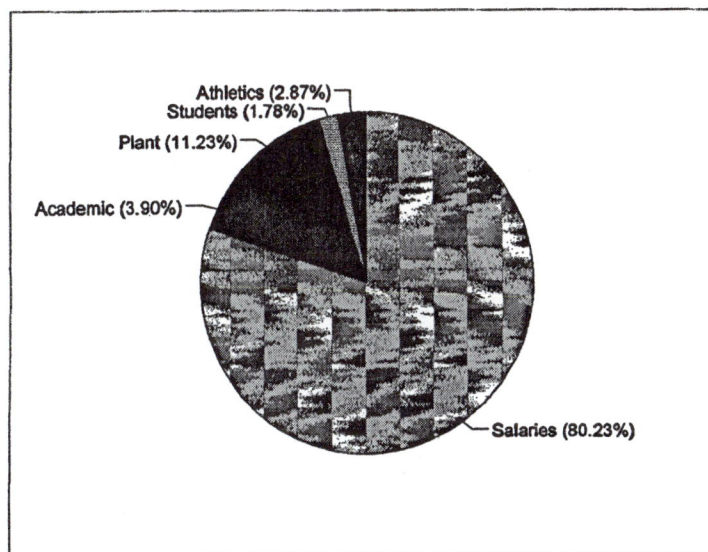

Pie chart labels: Athletics (2.87%), Students (1.78%), Plant (11.23%), Academic (3.90%), Salaries (80.23%)

42. a. 121 kwh   **b.** 44 kwh   **c.** 209 kwh   **d.** 55 kwh

43.

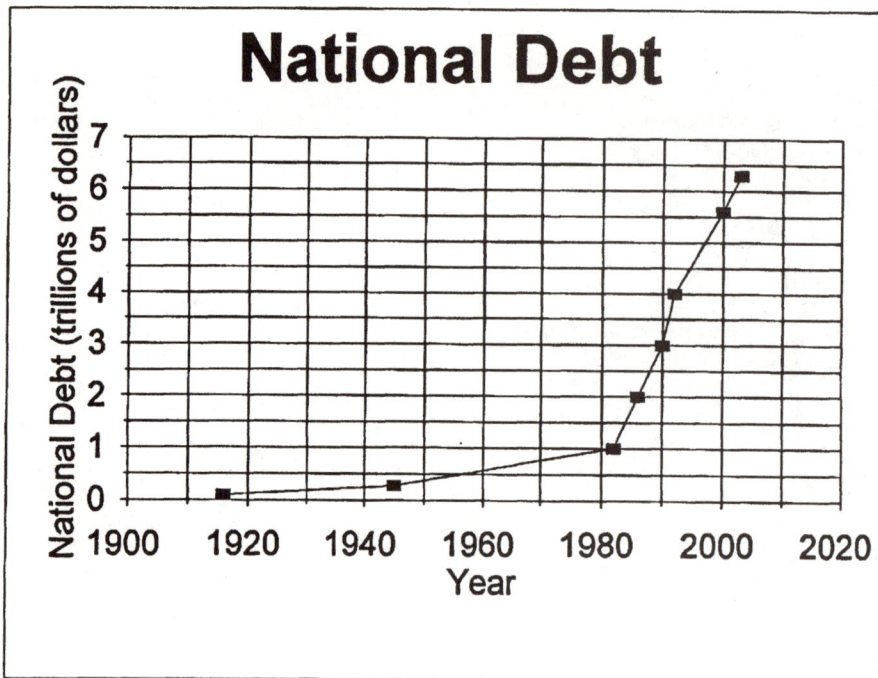

44. Each figure represents 30 managing directors.
Paine Webber: 46 of 465, or about 10%

Goldman, Sachs: 9 out of 173, or about 5%

Merrill Lynch: 76 out of 694, or about 11%

45. Answers vary; graph is meaningless without a scale.

46. Answers vary; the graphs are based on height, but the impression created is that of area, and the area of the second tree is much more than 65% greater (even though the height is 65% greater).

47. The graph is appropriate and accurate.

48. Three-dimensional bars are used to represent linear data, which as we know, exaggerates the difference. However, since the width and the length of each bar remains constant, the actual volume changes are faithful to the differences in height.

Level 3, page 621

49. The advertisement says that the car is 57 in. wide on the outside, but a full 5 ft wide across on the inside.   **50.** $50 billion   **51.** 3.4 persons   **52.** Persons who were born shortly after World War II (1945-1950) from Hirsh, *Dictionary of Cultural Literature*   **53.** 2036   **54.** $800 billion

55. The line graph at the right begins at 50 rather than 0.   **56.** Three-dimensional objects (people) are used to represent linear data (number of workers).

57.

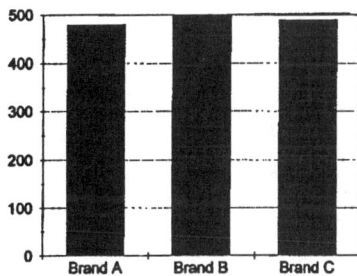

58.

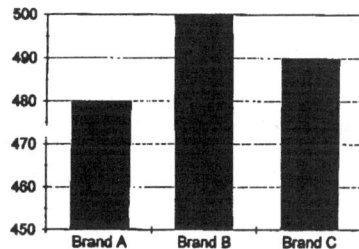

59. The graph in Figure 12.25b.

Level 3 Problem Solving, page 621

60. a. salt **b.** dime and penny plus two other coins **c.** die **d.** beach sand **e.** soda straw

12.2 Descriptive Statistics, page 622

Level 1, page 630

1. Answers vary; an *average* is a measure of central tendency and includes the mean, median, and mode.   **2.** Answers vary; a *measure of central tendency* is a single number that is used to represent the central location for a set of data.   **3.** Answers vary; a *measure of dispersion* is a single number that tells how closely the data is related to a measure of central tendency and includes the range, standard deviation, and the variance.   **4.** Answers vary; the *mean* is the most sensitive average. It reflects the entire distribution and is the most common average. The *median* gives the middle value. It is useful when there are a few extraordinary values to distort the mean. The *mode* is the average that measures "popularity." It is possible to have no mode or more than one mode.   **5.** As the example shows, Andy was better than Dave in both 1989 and 1990. However, the combined batting average for Andy is 253 hits for 969 at bats for an average of 0.261; the combined batting average for Dave is 136 hits for 490 at bats for an average of 0.278; as you can see Dave beat Andy for the combined season.   **6.** No; the square root of a number between 0 and 1 is larger than the original number.   **7.** The data values are all the same.   **8.** Answers vary. First student's scores: 80, 80, 80, 80, and 80. Second student's scores: 85, 85, 85, 85, and 60.   **9.** First student's scores: 80, 80, 80, 80, 80, and 80. Second student's scores: 100, 100, 100, 60, 50, and 70.   **10.** Answers vary; nothing to be reported; half should always be below the median (average) and half should be above — this is what median means. Can you say the same about the mean? How about the mode?

11. The person making the comment scored better in English than 78% of the people taking the test and received the median score in math.   **12.** mean = 3; median = 3; no mode   **13.** mean = 19; median = 19; no mode   **14.** mean = 105; median = 105; no mode

15. mean = 767; median = 767; no mode   **16.** mean = 10; median = 8; no mode

17. mean = 11; median = 9; no mode   **18.** mean = 91; median = 95; mode = 95

19. mean = 82; median = 81; no mode   **20.** mean = 3; median = 3; mode = 3

21. mean = 6; median = 2.5; mode = 1   **22.** range = 4; var. = 2.5; $s \approx 1.58$

23. range = 4; var. = 2.5; $s \approx 1.58$ **24.** range = 4; var. = 2.5; $s \approx 1.58$

25. range = 4; var. = 2.5; $s \approx 1.58$ **26.** range = 18; var. = 52; $s \approx 7.21$

27. range = 24; var. = 93.5; $s \approx 9.67$ **28.** range = 17; var. = 50.5; $s \approx 7.11$

29. range = 25; var. = 83; $s \approx 9.11$ **30.** range = 4; var. = $\frac{5}{3}$; $s \approx 1.29$

31. range = 21; var. = 62.86; $s \approx 7.93$

32. a. mean = 217,851   **b.** $s \approx 1.58$

33. Q_2 = median = 432.51; Q_1 = 427.48; Q_3 = 442.28

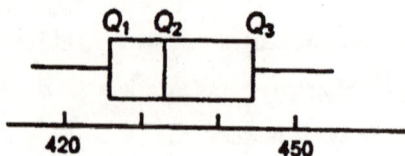

Level 2, page 631

34. mean = \$28,900, median = \$28,000, mode = \$25,000　　**35.** mean = 13; median = $\frac{10 + 12}{2} = 11$;

mode = 10　　**36.** mean $\approx$ 68.1; median = 70; mode = 70　　**37.** mean = 68; median = 70;

mode = 70; range = 50　　**38.** mean = 56; median = 65; mode = 70; range = 90

39. mean $\approx$ 366; median = 365; no mode　　**40.** mean = \$78,000; median = \$47,000; bimodal at

\$40,000 and \$30,000. The median is most descriptive.　　**41.** *A* (since the mode is the largest value,

namely 10)　　**42.** *A* (since the mode is the largest value, namely 52)　　**43.** C (it is a symmetric

distribution)　　**44.** A (since the mode is the smallest value)　　**45.** This problem can be calculated

directly, or by using a calculator.

Data lists:　　　　　　　　　　　　　　Calculator output:

a. mean = 105.25　**b.** median = 12　**c.** mode = 42

46.　a. $Q_1 = 42$, $Q_2 = 45$ (this is the median), $Q_3 = 57$

　b.

$Q_1 = 90$, Q_2 = median = 110, $Q_3 = 120$, maximum value is 160, and minimum value is 35.

　c. mean = 238; standard deviation $\approx$ 35

47. $s \approx 5$ (4.6)　　　　　　　　**48.** $s \approx 6{,}008$ (6,008.3)

49. $s \approx 12$ (11.9)　　　　　　　**50.** $s \approx 10$ (9.8)

51. $s \approx 15$ (15.1)　　　　　　　**52.** $s \approx 27$ (26.7)

53. mean = 19,200; range = 4,000; $s \approx 1{,}483$

54. Answers vary.   55. Answers vary.

56. **a.** She was 23rd from the top, so $\dfrac{240 - 23}{240} = 0.904$; the percentile ranking is the 90th

b. $0.85(50) = 42.5$; this means that there are 42 students below Lee, so his rank is 8.

Level 3, page 632

57. **a.** $\overline{x} = 6.5$; H.M. $= 4.7$ **b.** $\overline{x} = 56.5$ mph; H.M. $= 56.1$ mph

58. **a.** $x = 6$; G.M. $= 5.3$ **b.** $\overline{x} = 1.47\%$; G.M. $= 1.39\%$

Level 3 Problem Solving, page 633

59. Answers vary. **a.** Population A is Player A and population B is Player B.

$r_1 = 0.223$, $r_2 = 0.284$, $r = 0.257$, $R_1 = 0.232$, $R_2 = 0.296$ and $R = 0.251$. Finally, $C_1 = $ against right-handed pitchers and $C_2 = $ against left-handed pitchers   **b.** Examples vary.

60. False; answers vary.

12.3 The Normal Curve, page 633

Level 1, page 640

1.   A cumulative frequency is the sum of all preceding frequencies in which some order has been established.

2.   Answers vary; grading on a curve means determining students' grades by giving Cs to those students within 1 standard deviation from the mean; B and D to those students from 1 to 2 standard deviations above and below the mean, respectively.

3.   Answers vary; a normal curve is shown in Figure 12.33.

4.   In a normal curve that is skewed to the right, the mean is to the right of the mode (see Figure 12.39).

5.   The z-score is used to determine how far a given score is from the mean of the distribution. The answer is given in terms of standard deviations. If x is a value from a distribution with mean μ and standard deviation σ, then its z-score is $z = \dfrac{x - \mu}{\sigma}$.

6.   The z-score is negative when the given value is less than the mean; that is, when $x - \mu$ is negative.

7.

	Cumulative
0	5%
1	16%
2	45%
3	79%
4	94%
5	99%
6 or more	100%

$$\overline{x} = [0(0.05) + 1(0.11) + 2(0.29)$$
$$+ 3(0.34) + 4(0.15) + 5(0.05)$$
$$+ 6(0.01)]/1$$
$$= 2.62$$

median = 3
mode = 3

8. Use the mean of each interval as a representative value.

		Cumulative
0.00-0.99	0.495	11%
1.00-1.74	1.37	17%
1.75-2.49	2.12	36%
2.50-3.24	2.87	77%
3.25-4.00	3.625	100%

$$\overline{x} = \frac{0.495(0.11) + 1.37(0.06) + 2.12(0.19) + 2.87(0.41) + 3.625(0.23)}{1}$$
$$\approx 2.55$$

median = 2.87
mode = 2.87

9.

	Cumulative
0	1%
1	12%
2	47%
3	68%
4	88%
5	94%
6	97%
7	99%
8	100%

$\overline{x} = 2.94$
median = 3
mode = 2

10.

	Cumulative
0	0.18
1	0.42
2	0.74
3	0.90
4	0.99
5	1.00

$\overline{x} = 1.77$
median = 2
mode = 2

11.

	Cumulative
2	0.28
6	0.40
8	0.75
12	0.81
16	1.00

$\overline{x} = 7.84$
median = 8
mode = 8

12.

	Cumulative
11	0.01
12	0.28
13	0.41
14	0.48
15	0.53
16	0.85
17	0.87
18	0.98
19	1.00

$\overline{x} = 14.59$
median = 15
mode = 16

For Problems 13-23, remember a normal curve is symmetric about the mean.

13. 41.92%   **14.** 11.79%   **15.** 49.25%   **16.** 46.86%   **17.** 49.99%   **18.** 22.57%

19. 49.01%   **20.** 19.15%   **21.** 17.72%   **22.** 38.30%   **23.** 48.68%   **24.** 49.99%

25. a. 34 people   **b.** 33 people   **26. a.** 24 people   **b.** 6 people   **27.** 91.92%   **28.** 25

29.

		Cumulative
155		0.1%
160	1	2.3%
165	7	15.9%
170	17	50.0%
175	17	84.1%
180	7	97.7%
185	1	99.9%
190		100.0%

30. a. 31   **b.** 1   **31.** 60 or above   **32.** 0.0228   **33.** 25

34.

	Cumulative
35	0.1%
40	2.3%
45	15.9%
50	50.0%
55	84.1%
60	97.7%
65	99.0%
70	100.0%

Level 2, page 641

35. A; 50% − 6% = 44%; Look at Table 12.9 to find

$$z = 1.56; \text{ use } z = \frac{x - \mu}{\sigma}$$

$$1.56 = \frac{x - 75}{8}$$

$$x = 87.48$$

Cutoff is 87.

36. B; 6% + 16% = 22% $z = 0.58$, so $0.58 = \dfrac{x - 75}{8}$

$$x = 79.64$$

The cutoff is 80 points.

37. C; 80 − 75 = 5 points, so the cutoff is 70; that is, C ranges from 70 to 79.

38. D; cutoff is 65 (Symmetric − see answer to Problem 36.) That is, D ranges from 65 to 69.

39.

		Cumulative
A	87^+	6%
B	80-86	22%
C	70-79	78%
D	65-69	94%
F	64^-	100%

40. Answers vary. The mean is the largest; see Figure 12.39.

41. Answers vary. The mode is the largest; see Figure 12.39.

42. $0.85 = \dfrac{x - 85.7}{4.85}$

$x = 89.8225$

43. $2.55 = \dfrac{x - 85.7}{4.85}$

$x = 98.0675$

44. $-1.25 = \dfrac{x - 85.7}{4.85}$

$x = 79.6375$

45. $-3.46 = \dfrac{x - 85.7}{4.85}$

$x = 68.919$

46. $z = \dfrac{130 - 100}{16} = 1.875 \approx 1.88$; from Table 12.9, we find 0.4699, so the probability is

$0.50 + 0.4699 = 0.9693$

47. One standard deviation is 0.3413, so greater than one standard deviation is

$0.50 - 0.3413 = 0.1587$

48. 2.3% is 2 standard deviations above the mean, so the number of inches is

$35.5 + 2(2.5) = 40.5$ inches

49. $z = \dfrac{36 - 35.5}{2.5} = 0.2$; from Table 12.9, we find 0.0793, so the probability is

$0.50 - 0.0793 = 0.4207$

50. $z = \dfrac{30.5 - 35.5}{2.5} = -2$; note that 30.5 is two standard deviations less than the mean, so the

percent is $0.50 + 0.4772 = 0.9772$

51. Variance of 9 is a standard deviation of $\sigma = 3$. We find the z-score: $z = \dfrac{159 - 165}{3} \approx -2$;

from the table in Table 12.9, we look at -2 to find the area of the left of $z = -2$ is

$0.5000 - 0.4772 = 0.0228$. The probability that it will be defective is about 2.28%.

52. Variance is 0.0004 is a standard deviation of $\sigma = 0.02$. The diameter of 0.44 is two standard

deviations above the mean, so the probability is $0.50 - 0.4772 = 0.0228$.

53. $z = \dfrac{0.41 - 0.4}{0.02} = 0.50$; from Table 12.9, the area is 0.1915, so the probability of exceeding this is

$0.50 - 0.1915 = 0.3085$.

54. **a.** 1   **b.** 22   **c.** 136   **d.** 341   **e.** 341   **f.** 136   **g.** 23

55. $\text{Var} = 0.04$; $\sigma = 0.2$

a. 50% are less than the mean   **b.** 2.3% is the percent that is greater than two standard

deviations above mean: $12 + 2(0.2) = 12.4$ oz

56. We recognize 2.3% to be two standard deviations, so 13.5 is two standard deviations; we find

$13.5 - 2(0.3) = 12.9$ oz

Level 3 page 642

57.			58.			59.		
	-4	0.00013		-4	0.00001		20	0.0004
	-3	0.00443		-3	0.00195		30	0.0054
	-2	0.05399		-2	0.06260		40	0.0242
	-1	0.24197		-1	0.50000		50	0.0399
	0	0.39894		0	1.00000		60	0.0242
	1	0.24197		1	0.50000		70	0.0054
	2	0.05399		2	0.06250		80	0.0004
	3	0.00443		3	0.00195			
	4	0.00013		4	0.00001			

Answers vary.

60. Answers vary; graph **a** has less variance and graph **b** has more variance.

12.4 Correlation and Regression, page 643

It is fairly easy to use a spreadsheet to find the regression line and the correlation coefficient.

```
Data Analysis

Item number, n     x              y              xy        x^2        y^2
1              [Enter data]   [Enter data]   B4*C4     B4^2       C4^2
A4+1           [Enter data]   [Enter data]   replicate replicate  replicate
replicate      [Enter data]   [Enter data]
               [Enter data]   [Enter data]
               [Enter data]   [Enter data]
               [Enter data]   [Enter data]
               [Enter data]   [Enter data]
               [Enter data]   [Enter data]
               [Enter data]   [Enter data]
               [Enter data]   [Enter data]
               [Enter data]   [Enter data]
               [Enter data]   [Enter data]
               [Enter data]   [Enter data]
               [Enter data]   [Enter data]
               [Enter data]   [Enter data]
               [Enter data]   [Enter data]
               @SUM(B4..B20)  @SUM(C4..C20 @SUM(D4..D20) @SUM(E4..E20) @SUM(F4..F20)

What is n? [Enter data]  r =  (B24*D22-B22*C22)/(sqrt(B24*E22-(B22)^2)*sqrt(B24*F22-(C22)^2))

                         m =  (B24*D22-B22*C22)/(B24*E22-(B22)^2)
                         b =  (C22-E26*B22)/B24
```

Level 1, page 648

1. Answers vary; *correlation* is a number which measures the relationship between two variables.

2. The *linear correlation coefficient*, r, is $r = \dfrac{n\Sigma xy - (\Sigma x)(\Sigma y)}{\sqrt{n(\Sigma x^2) - (\Sigma x)^2}\sqrt{n(\Sigma y^2) - (\Sigma y)^2}}$

3. To decide if there is a correlation between two variables, use the linear correlation coefficient, r, and Table 12.10. If r is greater than the number shown in the table, then there is a linear correlation.

4. The *least squares* (or *regression*) *line* is $y' = mx + b$ where

$$m = \frac{n(\Sigma xy) - (\Sigma x)(\Sigma y)}{n(\Sigma x^2) - (\Sigma x)^2} \qquad b = \frac{\Sigma y - m(\Sigma x)}{n}$$

This is the line of *best fit.*   5. Strong positive correlation   6. Strong positive correlation

7. no   8. no   9. yes   10. yes   11. no   12. no   13. yes   14. yes   15. yes   16. no

17. no   18. yes

19. $r = -0.765$; not significant

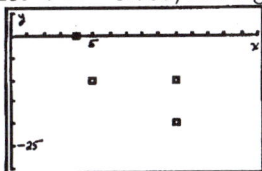

20. $r = 0.995$, significant at 1%

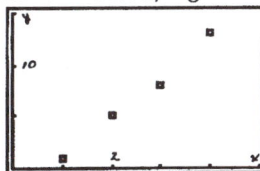

21. $r = -0.954$, significant at 5%

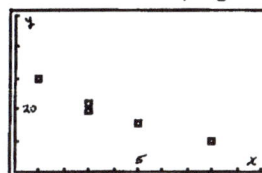

22. $r = -0.985$, significant at 1%

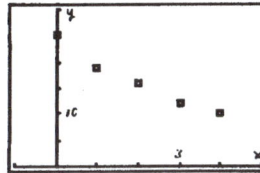

23. $r = -0.890$, significant at 5%

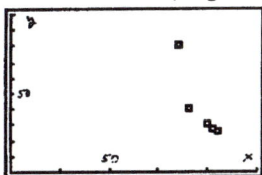

24. $r = 0.911$, significant at 5%

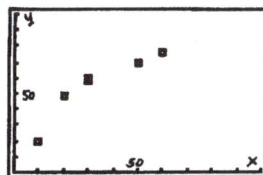

25. $y' = -1.95x + 4.146$ 26. $y' = 3.9x - 3$ 27. $y' = -2.7143x + 30.0571$

28. $y' = -3.7x + 23.8$ 29. $y' = -2.380x + 270.00$ 30. $y' = 0.96x + 23.12$

31. D   32. E   33. A   34. B   35. C   36. F

Level 2, page 649

37. $\quad m = \dfrac{8(240) - 40(48)}{8(210) - (40)^2} = 0 \qquad b = \dfrac{48 - m(40)}{8} = 6;$ Least squares line is $y' = 6$

$\quad r = \dfrac{8(240) - 40(48)}{\sqrt{8(210) - (40)^2}\sqrt{8(288) - (48)^2}} = 0;$ Correlation coefficient is 0

38. $\quad m = \dfrac{10(60) - 30(20)}{10(90) - (30)^2} = 0 \qquad b = \dfrac{20 - m(30)}{10} = 2;$ Least squares line is $y' = 2$

Correlation coefficient is 0

39. $\quad m = \dfrac{6(322) - 36(48)}{6(250) - (36)^2} = 1 \qquad b = \dfrac{48 - m(36)}{6} = 2;$ Least squares line is $y' = x + 2$

$\quad r = \dfrac{6(322) - 36(48)}{\sqrt{6(250) - (36)^2}\sqrt{6(418) - (48)^2}} = 1;$ Correlation coefficient is 1

40. $\quad m = \dfrac{10(-13) - 57(1)}{10(353) - (57)^2} = -0.67 \qquad b = \dfrac{1 - m(57)}{10} = 3.89;$

Least squares line is $y' = -0.67x + 3.89$

$\quad r = \dfrac{10(-13) - 57(1)}{\sqrt{10(353) - (57)^2}\sqrt{10(45) - (1)^2}} = -0.53;$ Correlation coefficient is -0.53

41. $r = 0.8830$, significant at 5%

42. $r = 0.64346$, significant at 5%

43. $r = 0.8421$, significant at 1%

44. $y' = 0.0224x + 0.31734$

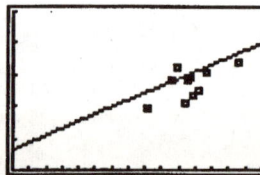

45. $y' = 2.4545x + 6.0909$

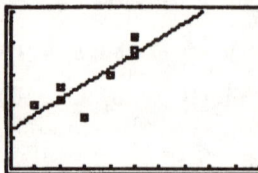

46. $r = 0.836$, significant at 1%

```
LinReg
 y=ax+b
 a=.3811708677
 b=6.481153444
 r=.8360941554
```

47. $r = 0.358$, no significant correlation

```
LinReg
 y=ax+b
 a=.46761113
 b=19.23530709
 r=.3584296566
```

48. $r = -0.154$, no significant correlation

```
LinReg
 y=ax+b
 a=-.1815718157
 b=14.69918699
 r=-.1537939615
```

49. $r = -0.936$, significant at 1%

```
LinReg
 y=ax+b
 a=-6.186388358
 b=87.17895046
 r=-.9356261053
```

50. $y' = 0.381x + 6.481$

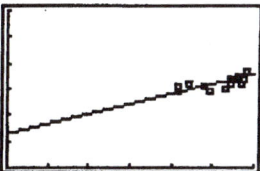

51. $y' = 0.468x + 19.235$

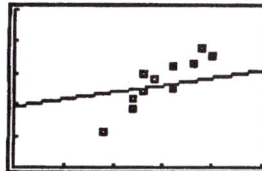

52. $y' = -0.182x + 14.699$

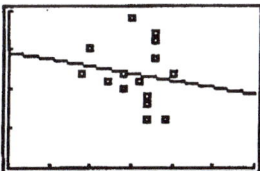

53. $y' = -6.186x + 87.179$

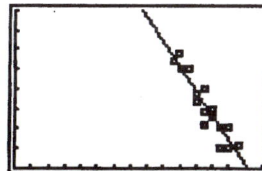

In Problems 54-57, don't forget to change minutes to parts of a hour.

54.   Note 2:57 is 2 hr, 57 min or $2 + \frac{57}{60} = 2.95$. Note the other *n*-values below are found similarly.

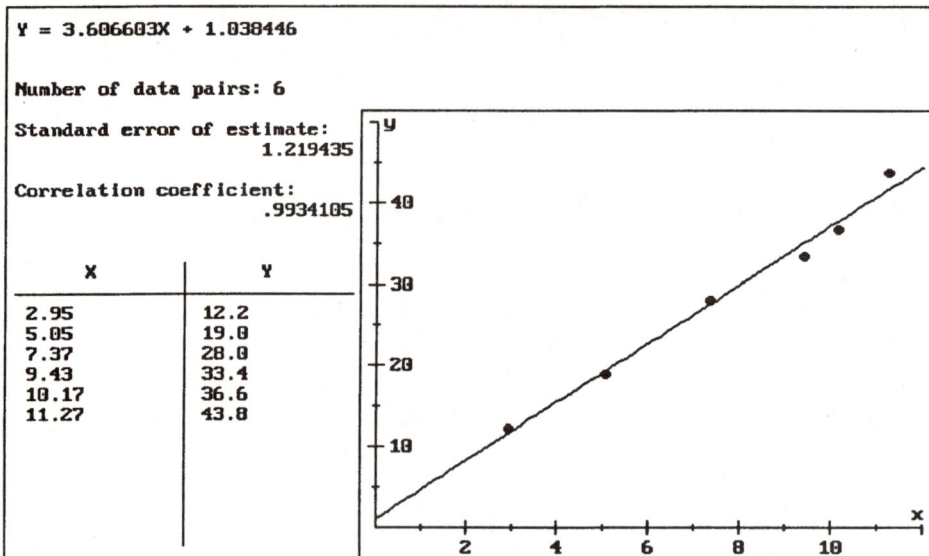

```
Y = 3.606603X + 1.038446

Number of data pairs: 6

Standard error of estimate:
                   1.219435

Correlation coefficient:
                   .9934185

      X              Y

    2.95           12.2
    5.05           19.0
    7.37           28.0
    9.43           33.4
   10.17           36.6
   11.27           43.0
```

Best fitting line is $y' = 3.6x + 1.04$ $51.7 = 3.6x + 1.04$
$$x = 14.072$$
Time is 14:04; 14 hr 4 min

55. $100.1 = 3.6x + 1.04$
$$x = 27.5167$$
Time is 27:31. The difference is 4 hr. 29 minutes. The reason is that he may be getting tired.

56.   Note 3:52 is 3 hr, 53 min or $3 + \frac{53}{60} \approx 3.88$. Note the other *n*-values below are found similarly.

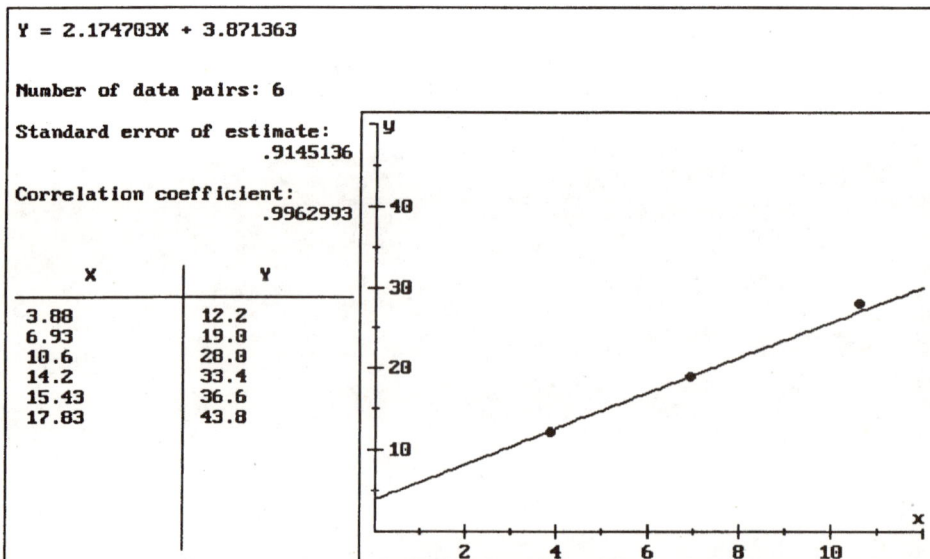

```
Y = 2.174703X + 3.871363

Number of data pairs: 6

Standard error of estimate:
                   .9145136

Correlation coefficient:
                   .9962993

      X              Y

    3.88           12.2
    6.93           19.0
   10.6            28.0
   14.2            33.4
   15.43           36.6
   17.83           43.0
```

Best fitting line is $y = 2.17x + 3.87$ $51.7 = 2.17x + 3.87$

$$x = 22.04$$

Time is 22:49; 22 hr, 49 min

Level 3, page 650

57. $100.1 = 2.17x + 3.87$

$x = 44.35$ *Change 0.35 to minutes; namely* $60(0.35) \approx 21$.

Time is 44:21 or 44 hr. 21 min, which is 3 hours and 29 minutes less than his actual time.

58. Correlation is $r = 0.36$. There is significant correlation at the 5% level, but not at the 1% level (Table 12.10).

59. Correlation is $r = 0.380$. There is significant correlation at the 5% level, but not at the 1% level (Table 12.10).

60. $y' = -0.29x + 2.57$ with correlation $r = -0.33$

This is significant at the 5% level. However, without the last data point, we obtain $r = -0.1794152$. This is not significant at the 1% or 5% levels.

12.5 Sampling, page 651

Level 1, page 655

1. Answers vary. Dictionary: (1) facts or data of a numerical kind, assembled, classified, and tabulated so as to present significant information about a given subject (2) the science of assembling, classifying, tabulating, and analyzing such facts and data (*Webster's New World Dictionary*)

2. Answers vary. Descriptive statistics is concerned with the accumulation of data, measures of central tendency, and dispersion. Inferential statistics is concerned with making generalizations or predictions about a population based on a sample from that population.

3. Answers vary. The fallacy of exception occurs when you form a conclusion by looking at a single case.

4. The target population is the ice cream customers, so C is the correct choice.

5. The target population is Pacific Bell's customers, so B is the correct choice.

6. The target population is those who come to the theater to watch movies, so C is the correct choice.

7. The target population is the registered voters, so B is the correct choice.

8. The target population are parents of mathematics students, so D is the correct choice. Choice C has the correct target population, but the first 1,000 names is not a random selection from the target population.

9. The target population are members of the environmental group, so D is the correct choice.

10. The target population are those living in a particular neighborhood, so D is the correct choice.

11. The target population is the customers of Betty's Koi Emporium, so A is the correct choice.

12. The target population is the residents of a particular city, so B is the correct choice.

13. The target population is the customers of an espresso coffee franchise, so D is the correct choice.

Level 2, page 657

Answers to Problems 14-23 may vary.

14. Survey customers whose names are randomly chosen from a listing of all customers.

15. Survey AT&T's customers by randomly choosing from a list of AT&T's customers.

16. Survey from a random sample of people whose names are taken from a list of cat owners.

17. Survey a selection of people whose names are randomly chosen from a local list of registered voters who are party members.

18. Survey a selection of people whose names are randomly chosen from a list of all club members.

19. Survey a selection of people whose names are randomly chosen from a list of all union members.

20. Survey a selection of people whose names are randomly chosen from the city's telephone directory.

21. Survey a random selection of students chosen from a list of all students.

22. Survey a random selection of people whose names appear on lists of persons who have purchased video recorder from that retailer.

23. Survey a selection of people whose names are randomly chosen from a list of all customers.

24. The four possibilities are:

    (1) You accept that 72 is the mean, and it is the mean.

    (2) You accept that 72 is the mean, and it is not the mean.

    This is Type II error (accept a false conclusion).

    (3) You do not accept that 72 is the mean, and it is the mean.

    This is Type I error (reject a true conclusion).

    (4) You do not accept that 72 is the mean, and it is not the mean.

25. The four possibilities are:

(1) You accept that 72 is the mean, and it is the mean.

(2) You accept that 72 is the mean, and it is not the mean.

This is Type II error (accept a false conclusion).

(3) You do not accept that 72 is the mean, and it is the mean.

This is Type I error (reject a true conclusion).

(4) You do not accept that 72 is the mean, and it is not the mean.

26. I would reject the hypothesis that the majority feel social security taxes should be raised.

(1) You accept that the majority feel social security taxes should be raised, and they do.

(2) You accept that the majority feel social security taxes should be raised, and they do not.

This is Type II error (accept a false conclusion).

(3) You reject that the majority feel social security taxes should be raised, and they do.

This is Type I error (reject a true conclusion).

(4) You reject that the majority feel social security taxes should be raised, and they do not.

Level 3 Problem Solving, page 657

27. Answers vary.   **28.** Answers vary.   **29.** Answers vary.

30. Suppose the coin lands heads 60% of the time. Then it lands tails 40% of the time.

$P(\text{heads/tails}) = 0.6 \times 0.4 = 0.24$; $P(\text{tails/heads}) = 0.4 \times 0.6 = 0.24$.

Chapter 12 Review Questions, page 659

1. a. Heads: ⊞⊞ ⊞⊞ ⊞⊞ ⦀ (18); Tails ⊞⊞ ⊞⊞ ⊞⊞ ⊞⊞ ⸾⸾ (22)

b.

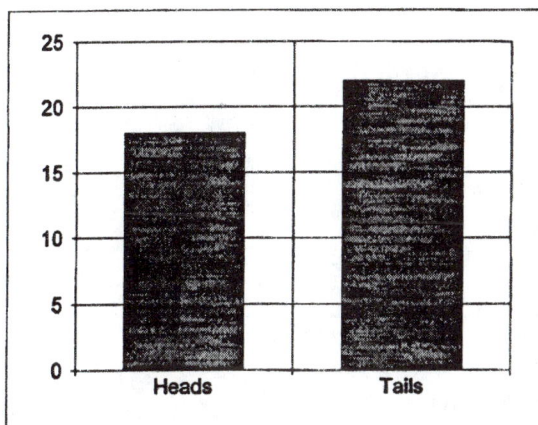

2. a.

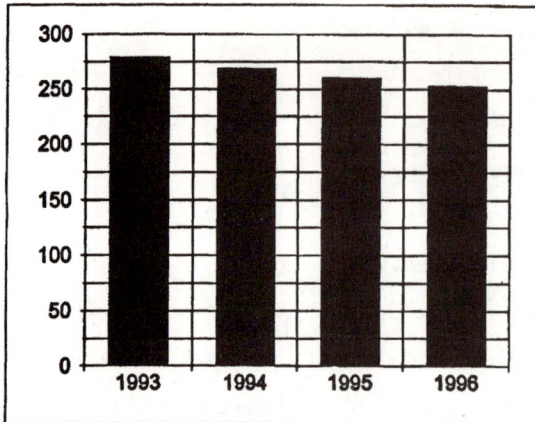

b.

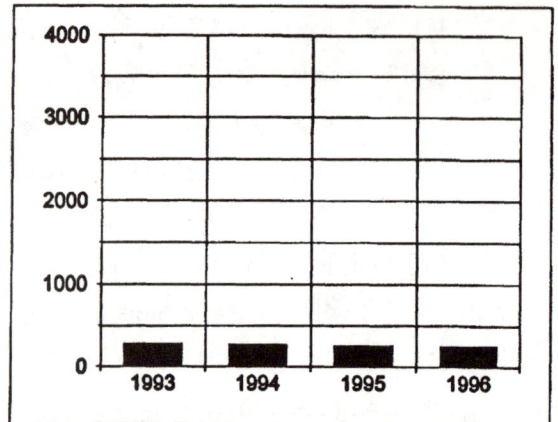

c.

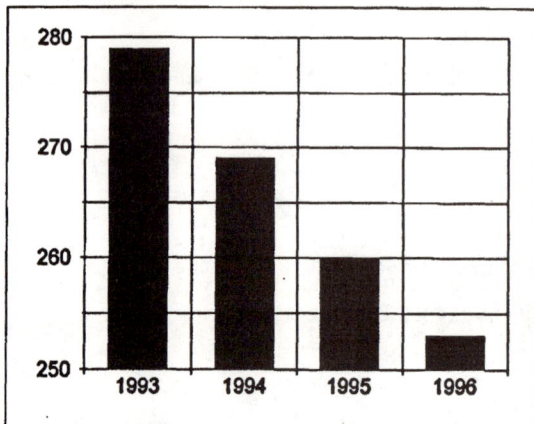

d. Answers vary; impressions can be greatly influenced by using faulty or inappropriate scale (or even worse, no scale at all).

3. $\quad \text{mean} = \dfrac{5 + 21 + 21 + 25 + 30 + 40}{6} = 23\frac{2}{3}$ **4.** The median is the mid value or the mean of the mid-values; $\dfrac{21 + 25}{2} = 23.$ **5.** The mode is 21.   **6.** The range is $40 - 5 = 35$

7. The standard deviation is found:

Score	(deviation from the mean)^2
5	348.44
21	7.11
21	7.11
25	1.78
30	40.11
40	266.78

$$\frac{344.44 + 7.11 + 7.11 + 1.78 + 40.11 + 266.78}{6 - 1} = \frac{671.328}{5} \approx 134.266$$

The standard deviation is $\sqrt{134.266} \approx 11.59$

8. mean = 11; median = 12; mode = 12; mode is the most appropriate measure

9. mean = 79; median = 74; no mode; mean is the most appropriate measure

10. mean = $\dfrac{\$2,460,172}{38} \approx \$64,741.37$; median = $\dfrac{\$67,691 + \$69,000}{2} = \$68,345.50$; no mode; the median is the most appropriate measure

11. **a.** yes

b. yes

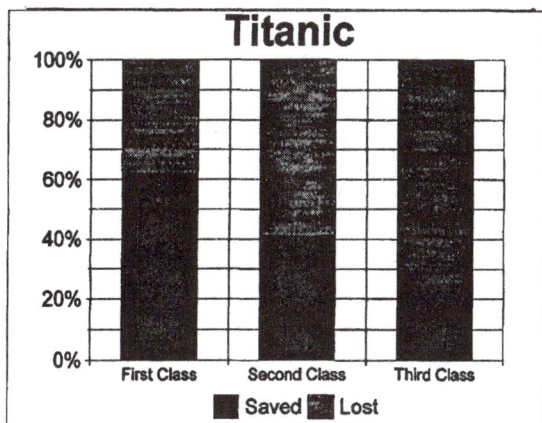

Survival rates of first, second
and third class passengers.

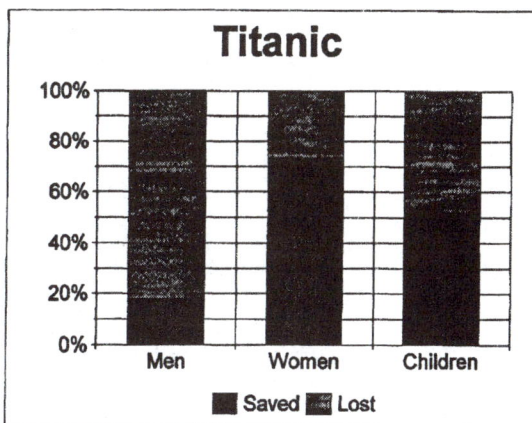

Survival rates of men, women,
and children.

12. **a.** 1982   **b.** 1994   **c.** $41,500

13. **a.**

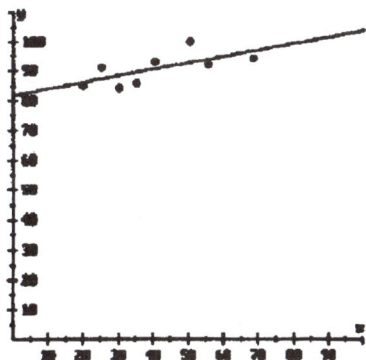

$$m = \frac{n(\Sigma xy) - (\Sigma x)(\Sigma y)}{n(\Sigma x^2) - (\Sigma x)^2}$$

$$= \frac{8(29,677) - (323)(725)}{8(14,899) - 104,329}$$

$$\approx 0.2180582655$$

$$b = \frac{\Sigma y - m(\Sigma x)}{n}$$

$$= \frac{725 - (0.2180582655)(323)}{8}$$

$$\approx 81.82089753$$

The best-fitting line is $y' = 0.22x + 81.8$.

b. $n = 8$ $\Sigma x^2 = 14,899$

$\Sigma x = 323$ $\Sigma y^2 = 65,907$

$\Sigma y = 725$ $(\Sigma x)^2 = 104,329$

$\Sigma xy = 29,677$ $(\Sigma y)^2 = 525,625$

$$r = \frac{n\Sigma xy - (\Sigma x)(\Sigma y)}{\sqrt{n(\Sigma x^2) - (\Sigma x)^2}\sqrt{n(\Sigma y^2) - (\Sigma y)^2}}$$

$$= \frac{8(29,677) - (323)(725)}{\sqrt{8(14,899) - 104,329}\sqrt{8(65,907) - 525,625}}$$

$$\approx 0.6582620404; \text{ not significant at 1\% level or 5\% level}$$

14. $314 - 266 = 48$; since the standard deviation is 16 days, we note this is 3 standard deviations greater than the mean. From the standard normal curve, this should happen about 0.001 (0.1 %) of the time. This means that approximately 1 in 1,000 pregnancies will have a 314-day duration.

15. a.

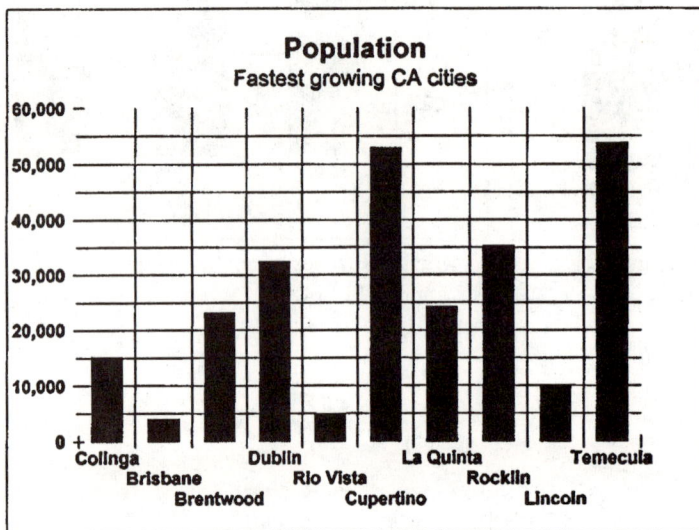

b. $\text{mean} = \dfrac{255,585}{10} = 25,558.5; \quad \text{median} = 23,675; \text{ no mode}$

Group Research Problems, pages 661-662

G44. a. $\bar{x} = \dfrac{2,264}{30} \approx 75.46$; the median is 76, and the data is bimodal: 76 and 87. The standard deviation is 14.40. The grading scale for the class of 30 students has a mean of 75.46 and a standard deviation of 14.4, so grades of A and F are more than two standard deviations away from the mean, with the grades of B and D one standard deviation from the mean:

Score	Grade	Number
104 or above	A	0
90-103	B	3
61-89	C	24
47-60	D	2
Below 47	F	1

b. Mean is 43.29 years; since there are 76 pieces of data, the median is found by looking at the stem-and-leaf plot: it is 42 years; the mode is 38. The ages range from 29 to 76 (47 years) and the standard deviation is 8.8 years.

G45. Answers vary; this is the famous Buffon's Needle Problem.

G46. Answers vary.

Individual Research Problems, pages 663-664

The projects of this section are all research papers, so the answers will, of course, vary. Be sure to check out the web resources at **mathnature.com**.

CHAPTER 13

THE NATURE OF GRAPHS AND FUNCTIONS

13.1 Cartesian Coordinates and Graphing Lines, page 666

Level 1, page 673

1. a. Aphrodite **b.** Maxwell Montes **c.** Atalanta Planitia **d.** Rhea Mons **e.** Lavina Planitia
2. To solve a linear equation in two variables means to find all ordered pairs that satisfy the given equation. Since there are generally infinitely many such order pairs, the solution is generally shown as a graph. **3.** The y-intercept of a line is the point $(0, b)$ where the line crosses the y-axis. The slope of a line is the steepness of a line. **4.** Find three ordered pairs that satisfy the given equation, plot those points, and use a straightedge to draw the line passing through the three points. Note: only 2 points are necessary, the third point being used as a check. **5.** Horizontal lines have slope 0. Vertical lines have no slope.
6.

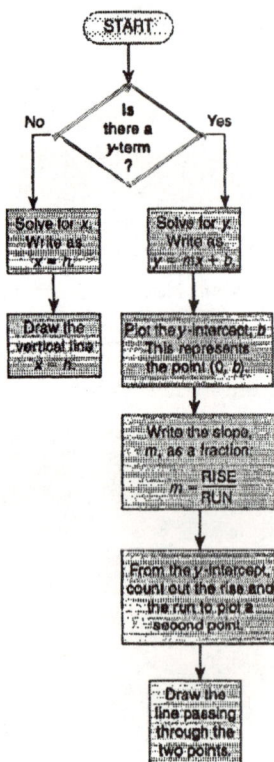

Ordered pairs in Problems 7-18 may vary.

7. $(0, 5), (1, 6), (2, 7)$

8. $(0, -1), (1, 1), (2, 3)$

9. $(0, 5), (1, 7), (-1, 3)$

10. $(0, -4), (1, -3), (4, 0)$

11. $(0, -1), (1, 0), (2, 1)$

12. $(0, 1), (-1, 0), (1, 2)$

13. $(0, 1), (1, -1), (-1, 3)$

14. $(0, 1), (1, -2), (2, -5)$

15. $(0, 1), (1, 3), (2, 5)$

16. $(0, 1), (1, -2), (2, -5)$

17. $(0, 2), (2, \frac{1}{2}), (4, -1)$

18. $(0, 2), (4, 0), (2, 1)$

19.   **20.**

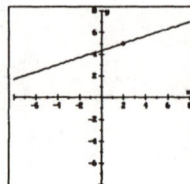   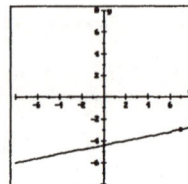

21.

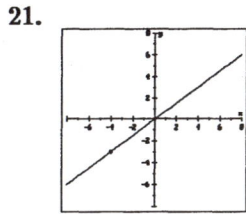

22.

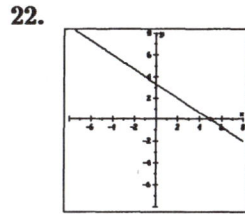

23.

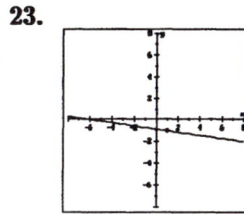

24.

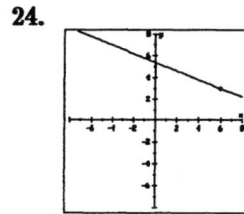

25.

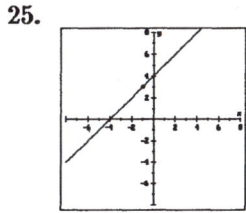

26.

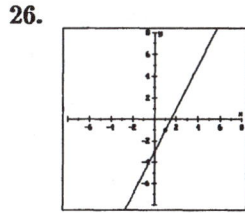

27.

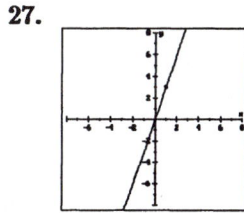

28.

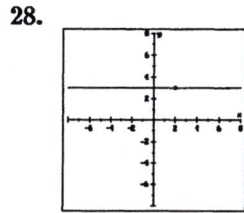

Level 2, page 674

29.

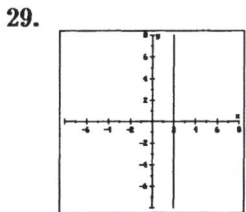

30.

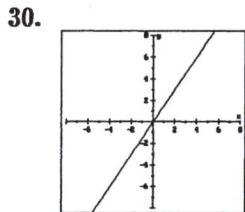

31.

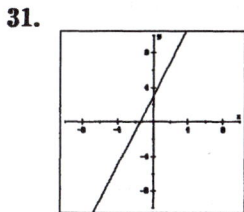

32.

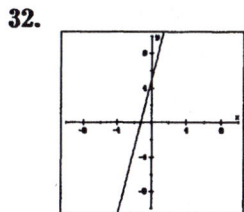

33.

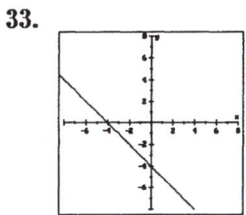

34.

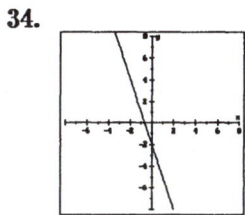

35.

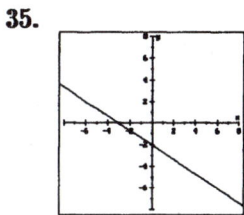

36.

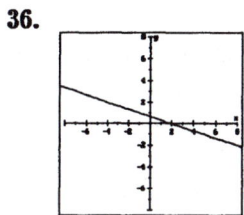

37.

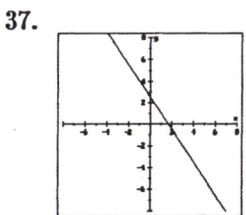

38.

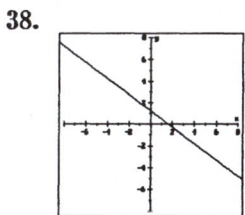

39.

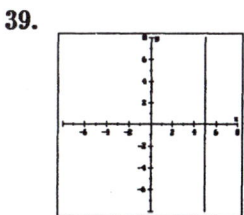

40.

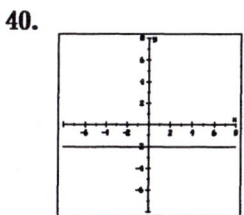

41.

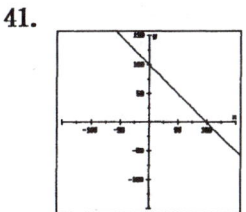

42.

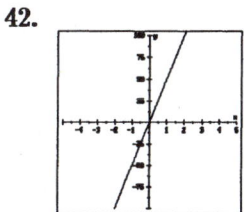

43.

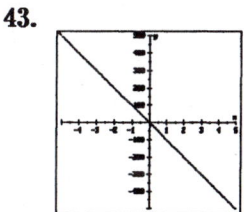

44.

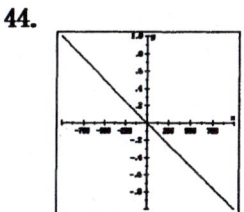

Level 3 Problem Solving, page 674

45. The y-intercept appears to be -2 and the slope $\frac{7}{10}$ (rather than $\frac{9}{10}$); C

46. The y-intercept appears to be -4 and the slope $\frac{7}{10}$ (rather than $\frac{1}{2}$); H

47. The y-intercept appears to be -4 and the slope $\frac{5}{10}$ (rather than $\frac{7}{10}$); A

48. The y-intercept appears to be 2 and the slope $-\frac{1}{2}$ (rather than $-\frac{9}{10}$); B

49. The y-intercept appears to be 2 and the slope $-\frac{9}{10}$ (rather than $-\frac{5}{10}$); E

50. The y-intercept appears to be 4 and the slope $-\frac{5}{10}$ (rather than $-\frac{7}{10}$); D

Level 3 Problem Solving, page 657

Problems 51-52 were inspired by the Media Clips feature of the December 1996 issue of The Mathematics Teacher, pp. 739-740. My thanks to Ron Lancaster, one of the editors of that column.

51.   **a.** It is declining at a rate that is nearly linear, but you can use slopes to prove that it is not linear. (Looking at a graph does not provide sufficient information to make a decision.)

    b. Answers vary; slope's vary from a low of -0.17 to a high of -0.14 with low estimate fertility rate of 2.2 to a high estimate of 5.8. Suppose we choose 1970 as the base year; that is, let $x = 0$ represent 1970. Use the data points $(0, 5.8)$ and $(10, 4.4)$ to find
$$m = \frac{4.4 - 5.8}{10 - 0} = -0.14$$
for equation $y = -0.14x + 5.8$.

    c. In 2005, $x = 35$, we predict $y = -0.14(35) + 5.8 = 0.9$.   (Predictions may vary.)

52.   Solve the two given equation for y to put them into slope-intercept form. Compare the slopes; if the slopes are equal, but the y-intercepts are different, then the lines are parallel. By the way, if the slopes are the same and the y-intercepts are also the same, then there is only one line.

53.   **a.**   **b.** 800   **c.** 10,000

$$A = 10{,}000(1 + 0.08t)$$
$$= 800t + 10{,}000$$

54.   **a.** $P = 20{,}000 - (18{,}000/8)t$
$$= -2{,}250t + 20{,}000$$

b.

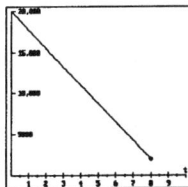

c. $-2,250$

55. a.

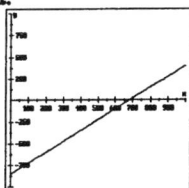

b. When $x = 0$, $P = -850$; if no items are sold there is a loss of $850.

c. 1.25; it is the profit increase corresponding to each unit increase in number of items sold

56. a.

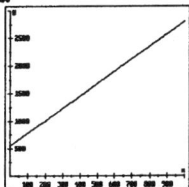

b. When $x = 0$, $C = 550$; the fixed costs are $550.

c. 2.25; it is the increase in cost corresponding to each unit increase in number of items

57.

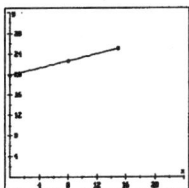

Let $x = 0$ be the base year 1990.

$$m = \frac{19.8 - 17.0}{8 - 0} = 0.35$$

$$y = 0.35x + 17.0$$

For 2005, $x = 15$ so $y = 0.35(15) + 17.0 = 22.25$.

The projected population is about 22.3 million.

58. $y = mx + b$ passes through (h, k) means $k = mh + b$ and $b = k - mh$; substitute back into the original equation:

$$y = mx + (k - mh)$$
$$y - k = mx - mh$$
$$y - k = m(x - h)$$

59. Given $y - k = m(x - h)$. Since the line passes through (x_1, y_1) and (x_2, y_2) the slope is

$m = \frac{y_2 - y_1}{x_2 - x_1}$. Let $(h, k) = (x_1, y_1)$ be the known point. Then, by substitution,

$$y - y_1 = \left(\frac{y_2 - y_1}{x_2 - x_1}\right)(x - x_1).$$

60. Since the line passes through $(0, b)$ and $(a, 0)$, $a \neq 0$, $b \neq 0$, we have from Problem 59:

$$y - 0 = \left(\frac{0 - b}{a - 0}\right)(x - a)$$

$$y = \frac{-b}{a}(x - a)$$

$$ay = -bx + ab$$

$$bx + ay = ab$$

$$\frac{x}{a} + \frac{y}{b} = 1 \qquad \text{Divide by } ab; \ (a \neq 0, b \neq 0).$$

13.2 Graphing Half-Planes, page 675

Level 1, page 676

1. A *linear inequality* in two variables, say x and y, is an inequality of the form

$Ax + By + C \leq 0$, $Ax + By + C < 0$, $Ax + By + C \geq 0$, $Ax + By + C > 0$.   **2.** *Step 1. Graph the boundary.* Replace the inequality symbol with an equality symbol and draw the resulting line. This is the boundary line. Use a solid line when the boundary is included ($\leq$ or $\geq$). Use a dashed line when the boundary is not included ($<$ or $>$). *Step 2. Test a point.* Choose any point in the plane that is not on the boundary line; the point $(0, 0)$ is usually the simplest choice. If this *test point* makes the *inequality* true, shade in the half-plane which contains the test point. That is, the shaded plane is the solution set. If the test point makes the *inequality* false, shade in the other half-plane for the solution.

3. F; the boundary is $2x + 5y = 2$.   **4.** F; can't choose a test point on the boundary line.   **5.** F; $(0, 0)$ does not work because it is on the boundary; choose a point not on the boundary.   **6.** T

7. T   **8.** F; $(0, 0)$ is not a test point because it lies on the boundary.   **9.** T   **10.** T   **11.** T

Level 2, page 677

12. 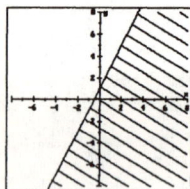   **13.** 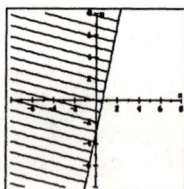   **14.** 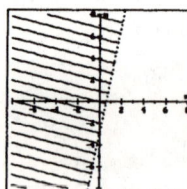   **15.**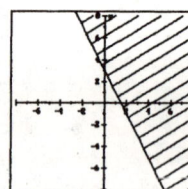

16. 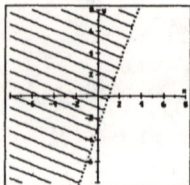   **17.** 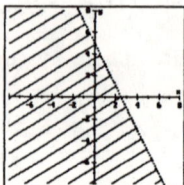   **18.** 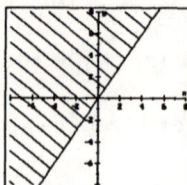   **19.**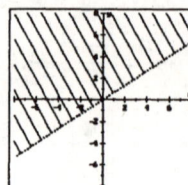

20.

21.

22.

23.

24.

25.

26.

27.

28.

29.

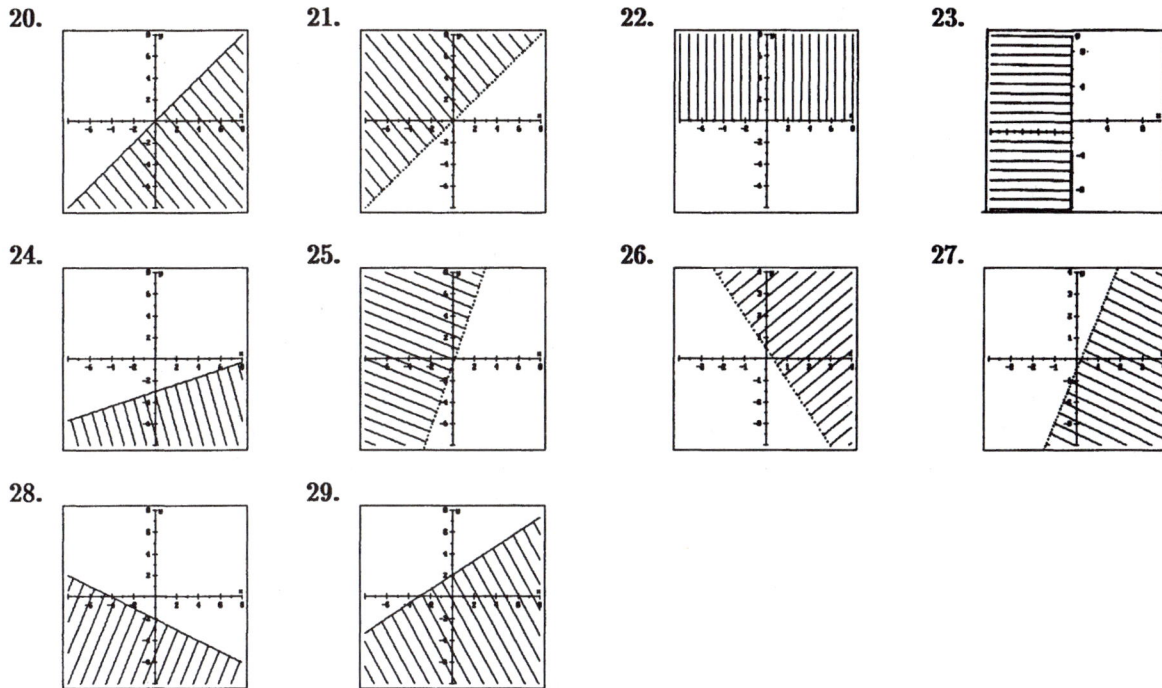

Level 3 Problem Solving, page 677

30. The batter will get a hit or will not get a hit. If the batter gets a hit, then we need to compare

$$\frac{h}{a} \text{ and } \frac{h+1}{a+1}$$

$$\frac{h}{a} \stackrel{?}{=} \frac{h+1}{a+1}$$

$$ha + h \stackrel{?}{=} ah + a$$

$$h \stackrel{?}{=} a$$

Equality holds if the number of hits is the same as the number of times at bat. We know, however, that $a \geq h$, so $h \leq a$. Thus,

$$\frac{h}{a} \leq \frac{h+1}{a+1}$$

with equality holding only if the number of hits is the same as the number of times at bat. Otherwise, the batting average must go up.

If the batter does not get a hit, then we need to compare

$$\frac{h}{a} \text{ and } \frac{h}{a+1}$$

$$\frac{h}{a} \overset{?}{=} \frac{h}{a+1}$$

$$ha + h \overset{?}{=} ah$$

$$h \overset{?}{=} 0$$

Equality holds if the number of hits is 0. Otherwise,

$$h > 0 \qquad \text{so} \qquad \frac{h}{a} > \frac{h}{a+1}$$

This says that the batting average must go down. The only time the batting average stays the same is if $a = h$ or if $h = 0$. That is, the batting average will stay the same if the player is batting a thousand and gets a hit or if the player has not had any hits and does not get a hit. Otherwise, the batting average goes up with every hit and goes down without a hit.

13.3 Graphing Curves, page 677

Level 1, page 681

1. A *parabola* is a curve that is described by the path of a projectile, as shown in Figure 13.16.

2. The symbol e represents an irrational number between 2.7 and 2.8. Using a calculator, we can find a better approximation: $e \approx 2.718281828$. In calculus, a limiting process is discussed and then used to define e as the limit of $\left(1 + \frac{1}{n}\right)^n$ as n becomes infinitely large.   3. Answers vary; use the formula $P = P_0 e^{rt}$.

4.
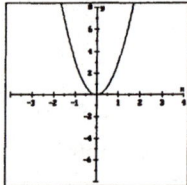

5.
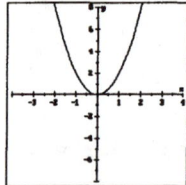

6.
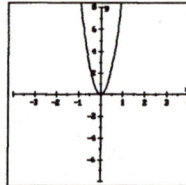

7.
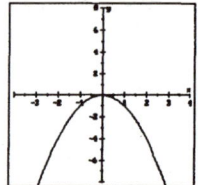

8.
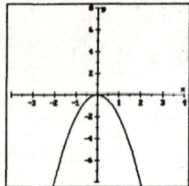

9.
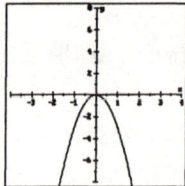

10.
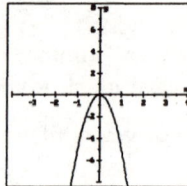

11.
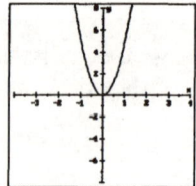

12.

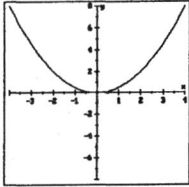

13.

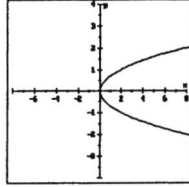

14.

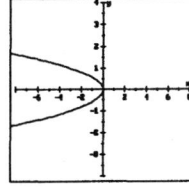

15.

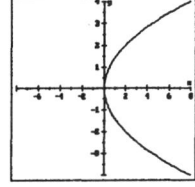

16.

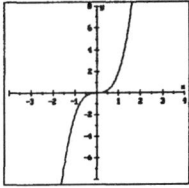

17.

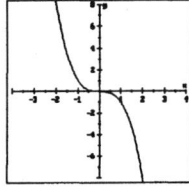

18.

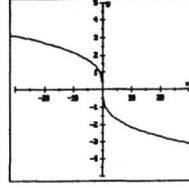

19.

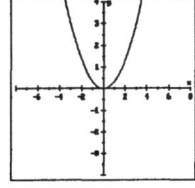

20.

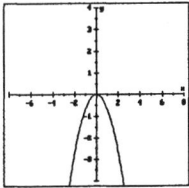

21.

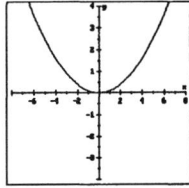

22.

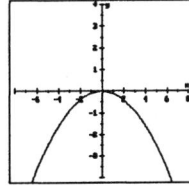

23.

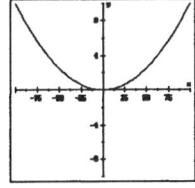

24.

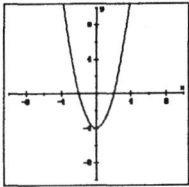

25.

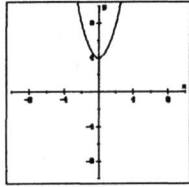

26.

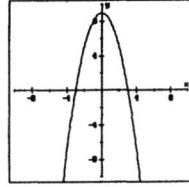

27.

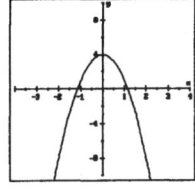

28.

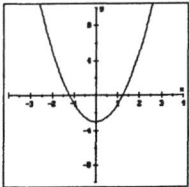

29.

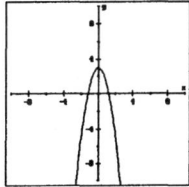

30.

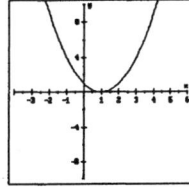

31.

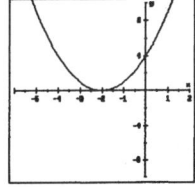

Level 2, page 682

32.

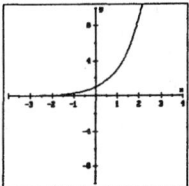

33.

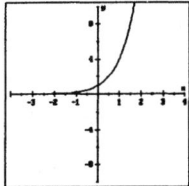

34.

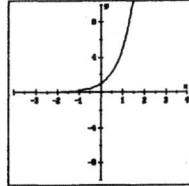

35.

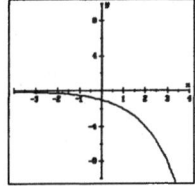

36.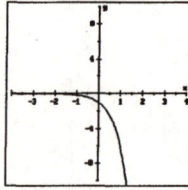

37.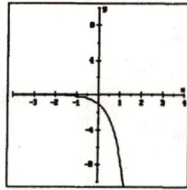

38.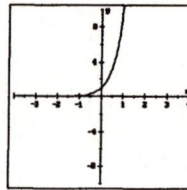

39.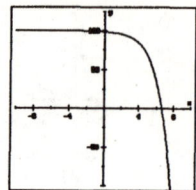

40.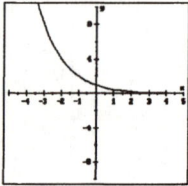

41.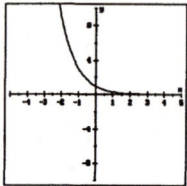

42.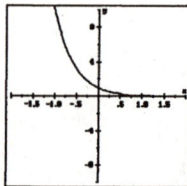

43.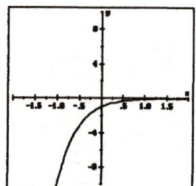

44. a. 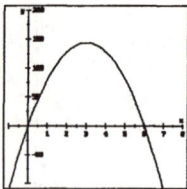　　**b.** downward　　**c.** no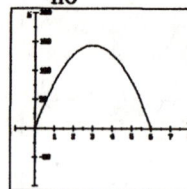

45. a. 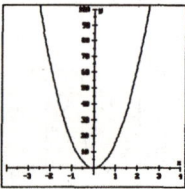　　**b.** upward　　**c.**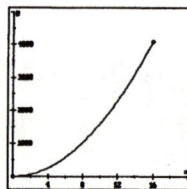

46.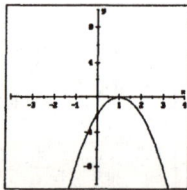

47.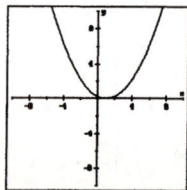

48.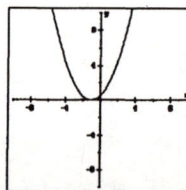

49.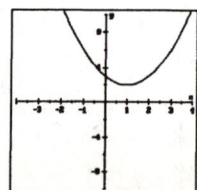

50.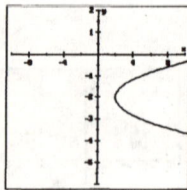

51.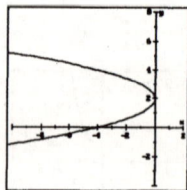

52.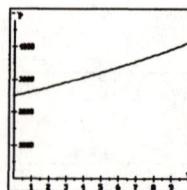

53.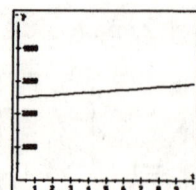

54.

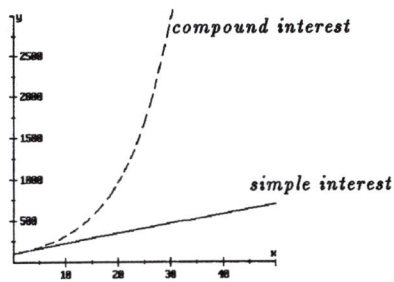

55.

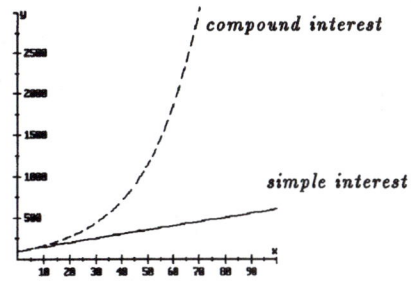

Level 3 Problem Solving, page 682

56.

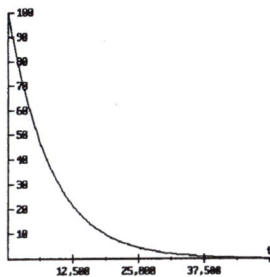

57.

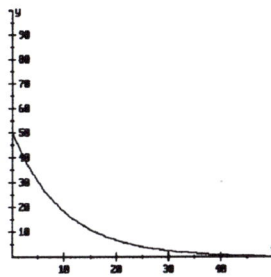

58.

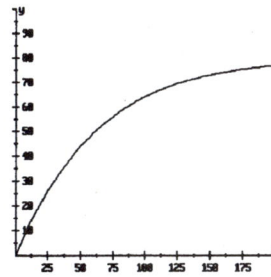

59.

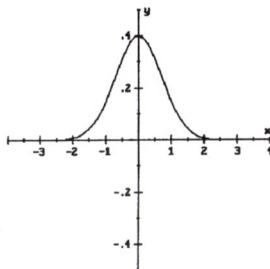

$$y = \frac{e^{-x^2/2}}{\sqrt{2\pi}}$$

60.

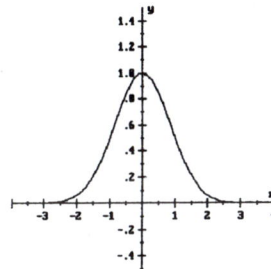

$$y = 2^{-x^2}$$

The shape of the two curves is the same. The *y*-intercepts are (0, 0.4), and (0, 1), respectively. Look at both graphs on the same axis:

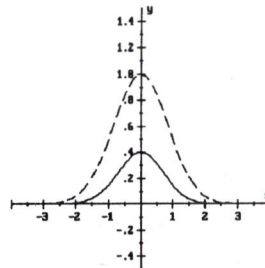

13.4 Conic Sections, page 682

Transparencies 29 and 30 can be used to graph circles, parabolas, ellipses, and hyperbolas. I demonstrate how to do this in class and then I pass out copies of these pages whenever I assign any of the Problems 14-23. These exercises not only help the students learn the definition of the conic sections, but they also help them to learn the correct shape for each of these conic sections.

Level 1, page 694

1. A *conic section* is the set of points which results from an intersection of a cone and a plane. It includes lines, parabolas, ellipses, circles, and hyperbolas.　　**2.** A *parabola* is the set of all points in a plane equidistant from a given point (called the focus) and a given line (called the directrix).　　**3.** An *ellipse* is the set of all points in a plane such that, for each point on the ellipse, the sum of its distances from two fixed points (called the foci) is a constant.　　**4.** A *hyperbola* is the set of all points in a plane such that, for each point on the hyperbola, the difference of its distances from two fixed points (called the foci) is a constant.　　**5.** Write the equation in standard form, so that there is a 1 on the right and the coefficients of the square terms are also 1. The center is $(0, 0)$; plot the intercepts on the x-axis and the y-axis. For the x-intercepts, plot $\pm$ the square root of the number under the x^2; for the y-intercepts, plot $\pm$ the square root of the number under the y^2. Finally, draw the ellipse using these intercepts.　　**6.** Write the equation in standard form, so that there is a 1 on the right and the coefficients of the square terms are also 1. The center is $(0, 0)$; plot the intercepts on either the x-axis and the y-axis. If the negative is in front of the x, then there are no x-intercepts and if the negative is in front of the y, then there are no y-intercepts. For the x-intercepts, plot $\pm$ the square root of the number under the x^2; for the y-intercepts, plot $\pm$ the square root of the number under the y^2. Plot the pseudovertices, draw the slant asymptotes, then the central rectangle. Finally, draw the hyperbola using these vertices and the slant asymptotes.　　**7. a.** Both variables are first-degree. **b.** One variable is first degree the other is second-degree.　**c.** Both variables are second degree and both have the same sign (in general form).　**d.** Both variables are second degree and they have opposite signs (in general form).　　**8. a.** $\epsilon = 1$　**b.** $0 \leq \epsilon < 1$; with a circle having eccentricity $\epsilon = 0$　**c.** $\epsilon > 1$　　**9.** A and C have opposite signs.　　**10.** $A = C$　　**11.** $A = 0$ and $C \neq 0$ or $A \neq 0$ and $C = 0$　　**12.** A and C have the same sign.　　**13.** $A = C = 0$

14.

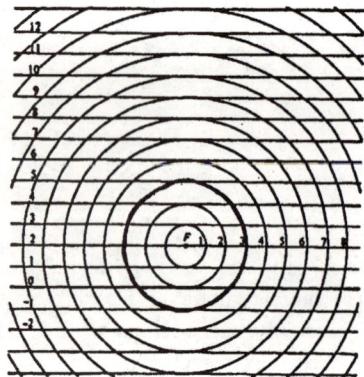

15.

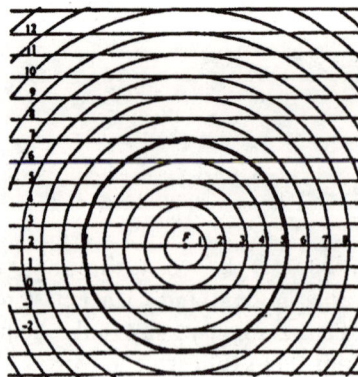

16.

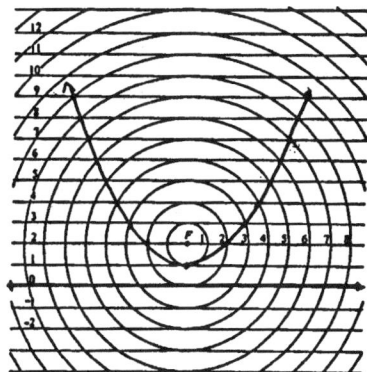

17.

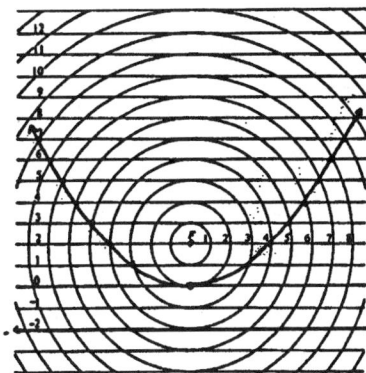

18.

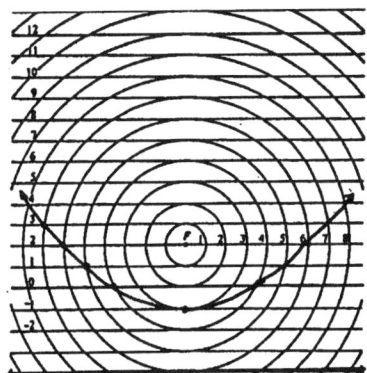

19.

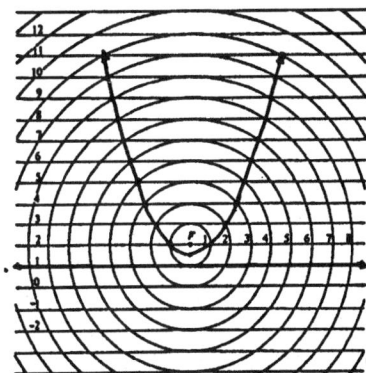

20.

21.

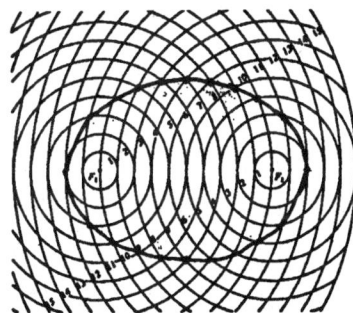

22.

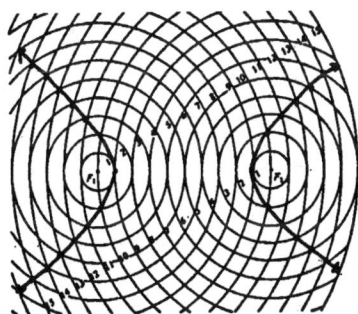

23.

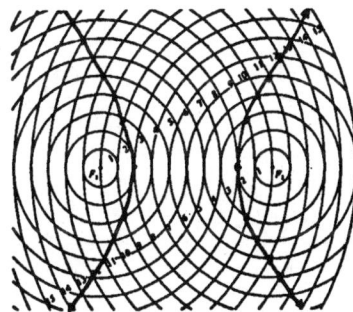

24. a. line **b.** parabola **c.** line   **25. a.** line **b.** ellipse **c.** parabola

26. a. circle **b.** parabola **c.** hyperbola   **27. a.** parabola **b.** hyperbola **c.** ellipse

Level 2, page 694

28.

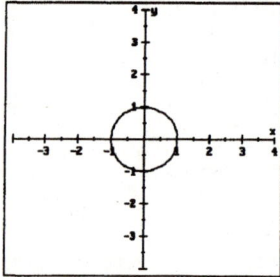

29.

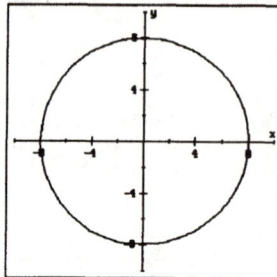

30.

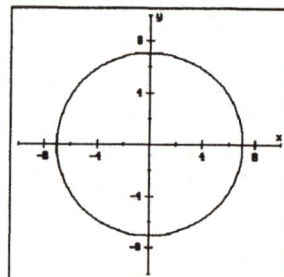

31.

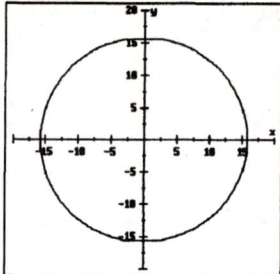

32.

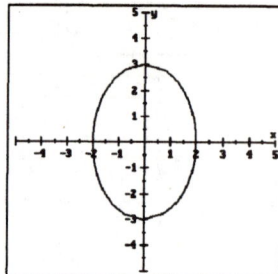

33.

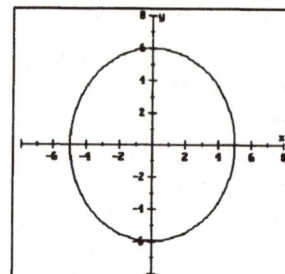

34.

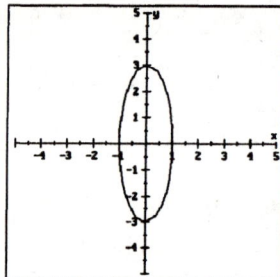

35.

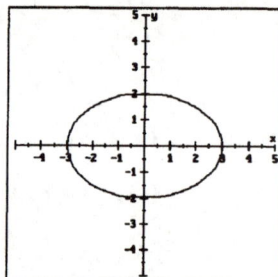

36.

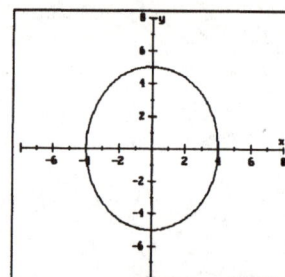

37.

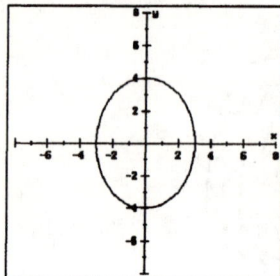

38.

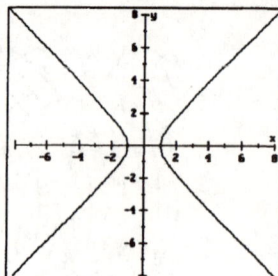

39.

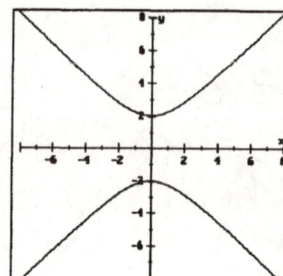

40.

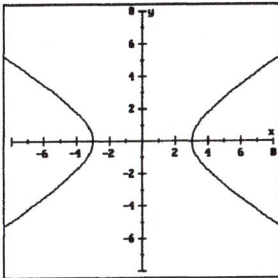

41.

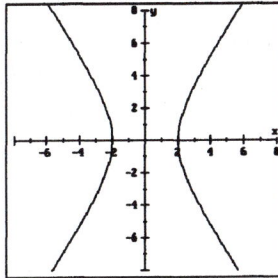

42.

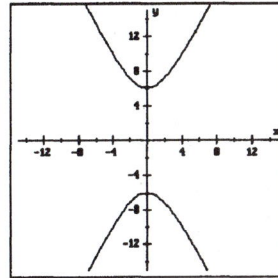

43.

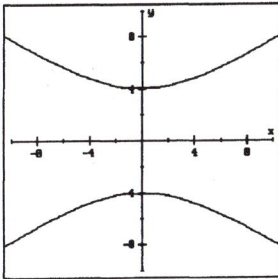

44.

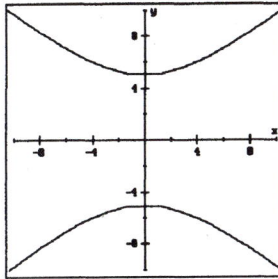

45.

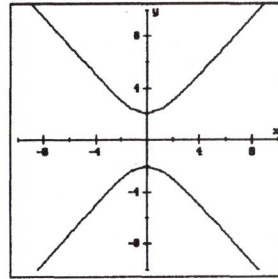

46.

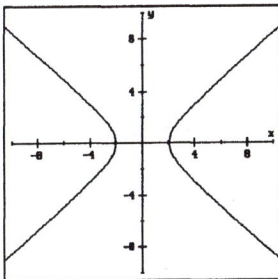

47.

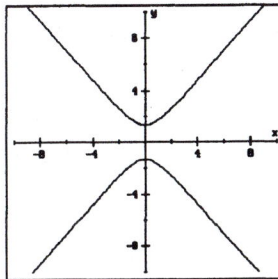

48.

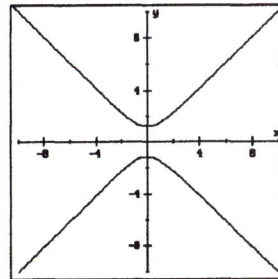

49.

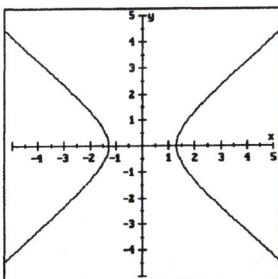

50.

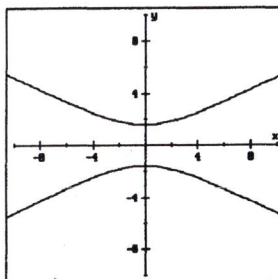

51.

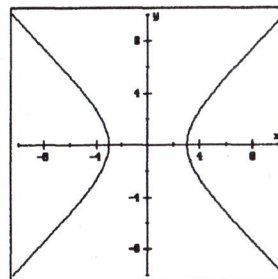

52. $2a = 186,000,000$

      $a = 93,000,000$

      $\epsilon = \dfrac{c}{a}$

    $0.017 = \dfrac{c}{93,000,000}$

      $c = 93,000,000(0.017)$

      $c = 1,581,000$

aphelion: $a + c = 9.3 \times 10^7 + 1.6 \times 10^6$ perihelion: $a - c = 9.3 \times 10^7 - 1.6 \times 10^6$

              $= 9.46 \times 10^7$ mi $= 9.14 \times 10^7$ mi

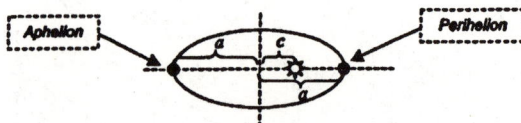

53. $a = 1.4 \times 10^8$, $\epsilon = 0.093$

      $\epsilon = \dfrac{c}{a}$

    $0.093 = \dfrac{c}{1.4 \times 10^8}$

      $c = 1.302 \times 10^7$

aphelion $= a + c = (1.4 \times 10^8) + (1.302 \times 10^7) = 1.5 \times 10^8$ mi

perihelion $= a - c = (1.4 \times 10^8) - (1.302 \times 10^7) = 1.3 \times 10^8$ mi

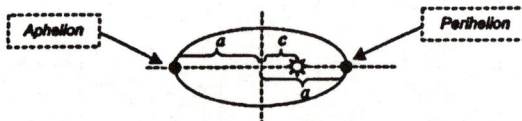

54.   perihelion: $a - c = 2.8 \times 10^7$

    $\epsilon = \dfrac{c}{a}$ so $\dfrac{1}{5} = \dfrac{c}{a}$ and $a = 5c$

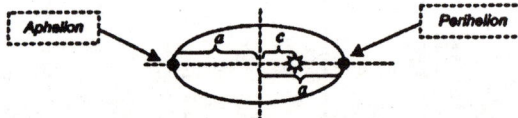

      $5c - c = 2.8 \times 10^7$ *Substitute $a = 5c$.*

        $4c = 2.8 \times 10^7$ *Simplify.*

         $c = 7 \times 10^6$ *Solve for c.*

         $a = 5(7 \times 10^6)$ *Use $a = 5c$ and substitute for c.*

         $a = 3.5 \times 10^7$

       $2a = 7 \times 10^7$ *The number $2a$ is the length of the major axis.*

The length of the major axis is 7×10^7 miles.

55.   apogee $= a + c = 199,000$

        $2a = 378,000$

         $a = 189,000$

$$a + c = 199,000$$

$$189,000 + c = 199,000 \qquad \textit{Substitute } a = 189,000.$$

$$c = 10,000 \qquad \textit{Solve for c.}$$

$$\epsilon = \frac{c}{a}$$

$$\epsilon = \frac{10,000}{189,000} = \frac{10}{189} \approx 0.053 \qquad \textit{Substitute values for c and a.}$$

Level 3 Problem Solving, page 695

56. Let the vertex be at $(0, -3)$. The parabola opens down and passes through $(6, -9)$.

$$(x - h)^2 = -4c(y - k)$$

$$(x - 0)^2 = -4c[y - (-3)]$$

$$x^2 = -4c(y + 3)$$

Substitute $(x, y) = (6, -9)$:

$$6^2 = -4c(-9 + 3)$$

$$36 = 24c$$

$$c = \frac{3}{2}$$

$$4c = 6$$

The equation of the parabola portion is $x^2 = -6(y + 3)$.

Since the widest portion of the parabola is 12 ft, the domain is $D = [-6, 6]$.

57. Let the vertex be $(0, 0)$. Since the parabola opens right, $y^2 = 4cx$. The parabola passes through $(4, 6)$.

$$y^2 = 4cx$$

Substitute $(4, 6)$ for (x, y):

$$6^2 = 4c(4)$$

$$36 = 16c$$

$$c = \frac{9}{4} \text{ or } 2.25$$

The focus is 2.25 m from the vertex on the axis of the parabola.

58. The vertex is at $V(0, 0)$. The parabola passes through $(8, 8)$.

$$x^2 = 4cy$$

$$8^2 = 4c(8)$$

$$64 = 32c$$

$$c = 2$$

The focus is 2 cm from the vertex on the axis of the parabola.

59. The vertex is at $V(0, 0)$. The parabola passes through $(4, 6)$.

$$x^2 = 4cy$$
$$4^2 = 4c(6)$$
$$\tfrac{2}{3} = c$$

The focus is 8 in. from the vertex on the axis of the parabola.

60. By inspection the center is $(0, 0)$, $a = 20$ and $b = 15$.

The equation is:
$$\frac{x^2}{400} + \frac{y^2}{225} = 1$$
$$\frac{y^2}{225} = 1 - \frac{x^2}{400}$$
$$y^2 = 225\left(1 - \frac{x^2}{400}\right)$$
$$y^2 = \frac{225}{400}(400 - x^2)$$
$$y = \frac{15}{20}\sqrt{400 - x^2} \qquad \textit{Use the square root property, and select}$$
$$\textit{nonnegative value.}$$
$$y = \frac{3}{4}\sqrt{400 - x^2}$$

for $x = 4$: $y = \frac{3}{4}\sqrt{400 - 4^2}$ for $x = 8$: $y = \frac{3}{4}\sqrt{400 - 8^2}$ for $x = 12$: $y = \frac{3}{4}\sqrt{400 - 12^2}$

$\qquad\qquad y = 14.7$ ft $\qquad\qquad\qquad\qquad y = 13.7$ ft $\qquad\qquad\qquad\qquad y = 12.0$ ft

for $x = 16$: $y = \frac{3}{4}\sqrt{400 - 16^2}$ for $x = 20$: $y = \frac{3}{4}\sqrt{400 - 20^2}$

$\qquad\qquad y = 9.0$ ft $\qquad\qquad\qquad\qquad y = 0.0$ ft

13.5 Functions, page 695

Level 1, page 701

1. A *function* is a set of ordered pairs in which the first component is associated with exactly one second component.   **2.** $f(x)$ is the second component of the ordered pair for a function f.

3. function   **4.** function   **5.** function   **6.** not a function   **7.** not a function

8. function   **9.** function   **10.** not a function   **11.** function   **12.** function

13. not a function   **14.** not a function

15. a. $f(4) = 12$ **b.** $f(6) = 14$ **c.** $f(-8) = 0$ **d.** $f(\tfrac{1}{2}) = 8\tfrac{1}{2}$ **e.** $f(x) = x + 8$

16.   a. $F(4) = -1$ **b.** $F(6) = 1$ **c.** $F(-8) = -13$ **d.** $F(\tfrac{1}{2}) = -4\tfrac{1}{2}$ **e.** $F(x) = x - 5$

17. a. $M(4) = 17$ **b.** $M(6) = 37$ **c.** $M(-8) = 65$ **d.** $M(\tfrac{1}{2}) = 1\tfrac{1}{4}$ **e.** $M(x) = x^2 + 1$

18. a. $S(4) = 12$ **b.** $S(6) = 18$ **c.** $S(-8) = -24$ **d.** $S(\tfrac{1}{2}) = \tfrac{3}{2}$ **e.** $S(x) = 3x$

19.   **a.** $g(4) = 7$ **b.** $g(6) = 11$ **c.** $g(-8) = -17$ **d.** $g(\tfrac{1}{2}) = 0$ **e.** $g(x) = 2x - 1$

20.   **a.** $c(4) = 4\tfrac{1}{2}$ **b.** $c(6) = 6\tfrac{1}{2}$ **c.** $c(-8) = -7\tfrac{1}{2}$ **d.** $c(\tfrac{1}{2}) = 1$ **e.** $c(x) = x + \tfrac{1}{2}$

21. **a.** 8   **b.** -16 **c.** $p - 7$ **22.** **a.** 200   **b.** -50 **c.** $2m$ **23.** **a.** -1 **b.** -31 **c.** $3a - 1$

24. **a.** 10   **b.** $1\tfrac{1}{4}$ **c.** $b^2 + 1$ **25.** **a.** 5   **b.** -2 **c.** $\tfrac{3}{2}$ **26.** **a.** 2.46   **b.** 1.38   **c.** 1.5

27. not a function; domain: $-1 \le x \le 1$; range: $-3 \le y \le 3$

28. not a function; domain: $-3 \le x \le 3$; range: $-3 \le y \le 3$

29. not a function; domain: $x \ge -3$; range: $\mathbb{R}$ **30.** function; domain: $\mathbb{R}$; range: $y \ge -80$

31. function; domain: $-2 \le x \le 3$; range: $-8 \le y \le 4$

32. function; domain: $\mathbb{R}$; range: $\mathbb{R}$

33. quadratic;

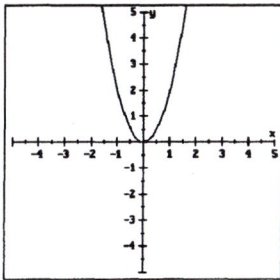

34. linear;

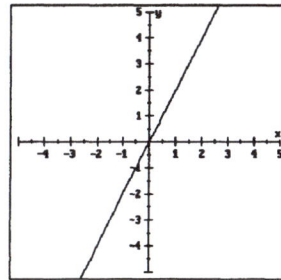

35. logarithmic;

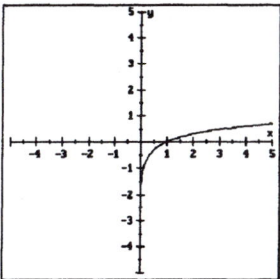

36. exponential;

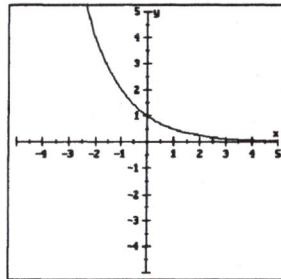

37. probability;

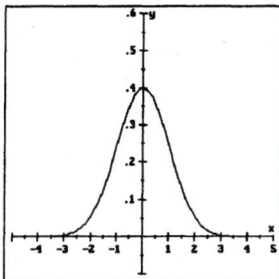

38. linear;

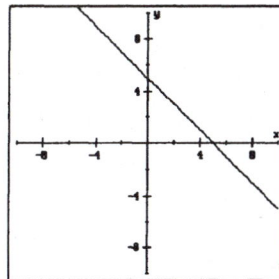

Page 305

Level 2, page 702

39.
$$\frac{f(x+h)-f(x)}{h} = \frac{3(x+h)-5-(3x-5)}{h}$$

$$= \frac{3x+3h-5-3x+5}{h}$$

$$= 3$$

40.
$$\frac{f(x+h)-f(x)}{h} = \frac{3(x+h)^2-5-(3x^2-5)}{h}$$

$$= \frac{3x^2+6xh+3h^2-5-3x^2+5}{h}$$

$$= 6x+3h$$

41.
$$\frac{f(x+h)-f(x)}{h} = \frac{(x+h)^3-x^3}{h}$$

$$= \frac{x^3+3x^2h+3xh^2+h^3-x^3}{h}$$

$$= \frac{h(3x^2+3xh+h^2)}{h}$$

$$= 3x^2+3xh+h^2$$

42.
$$\frac{f(x+h)-f(x)}{h} = \frac{5(x+h)^3-5x^3}{h}$$

$$= \frac{5(x^3+3x^2h+3xh^2+h^3)-5x^3}{h}$$

$$= \frac{15x^2h+15xh^2+5h^3}{h}$$

$$= 15x^2+15xh+5h^2$$

43.
$$\frac{f(x+h)-f(x)}{h} = \frac{\frac{1}{x+h}-\frac{1}{x}}{h}$$

$$= \frac{x-(x+h)}{x(x+h)h}$$

$$= \frac{-h}{x(x+h)h}$$

$$= \frac{-1}{x(x+h)}$$

44.

$$\frac{f(x+h) - f(x)}{h} = \frac{\frac{1}{(x+h)^2} - \frac{1}{x^2}}{h}$$

$$= \frac{x^2 - (x+h)^2}{(x+h)^2 x^2 h}$$

$$= \frac{x^2 - x^2 - 2xh - h^2}{(x+h)^2 x^2 h}$$

$$= \frac{-2x - h}{(x+h)^2 x^2}$$

Level 3, page 702

45. **a.** 64 **b.** 96 **c.** 128 **d.** 256 **e.** 512

46. $(0, 0)$, $(8, 256)$, $(16, 512)$

47. **a.** 1,430 **b.** 1,050 **c.** 670 **d.** 290 **e.** 100

48. $(0, 2000)$, $(5, 1050)$, $(10, 100)$

49. A prime number is a counting number with exactly two factors. The primes less than 100 are:

2, 3, 5, 7, 11, 13, 17, 19, 23, 29, 31, 37, 41, 43, 47, 53, 59, 61, 67, 71, 73, 79, 83, 89, and 97.

a. $P(10) = 4$ **b.** $P(-10) = 0$ **c.** $P(100) = 25$

50. a. $S(32) = 5$ **b.** $S(\frac{1}{8}) = -3$ **c.** $S(\sqrt{2}) = \frac{1}{2}$

51. $y = 3x - 5$; domain: $\mathbb{R}$ **52.** $y = x^2 - 5x$; domain; $\mathbb{R}$ **53.** $y = \sqrt{5 - x}$; domain; $x \leq 5$

54. $y = \frac{5 + 5x}{x + 1}$; domain; $\mathbb{R}$, $x \neq -1$

Level 3 Problem Solving, page 703

55. a. $\{(a, e), (e, a), (b, e), (e, b), (b, c), (c, b), (c, e), (e, c), (e, d), (d, e)\}$; it is not a function

b. $\{(a, e), (e, b), (b, c), (c, e), (d, e)\}$; it is a function

56. $f(x) = (x + 5)^2 - x^2$

$$= x^2 + 10x + 25 - x^2$$

$$= 10x + 25$$

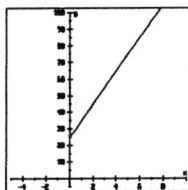

57. They could toss a rock into the well and measure the time it takes to hit the bottom. We know 1 mi = 5,280 ft, so

$$5{,}280 = 16t^2$$

$$330 = t^2$$

$$t \approx \pm 18.1$$

Reject the negative solution, and conclude that it would be about 18 seconds to the bottom.

58. **a.** $P(9) = 20 - \dfrac{6}{9+1} = 19.4$; the expected population is 19,400

b. $P(10) = 20 - \dfrac{6}{10+1} \approx 19.455$; the increase in population is $P(10) - P(9) = 0.055$

In the ninth year, the population would increase by about 55 people.

c. $P(t) = 20 - \dfrac{6}{t+1}$; as t becomes large, $\dfrac{6}{t+1}$ is getting small, so it looks like the population is approaching 20,000.

59. $A = s^2$ and $P = 4s$, so $A = \left(\dfrac{P}{4}\right)^2$.

60. $A = \pi r^2$ and $C = 2\pi r$, so $A = \pi\left(\dfrac{C}{2\pi}\right)^2 = \dfrac{C^2}{4\pi}$.

Chapter 13 Review Questions, page 704

1. Solve for y:
$$5x - y = 15$$
$$5x - 15 = y$$

The y-intercept is -15; the slope is 5; rise of 5 and run of 1.

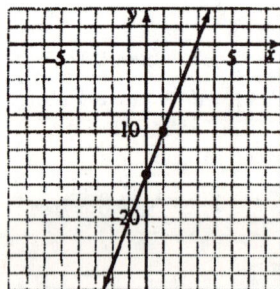

2. $y = -\frac{4}{5}x - 3$

The y-intercept is -3 and the slope is $-\frac{4}{5}$; rise of -4, and run of 5.

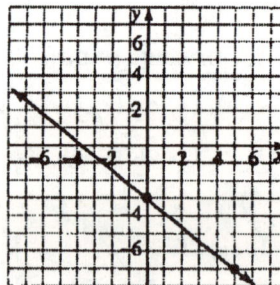

3. Solve for y: $2x + 3y = 15$

$3y = -2x + 15$

$y = -\frac{2}{3}x + 5$

The y-intercept is 5; the slope is $-\frac{2}{3}$;

rise of -2 and run of 3.

4. Solve for y: $x = -\frac{2}{3}y + 1$

$3x = -2y + 3$

$2y = -3x + 3$

$y = -\frac{3}{2}x + \frac{3}{2}$

y-intercept is $\frac{3}{2}$; the slope is $-\frac{3}{2}$;

rise of -3 and run of 2.

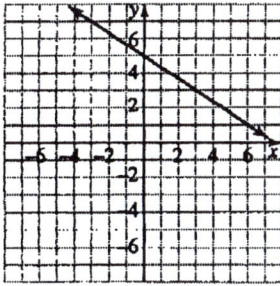

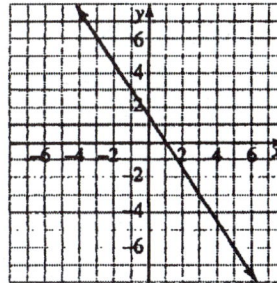

5. $x = 150$; recognize this as a vertical line; watch the scale.

6. $x < 3y$

Graph the boundary $x = 3y$.

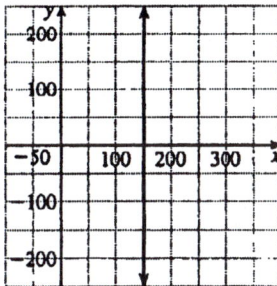

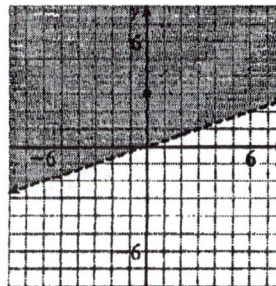

Test point: (0, 3)
$x < 3y$
$0 < 3(3)$ is true

Shade the half-plane
on the same side as the
test point.

7. $y = 1 - x^2$

x	y
1	0
-1	0
2	-3
3	-8

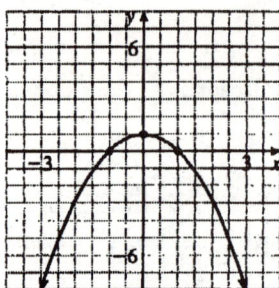

8. $y = -2^x$

x	y
0	-1
1	-2
2	-4
3	-8
-1	$-1/2$
-2	$-1/4$
-3	$-1/8$

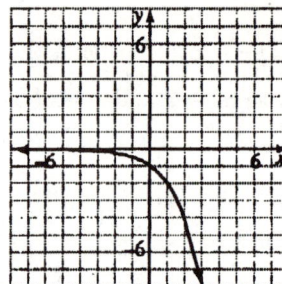

9.

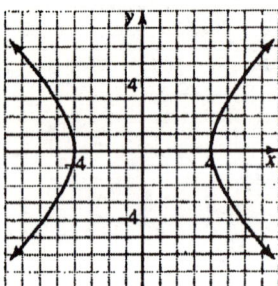

10.

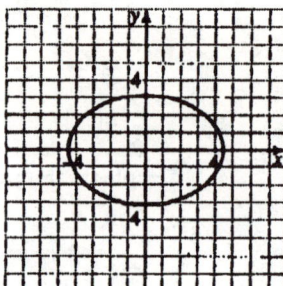

11.

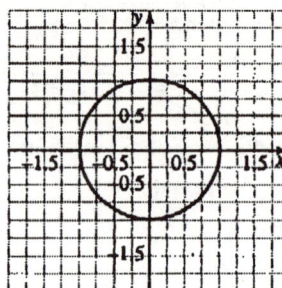

12.

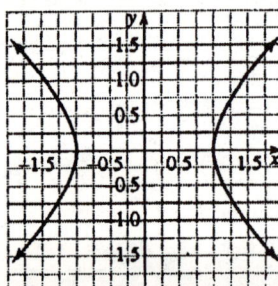

13. Yes, it is a function because each x-value (namely, 4, 5, and 6) is associated with exactly one y-value.

14. $f(6) = 3(6) + 2 = 20$

15. $g(0) = 0^2 - 3 = -3$

16. $F(10) = 5(10) + 25 = 75$

17. $m(10) = 5$

18. $A = 1.7P$; label the horizontal axis P and the vertical axis A. The intercept is 0 and the slope is 1.7 — that is, a rise of 1.7 and a run of 1 (or a rise of 17 and a run of 10). Note that since P represents an amount of money, it is nonnegative.

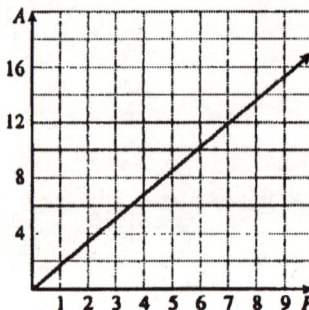

19. Form a table of values for $y = 128t - 16t^2$:

t	0	1	2	3	4	5	6	7	8
y	0	112	192	240	256	240	192	112	0

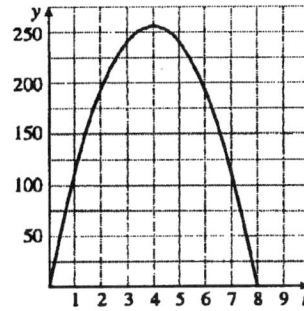

20. Form a table of values for $P = 53,000e^{0.013t}$; label the horizontal axis t and the vertical axis P (in units of thousands)

t	0	10	20	30	40	50
P	53,000	60,358	68,737	78,280	89,147	101,524

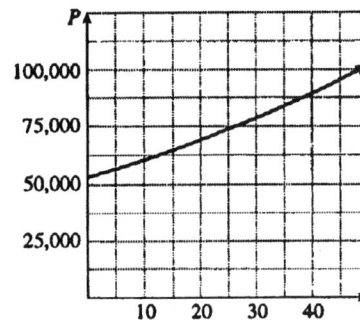

Group Research Problems, pages 705-706

G47. Consider the equation $y - k = a(x - h)^2$.

The vertex must occur halfway to the place where it hits the ground; thus, the vertex is $(100, 50)$ so

$$y - 50 = a(x - 100)^2$$

The parabola passes through $(0, 0)$ so

$$0 - 50 = a(0 - 100)^2$$
$$-50 = 100^2 a$$
$$-\frac{50}{100^2} = a$$
$$-\frac{1}{200} = a$$

Thus, the equation is $y - 50 = -\frac{1}{200}(x - 100)^2$. This can also be written $y = -\frac{1}{200}x^2 + x$.

G48. a. We use 1987 to 1988, we estimate the rate to be $r \approx 0.1282284727$ and find

$$A = 21,978e^{8r} \approx 61,305.5106$$

We estimate the number of deaths in 1996 to be about 61,000.

b.   Let $1988 = 0$.

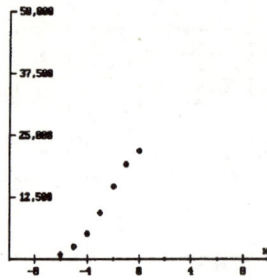

If we use a normal curve to approximate these data points we find

$$y = \frac{56{,}000\,e^{-x^2/10}}{\sqrt{2\pi}}.$$ The graph is shown:

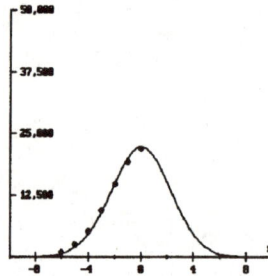

If we let $x = 8$, then $y \approx 37$; the predicted number is 37,000.

c.   The model from part **a** makes a better model, by we might explore some other models (by using a computer program).

Here is a linear model:

$y = 3{,}791.572x + 21{,}957$

If $x = 8$, then $y \approx 52{,}000$

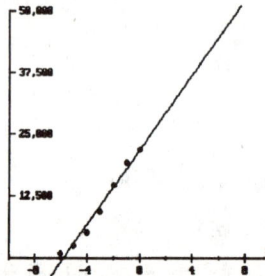

Here is an exponential model:

$y = 32{,}787.75\,e^{0.5209402x}$

If $x = 8$, then $y \approx 2{,}100{,}000$

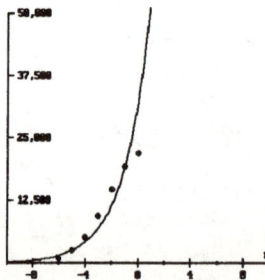

Here is a polynomial model:

$$y = 5.6x^6 + 107.3x^5 + 739.7x^4$$
$$+\, 2{,}059.8x^3 + 1{,}445.2x^2$$
$$+\, 2{,}668.4x + 21{,}978$$

If $x = 8$, then $y \approx 9{,}200{,}000$

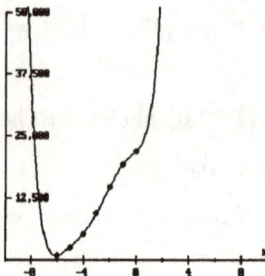

Page 312

Compare these models with actual data obtained from the Centers of Disease Control. New drug "cocktails" have cut back the death rate in 1997.

G49. Answers vary.

G50. In the article, "Why Were So Few Mathematicians Female," Loretta Kelley answers: "My answer is twofold. First, many women *have* become mathematicians. Second, when I study their lives, I do not wonder why more women did not choose this career. Considering the little support that they have received and the many barriers that have been placed in their way, it is remarkable that so many women have accomplished so much in mathematics." Here is a list of some of the more prominent women mathematicians: Emilie Du Châtelet (1706-1749), Winifred Haring Edgerton (1862-1951), Sophie Germain (1776-1831), Grace Hopper (1906-1992), Hypatia (370-415), Sofia Kovalevskaia (1850-1891), Ada Byron Lovelace (1815-1852), Maria Goeppert Mayer (1906-1972), Emmy Noether (1882-1935), Mary Frances Winston Newson (1869-1959), Ellen Richards (1842-1911), Julia Robinson (1919-1985), Charlotte Angas Scott (1858-1931), Mary Fairfax Somerville (1780-1872). The author would appreciate additional references for distinguished women mathematicians. (email: SMITHKJS@mathnature.com)

G51. Benjamin Banneker, Joseph Battle, Marjorie Lee Browne, Elbert Cox, Jean-Baptiste Lislet-Geoffray, Evelyn Boyd Glanville, Eleanor Green Dawley Jones, Lee Lorch, Narbert Rillieux, André Rebouças Granville T. Woods. The author would appreciate additional references for distinguished black mathematicians. (email: SMITHKJS@mathnature.com)

G52. Karl Gauss, Karl Jacobi, Karl Pierson, Karl Weierstrass. The author would appreciate additional references for distinguished mathematicians with the first name of Karl. (email: SMITHKJS@mathnature.com)

G53. Answers vary.

Individual Research Problems, pages 706-707

P13.1. Answers vary.

P13.2. 33,547,618; answers vary.

P13.3. 7,975; answers vary.

P13.4. Answers vary.

P13.5. From the definition of a hyperbola, we can now answer the question — the path is that of a hyperbola. Let us continue with this application. Suppose the difference in the arrival of the time signals is 300 μsec. (*Note*: Remember, μsec is a microsecond — that is, one millionth of a second.) Also, suppose that F_1 and F_2 are 100 miles apart. Finally, suppose that signals travel at 980 ft/μsec. Since the time difference is 300 μsec, the difference of the ship's distance of the ship to each station is

$$d = rt = (980 \text{ ft}/\mu\text{sec})(300 \ \mu\text{sec}) = 294{,}000 \text{ ft}$$

We change feet to miles by dividing by 5,280 to find

$$d = \frac{294{,}000}{5{,}280} = 55.68\overline{18} \text{ mi}/\mu\text{sec}$$

The difference of the ship's distances from each station is $2a$ (this is true for any hyperbola); so $2a = 55.68\overline{18}$. To describe the path of the ship, we superimpose a coordinate system with F_1 and F_2 on the x-axis. Let the coordinates of the ship be $P(x, y)$. Then the center of the coordinate system is the midpoint of the segment $\overline{F_1 F_2}$ which is at $(0, 0)$. From the definition of a hyperbola,

$$\frac{x^2}{a^2} - \frac{y^2}{b^2} = 1$$

where $a = 27.84\overline{09}$. Also, c is the distance from the center to a focus, so $c = 50$, and we find b:

$$b^2 = c^2 - a^2 = 50^2 - (27.84\overline{09})^2 \approx 1{,}724.883781$$

Thus, the ship's path is described by the equation of a hyperbola where $a^2 \approx 775$ and $b^2 \approx 1{,}725$:

$$\frac{x^2}{775} - \frac{y^2}{1725} = 1$$

The graph of the ship's position is shown:

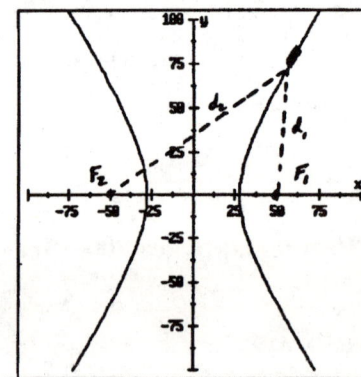

P13.6. Answers vary.

CHAPTER 14

THE NATURE OF MATHEMATICAL SYSTEMS

14.1 Systems of Linear Equations, page 710

Level 1, page 713

1. A *system of linear equations* is two or more first degree equations considered at the same time.

2. An *inconsistent system* is a system for which there is no solution and a *dependent system* is a system for which there are infinitely many solutions.   3. Answers vary; graph each of the equations independently and look for the points of intersection.   4. *Solve* one of the equations for one of the variables. *Substitute* the expression that you obtain into the other equation. *Solve* the resulting equation. *Substitute* that solution into either of the original equations to find the value of the other variable. *State* the solution.   5. *Multiply* one or both of the equations by a constant or constants so that the coefficients of one of the variables become opposites. *Add* corresponding members of the equations to obtain a new equation in a single variable. *Solve* the derived equation for that variable. *Substitute* the value of the found variable into either of the original equations, and solve for the second variable. *State* the solution.

6. $(3, 1)$ **7.** $(4, -1)$ **8.** $(-2, -5)$ **9.** inconsistent

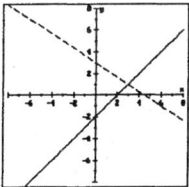   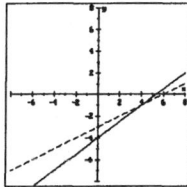   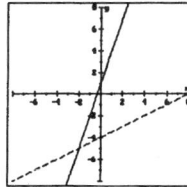   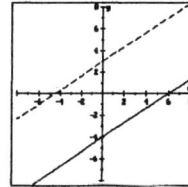

10. $(3, -3)$ **11.** $(-1, 1)$ **12.** $(-1, -1)$ **13.** $(1, 1)$

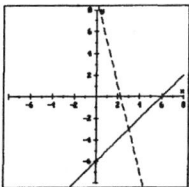   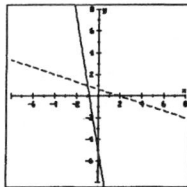   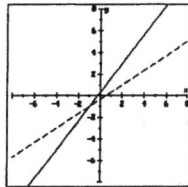   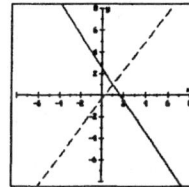

14. $(23, -43)$ **15.** $(-5, -3)$ **16.** $(2, -1)$ **17.** $(-1, -2)$ **18.** inconsistent

19. $(2, 10)$ **20.** $(-\frac{3}{5}, \frac{5}{4})$ **21.** $(9, -4)$ **22.** $(t_1, t_2) = (57, 274)$ **23.** $(4, -1)$ **24.** $(-1, 1)$

25. $(6, -2)$ **26.** $(13, 3)$ **27.** $(525, 35)$ **28.** $(8, -2)$ **29.** $(5, -2)$ **30.** dependent

31. dependent   **32.** $(a_1, a_2) = (-7, 3)$ **33.** $(s_1, s_2) = (3, 4)$ **34.** $(s, t) = (\frac{20}{3}, \frac{16}{3})$

35. $(u, v) = (5, -2)$ **36.** $(-14, 7)$ **37.** $(\frac{3}{5}, \frac{1}{2})$

Level 2, page 714

38. $(3, 4)$ **39.** $(5, -3)$ **40.** $(4, 3)$ **41.** $(3, 1)$ **42.** $(-2, -5)$ **43.** $(3, 1)$ **44.** $(3, 1)$

45. $(4, -1)$ **46.** $(\frac{3}{2}, -3)$ **47.** $(\frac{1}{3}, -\frac{2}{3})$ **48.** $(3, 8)$ **49.** $(1, 1)$ **50.** $(2, 200)$

51. inconsistent   **52.** $(q, d) = (63, 84)$ **53.** $(8, 2)$ **54.** $(3, 15)$ **55.** $(-\frac{8}{5}, -\frac{21}{5})$

Level 3 Problem Solving, page 714

56. 1,301 days after the dog's birth, namely November 9, 1995.   **57.** 2,601 days after the cat's birth, namely June 2, 1999.   **58.** Answers vary.

59. $\left(\dfrac{a + \sqrt{a^2 - 4}}{2}, \dfrac{a - \sqrt{a^2 - 4}}{2} \right), \left(\dfrac{a - \sqrt{a^2 - 4}}{2}, \dfrac{a + \sqrt{a^2 - 4}}{2} \right)$

60. $\begin{cases} x + \frac{2}{3}y = 29 \\ \frac{3}{4}x + y = 29 \end{cases}$ or $\begin{cases} 3x + 2y = 87 \\ 3x + 4y = 116 \end{cases}$

Solution: $(\frac{58}{3}, \frac{29}{2})$

The first person has $19\frac{1}{3}$ pistoles and the second has 14.5 pistoles.

14.2 Problem Solving with Systems, page 714

Level 1, page 722

1. The box has 31 nickels and 56 dimes.   **2.** There are 99 dimes in the box.   **3.** There are 18 nickels.   **4.** The boat's speed in still water is 20 mph.   **5.** The plane's speed in still air is 175 mph. **6.** The speed of the plane in still air is 390 mph.   **7.** The equilibrium point for the system is $(50, 250\,000)$.   **8.** The optimum price for the calculators is $25.   **9.** The optimum price for the items is $3.   **10.** Dom DeLuise was born in 1933.   **11.** Debra Winger was born in 1955. **12.** José Feliciano was born in 1945.   **13.** Mixture **a** has 7 lb of micoden.   **14.** Mixture **b** has $0.2x + 0.2y$ lb of water.   **15.** Mixture **c** has $0.5p$ L of bixon.   **16.** Mixture **a** contains 20% water.

17. Mixture **a** contains $33\frac{1}{3}\%$ bixon.   **18.** Mixture **a** contains $46\frac{2}{3}\%$ micoden.   **19.** There are 40 oz of the base metal.   **20.** Add 9 oz of water.   **21.** Mix 45 gal of milk with 135 gal of cream.

Level 2, page 723

22. Both rates are the same if you drive 150 miles.   **23.** Both rates are the same if you drive 160 miles.   **24.** The equilibrium point is (58, 9 200).   **25.** The equilibrium point is (2, 10 000).
26. There are 15 dimes and 9 quarters.   **27.** There are 14 dimes and 28 quarters.   **28.** The rate of the current is 1.5 mph.   **29.** The wind speed is 22 mph.   **30.** Clint Eastwood was born in 1931.
31. Kristy McNichol was born in 1967.   **32.** The mixture should consist of 25 oz silver and 75 oz silver alloy.   **33.** Add $\frac{1}{3}$ gal water.   **34.** Should mix 1 gal of antifreeze.   **35.** NY is 47,831 sq mi and CA is 156,361 sq mi.   **36.** TX is 262,134 sq mi and FL is 54,090 sq mi.   **37.** There are 37 quarters.   **38.** There are 26 pennies.   **39.** There are 30 dimes.   **40.** The box contains 54 nickels, 54 dimes, and 30 pennies.   **41.** The robbery included 14 $5 bills, 70 $10 bills, and 31 $20 bills.
42. The plane's speed without the wind is 304.5 mph.   **43.** The wind speed is 22.5 mph.   **44.** Mix together 6 L of a 50% solution with 4 L of a 75% solution.   **45.** Mix together 28.5 gal of milk with 1.5 gal of cream.   **46.** Mix together 237.5 gal of Lot I with 262.5 gal of Lot II.   **47.** Mix together 38.46 mg of 12% aspirin with 61.54 mg of 25% aspirin.   **48.** The height of the Transamerica Tower is 853 ft and the height of the Bank of America Building is 779 ft.   **49.** The Sears Tower is 1,454 ft and the Standard Oil building is 1,136 ft.   **50.** The Golden Gate bridge is 4,200 ft.   **51.** The Verrazano-Narrows bridge is 4,260 ft.   **52.** The equilibrium point is $3.50 for 4,000 items.
53. The equilibrium point is $2 for 40,000 items.   **54.** The mixture should contain 20 g of pure silver and 80 g of sterling silver.   **55.** Replace 9 qt of the liquid with antifreeze.   **56.** Mix 25 gal of butterfat cream.   **57. a.** 125 items would be supplied; 75 would be demanded   **b.** No items would be supplied at $200.   **c.** No items would be demanded at $400.   **d.** The equilibrium price is $\frac{700}{3} \approx$ $233.33.   **e.** The number of items produced at the equilibrium price is $\frac{250}{3} \approx 83$.
58. a. 37,000 items would be supplied and 9,000 would be demanded.   **b.** No items would be supplied at $0.20.   **c.** No items would be demanded at $21.   **d.** The equilibrium price is $8.   **e.** The number of items produced at the equilibrium price is 19,500.

Level 3 Problem Solving, page 724

59. False; the cork costs 5¢ and the bottle $1.05.   **60.** False; same amount

14.3 Matrix Solution of a System of Equations, page 724

Level 1, page 732

1. A matrix is a rectangular array of entries.　　2. *RowSwap* Interchange any two rows.
Row+ Row addition — add a row to any other row.　*Row* Scalar multiplication — multiply (or
divide) all the elements of a row by the same nonzero real number.　*Row+* Multiply all the entries of
a row (pivot row) by a nonzero real number and add each resulting product to the corresponding entry
of another specified row (target row).　　3. Select as the first pivot the element in the first row, first
column, and pivot. The next pivot is the element in the second row, second column; pivot. Repeat the
process until you arrive at the last row, or until the pivot element is a zero. If it is a zero and you
can interchange that row with a row below it, so that the pivot element is no longer a zero, do so and
continue. If it is zero and you cannot interchange rows so that it is not a zero, continue with the next
row. The final matrix is called the *row-reduced form*.　　4. true　　5. false; it is 4×5　　6. false; it
is the scalar　　7. true　　8. false; it is Row+([A],3,5)　9. true　　10. true　　11. false; multiply
row 4 by $\frac{1}{3}$ and add to row 2　　12. false; the target row is 5　　13. false; RowSwap([A],1,2)

14. a. $\begin{bmatrix} 4 & 5 & | & -16 \\ 3 & 2 & | & 5 \end{bmatrix}$　　b. $\begin{bmatrix} 1 & 1 & 1 & | & 4 \\ 3 & 2 & 1 & | & 7 \\ 1 & -3 & 2 & | & 0 \end{bmatrix}$　　c. $\begin{bmatrix} 1 & 2 & -5 & 1 & | & 5 \\ 1 & 0 & -3 & 6 & | & 0 \\ 0 & 0 & 1 & -3 & | & -15 \\ 0 & 1 & -5 & 5 & | & 2 \end{bmatrix}$

15. a. $\begin{cases} 6x + 7y + 8z = 3 \\ x + 2y + 3z = 4 \\ y + 3z = 4 \end{cases}$　　b. $\begin{cases} x_1 = 3 \\ x_2 + 2x_3 = 4 \end{cases}$　　c. $\begin{cases} x_1 = 32 \\ x_2 = 27 \\ x_3 = -5 \\ 0 = 3 \end{cases}$

16. $\begin{bmatrix} 1 & 3 & -4 & | & 9 \\ 0 & 2 & 4 & | & 5 \\ 3 & 1 & 2 & | & 1 \end{bmatrix}$
RowSwap([A], 1, 3)

17. $\begin{bmatrix} 1 & 0 & 2 & | & -8 \\ -2 & 3 & 5 & | & 9 \\ 0 & 1 & 0 & | & 5 \end{bmatrix}$
RowSwap([B], 1, 2)

18. $\begin{bmatrix} 1 & 2 & 5 & | & -6 \\ 6 & 3 & 4 & | & 6 \\ 10 & -1 & 0 & | & 1 \end{bmatrix}$
*Row(1/2, [C], 1)

19. $\begin{bmatrix} 1 & 4 & 3 & | & \frac{6}{5} \\ 7 & -5 & 3 & | & 2 \\ 12 & 0 & 1 & | & 4 \end{bmatrix}$
*Row(1/5, [A], 1)

20.
$$\begin{bmatrix} 1 & 2 & -3 & | & 0 \\ 0 & 3 & 1 & | & 4 \\ 0 & 1 & 7 & | & 6 \end{bmatrix}$$
*Row+(−2, [A], 1, 3)

21.
$$\begin{bmatrix} 1 & 3 & -5 & | & 6 \\ 0 & 13 & -14 & | & 20 \\ 0 & 5 & 1 & | & 3 \end{bmatrix}$$
*Row+(3, [B], 1, 2)

22.
$$\begin{bmatrix} 1 & 2 & 4 & | & 1 \\ 0 & 9 & 8 & | & 4 \\ 0 & 13 & 17 & | & 7 \end{bmatrix}$$
*Row+(2, [C], 1, 2)
*Row+(4, [Ans], 1, 3)

23.
$$\begin{bmatrix} 1 & 5 & 3 & | & 2 \\ 0 & -7 & -7 & | & 0 \\ 0 & -13 & -8 & | & -6 \end{bmatrix}$$
*Row+(−2, [A], 1, 2)
*Row+(−3, [Ans], 1, 3)

24.
$$\begin{bmatrix} 1 & 3 & 5 & | & 2 \\ 0 & 1 & 3 & | & -4 \\ 0 & 3 & 4 & | & 1 \end{bmatrix}$$
*Row(1/2, [A], 2)

25.
$$\begin{bmatrix} 1 & 5 & -3 & | & 5 \\ 0 & 1 & 3 & | & -5 \\ 0 & 2 & 1 & | & 5 \end{bmatrix}$$
*Row(1/3, [B], 2)

26.
$$\begin{bmatrix} 1 & 4 & -1 & | & 6 \\ 0 & 1 & -5 & | & -2 \\ 0 & 4 & 6 & | & 5 \end{bmatrix}$$
*Row+(−1, [C], 3, 2)

27.
$$\begin{bmatrix} 1 & 3 & -2 & | & 0 \\ 0 & 1 & -4 & | & 8 \\ 0 & 3 & 6 & | & 1 \end{bmatrix}$$
*Row+(−1, [A], 3, 2)

28.
$$\begin{bmatrix} 1 & 0 & -23 & | & -23 \\ 0 & 1 & 4 & | & 5 \\ 0 & 0 & -8 & | & -13 \end{bmatrix}$$
*Row+(−5, [A], 2, 1)
*Row+(−3, [Ans], 2, 3)

29.
$$\begin{bmatrix} 1 & 0 & 12 & | & 27 \\ 0 & 1 & -2 & | & -5 \\ 0 & 0 & -2 & | & -4 \end{bmatrix}$$
*Row+(−3, [B], 2, 1)
*Row+(2, [Ans], 2, 3)

30.
$$\begin{bmatrix} 1 & 0 & -45 & -14 & | & 1 \\ 0 & 1 & 7 & 3 & | & 0 \\ 0 & 0 & -17 & -9 & | & -2 \\ 0 & 0 & 17 & 7 & | & 0 \end{bmatrix}$$
*Row+(−6, [C], 2, 1)
*Row+(−3, [Ans], 2, 3)
*Row+(2, [Ans], 2, 4)

31.
$$\begin{bmatrix} 1 & 0 & -26 & -8 & | & 8 \\ 0 & 1 & 5 & 2 & | & 0 \\ 0 & 0 & -1 & -2 & | & 5 \\ 0 & 0 & -13 & -3 & | & 7 \end{bmatrix}$$
*Row+(−5, [D], 2, 1)
*Row+(−1, [Ans], 2, 3)
*Row+(−2, [Ans], 2, 4)

32. $\begin{bmatrix} 1 & 0 & 4 & | & -5 \\ 0 & 1 & -3 & | & 6 \\ 0 & 0 & 1 & | & 2 \end{bmatrix}$

*Row(1/5, [A], 3)

33. $\begin{bmatrix} 1 & 0 & -4 & | & -5 \\ 0 & 1 & 3 & | & 6 \\ 0 & 0 & 1 & | & 1.5 \end{bmatrix}$

*Row(1/8, [B], 3)

34. $\begin{bmatrix} 1 & 0 & -2 & 1 & | & 1 \\ 0 & 1 & 6 & 2 & | & 0 \\ 0 & 0 & 1 & 0 & | & -0.5 \\ 0 & 0 & 2 & 1 & | & 0 \end{bmatrix}$

*Row(1/4, [C], 3)

35. $\begin{bmatrix} 1 & 0 & -8 & 2 & | & 8 \\ 0 & 1 & -1 & 3 & | & 2 \\ 0 & 0 & 1 & 0 & | & 5 \\ 0 & 0 & -2 & 1 & | & 6 \end{bmatrix}$

*Row(1/2, [D], 3)

36. $\begin{bmatrix} 1 & 0 & 0 & | & 9 \\ 0 & 1 & 0 & | & -7 \\ 0 & 0 & 1 & | & 4 \end{bmatrix}$

*Row+(−2, [A], 3, 2)
*Row+(1, [Ans], 3, 1)

37. $\begin{bmatrix} 1 & 0 & 0 & | & 7 \\ 0 & 1 & 0 & | & -7 \\ 0 & 0 & 1 & | & 3 \end{bmatrix}$

*Row+(−4, [B], 3, 2)
*Row+(3, [Ans], 3, 1)

38. $\begin{bmatrix} 1 & 0 & 0 & 4 & | & -1 \\ 0 & 1 & 0 & 3 & | & 8 \\ 0 & 0 & 1 & 0 & | & -2 \\ 0 & 0 & 0 & 1 & | & -6 \end{bmatrix}$

*Row+(1, [C], 3, 1)
*Row+(−4, [Ans], 3, 2)
*Row+(3, [Ans], 3, 4)

39. $\begin{bmatrix} 1 & 0 & 0 & 2 & | & 24 \\ 0 & 1 & 0 & 2 & | & -8 \\ 0 & 0 & 1 & 0 & | & 2 \\ 0 & 0 & 0 & 1 & | & 9 \end{bmatrix}$

*Row+(8, [D], 3, 1)
*Row+(−4, [Ans], 3, 2)
*Row+(1, [Ans], 3, 4]

Level 2, page 734

40. (3, 4) **41.** (5, −3) **42.** (4, 3) **43.** (3, 1) **44.** (1, 1) **45.** (4, −1) **46.** (3, 1)

47. (4, −1) **48.** $(\frac{3}{2}, \frac{1}{2}, -\frac{1}{2})$ **49.** (2, 0, 1) **50.** (−1, −2, 2) **51.** (3, 2, 5) **52.** (−1, 2, −3)

53. (2, −3, −1) **54.** (1, −3, 5) **55.** (2, −3, 2) **56.** (5, 3, 2) **57.** $(1, 2, -\frac{1}{2})$

Level 3 Problem Solving, page 734

58.

	Spray I	Spray II	Total
A	5	2	23
B	2	7	34

Let x = amount of Spray I
y = amount of Spray II

$$\begin{cases} 5x + 2y = 23 \\ 2x + 7y = 34 \end{cases}$$

Solution: (3, 4)

Mix 3 containers of Spray I with 4 containers of Spray II.

59.

Candy:	I	II	III	Total
Chocolate	7	3	4	67
Milk	5	2	3	48
Mint	1	2	3	32

Let x = amount of Candy I
y = amount of Candy II
z = amount of Candy III

$$\begin{cases} 7x + 3y + 4z = 67 \\ 5x + 2y + 3z = 48 \\ x + 2y + 3z = 32 \end{cases}$$

Solution: (4, 5, 6)
Produce 4 units of candy I, 5 units of candy II, and 6 units of candy III.

60. Produce 3 units of candy I, 7 units of candy II, and 5 units of candy III.

14.4 Inverse Matrices, page 734

Level 1, page 745

1. [M] = [N] if and only if matrices [M] and [N] are the same order and the corresponding entries are the same.   **2.** [M] + [N] = [S] if and only if [M] and [N] are the same order and the entries of [S] are found by adding the corresponding entries of [M] and [N].   **3.** Let [M] be an $m \times r$ matrix and [N] an $r \times n$ matrix. The product matrix [M][N] = [P] is an $m \times n$ matrix. The entry in the ith row and jth column of [M][N] is *the sum of the products formed by multiplying each entry of the ith row of* [M] *by the corresponding element in the jth column of* [N].   **4.** Inverse for addition: [M] + [−M] = [−M] + [M] = [0]; inverse for multiplication: $[M][M]^{-1} = [M]^{-1}[M] = [I]$

5. A matrix is nonsingular if it has an inverse for multiplication.   **6.** To find the inverse of a *square* matrix A: Augment [A] with [I]; that is, write [A | I], where [I] is the matrix of the same order as [A]. Perform elementary row operations using Gauss-Jordan elimination to change the matrix A into the identity matrix (if possible). If at any time you obtain all zeros in a row or column to the left of the dashed line, then there will be no inverse. If steps 1 and 2 can be performed, the result in the augmented part is the inverse of [A].   **7.** If $[A][X] = [B]$, then $[X] = [A]^{-1}[B]$.   **8.** A *communication matrix* is a square matrix in which the entries symbolize the occurrence of some facet or event with a 1 and the nonoccurrence with a 0.

9. **a.** $\begin{bmatrix} 2 & 4 & 2 \\ 6 & -2 & 4 \\ 2 & 2 & 5 \end{bmatrix}$ **b.** $\begin{bmatrix} -6 & -1 & -2 \\ 3 & -7 & -3 \\ 4 & -7 & -2 \end{bmatrix}$

10. **a.** $\begin{bmatrix} -3 & 6 & 10 \\ -4 & 15 & 8 \\ 7 & 15 & 2 \end{bmatrix}$ **b.** $\begin{bmatrix} 16 & 19 & 10 \\ 29 & 16 & 15 \\ 35 & 9 & 31 \end{bmatrix}$

11. **a.** $\begin{bmatrix} 20 & 21 & 34 \\ 29 & 16 & 15 \\ 7 & 48 & 5 \end{bmatrix}$ **b.** $\begin{bmatrix} -1 & 37 & 32 \\ 4 & 14 & 45 \\ 27 & 9 & 28 \end{bmatrix}$

12. **a.** $\begin{bmatrix} 2 & 8 \\ 16 & 8 \\ -7 & 7 \\ 7 & 7 \end{bmatrix}$ **b.** $\begin{bmatrix} 14 & 4 & 8 & 30 \\ -7 & -1 & -9 & 3 \end{bmatrix}$

13. **a.** $\begin{bmatrix} 14 & 14 \\ -7 & 7 \end{bmatrix}$ **b.** $\begin{bmatrix} 1 & 0 & 0 & 0 \\ 0 & 1 & 0 & 0 \\ 0 & 0 & 1 & 0 \\ 0 & 0 & 0 & 1 \end{bmatrix}$

14. **a.** $\begin{bmatrix} 13 & 25 & 20 \\ 25 & 31 & 23 \\ 42 & 24 & 33 \end{bmatrix}$ **b.** $\begin{bmatrix} 13 & 25 & 20 \\ 25 & 31 & 23 \\ 42 & 24 & 33 \end{bmatrix}$

15. **a.** not conformable   **b.** not conformable

16. **a.** $\begin{bmatrix} 34 & 117 & 44 \\ 45 & 143 & 97 \\ 109 & 100 & 151 \end{bmatrix}$ **b.** $\begin{bmatrix} 34 & 117 & 44 \\ 45 & 143 & 97 \\ 109 & 100 & 151 \end{bmatrix}$

17. $\begin{cases} x + 2y + 4z = 13 \\ -3x + 2y + z = 11 \\ 2x + z = 0 \end{cases}$ 18. $\begin{cases} 4x + y = 2 \\ 3x - y + 2z = 11 \\ 2x + 3y + z = -1 \end{cases}$

19. $[A][B] = \begin{bmatrix} 8-7 & -14+14 \\ 4-4 & -7+8 \end{bmatrix}$ $[B][A] = \begin{bmatrix} 8-7 & 28-28 \\ -2+2 & -7+8 \end{bmatrix}$

$= \begin{bmatrix} 1 & 0 \\ 0 & 1 \end{bmatrix}$ $= \begin{bmatrix} 1 & 0 \\ 0 & 1 \end{bmatrix}$

[A] and [B] are inverses since [A][B] = [B][A] = [I].

20. $[A][B] = \begin{bmatrix} -16-4+21 & -32-10+42 & 16-2-14 \\ 7+2-9 & 14+5-18 & -7+1+6 \\ -3+0+3 & -6+0+6 & 3+0-2 \end{bmatrix}$

$= \begin{bmatrix} 1 & 0 & 0 \\ 0 & 1 & 0 \\ 0 & 0 & 1 \end{bmatrix}$

$[B][A] = \begin{bmatrix} -16+14+3 & -2+2+0 & 7-6-1 \\ -32+35-3 & -4+5+0 & 14-15+1 \\ -48+42+6 & -6+6+0 & 21-18-2 \end{bmatrix}$

$= \begin{bmatrix} 1 & 0 & 0 \\ 0 & 1 & 0 \\ 0 & 0 & 1 \end{bmatrix}$

[A] and [B] are inverses since [A][B] = [B][A] = [I].

Level 2, page 746

21. $\begin{bmatrix} 2 & 7 \\ 1 & 4 \end{bmatrix}$ 22. $\begin{bmatrix} \frac{1}{11} & -\frac{3}{22} \\ \frac{1}{22} & \frac{2}{11} \end{bmatrix}$ 23. $\begin{bmatrix} 9 & -4 & -2 \\ -18 & 9 & 4 \\ -4 & 2 & 1 \end{bmatrix}$ 24. $\begin{bmatrix} 3 & -17 & -20 \\ 3 & -18 & -20 \\ -1 & 6 & 7 \end{bmatrix}$

25. $\begin{bmatrix} 1 & 0 & -1 & 0 \\ 0 & \frac{1}{2} & 0 & 0 \\ -2 & 0 & 2 & 1 \\ 0 & 0 & 1 & 0 \end{bmatrix}$ **26.** $\begin{bmatrix} -\frac{6}{5} & 0 & \frac{4}{5} & \frac{1}{10} \\ \frac{3}{5} & 0 & -\frac{2}{5} & \frac{1}{5} \\ \frac{1}{5} & 0 & \frac{1}{5} & -\frac{1}{10} \\ 0 & 1 & 0 & 0 \end{bmatrix}$ **27.** $(3, 2)$ **28.** $(-4, 7)$

29. $(5, -4)$ **30.** $(25, 14)$ **31.** $(38, 21)$ **32.** $(50, 29)$ **33.** $(3, -2)$ **34.** $(-1, 4)$

35. $(3, -5)$ **36.** $(-5, 2)$ **37.** $(-4, 1)$ **38.** $(-3, -2)$

Problems 39-44 all use the same inverse: $\begin{bmatrix} \frac{2}{5} & \frac{1}{5} \\ \frac{1}{15} & -\frac{2}{15} \end{bmatrix}$

39. $(3, 1)$ **40.** $(4, -2)$ **41.** $(-2, 2)$ **42.** $(12, -5)$ **43.** $(1, -8)$ **44.** $(0, 4)$

45. $(5, 6, 1)$ **46.** $(-2, 4, 3)$ **47.** $(-26, 52, 15)$ **48.** $(3, -5, 2)$ **49.** $(88, -176, -38)$

50. $(-5, 18, 5)$ **51.** $(1, 1, 1)$ **52.** $(5, 4, -1)$ **53.** $(2, -4, -1)$ **54.** $(54, 53, -16)$

Level 3, page 747

55. **a.** $P^{-1} = \begin{bmatrix} 1 & 0 & 0 & 0 & 0 \\ -1 & 1 & 0 & 0 & 0 \\ 1 & -2 & 1 & 0 & 0 \\ -1 & 3 & -3 & 1 & 0 \\ 1 & -4 & 6 & -4 & 1 \end{bmatrix}$

b. It is the same as the Pascal's matrix except that the signs of the terms alternate.

56. $(P + I)^{-1} = \begin{bmatrix} \frac{1}{2} & 0 & 0 & 0 & 0 \\ -\frac{1}{4} & \frac{1}{2} & 0 & 0 & 0 \\ 0 & -\frac{1}{2} & \frac{1}{2} & 0 & 0 \\ \frac{1}{8} & 0 & -\frac{3}{4} & \frac{1}{2} & 0 \\ 0 & \frac{1}{2} & 0 & -1 & \frac{1}{2} \end{bmatrix}$

57. $[A]^3 = \begin{bmatrix} 2 & 4 & 1 & 3 \\ 4 & 2 & 3 & 4 \\ 1 & 3 & 0 & 1 \\ 3 & 4 & 1 & 2 \end{bmatrix}$ For example, the U.S. can talk to Cuba through two intermediaries in one way, namely, U.S. to Mexico to Russia to Cuba.

58. **a.** $[T] = \begin{bmatrix} 0 & 1 & 0 & 1 \\ 1 & 0 & 1 & 0 \\ 0 & 1 & 0 & 1 \\ 1 & 0 & 1 & 0 \end{bmatrix}$

b. $[T]^2 = \begin{bmatrix} 2 & 0 & 2 & 0 \\ 0 & 2 & 0 & 2 \\ 2 & 0 & 2 & 0 \\ 0 & 2 & 0 & 2 \end{bmatrix}$

There are no ways of going from KC to SF with one stop.

c. $[T]^3 = \begin{bmatrix} 0 & 4 & 0 & 4 \\ 4 & 0 & 4 & 0 \\ 0 & 4 & 0 & 4 \\ 4 & 0 & 4 & 0 \end{bmatrix}$

There are four ways of going from SF to KC with two stops.

Level 3 Problem Solving, page 747

59. **a.** $[C][W] = \begin{bmatrix} 1 & 4 & 6 \end{bmatrix} \begin{bmatrix} 2 & 1 & 3 \\ 4 & 3 & 6 \\ 1 & 2 & 4 \end{bmatrix}$

$$= [2+16+6 \quad 1+12+12 \quad 3+24+24] = [24 \quad 25 \quad 51]$$

Riesling costs 24; Charbono, 25; and Rosé, 51

b. $[W][D] = \begin{bmatrix} 2 & 1 & 3 \\ 4 & 3 & 6 \\ 1 & 2 & 4 \end{bmatrix} \begin{bmatrix} 40 \\ 60 \\ 30 \end{bmatrix} = \begin{bmatrix} 80 + 60 + 90 \\ 160 + 180 + 180 \\ 40 + 120 + 120 \end{bmatrix} = \begin{bmatrix} 230 \\ 520 \\ 280 \end{bmatrix}$

Outside bottling, 230; produced and bottled at winery, 520; estate bottles, 280

c. $([C][W])[D] = \begin{bmatrix} 24 & 25 & 51 \end{bmatrix} \begin{bmatrix} 40 \\ 60 \\ 30 \end{bmatrix} = [960+1,500+1,530] = [3,990]$

Page 325

$$[C]([W][D]) = \begin{bmatrix} 1 & 4 & 6 \end{bmatrix} \begin{bmatrix} 230 \\ 520 \\ 280 \end{bmatrix} = [230+2,080+1,680] = [3,990]$$

It is the total cost of production of all three wines.

60. Pronounce: "A big wheel," "A wheel," "Up and comer," and "Pea-on."

  a.

	Abigweel	Aweel	Upancomer	Pon
Abigweel	0	1	1	1
Aweel	0	0	1	1
Upancomer	0	0	0	1
Pon	0	0	0	0

  b.

$$\begin{bmatrix} 0 & 1 & 1 & 1 \\ 0 & 0 & 1 & 1 \\ 0 & 0 & 0 & 1 \\ 0 & 0 & 0 & 0 \end{bmatrix} \begin{bmatrix} 0 & 1 & 1 & 1 \\ 0 & 0 & 1 & 1 \\ 0 & 0 & 0 & 1 \\ 0 & 0 & 0 & 0 \end{bmatrix} = \begin{bmatrix} 0 & 0 & 1 & 2 \\ 0 & 0 & 0 & 1 \\ 0 & 0 & 0 & 0 \\ 0 & 0 & 0 & 0 \end{bmatrix}$$

  c.

$$[D] + [D]^2 = \begin{bmatrix} 0 & 1 & 1 & 1 \\ 0 & 0 & 1 & 1 \\ 0 & 0 & 0 & 1 \\ 0 & 0 & 0 & 0 \end{bmatrix} + \begin{bmatrix} 0 & 0 & 1 & 2 \\ 0 & 0 & 0 & 1 \\ 0 & 0 & 0 & 0 \\ 0 & 0 & 0 & 0 \end{bmatrix}$$

$$= \begin{bmatrix} 0 & 1 & 2 & 3 \\ 0 & 0 & 1 & 2 \\ 0 & 0 & 0 & 1 \\ 0 & 0 & 0 & 0 \end{bmatrix}$$

The power of Abigweel is 6; Aweel, 3; Upancomer, 1; and Pon is 0.

  d. Ranking: Abigweel, Aweel, Upancomer, and Pon

14.5 Systems of Inequalities, page 748

Level 1, page 749

1. A *system of linear inequalities* is two or more first degree inequalities considered at the same time.

2. Graph the boundary lines, use a test point to determine the half planes, and shade the intersecting of the half-planes.

3. a. 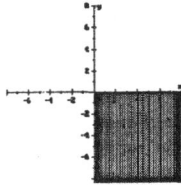　b. 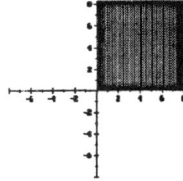　4. a. 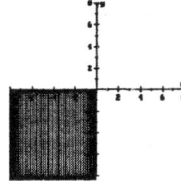　b.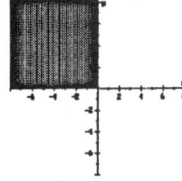

5. 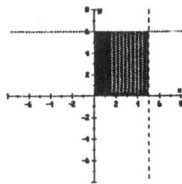　6. 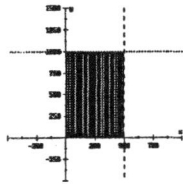　7. 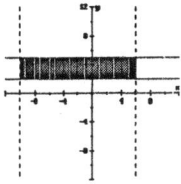　8.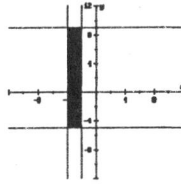

9. 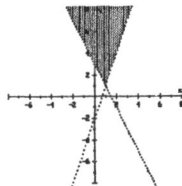　10. 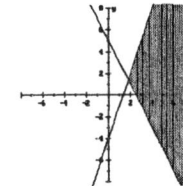　11. 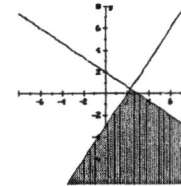　12.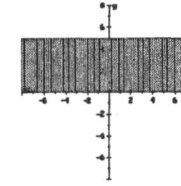

13. 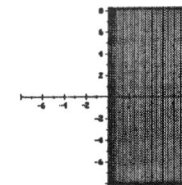　14. 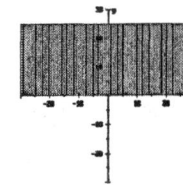　15. 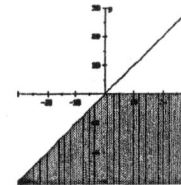　16.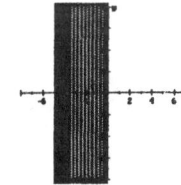

Level 2, page 749

17. 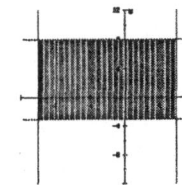　18. 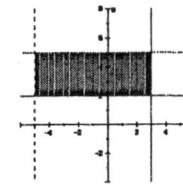　19. 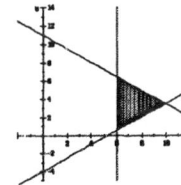　20.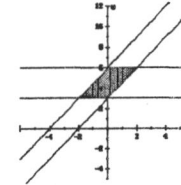

21.
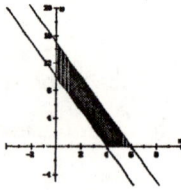

22.
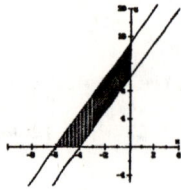

23.
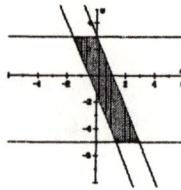

24.
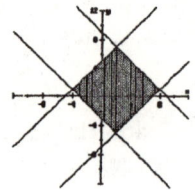

25.
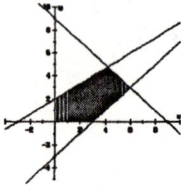

26.
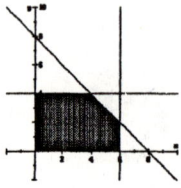

27.
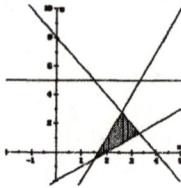

28.
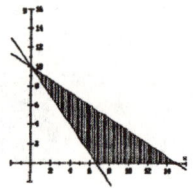

29.
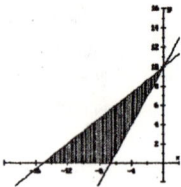

30.
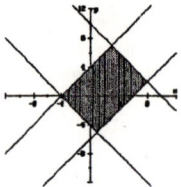

14.6 Modeling with Linear Programming, page 749

Level 1, page 755

1. A *linear programming problem* is a problem which seeks the maximum or minimum value of an expression, called the objective function. This solution is subject to a set of constraints, which take the form of linear inequalities.

2. An *objective function* is the function which needs to be maximized or minimized in a linear programming problem. A system of inequalities that must be satisfied in a linear programming problem are called the *constraints*.

3. a. no b. yes c. no d. no e. no f. no

4. a. no b. yes c. no d. yes e. yes f. no

Level 2, page 755

5. $(0, 0)$, $(0, \frac{9}{2})$, $(5, 2)$, $(6, 0)$ 6. $(0, 0)$, $(0, 4)$, $(5, 2)$, $(6, 0)$ 7. $(0, 0)$, $(0, 4)$, $(2, 3)$, $(4, 0)$

8. $(0, 6)$, $(0, 8)$, $(10, 8)$, $(10, 0)$, $(4, 0)$ 9. $(0, 0)$, $(0, 4)$, $(4, 4)$, $(6, 2)$, $(6, 0)$

10. $(0, 6)$, $(0, 9)$, $(\frac{25}{2}, 9)$, $(8, 0)$, $(6, 0)$ 11. $(0, 0)$, $(0, \frac{8}{5})$, $(\frac{24}{13}, \frac{16}{13})$, $(\frac{8}{3}, 0)$

12. $(0, 2)$, $(0, 10)$, $(\frac{15}{2}, \frac{5}{2})$, $(6, 2)$ 13. $(50, 0)$, $(\frac{200}{7}, \frac{60}{7})$, $(8, 24)$, $(0, 40)$

14. $(0, 0)$, $(0, 2)$, $(\frac{21}{5}, \frac{24}{5})$, $(6, 3)$, $(3, 0)$ 15. $(3, 2)$, $(5, 5)$, $(7, 5)$, $(\frac{10}{3}, \frac{4}{3})$

16. $(4, 0)$, $(\frac{5}{3}, \frac{14}{3})$, $(9, 12)$, $(9, 1)$, $(7, 0)$ 17. maximum $W = 190$ at $(5, 2)$

18. maximum $T = 600$ at $(6, 0)$ 19. maximum $P = 500$ at $(2, 3)$

20. minimum $K = 24$ at $(4, 0)$ 21. minimum $A = -12$ at $(0, 4)$

Level 3, page 756

22. Let $x =$ number of regular widgets
    $y =$ number of deluxe widgets
    Maximize $P = 25x + 30y$
    Subject to $\begin{cases} x \geq 0, y \geq 0 \\ 3x + 2y \leq 8 \\ 2x + 4y \leq 8 \end{cases}$

23. Let $x =$ number of units of food A
    $y =$ number of units of food B
    Minimize $C = 0.29x + 0.15y$

    Subject to $\begin{cases} x \geq 0, y \geq 0 \\ 10x + 5y \geq 200 \\ 2x + 5y \geq 100 \\ 3x + 4y \geq 20 \end{cases}$

24. Let $x =$ number of days Gainesville operates
    $y =$ number of days Sacramento operates
    Minimize $C = 20{,}000x + 15{,}000y$

    Subject to $\begin{cases} x \geq 0, x \leq 30 \\ y \geq 0, y \leq 30 \\ 600x + 300y \geq 24{,}000 \\ 100x + 100y \geq 5{,}000 \end{cases}$

25. Let $x =$ amount invested in stock (in millions of dollars)
    $y =$ amount invested in bonds (in millions of dollars)
    Maximize $T = 0.12x + 0.08y$

    Subject to $\begin{cases} x \geq 0 \\ x \leq 8, y \geq 2 \\ x + y \leq 10 \\ x \leq 3y \end{cases}$

Level 3 Problem Solving, page 756

26. Maximum profit $= 8{,}000$ at $(\frac{400}{11}, 0)$; this means that the maximum profit is $8,000 with

$\frac{400}{11} \approx 36.4$ acres of corn planted and the $\frac{700}{11} \approx = 63.6$ acres remaining are unplanted.

27. The maximum profit $P = \$14,300$ is achieved with all 100 acres planted in corn.

28. The maximum profit of $156,000 is achieved with 400 standard models and 4,600 economy models produced each week.

29. The minimum cost of $1.19 is obtained with 17 oz of Corn Flakes and no Honeycombs.

30. Invest $70,000 in stocks and $30,000 in bonds for a yield of $17,000.

Chapter 14 Review Questions, page 758

1. $[C][B] - 3[D] = \begin{bmatrix} 2 & 0 \\ 1 & 2 \\ -1 & 1 \end{bmatrix} \begin{bmatrix} 2 & -1 & 0 \\ 1 & 0 & 1 \end{bmatrix} - 3 \begin{bmatrix} 0 & 1 & 0 \\ -1 & 0 & 0 \\ 0 & 1 & -1 \end{bmatrix}$

$= \begin{bmatrix} 4 & -2 & 0 \\ 4 & -1 & 2 \\ -1 & 1 & 1 \end{bmatrix} + \begin{bmatrix} 0 & -3 & 0 \\ 3 & 0 & 0 \\ 0 & -3 & 3 \end{bmatrix} = \begin{bmatrix} 4 & -5 & 0 \\ 7 & -1 & 2 \\ -1 & -2 & 4 \end{bmatrix}$

2. $[A][B][C] = \begin{bmatrix} 1 & 0 \\ 2 & -1 \end{bmatrix} \begin{bmatrix} 2 & -1 & 0 \\ 1 & 0 & 1 \end{bmatrix} \begin{bmatrix} 2 & 0 \\ 1 & 2 \\ -1 & 1 \end{bmatrix}$

$= \begin{bmatrix} 2 & -1 & 0 \\ 3 & -2 & -1 \end{bmatrix} \begin{bmatrix} 2 & 0 \\ 1 & 2 \\ -1 & 1 \end{bmatrix} = \begin{bmatrix} 3 & -2 \\ 5 & -5 \end{bmatrix}$

3. $[B]$ is a 2×3 matrix and $[A]$ is 2×2 so these matrices are not conformable for multiplication.

4. $[C][A][B] = \begin{bmatrix} 2 & 0 \\ 1 & 2 \\ -1 & 1 \end{bmatrix} \begin{bmatrix} 1 & 0 \\ 2 & -1 \end{bmatrix} \begin{bmatrix} 2 & -1 & 0 \\ 1 & 0 & 1 \end{bmatrix}$

$= \begin{bmatrix} 2 & 0 \\ 5 & -2 \\ 1 & -1 \end{bmatrix} \begin{bmatrix} 2 & -1 & 0 \\ 1 & 0 & 1 \end{bmatrix} = \begin{bmatrix} 4 & -2 & 0 \\ 8 & -5 & -2 \\ 1 & -1 & -1 \end{bmatrix}$

5. $\begin{bmatrix} 2 & 1 & | & 1 & 0 \\ -\frac{3}{2} & -\frac{1}{2} & | & 0 & 1 \end{bmatrix} \rightarrow \begin{bmatrix} 1 & \frac{1}{2} & | & \frac{1}{2} & 0 \\ -\frac{3}{2} & -\frac{1}{2} & | & 0 & 1 \end{bmatrix} \rightarrow \begin{bmatrix} 1 & \frac{1}{2} & | & \frac{1}{2} & 0 \\ 0 & \frac{1}{4} & | & \frac{3}{4} & 1 \end{bmatrix}$

$\rightarrow \begin{bmatrix} 1 & \frac{1}{2} & | & \frac{1}{2} & 0 \\ 0 & 1 & | & 3 & 4 \end{bmatrix} \rightarrow \begin{bmatrix} 1 & 0 & | & -1 & -2 \\ 0 & 1 & | & 3 & 4 \end{bmatrix}$ Inverse is $\begin{bmatrix} -1 & -2 \\ 3 & 4 \end{bmatrix}$.

6. $\begin{bmatrix} 1 & 3 & 3 & | & 1 & 0 & 0 \\ 1 & 4 & 3 & | & 0 & 1 & 0 \\ 1 & 3 & 4 & | & 0 & 0 & 1 \end{bmatrix} \rightarrow \begin{bmatrix} 1 & 3 & 3 & | & 1 & 0 & 0 \\ 0 & 1 & 0 & | & -1 & 1 & 0 \\ 0 & 0 & 1 & | & -1 & 0 & 1 \end{bmatrix}$

$\rightarrow \begin{bmatrix} 1 & 0 & 3 & | & 4 & -3 & 0 \\ 0 & 1 & 0 & | & -1 & 1 & 0 \\ 0 & 0 & 1 & | & -1 & 0 & 1 \end{bmatrix} \rightarrow \begin{bmatrix} 1 & 0 & 0 & | & 7 & -3 & -3 \\ 0 & 1 & 0 & | & -1 & 1 & 0 \\ 0 & 0 & 1 & | & -1 & 0 & 1 \end{bmatrix}$

Inverse is $\begin{bmatrix} 7 & -3 & -3 \\ -1 & 1 & 0 \\ -1 & 0 & 1 \end{bmatrix}$.

7. Graph $2x - y = 2$ by writing $y = 2x - 2$; y-intercept is -2 and slope is $2 = \frac{2}{1}$.

   Graph $3x - 2y = 1$ by writing $y = \frac{3}{2}x - \frac{1}{2}$; y-intercept is $-\frac{1}{2}$ and slope is $\frac{3}{2}$.

   The intersection point appears to be $(3, 4)$. Check:

   $2(3) - 4 = 6 - 4 = 2$

   $3(3) - 2(4) = 9 - 8 = 1$

8. $\quad 2 \begin{cases} x + 3y = 3 \\ 4x - 6y = -6 \end{cases} + \begin{cases} 2x + 6y = 6 \\ 4x - 6y = -6 \end{cases}$

   $$6x = 0$$
   $$x = 0 \qquad \text{If } x = 0, \text{ then } 0 + 3y = 3 \text{ or } y = 1.$$

   The solution is $(0, 1)$.

9. Substitute the first equation into the second: $5x + 2(1 - 2x) = 1$

   $$5x + 2 - 4x = 1$$
   $$x = -1$$

   If $x = -1$, then $y = 1 - 2(-1) = 1 + 2 = 3$; the solution is $(-1, 3)$.

10. $\begin{bmatrix} 1 & 1 & 1 & | & 2 \\ 1 & 2 & -2 & | & 1 \\ 1 & 1 & 3 & | & 4 \end{bmatrix} \rightarrow \begin{bmatrix} 1 & 1 & 1 & | & 2 \\ 0 & 1 & -3 & | & -1 \\ 0 & 0 & 2 & | & 2 \end{bmatrix} \rightarrow \begin{bmatrix} 1 & 0 & 4 & | & 3 \\ 0 & 1 & -3 & | & -1 \\ 0 & 0 & 2 & | & 2 \end{bmatrix}$

   $\rightarrow \begin{bmatrix} 1 & 0 & 4 & | & 3 \\ 0 & 1 & -3 & | & -1 \\ 0 & 0 & 1 & | & 1 \end{bmatrix} \rightarrow \begin{bmatrix} 1 & 0 & 0 & | & -1 \\ 0 & 1 & 0 & | & 2 \\ 0 & 0 & 1 & | & 1 \end{bmatrix}$ The solution is $(-1, 2, 1)$.

11. Let $[A] = \begin{bmatrix} 2 & 1 \\ -\frac{3}{2} & -\frac{1}{2} \end{bmatrix}$, $[X] = \begin{bmatrix} x \\ y \end{bmatrix}$, $[B] = \begin{bmatrix} 13 \\ 10 \end{bmatrix}$; The solution is $[X] = [A]^{-1}[B]$.

We found the inverse in Problem 5: $[X] = \begin{bmatrix} -1 & -2 \\ 3 & 4 \end{bmatrix}\begin{bmatrix} 13 \\ 10 \end{bmatrix} = \begin{bmatrix} -33 \\ 79 \end{bmatrix}$.

The solution is $(-33, 79)$.

12.

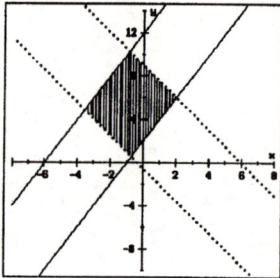

13. Let x = number of bars of product I.

y = number of bars of product II.

	A	B
Product I:	$3x$	$5x$
Product II:	$4y$	$7y$
Total:	33	56

Thus,

$$\begin{cases} 3x + 4y = 33 \\ 5x + 7y = 56 \end{cases}$$

The solution to this system is $(7, 3)$. This means she uses 7 bars of Product I and 3 bars of Product II.

14. Let x, y, and z be the number of products of I, II, and III that can be manufactured. This leads to the following system:

$$\begin{cases} 2x + 2y + 3z = 1{,}250 \\ x + 2y + 2z = 900 \\ x + y + 2z = 750 \end{cases}$$

The solution to this system is $(100, 150, 250)$. This means the number of product I to be manufactured is 100; product II, 150; and product III, 250.

15. Let x = number of acres of corn; y = number of acres of wheat.

Maximize profit, $P = 2.10(100)x + 2.50(40)y = 210x + 100y$. The constraints are:

$$\begin{cases} x \geq 0 \\ y \geq 0 \\ x + y \leq 500 \\ 100x + 40y \leq 18{,}000 \\ 120x + 60y \leq 24{,}000 \end{cases}$$

The corner points are (100, 200), (0, 400), and (180, 0).

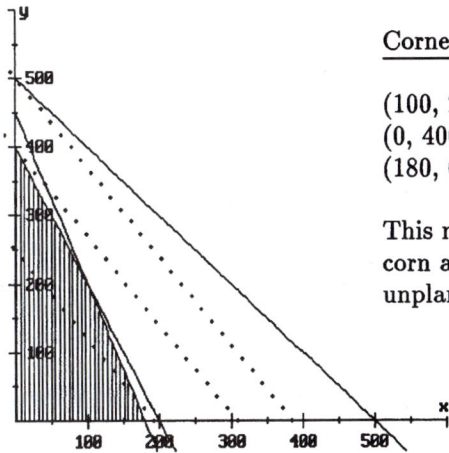

Corner point	Values of objective function
(100, 200)	41,000
(0, 400)	40,000
(180, 0)	37,800

This means the farmer will maximize the profit if 100 acres of corn and 200 acres of wheat are planted (200 acres left unplanted).

Group Research Problems, pages 759-760

G54.

$$[M] = \begin{array}{c} \\ A \\ B \\ C \\ D \\ E \\ F \end{array}
\begin{array}{cccccc}
A & B & C & D & E & F \\
0 & 0 & 0.5 & 0 & 0 & 1 \\
1 & 0 & 1 & 0 & 0 & 1 \\
0 & 0 & 0 & 0 & 1 & 1 \\
0.5 & 0 & 0 & 0 & 0 & 1 \\
1 & 0 & 0 & 0 & 0 & 1 \\
0 & 0 & 1 & 0.5 & 0 & 0
\end{array}$$

$$[M]^2 = \begin{bmatrix}
0 & 0 & 1 & 0.5 & 0.5 & 0.5 \\
0 & 0 & 1.5 & 0.5 & 1 & 2 \\
1 & 0 & 1 & 0.5 & 0 & 1 \\
0 & 0 & 1.25 & 0.5 & 0 & 0.5 \\
0 & 0 & 1.5 & 0.5 & 0 & 1 \\
0.25 & 0 & 0 & 0 & 1 & 1.5
\end{bmatrix} \qquad [T] = \begin{bmatrix} 1 \\ 1 \\ 1 \\ 1 \\ 1 \\ 1 \end{bmatrix}$$

Page 333

$$([M] + [M]^2)[T] = \begin{bmatrix} 4 \\ 8 \\ 5.5 \\ 3.75 \\ 5 \\ 4.25 \end{bmatrix} \quad \text{Rank: B, C, E, F, A, D}$$

55. Let s = number of steps on a stopped escalator, and let r = rate of the escalator (in steps per second). This leads to the system

$$\begin{cases} s - 20 = 20r \\ s - 32 = 16r \end{cases}$$

Solve this system to find $s = 80$. There are 80 steps on the escalator.

56. Let n be the number of cattle, so that n^2 is the number of dollars obtained from the sale. The number of 10's in n^2 also represents the number of sheep they bought. Since the sheep could not be evenly divided between the two ranchers, there must be an odd number of 10's in n^2. What could that number be? Here are some possibilities: 16, 36, 196, 256, 576, 676, 1,156, 1,296, $\cdots$. Notice that each number ends in the same digit. If this were not the case, the problem would not have a unique answer. Choose any one of them for n^2 (say 36; in this case, the ranchers will have bought three $10 sheep and one $6 lamb). The rancher that took the two sheep must have compensated his partner $2. Thus, the value of the watch is $2. Note that it is not necessary to find the number of sheep or cattle, nor it is necessary to find the total number of dollars.

57. Minimize the objective function $C = 0.14x + 0.18y$ where x = number of barrels (in millions) produced by the four-field precipitator and y = number of barrels (in millions) produced by the five-field precipitator. The constraints are $x \geq 0$, $y \geq 0$, $x + y \leq 2,500,000$, $1.5x + 1.8y \geq 4,200,000$. Graph the constraints to find corner points $(0, 2.5)$, $(0, 7/3)$, and $(1, 1.5)$. The cost is minimized for $(1, 1.5)$. This means that the least cost solution is to install the four-field precipitator on kilns producing 1,000,000 barrels and the five-field precipitator on kilns producing 1,500,000 barrels. The minimum cost is:

$$C = 0.14(1,000,000) + 0.18(1,500,000) = \$410,000$$

58. a. 7   b. 21   See **www.mathnature.com**, Section 14.6 for an excellent site with links.

CHAPTER 15
THE NATURE OF NETWORKS AND GRAPH THEORY

15.1 Euler Circuits and Hamiltonian Cycles, page 762

I use Transparency 31 to introduce this section. This is Problem 15 from the Prologue Problem Set. I happen to live in Santa Rosa, so I have selected landmarks in Santa Rosa surrounding my college. The map from the Santa Rosa Street Problem is topologically equivalent to the Königsberg Bridge problem. You could do the same for a map around your school.

Level 1, page 770

1. In the 18th century, in the German town of Königsberg (now a Russian city), a popular pastime was to walk along the bank of the Pregel River and cross over some of the seven bridges that connected two islands, as shown. The problem was to take a walk and cross each of the seven bridges once and only once, and return to the starting location.

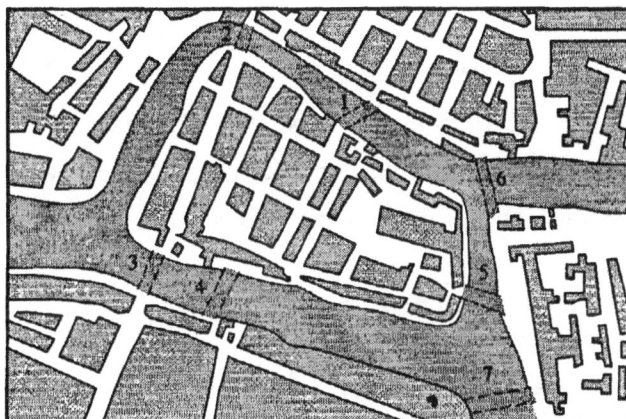

2. The floor problem is to take a trip through all the rooms of a floor plan so that you pass through each door once and only once. 3. Answers vary; see Table 15.1. If there are no odd vertices, the network is traversable and any point may be a starting point. The point selected will also be the ending point. If there is one odd vertex, the network is not traversable. A network cannot have only one starting or ending point without the other. If there are two odd vertices, the network is traversable; one odd vertex must be a starting point and the other odd vertex must be the ending point. If there are more than two odd vertices, the network is not traversable. A network cannot

have more than one starting point and one ending point.   **4.** A salesperson starts at home and wants to visit several cities without going through the same city again, and then return to the starting city.

5. The roles of vertices and paths reverse with the Euler circuits and Hamiltonian cycles. That is, with the Euler circuit problem we travel each path exactly once, and with a Hamiltonian cycle we visit each vertex exactly once.   *Results shown for Problems 6-11 may vary.*   **6.** Euler circuit; traversable ($A \to B \to C \to D \to E \to F \to A$)   **7.** Traversable ($A \to B \to C \to D \to A \to F \to E \to D$), but not an Euler circuit because we do not return to the starting point.   **8.** Not an Euler circuit; not traversable (more than 2 odd vertices).   **9.** Not an Euler circuit; not traversable (more than 2 odd vertices).   **10.** Not an Euler circuit; not traversable (more than 2 odd vertices).   **11.** Not an Euler circuit; not traversable (more than 2 odd vertices).   *Routes shown for Problems 12-17 may vary.*

12. Hamiltonian cycle: $A \to B \to E \to C \to D \to A$ **13.** Hamiltonian cycle: $A \to B \to C \to D \to A$

14. Hamiltonian cycle: $A \to E \to B \to C \to D \to A$

15. Hamiltonian cycle: $A \to B \to C \to G \to H \to F \to E \to D \to A$

16. Hamiltonian cycle: $A \to B \to E \to C \to D \to A$

17. Hamiltonian cycle: $A \to B \to C \to D \to A$

Possible paths in Problems 18-23 may vary.

18. traversable; $2 \to 5 \to 6 \to 3 \to 2 \to 1 \to 4 \to 5$ **19.** There are five rooms (vertices), three of which are odd, so it is not traversable.   **20.** traversable; $3 \to 4 \to 2 \to$ outside $\to 1 \to$ outside $\to 3 \to$ outside $\to 4 \to 2$ **21.** There are five rooms (vertices) three of which are odd, so it is not traversable.   **22.** There are six rooms, four of which are odd, it it is not traversable.   **23.** There are five rooms (vertices), two of which are odd, so it is traversable. Here is one possibility:

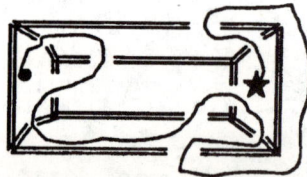

Level 2, page 770

24. Euler circuit; $A \to B \to C \to G \to F \to H \to C \to D \to E \to F \to A$

25. There are more odd vertices, so it is not an Euler circuit, since there are more than two odd vertices, it is not traversable.   **26.** There are odd vertices, so it is not an Euler circuit since it is not traversable.   **27.** There are odd vertices, so it is not an Euler circuit since it is not traversable.   **28.** not possible   **29.** not possible   **30.** *Answers may vary.* Hamiltonian cycle: $A \to E \to B \to G \to C \to F \to D \to A$ **31.** not possible   **32.** Transform the problem into a network; it is now traversable by starting at an odd vertex. Actual routes may vary.   **33.** *Answers may vary.* $1 \rightarrow 2 \rightarrow 3 \rightarrow 10 \rightarrow 11 \rightarrow 12 \rightarrow 4 \rightarrow 5 \rightarrow 14 \rightarrow 13 \rightarrow 19 \rightarrow 18 \rightarrow 17 \rightarrow 9 \rightarrow 8 \rightarrow 7 \rightarrow 16 \rightarrow 20 \rightarrow 15 \rightarrow 6 \rightarrow 1$ **34.** Not an Euler circuit; not traversable since there are more than two odd vertices.   **35.** Transform the problem into a network; it is not traversable since there are more than two odd vertices.   **36.**   From Point *A* cross Bridge 1 (George Washington Bridge), then Bridge 2 (Triborough Bridge), Bridge 3 (Queesboro or 59th St. Bridge), Bridge 4 (Williamsburg Bridge), Bridge 5 (Manhattan Bridge), and finally across Bridge 6 (Brooklyn Bridge) to reach point *B*. (There are other possible paths).   **37.** There are two odd vertices, so it is traversable; begin at either Queens or Manhattan.   **38.** It is possible; Bank and South Kensington are the only odd vertices.   **39.** Yes; 2 odd vertices.   **40.** No; there are now 4 odd vertices.

Level 3, page 772

41. No; there are 8 odd vertices.

42. Brute force: NYC $\rightarrow$ Boston $\rightarrow$ D.C. $\rightarrow$ Cleveland $\rightarrow$ NYC; $216 + 441 + 375 + 481 = 1,513$

NYC $\rightarrow$ Boston $\rightarrow$ Cleveland $\rightarrow$ D.C. $\rightarrow$ NYC; $216 + 667 + 375 + 235 = 1,493$

NYC $\rightarrow$ Cleveland $\rightarrow$ Boston $\rightarrow$ D.C. $\rightarrow$ NYC; $481 + 667 + 441 + 235 = 1,824$

NYC $\rightarrow$ Cleveland $\rightarrow$ D.C. $\rightarrow$ Boston $\rightarrow$ NYC; $481 + 375 + 441 + 216 = 1,513$

NYC $\rightarrow$ D.C. $\rightarrow$ Boston $\rightarrow$ Cleveland $\rightarrow$ NYC; $235 + 441 + 667 + 481 = 1,824$

NYC $\rightarrow$ D.C. $\rightarrow$ Cleveland $\rightarrow$ Boston $\rightarrow$ NYC; $235 + 375 + 667 + 216 = 1,493$

There is one route (along with its reverse route) with the mileage of 1,493 mi.

43. a. NYC to Boston to Washington, D.C. to NYC; $216 + 441 + 235 = 892$ mi; forms a loop without including Cleveland.   **b.** NYC to Boston to Washington, D.C. to Cleveland to NYC; $216 + 441 + 375 + 481 = 1,513$ mi   **44. a.** Denver to St. Louis to New Orleans, to Denver; $879 + 677 + 1,344 = 2,900$ mi; forms a loop that does not include L.A.   **b.** Denver to St. Louis to New Orleans to L.A. to Denver; $879 + 677 + 2,009 + 1,062 = 4,627$ mi

45. Brute force: Denver $\rightarrow$ S.L. $\rightarrow$ LA $\rightarrow$ N.O. $\rightarrow$ Denver; $879 + 1,844 + 2,009 + 1,344 = 6,076$

Denver $\rightarrow$ S.L. $\rightarrow$ N.O. $\rightarrow$ LA $\rightarrow$ Denver; $879 + 677 + 2,009 + 1,062 = 4,627$

Denver $\rightarrow$ LA $\rightarrow$ S.L. $\rightarrow$ N.O. $\rightarrow$ Denver; $1,062 + 1,844 + 677 + 1,344 = 4,927$

Denver $\rightarrow$ LA $\rightarrow$ N.O. $\rightarrow$ S.L. $\rightarrow$ Denver; $1,062 + 2,009 + 677 + 879 = 4,627$

Denver $\rightarrow$ N.O. $\rightarrow$ S.L. $\rightarrow$ LA $\rightarrow$ Denver; $1,344 + 677 + 1,844 + 1,062 = 4,927$

Denver $\rightarrow$ N.O. $\rightarrow$ LA $\rightarrow$ S.L. $\rightarrow$ Denver; $1,344 + 2,009 + 1,844 + 879 = 6,076$

There is one route (along with its reverse route) with the mileage 4,627 mi. Note this is the same as the sorted-edge solution from Problem 44b.

46.

	V	E	R	$V + R - 2$
(6)	6	6	2	6
(7)	6	7	3	7
(8)	7	10	5	10
(9)	4	6	4	6
(10)	5	12	9	12
(11)	12	16	6	16
(12)	5	8	5	8
(13)	4	6	4	6
(14)	5	6	3	6
(15)	8	11	5	11
(16)	5	10	7	10
(17)	4	8	6	8

$$V + R - 2 = E$$

Level 3 Problem Solving, page 772

47. Use the sorted edge method: $NYC \to B \to DC \to C \to A \to NYC$;

$216 + 441 + 375 + 780 + 887 = 2,699$ mi

48. $C \to M \to D \to A \to B \to C$; $420 + 963 + 691 + 1,115 + 1,013 = 4,202$ mi

49. A tetrahedron has four triangular faces; the sum of the measures of the angles on each face is 180°, so the sum of the measures of the face angles of a tetrahedron is 720°.

50. A pentagonal prism has five faces that are rectangles and two faces that are pentagons. The sum of the measures of the angles on each rectangular face is 360°, and the sum of the measures of the angles on each pentagonal face is 540°. Thus, the sum of the measures of the face angles of the pentagonal prism is $5(360°) + 2(540°) = 2,880°$.

51. Luke in this room: 1   2   3   4   5   6   7   8   9   10   11   12   13
No. of routes: 1   2   3   5   8   13   21   34   55   89   144   233   377

a. Room 1: 1 path; room 2: 2 paths; room 3: 3 paths; room 4: 5 paths   **b.** 89   **c.** 377

52. Look for a pattern; the number of paths to Room n is found by the sum of the previous two numbers.

53. **a.** Start at the top (or at the right) at one of the two odd vertices.
b. The chance that you will choose one of the two odd vertices at random is small.

54. One edge

55. The result is a twisted band twice as long and half as wide as the band with which you began.

56. Two interlocking pieces

57. Two interlocking pieces, one twice as long and half as wide as the other

58. Yes

59. One edge; one side; after cutting, you will have a single loop with a knot.

60. A Klein bottle is a bottle with only one side. It is impossible for a real Klein bottle in a three dimensional world.

15.2 Trees and Minimum Spanning Trees, page 774

Level 1, page 781

1.   A *tree* is a graph which is connected and has no circuits.   2.   A *spanning tree* is a tree that is created from another graph by removing edges, but keeping a path to each vertex.

3.   To construct the minimum spanning tree from a weighted graph: 1. Select any edge with minimum weight.   2. Select the next edge with minimum weight among those not yet selected.   3. Continue to choose edges of minimum weight from those not yet selected, but make sure not to select any edge that forms a circuit.   4. Repeat this process until the tree connects all of the vertices of the original graph.

4. This is a tree.   5. This is not a tree.   6. This is a tree.   7. This is not a tree because it has a circuit.   8. This is not a tree because it has a circuit.   9. This is not a tree because it has a circuit and is not connected.   10. This is not a tree because it is not connected.   11. This is a tree.

Answers to Problems 12-19 vary.

12.

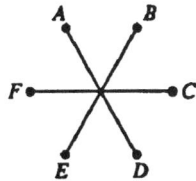 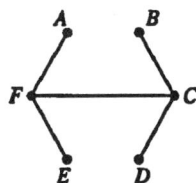

13.

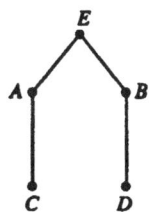 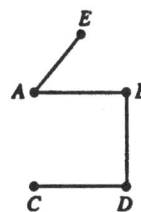

14.

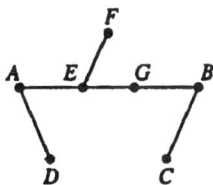 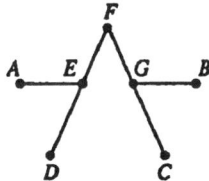

15.

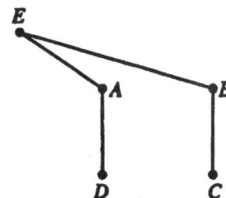 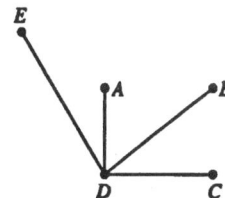

16.

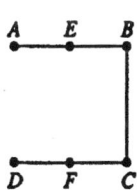 

17.

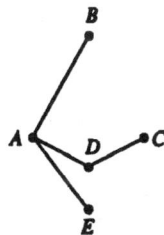 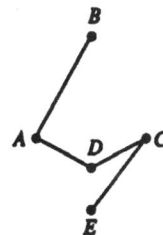

18.

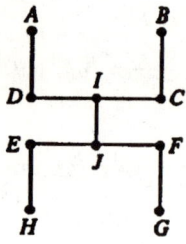 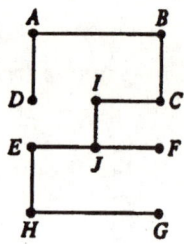

19.

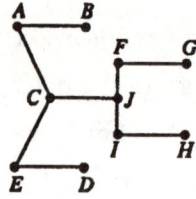 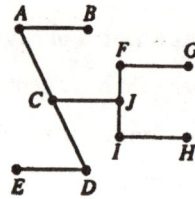

20.

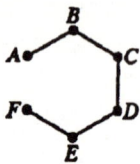

21.

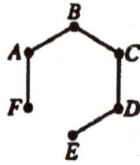

22.

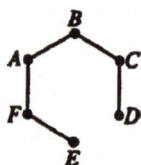

23.

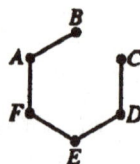

24.

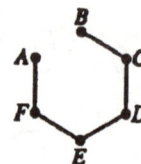

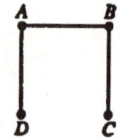

25.

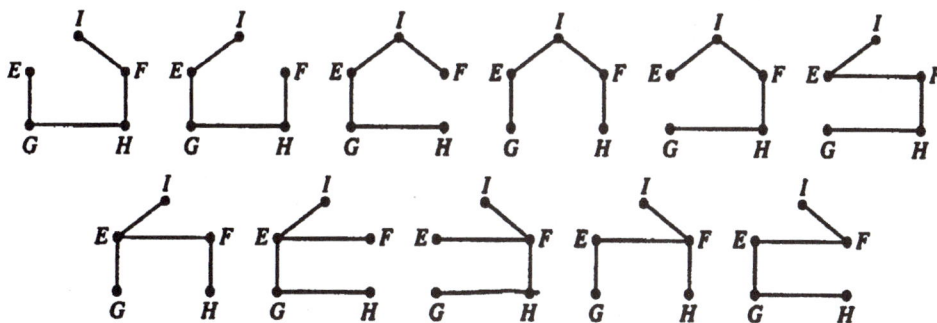

26. a tree   **27.** a tree   **28.** a tree

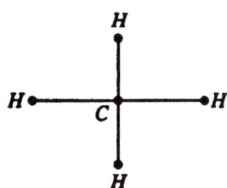

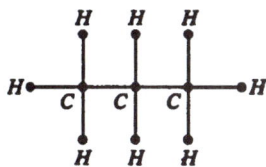

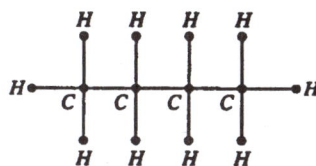

29. a tree   **30.** Not a tree because   **31.** Not a tree because
there is a circuit.   there is a circuit.

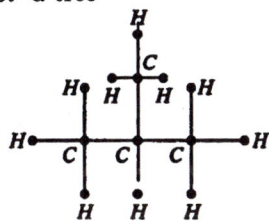

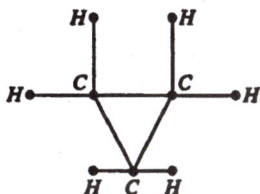

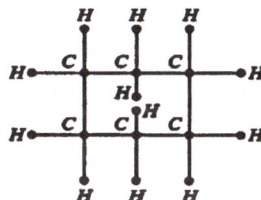

32.

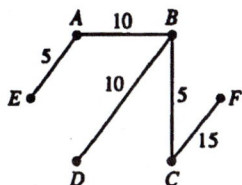

33.

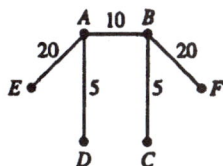

34.

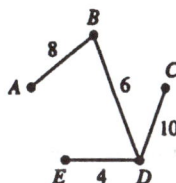

Minimum value:

$5 + 10 + 10 + 5 + 15$

$= 45$

Minimum value:

$20 + 10 + 5 + 5 + 20$

$= 60$

Minimum value:

$8 + 6 + 4 + 10$

$= 28$

35.

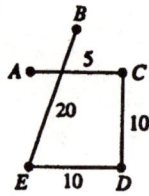

36.

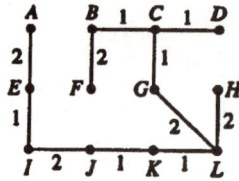

37.

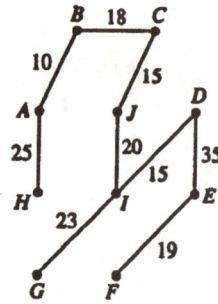

Minimum value, 45   Minimum value, 16   Minimum value, 180

38.

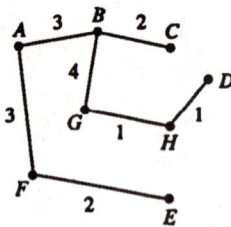

39.

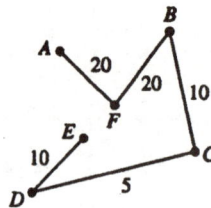

40.

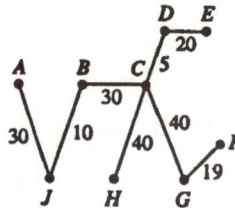

Minimum value

$2 + 3 + 3 + 4 + 1$
$+ 1 + 2$
$= 16$

Minimum value:

$20 + 20 + 10 + 5 + 10$
$= 65$

Minimum value:

$30 + 10 + 30 + 40 + 40$
$+ 20 + 19 + 5$
$= 194$

41.

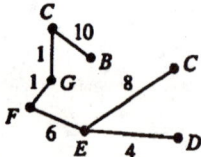

Minimum value: $10 + 1 + 1 + 6 + 4 + 8 = 30$

Level 2, page 783

42. From the number of edges and vertices in a tree theorem, and since the number of vertices is 39, the number of edges (or bounds) is $39 - 1 = 38$.   **43.** From the number of edges and vertices in a tree theorem, and since the number of vertices is 65, the number of edges (or bounds) is $65 - 1 = 64$.

44. No, because there would necessarily be some circuits.   **45.** No, because some sites lead you back to previously visited sites, which completes a circuit.

46.

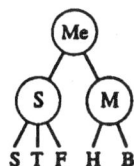

47.

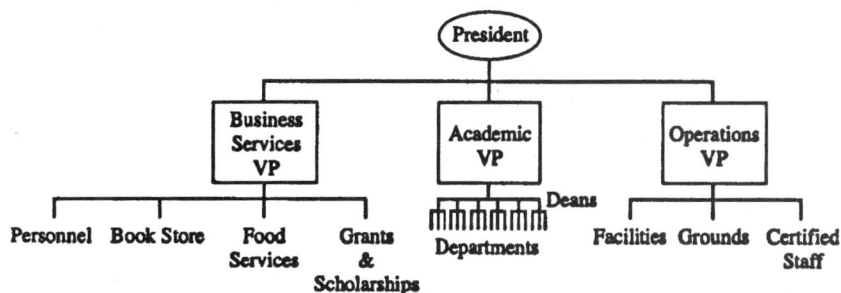

48. From the number of vertices and edges in a tree theorem, and since the number of vertices is 15, the number of edges is $15 - 1 = 14$.   **49.** From the number of vertices and edges in a tree theorem, and since the number of edges is 48, the number of vertices is $48 + 1 = 49$.   **50.** Use Kruskal's algorithm; there are six edges. Begin with the two of length 30 ft; then connect the ones of lengths 31 and 32 ft; finally, connect the last shrinker head using the edge of length 35 ft. The smallest number is $30 + 30 + 31 + 32 + 35 = 158$; that is 158 ft of drip hose is needed.   **51.** Use Kruskal's algorithm; there are six buildings using the walkways of length 10, 15, 25, 30, and 30. The total is 110 ft. Since the cost is \$350/ft, we estimate the minimum cost for the project to be (\$350/ft)(100 ft) = \$35,000. The actual amount is (\$350/ft)(110 ft) = \$38,500.   **52.** Use Kruskal's algorithm; there are five vertices; add edge AE, then BD, AB, and finally DC to connect complete the tree. The minimum cost is $\$650 + \$700 + \$860 + \$900 = \$3,110$.   **53.** The minimum spanning tree connects San Francisco to Oakland (smallest distance, 13 miles); Oakland to San Jose (next smallest distance, 46 miles); Manteca to Merced (54 miles); Merced to Fresno (56 miles); Santa Rosa to San Francisco (58 miles); Oakland to Manteca (66 miles), and then complete the graph with Merced to Yosemite Village (82 miles). The minimum spanning tree is 375 miles.   **54.** The minimum spanning tree connects Reno to Carson City (smallest distance, 30 miles); Beatty to Death Valley (next smallest distance, 47 miles); Bishop to Lone Pine (58 miles); Reno to Fallon (61 miles); Lee Vining to Bishop (66 miles); Beatty to Tonepah (93 miles); Death Valley to Lone Pine (106 miles); Carson City to Lee Vining (107 miles); and complete the tree by including Fallon to Austin (111 miles). The minimum spanning tree is 679 miles.

Level 3, page 784

55. a.

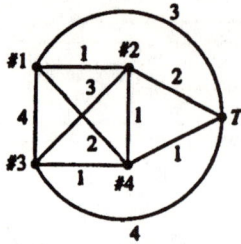

b.

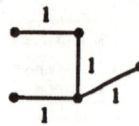

c. The minimum cost is $1 + 1 + 1 + 1 = 4$; that is, \$4,000,000.

56. a.

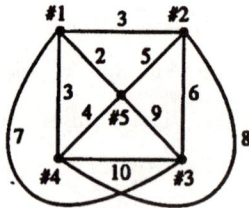

b.

c. The minimum cost is $3 + 2 + 3 + 6 = 14$.

57. a.

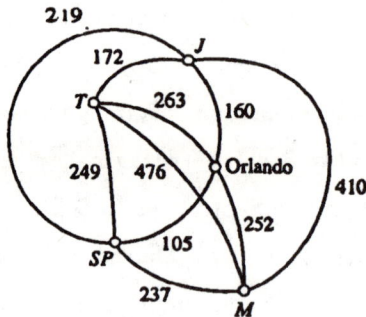

b.

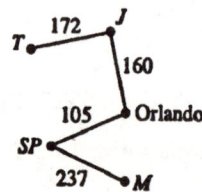

c. The minimum distance is
$172 + 160 + 105 + 237 = 674$ miles.
The minimum cost is $\$85(674) = \$57,290$.

58. a.

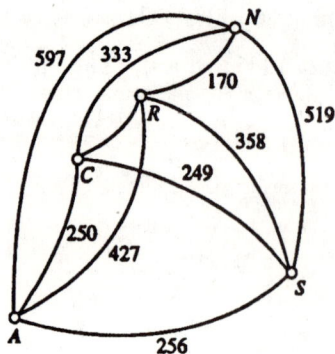

b.

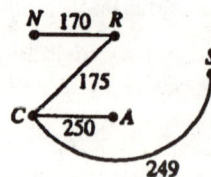

c. The minimum distance is
$170 + 175 + 250 + 249 = 844$ miles.
The minimum cost is $\$205(844) = \$173,020$.

Level 3 Problem Solving, page 785

59. a. If a graph is a tree with n vertices, then the number of edges is $n - 1$.

  b. There are no edges; yes, the property holds because $1 - 1 = 0$.

  c. Can only have one edge to have a tree. If there are two vertices and two or more edges, then there is a circuit, which is not permitted.

  d. If there are three vertices, then it takes two edges to form a tree. If there are three or more edges, then there is a circuit.

60. a. If $n = 3$, then $3^{3-2} = 3^1 = 3$

  These are shown below:

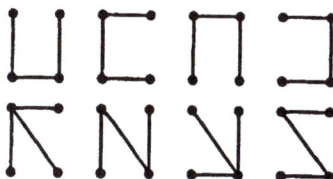

  b. If $n = 4$, then $4^{4-2} = 4^2 = 16$

  These are shown at the right:

  c. If $n = 5$, then $5^{5-2} = 5^3 = 125$

  d. If $n = 6$, then $6^{6-2} = 6^4 = 1{,}296$

15.3 Topology and Fractals, page 785

Level 1, page 791

1. *Topology* is that branch of mathematics that is concerned with discovering and analyzing the essential similarities and differences between sets and figures. An important idea of topology is that of elastic motion.   2.   The *four-color problem* states that any map on a plane or a sphere can be colored with at most four colors so that any two countries that share a common boundary are colored differently. 3. A and F; B and G; C, D, and H   4. F and H; B, C, and D; as well as A and G are topologically equivalent.   5. A and F are simple closed curves.   6. F and H are simple closed curves.   7. C, E, F, G, H, I, J, K, L, M, N, S, T, U, V, W, X, Y, and Z are the same class; A, D, O, P, Q, and R are another class; B is a third class. Note; if the Q has a visible loop as shown here, then it is in the same class as B. However, as shown in the text, it is categorized with the letter A.

8. A, B, C, D, E, and F are the same class; G is different.

9. A, C, E, and F are the same class; B and D are the same class; G is different.

Level 2, page 791

10. A and C are inside; B is outside.   **11.** B and C are inside; A is outside.   **12.** A and C are inside; B is outside.   **13.** A and B are outside; C is inside.   **14.** A and B are inside; C is outside.

15.
  a. It will cross an odd number of times for $\overline{AX}$ and $\overline{CX}$, and an even number of times for $\overline{BX}$.

  b. It will cross an odd number of times for $\overline{BX}$ and $\overline{CX}$, and an even number of times for $\overline{AX}$.

  c. It will cross an odd number of times for $\overline{AX}$ and $\overline{CX}$, and an even number of times for $\overline{BX}$.

  d. If $\overline{PX}$ crosses the curve an even number of times, then P is outside; an odd number of times, P is inside (for some point X which is outside).

16. two colors   **17.** Answers vary; map in Problem 16 is an example.   **18.** Answers vary; first region is one color, second another; then the third region (that lies on each of those) needs a third color.   **19.** Answers vary.

Level 3, page 792

20.
  a. 4 sides of $\frac{1}{3}$ unit each has a total length of $\frac{4}{3}$ unit.

  b. 16 sides of $\frac{1}{9}$ unit each has a total length of $\frac{16}{9}$ unit.

  c. 64 sides of $\frac{1}{27}$ unit each has a total length of $\frac{64}{27}$ unit.

21. Answers vary; construct a fractal curve.   **22.** Construction varies.   **23.** Construction varies.   **24.**

25. a. Answers vary (there are two possibilities)

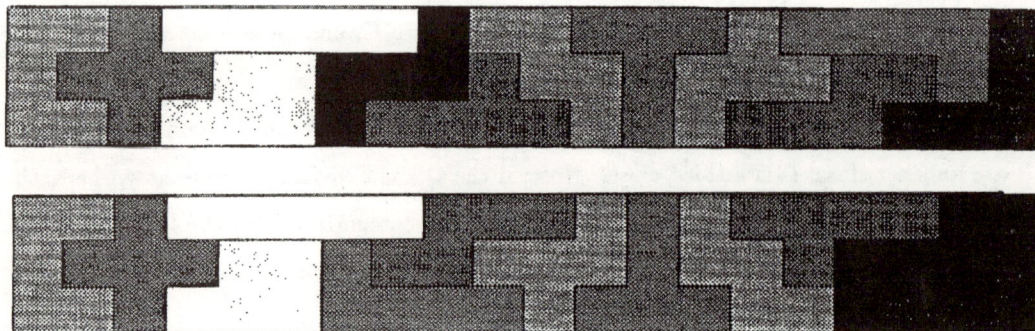

b. Answers vary; there are 364 possible designs. Here is one of them.

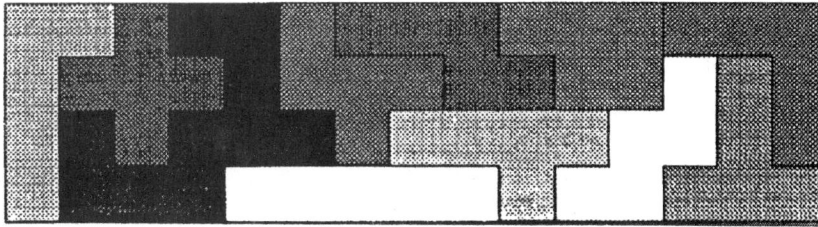

26. Answers vary.   **27.** Answers vary.   **28.** The images are topologically equivalent.

Level 3 Problem Solving, page 794

29. a. No, it will now fit the left hand.   **b.** Yes   **30.** It is not the first five-color map; simply begin over by starting with the inside regions and work your way out.

Chapter 15 Review Questions, page 796

1. Classify the vertices; they are all odd (degree 3), so there are more than 2 odd vertices. Thus it is not traversable.

2. Classify the vertices; the outside corners and the corner at the bottom of the inner rectangle are even; the intersecting lines at the top show an even vertex, and the two vertices at the top of the inner rectangle are odd (degree 5). Start at either of these odd vertices, and end at the other to show that the network is traversable.

3. There are 7 rooms, including the exterior. There are two rooms with an odd number of doors, so begin in either of those rooms and end in the other — it is possible.

4. Answers vary.

a.

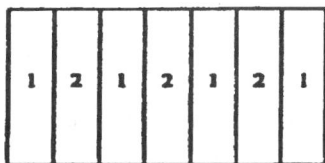

b.

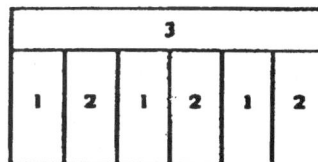

c.

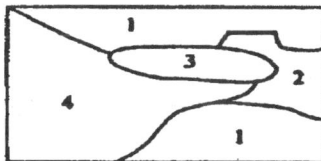

d. No such map exists.

5. **a.** Using the formula from Example 7 in Section 15.1, we see there are 8 cities, so there are
$\frac{8 \cdot 7 \cdot 6 \cdot 5 \cdot 4 \cdot 3 \cdot 2 \cdot 1}{2} = 20{,}160$ different routes.

   b. SF → LA → SD → P → SF; $369 + 133 + 353 + 755 = 1{,}610$

   SF → LA → P → SD → SF; $369 + 372 + 353 + 502 = 1{,}596$

   SF → SD → LA → P → SF; $502 + 133 + 372 + 755 = 1{,}762$

   SF → SD → P → LA → SF; $502 + 353 + 372 + 369 = 1{,}596$

   SF → P → SD → LA → SF; $755 + 353 + 133 + 369 = 1{,}610$

   SF → P → LA → SD → SF; $755 + 372 + 133 + 502 = 1{,}762$

   It looks like either of the routes shown in boldface is the best brute-force route.

6. **a.** SF → LA → SD → LA; $369 + 133 + 133 = 635$, but this does not include Phoenix, so this method does not give a best route.

   b. The weighted graph is shown:

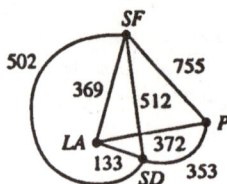

   The minimum spanning tree is:

   The minimum distance is $369 + 133 + 353 = 855$ miles.

7. **a.** Yes, there are many possibilities, for example, Minnesota, Wisconsin, Iowa, Illinois, Indiana, Michigan, Ohio, and Pennsylvania.

   b. Actually no, but if you use the ASC ferry system from Wisconsin to Pennsylvania, then the answer is yes; Iowa, Minnesota, Wisconsin, Pennsylvania, Ohio, Michigan, Indiana, Illinois, and Iowa. (There are other possibilities if you use the ferry system.)

8. **a.** $A → B → G → L → K → F → A$; the cost of this trip is 20.

   b. $M → H → I → N → O → J → E → D → C → B → G → L → K → F → A → B$; the cost of this trip is 49.

   c. There are 15 vertices, so the number of possibilities is $\frac{14 \cdot 13 \cdots 3 \cdot 2 \cdot 1}{2} = 4.3589 \times 10^{10}$.

d.   The cost of this trip is 39.

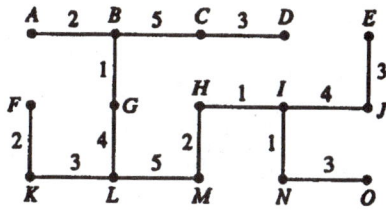

9. **a.** No, because there are more than two odd vertices.

    b. There are many possibilities; $A \to B \to C \to D \to I \to E \to F \to G \to H \to A$

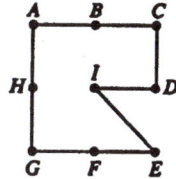

    c. Answers vary; here is one:

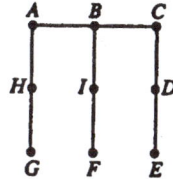

10. There are two edges and two sides; the result after cutting is two interlocking loops.

Group Research Problems, page 799

G59. The perimeters are $3a$, $4a$, $\frac{16}{3}a$, $\cdots 3a\left(\frac{4}{3}\right)^{n-1}, \cdots$

The perimeter for the snowflake increases without limit (geometric series with $r = \frac{4}{3} > 1$). There is no perimeter.

The area is $\dfrac{a^2\sqrt{3}}{4} + \dfrac{a^2\sqrt{3}}{12} + \dfrac{a^2\sqrt{3}}{12}\left(\dfrac{4}{9}\right) + \dfrac{a^2\sqrt{3}}{12}\left(\dfrac{4}{9}\right)^2 + \cdots$

Using the formula for the sum of a geometric series,

$$A = \frac{a^2\sqrt{3}}{4} + \frac{3a^2\sqrt{3}}{20} = \frac{2a^2\sqrt{3}}{5}$$

The snowflake curve has finite area and infinite perimeter.

Individual Research Problems, page 800

P15.1 to P15.4 Answers vary.   P15.5

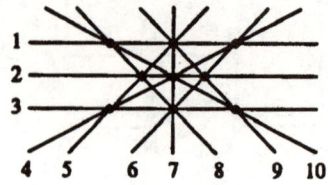

CHAPTER 16
THE NATURE OF VOTING AND APPORTIONMENT

16.1 Voting, page 802

I begin this chapter by asking the class to vote on some topic of current interest to the class. It might involve something happening in the community such as a freeway bypass, or a presidential election, or even to vote on an item of concern to the class itself, like how they would like to spend a day off. We then talk about how to tally and record the vote.

Level 1, page 811

1. **a.** yes **b.** yes **c.** no **d.** yes   *The answers to Problems 2-12 vary.*   **2.** Each voter votes for one candidate. The candidate receiving the most votes is declared the winner.   **3.** Each voter ranks the candidates. If there are *n* candidates, then *n* points are assigned to the first choice for each voter, with $n - 1$ points for the next choice, and so on. The points for each candidate are added and if one has more votes, that candidate is declared the winner.   **4.** Each voter votes for one candidate. If a candidate receives a majority of the votes, that candidate is declared to be a *first round winner*. If no candidate receives a majority of the votes, then with the **Hare method**, also known as the *plurality with an elimination runoff method*, then the candidate(s) with the fewest number of first-place votes is (are) eliminated. Each voter votes for one candidate in the second round. If a candidate receives a majority, that candidate is declared to be a *second round winner*. If no candidate receives a majority of the second round votes, then eliminate the candidate(s) with the fewest number of votes. Repeat this process until a candidate receives a majority.   **5.** In the **pairwise comparison method** of voting, the voters rank the candidates. A series of comparisons in which each candidate is compared to each of the other candidates. If choice *A* is preferred to choice *B*, then *A* receives 1 point. If *B* is preferred to *A*, then *B* receives 1 point. If the candidates tie, each receives $\frac{1}{2}$ point. The candidate with the most points is the winner.   **6.** With the **tournament method** candidates are teamed head-to-head or one-to-one, with the winner of one pairing facing a new opponent for the next election.   **7.** The **approval voting method** allows each voter to cast one vote for each candidate that meets with his or her approval. The candidate with the most votes is declared the winner.   **8.** class president
9. winning football team in an intramural competition   **10.** state convention voting for their

president   **11.** tennis tournament   **12.** selecting priorities in the master plan for an organization.
13. $8 + 4 + 3 + 0 + 2 + 5 = 22$ **14.** Since the total number of votes is even, a majority is
$\frac{22}{2} + 1 = 12$.   **15. a.** "(CAB)" means that the voter ranks three candidates in the order of C first, A
next, and candidate B last.   **b.** It means that in the election there were 5 voters who ranked the
candidates in the order CBA.   **16. a.** "(ACB)" means that the voter ranks three candidates in the
order of A first, C next, and in last place is candidate B.   **b.** In the election there were 4 voters who
ranked the candidates in the order ACB.   **17. a.** (ABC)   **b.** In the election there were 8 voters who
ranked the candidates in the order ABC.   **18. a.** "(BCA)" means that the voter preferences are for
B as the first choice, C next, with the last choice being A.   **b.** No people voted for this ranking.
19. $8 + 5 + 3 + 1 + 2 = 19$ votes   **20.** Since the total number of votes is odd, a majority is
$\frac{19 + 1}{2} = 10$.   **21. a.** "(DACB)" means the voters picks D as the first choice, followed by A, then C,
with B in last place.   **b.** It means that in the election there were 5 voters who ranked the candidates
in the order DACB.   **22. a.** "(ADBC)" means the voter picks A as the first choice, followed by D,
then C, and B is in last place.   **b.** It means that in the election there were 8 voters who ranked the
candidates in the order ADBC.   **23. a.** (BCDA)   **b.** 0   **24. a.** $4! = 24$
b. (ABCD), (ABDC), (ACBD), (ACDB), (ADCB), (BACD), (BCAD), (BCDA), (BDAC), (CABD),
(CADB), (DBAD), (CBDA), (CDAB), (CDBA), (DABC), (DBAC), (DBCA), (DCBA)
25. a. $3! = 6$ **b.** $4! = 24$ **c.** $5! = 120$

Level 2, page 811

26. a. First choice gets 5 points, second choice is 4 points, third choice is 3 points, then 2, and finally
one point. This means that each voter can cast $5 + 4 + 3 + 2 + 1 = 15$ votes. Consequently, if there
are 10 voters the total Borda count is 150.   **b.** Four candidates have a total count of 10 per voter, so
the total is $20(10) = 200$.   **c.** Three candidates have a total count of 6 per voter, so the total is
$200(6) = 1,200$.   **27.** $n!$ ways   **28.** The number of votes per candidate is $\frac{m(m + 1)}{2}$ and the number
of voters is n so the product is $n\left[\frac{m(m + 1)}{2}\right] = \frac{1}{2}mn(m + 1)$ **29.** The votes are A: $8 + 4 = 12$;
B: $3 + 0 = 3$, and C: $2 + 5 = 7$. The total vote is 22, and this is even so $\frac{22}{2} + 1 = 12$, so 12 votes are
necessary for a majority; thus, A wins according the to majority rule.   **30.** Any candidate who is a

majority winner must also be the plurality winner, so A is the winner.

31. The Borda count is:

Outcome	A	B	C		A	B	C
(ABC), 8:	8(3	2	1) =	24	16	8
(ACB), 4:	4(3	1	2) =	12	4	8
(BAC), 3:	3(2	3	1) =	6	9	3
(BCA), 0:	0(1	3	2) =	0	0	0
(CAB), 2:	2(2	1	3) =	4	2	6
(CBA), 5:	5(1	2	3) =	5	10	15
TOTAL:				51	41	40

A is the winner $_____$ ↑

32. A is the first round winner since A has received a majority of the votes.   **33.** AB → A; 1 point; AC → A; 1 point BC → tie; $\frac{1}{2}$ point for each; A wins   **34.** (1) First pair (A and B) → A; then pair the winner, A, with C: (A and C) → A; A wins. (2) First pair (A and C) → A; then pair the winner, A, with B: (A and B) → A; A wins. (3) First pair (B and C) → tie; if a tie-breaker is held and B is the winner, then (B with A) → A; A wins. On the other hand, if a tie-breaker is held and C is the winner, then (C with A) → A; A wins. In all pairings A wins.   **35.** snacks has 10 votes; drinks has 8 votes; travel slides has 3 votes; and a guest speaker has 9 votes. Snacks is the winner.   **36.** A has 10 votes; B has 10 votes; C has 11 votes; D has 9 votes; and E has 8 votes; C wins.   **37.** A majority winner is $\frac{293{,}472}{2} + 1 = 146{,}737$ Howard Dean is the majority winner.   **38.** A majority winner is $\frac{221{,}107 + 1}{2} = 110{,}554$ to win a majority. There is no majority winner in this race.

39. A: 2 + 4 = 6; B: 2 + 1 = 3; C: 2 + 1 = 3; there is no winner by using the majority rule.

40. Plurality vote chooses A.

41.

Outcome	A	B	C		A	B	C
(ABC), 2:	2(3	2	1) =	6	4	2
(ACB), 4:	4(3	1	2) =	12	4	8
(BAC), 2:	2(2	3	1) =	4	6	2
(BCA), 1:	1(1	3	2) =	1	3	2
(CAB), 2:	2(2	1	3) =	4	2	6
(CBA), 1:	1(1	2	3) =	1	2	3
TOTAL:				28	21	23

A is the winner $_____$↑

42. A has 6 first-place votes, B has 3 first-place votes, and C has 3 first-place votes. B and C have a tie for the fewest number of first-place votes. Some method should be specified for breaking this tie. Perhaps they could look at the Borda count (see Problem 41) and eliminate B, or they might flip a coin and eliminate C. Let us consider both of those possibilities. If they eliminate B, then the vote is: A: $2 + 4 + 2 = 8$ and C: $1 + 2 + 1 = 4$, so A wins in this case. If they eliminate C, then the vote is: A: $2 + 4 + 2 = 8$ and B: $2 + 1 + 1 = 4$, so A also wins in this case. This means that (in either case) A is the winner using the Hare method.   **43.** (AB): $2 + 4 + 2 = 8$ and (BA): $2 + 1 + 1 = 4$; A gets 1 point. (AC): $2 + 4 + 2 = 8$ and (CA): $1 + 2 + 1 = 4$; A gets 1 point. (BC): $2 + 2 + 1 = 5$ and (CB): $4 + 2 + 1 = 7$; C gets 1 point. This means that A has 2 points, B has no points, and C has 1 point. A is the winner. **44.** With the tournament method we look at the different pairings:

AB → A wins (8 to 4); then AC → A wins (8 to 4)

AC → A wins (8 to 4); then AB → A wins (8 to 4)

BC → C wins (5 to 7); then CA → A wins (8 to 4)

In all cases A wins.   **45.** We note that A has $1 + 3 = 4$ votes; B has $4 + 3 = 7$ votes; and C has $5 + 1 = 6$ votes. A majority requires $\frac{17 + 1}{2} = 9$ votes; there is no winner by using the majority rule.

46. Plurality vote chooses B.

47.

Outcome	A B C	A B C
(ABC), 1:	1(3 2 1) =	3 2 1
(ACB), 3:	3(3 1 2) =	9 3 6
(BAC), 4:	4(2 3 1) =	8 12 4
(BCA), 3:	3(1 3 2) =	3 9 6
(CAB), 5:	5(2 1 3) =	10 5 15
(CBA), 1:	1(1 2 3) =	1 2 3
TOTAL:		34 33 35; C is the winner

48. Since A obtains the fewest first-place votes, eliminate A from the running. The outcome now is:

(BC) (CB) (BC) (BC) (CB) (CB)

1   3   4   3   5   1

B: $1 + 4 + 3 = 8$

C: $3 + 5 + 1 = 9$; C is the winner

49. AB → A (9 to 8); AC → C (9 to 8); BC → C (9 to 8). This means that A has 1 point, B has no points, and C has 2 points. C is the winner.   **50.** With the tournament method we look at the different pairings:

AB → A wins (9 to 8); then AC → C wins (9 to 8)

AC → C wins (9 to 8); then CB → C wins (9 to 8)

BC → C wins (9 to 8); then AC → C wins (9 to 8)

In all cases C wins.

Level 3, page 812

51. a. The most common grade is a B. **b.** plurality vote   **52. a.** A grade is 4 points, B grade is 3 points, C is 2 points, D is 1 point, and F is 0 points. The GPA is:

$2 \times 4 + 6 \times 3 + 5 \times 2 + 1 \times 1 = 37$ divided by the number of courses 14: $\frac{37}{14} \approx 2.64$. (Note: if the courses were not all three unit courses, then we would need to multiply the number of units times the letter value, but then we would need to divide by the total number of units instead of the number of courses.) **b.** This is a Borda count.   **53. a.** The most common grade is a C. **b.** plurality vote

54. a. A grade is 4 points, B grade is 3 points, C is 2 points, D is 1 point, and F is 0 points. The GPA is: $14 \times 4 + 21 \times 3 + 35 \times 2 + 5 \times 1 + 2 \times 0 = 194$ divided by the number of courses 77: $\frac{194}{77} \approx 2.52$. (Note: if the courses were not all three unit courses, then we would need to multiply the number of units times the letter value, but then we would need to divide by the total number of units, instead of the number of courses.) **b.** This is a Borda count.   **55.** There are five candidates, so the number of possibilities is $5! = 120$.   **56.** The results are A: 360; B: 240; C: 200; D: 180; E: $80 + 40 = 120$. **a.** There were 1,100 votes cast, so a majority would require $\frac{1,100}{2} + 1 = 551$ votes; there is no majority winner. **b.** There is a plurality winner, and that is A.   **57.** Drop the person with the fewest first-place votes, which is E. Now the votes are:

    A: 360 = 360
    B: 240 + 80 = 320
    C: 200 + 40 = 240
    D: 180 = 180

No majority winner, so we now, eliminate D for the third ballot:

    A: 360 + 0 = 360
    B: 240 + 80 + 0 = 320
    C: 200 + 180 + 40 = 420

For the fourth ballot, eliminate B:

    A: 360
    C: 240 + 200 + 180 + 80 + 40 = 740; C is declared the winner.

This is commonly called the Hare method.

58.   There are a total of 1,100 votes cast, so since this is even, a majority is $\frac{1,100}{2} + 1 = 551$.   There is no majority, so the tally for the last place votes is:

A: $240 + 200 + 180 + 80 + 40 = 740$

C: 360

B, D, and E do not have any last place votes, so eliminate A.   Now the votes are:

A: 0
B: 240
C: 200
D: $360 + 180 = 540$
E: $80 + 40 = 120$

There is not yet a majority, so we once again count the number of **last place votes**:

B: $180 + 40 = 220$
C: $360 + 240 + 80 = 680$
D: 200
E: 0

At this point we need to eliminate C.   Now the votes are:

B: $240 + 200 = 440$
D: $360 + 180 = 540$
E: $80 + 40 = 120$

We need 551 for a majority, it looks like we need to go one more step.   Here is the count of **last place votes** (after eliminating both A and C):

B: $360 + 180 + 40 = 580$
D: $240 + 200 + 80 = 520$
E: 0

Eliminate B.   Here is the count after eliminating (A, B, and C):

D: $360 + 180 = 540$
E: $240 + 200 + 80 + 40 = 560$

E now has the required numbers to be declared the majority winner.

59.

	A	B	C	D	E		A	B	C	D	E
360(	5	2	1	4	3)	=	1,800	720	360	1,440	1,080
240(	1	5	2	3	4)	=	240	1,200	480	720	960
200(	1	4	5	2	3)	=	200	800	1,000	400	600
180(	1	2	4	5	3)	=	180	360	720	900	540
80(	1	4	2	3	5)	=	80	320	160	240	400
40(	1	2	4	3	5)	=	40	80	160	120	200
TOTAL:							2,540	3,480	2,880	3,820	3,780

The winner is D.

60.   As you can see (by looking at Problems 56-59 the winner by plurality is A, by the Hare method it is C, by runoff with elimination it is E and the Borda count gives D.   This can be a bit troubling, and leads us to the discussion about voting dilemmas, which is discussed in the next section.

16.2 Voting Dilemmas, page 813

Level 1, page 827

1. The *majority criterion* says that if a candidate receives a majority of the first-place votes, then that candidate should be declared the winner.   **2.** The *Condorcet criterion* says that if a candidate is favored when compared one-on-one with every other candidate, then that candidate should be declared the winner.   **3.** The *monotonicity criterion* says that a candidate who wins a first election and then gains additional support, without losing any of the original support, should also win a second election.   **4.** The *criterion of irrelevant alternatives* says that if a candidate is declared the winner of an election, and in a second election one or more of the other candidates is removed, then the previous winner should still be declared the winner.   **5.** A voting plan satisfies the *fairness criteria* if it satisfies all of the following properties: majority criterion, Condorcet criterion, monotonicity criterion and *irrelevant alternatives criterion*.   **6.** Answers vary; this is almost like asking you which of your children you like best.   **7.** Answers vary.   **8.** Insincere voting is when a person votes for a candidate which the voter does not wish to win, with the purpose of forcing the front runner to falter. Insincere voting also includes offering an amendment with the purpose of changing the final outcome of the election.   **9.** It tells us that we will not be able to make up, invent, or discover a voting plan without violating at least one of these conditions: unrestricted domain, decisiveness, symmetry and transitivity, independence of irrelevant alternatives, Pareto principle, or dictatorship.   **10.** The Pareto principle says that if each voter prefers A over B, then the group should choose A over B.

11. **a.** The vote for winner is:

A: $11 + 1 = 12$
B: $3 + 6 = 9$
C: $3 + 7 = 10$

There is no majority. Proceed with the Hare method of voting. The fewest first-place votes were cast for B, so eliminate B. The vote is now:

A: $11 + 1 + 3 = 15$
C: $6 + 3 + 7 = 16$

The winner is C, so the California Teachers Association will be the collective bargaining for the faculty association.

  b. No, this does not violate the majority criterion because there is no majority winner.

12.　**a.** The vote for winner is:

A: 5
B: 4
C: 3

Since there are 12 voters, a majority would be $\frac{12}{2} + 1 = 7$ votes.　There is no majority winner, but the plurality vote goes to A.

b.

Outcome	A B C	A	B	C
(ABC), 5:	5(3　2　1) =	15	10	5
(ACB), 0:	0(3　1　2) =	0	0	0
(BAC), 0:	0(2　3　1) =	0	0	0
(BCA), 4:	4(1　3　2) =	4	12	8
(CAB), 0:	0(2　1　3) =	0	0	0
(CBA), 3:	3(1　2　3) =	3	6	9
TOTAL:		22	28	22; B is the winner.

c.　The Borda count method declares B to be the winner and there is no majority winner, so it does not violate the majority criteria.　Notice, however, that the plurality winner is A, so this example violates what might be called a "plurality criteria."　This problem could serve as an example that the Borda count can violate the plurality criteria.

13.　**a.** The vote for winner is:

A: 6
B: 5
C: 3

Since there are 14 voters, a majority would be 8 votes.　There is no majority winner, but the plurality vote goes to A.

b.

Outcome	A B C	A	B	C
(ABC), 0:	0(3　2　1) =	0	0	0
(ACB), 6:	6(3　1　2) =	18	6	12
(BAC), 5:	5(2　3　1) =	10	15	5
(BCA), 0:	0(1　3　2) =	0	0	0
(CAB), 3:	3(2　1　3) =	6	3	9
(CBA), 0:	0(1　2　3) =	0	0	0
TOTAL:		34	24	26; A is the winner.

The Borda count method declares A to be the winner.

c.　No, this does not violate the majority criteria, since there is no majority winner.

14. **a.** The vote for winner is:

A: 16
B: 5
C: 9

Since there are 30 voters, a majority would be 16 votes.   The majority rules, A is the winner.

b.

Outcome	A	B	C	A	B	C
(ABC), 16:	16(3	2	1) =	48	32	16
(ACB), 0:	0(3	1	2) =	0	0	0
(BAC), 0:	0(2	3	1) =	0	0	0
(BCA), 5:	5(1	3	2) =	5	15	10
(CAB), 0:	0(2	1	3) =	0	0	0
(CBA), 9:	9(1	2	3) =	9	18	27
TOTAL:				62	65	53; B is the winner.

c. The Borda count method declares B to be the winner, even though A is the majority winner. This problem could serve as an example that the Borda count can, indeed, violate the majority rule.

15. **a.** The vote for winner is:

A: 2
B: 2
C: 1

The candidate with the fewest first-place votes is C. Delete C and have a runoff election between A and B:

A: 2 + 0 + 1 = 3
B: 0 + 2 + 0 = 2

A wins the runoff election.

b. If B withdraws before the election the votes would be:

(AC)   (CA)

2 + 0 + 0 = 2   2 + 1 + 0 = 3

There are 5 votes, so a majority winner is C.

c. Yes, this violates the irrelevant alternative criterion. Candidate B was not the winner in part a, where A won the election. However, if candidate B were to withdraw from the election, that would affect the outcome and cause C to win.

16. **a.** The vote for winner is:

A: 9
B: 9
C: 7

The candidate with the fewest first-place votes is C. Delete C and have a runoff election

between A and B:

A: $9 + 0 = 9$
B: $9 + 0 + 7 = 16$

B wins the runoff election.

b. If C withdraws before the election the votes would be:

(AB)   (BA)

$9 + 0 + 0 = 9$ $9 + 0 + 7 = 16$

There are 25 votes, so a majority winner is B.

c. No, none of the fairness criteria have been violated.

17. **a.** A Condorcet candidate is the one that wins all the one-on-one match-ups.

A over B: $8 + 6 = 14$ and B over A: $5 + 8 = 13$; A wins

A over C: $8 + 8 = 16$ and C over A: $5 + 6 = 11$; A wins

B over C: $8 + 5 + 8 = 21$ and C over B: 6; B wins

	A	B	C
A	–	A	A
B	A	–	B
C	A	B	–

We can see from the table that A is the Condorcet candidate.

b. Now, look at the original vote; the results are:

A: 8
B: $5 + 8 = 13$
C: 6

The total number of votes is 27, so there is no majority. The plurality vote goes to to B with

a total of 13 first-place votes. This does violate the Condorcet criterion.

18. **a.** A Condorcet candidate is the one that wins all the one-on-one match-ups.

A over B: $8 + 6 = 14$ and B over A: $5 + 8 = 13$; A wins
A over C: $8 + 8 = 16$ and C over A: $5 + 6 = 11$; A wins
B over C: $8 + 5 + 8 = 21$ and C over B: 6; B wins

	A	B	C
A	–	A	A
B	A	–	B
C	A	B	–

We can see from the table that A is the Condorcet candidate.

b. We do the Borda count:

Outcome	A B C		A	B	C
(ABC), 8:	8(3 2 1) =		24	16	8
(ACB), 0:	0(3 1 2) =		0	0	0
(BAC), 8:	8(2 3 1) =		16	24	8
(BCA), 5:	5(1 3 2) =		5	15	10
(CAB), 6:	6(2 1 3) =		12	6	18
(CBA), 0:	0(1 2 3) =		0	0	0
TOTAL:			57	61	44; B is the winner.

We see that the Borda count violates the Condorcet criterion.

19. a. The vote is:

A: 7
B: 3
C: 5
D: 2

There is no majority winner, but A wins by the plurality method.

b. If Carol drops out the voting preferences are:

(ADB) (BAD) (BDA) (DBA)

  7   5   3   2

The vote is:

A: 7
B: $5 + 3 = 8$
D: 2

The plurality winner is now B.

c. Yes, the irrelevant alternatives criterion has been violated because when Carol was removed the outcome was changed.

Level 2, page 828

20. We construct a table showing the one-on-one match-up winners.

A over B: $30 + 25 = 55$

B over A: $22 + 20 + 11 = 53$

A wins.

A over C: $30 + 25 = 55$

C over A: $22 + 20 + 11 = 53$

A wins.

A over D: $25 + 22 + 11 = 58$

D over A: $30 + 20 = 50$

A wins.

B over C: $30 + 22 = 52$

C over B: $25 + 20 + 11 = 56$

C wins.

	A	B	C	D
A	–	A	A	A
B	A	–	C	B
C	A	C	–	C
D	A	B	C	–

B over D: $25 + 22 + 20 + 11 = 78$

D over B: 30

B wins

C over D: $25 + 22 + 20 + 11 = 78$

D over C: 30

C wins

We see that A wins over all the other candidates on a one-to-one matching, so the Condorcet candidate is A.

21. Here is the vote count:

A: 25
B: 22
C: $20 + 11 = 31$
D: 30

There is no majority winner, but C is the plurality winner. This violates the Condorcet criterion because the Condorcet candidate is A.

22.

	A	B	C	D		A	B	C	D
30(	3	2	1	4)	=	90	60	30	120
25(	4	2	3	1)	=	100	50	75	25
22(	2	4	3	1)	=	44	88	66	22
20(	1	3	4	2)	=	20	60	80	40
11(	2	3	4	1)	=	22	33	44	11
TOTAL:						276	291	295	218

We see that C wins the Borda count. This violates the Condorcet criterion since the Condorcet candidate is A.

23. A majority is $\frac{108}{2} + 1 = 55$ votes.

A: 25
B: 22
C: $20 + 11 = 31$
D: 30

The least number of first-place votes are for B, so we eliminate B. The results of the second round of voting are:

A: 25
C: $22 + 20 + 11 = 53$
D: 30

There still is not majority winner, so we eliminate A. The results of the third round of voting are:

C: $25 + 22 + 20 + 11 = 78$
D: 30

The winner is C. This violates the Condorcet criterion, since the Condorcet candidate is A.

24. (See the solution to Problem 20 for details.)

A wins over B, so A gets one point.   C wins over B, so C gets one point.
A wins over C, so A gets one point.   B wins over D, so B gets one point.
A wins over D, so A gets one point.   C wins over D, so C gets one point.

The score is: A 3 points; B 1 point, C 2 points, and D 1 point. A wins the pairwise comparison method. No, this does not violate the Condorcet criterion.

25. We construct a table showing the one-on-one match-up winners.

A over B: $80 + 45 = 125$

B over A: $30 + 50 + 10 = 90$

A wins.

A over C: $80 + 45 = 125$

C over A: $30 + 10 + 50 = 90$

A wins.

A over D: $45 + 30 + 50 = 125$

D over A: $80 + 10 = 90$

A wins.

B over C: $80 + 30 = 110$

C over B: $45 + 10 + 50 = 105$

B wins.

	A	B	C	D
A	–	A	A	A
B	A	–	B	B
C	A	B	–	C
D	A	B	C	–

B over D: $45 + 30 + 10 + 50 = 135$

D over B: 80

B wins

C over D: $45 + 30 + 10 + 50 = 135$

D over C: 80

C wins

We see that A wins over all the other candidates on a one-to-one match-up so A is the Condorcet candidate.

26. A majority is $\dfrac{215 + 1}{2} = 108$ votes.

Here is the vote count:

A: 45
B: 30
C: 10 + 50 = 60
D: 80

There is no majority candidate. The plurality winner is D,

which violates the Condorcet criterion because the Condorcet candidate is A.

27.

	A	B	C	D		A	B	C	D
80(	3	2	1	4)	=	240	160	80	320
45(	4	2	3	1)	=	180	90	135	45
30(	2	4	3	1)	=	60	120	90	30
10(	1	3	4	2)	=	10	30	40	20
50(	2	3	4	1)	=	100	150	200	50
TOTAL:						590	550	545	465

We see that A wins the Borda count, which does not violates the Condorcet criterion.

28.   A majority is $\dfrac{215 + 1}{2} = 108$ votes.,

Here is the initial vote count:

A: 45
B: 30
C: 10 + 50 = 60
D: 80

The least number of first-place votes is B, so we eliminate B. The results of the second round of

voting are:

A: 45
C: 30 + 10 + 50 = 90
D: 80

There still is not majority winner, so we eliminate A. The results of the third round of voting

are:

C: 45 + 30 + 10 + 50 = 135
D: 80

The winner is C. This violates the Condorcet criterion, since the Condorcet candidate is A.

29.   (See the solution for Problem 25 for details.)

A wins over B, so A gets one point.   B wins over C, so B gets one point.

A wins over C, so A gets one point.   B wins over D, so B gets one point.

A wins over D, so A gets one point.   C wins over D, so C gets one point.

The score is A 3 points; B 2 points, C one point, and D zero points. A wins the pairwise

comparison method. No, this does not violate the Condorcet criterion.

30.   E over M: 15 and M over E: $6 + 6 + 6 = 18$; M wins, so M gets one point.

E over H: $15 + 6 + 6 = 27$ and H over E: 6; E wins, so E gets one point.

E over I: $15 + 6 + 6 = 27$ and I over E: 6; E wins, so E gets one point.

E over T: $15 + 6 + 6 = 27$ and T over E: 6; E wins, so E gets one point.

M over H: $6 + 6 = 12$ and H over M: $15 + 6 = 21$; H wins, so H gets one point.

M over I: $6 + 6 + 6 = 18$; I over M: 15; M wins, so M gets one point.

M over T: $6 + 6 = 12$; T over M: $15 + 6 = 21$; T wins, so T gets one point.

H over I: 6 and I over H: $15 + 6 + 6 = 27$; I wins, so I gets one point.

H over T: $15 + 6 + 6 = 27$ and T over H: 6; H wins, so H gets one point.

I over T: $15 + 6 = 21$ and T over I: $6 + 6 = 12$; I wins, so I gets one point.

Here is the score: E has 3 points; M has 2 points; H has 2 points; I has 2 points; and T has 1 point. E is the winner.

31.   Here is the Borda count:

		E	M	H	I	T		E	M	H	I	T
										Totals		
(EIHTM)	15(	5	1	3	4	2) =		75	15	45	60	30
(MIEHT)	6(	3	5	2	4	1) =		18	30	12	24	6
(HMETI)	6(	3	4	5	1	2) =		18	24	30	6	12
(TMEIH)	6(	3	4	1	2	5) =		18	24	6	12	30
TOTAL:								129	93	93	102	78

E is the Borda count winner.

32.   E over M: 15 and M over E: $6 + 6 + 6 = 18$; M wins, so M gets one point.

E over I: $15 + 6 + 6 = 27$ and I over E: 6; E wins, so E gets one point.

M over I: $6 + 6 + 6 = 18$; I over M: 15; M wins, so M gets one point.

Here is the score: E has 1 point; M has 2 points; and I has no points. M is the winner.

This does violate the irrelevant alternatives criterion. The original winner was E and it changed to M after one or more candidates was removed.

33.   Here is the Borda count:

		E	M	I		E	M	I
						Totals		
(EIM)	15(	3	1	2) =		45	15	30
(MIE)	6(	1	3	2) =		6	18	12
(MEI)	12(	2	3	1) =		24	36	12
TOTAL:						75	69	54

E is the Borda count winner. No, the Borda count method does not violate the irrelevant alternatives criterion.

34.   E over M: 15 and M over E: 6 + 6 + 6 = 18; M wins, so M gets one point.

M is the winner. Yes, this violates the irrelevant alternatives criterion.

35.   Here is the Borda count: Totals

		E	M		E	M
(EM)	15(	2	1	) =	30	15
(ME)	18(	1	2	) =	18	36
TOTAL:					48	51

M is the Borda count winner. Yes, the Borda count method violates the

irrelevant alternatives criterion because the pairwise winner is E.

36.   A majority is $\frac{93+1}{2} = 47$ votes.

Here is the result of the first day ballot:

A: 20 + 4 = 24
L: 35
T: 31
S: 3

With the Hare method, S is eliminated for the second ballot:

A: 20 + 4 = 24
L: 35
T: 31 + 3 = 34

A is eliminated from the second ballot. The third ballot results are:

T: 31 + 4 + 3 = 38
L: 35 + 20 = 55

Lillehammer (L) is the winner of the first day (straw) vote.

37.   A majority vote $\frac{93+1}{2} = 47$.

Here is the result of the second (binding) day ballot:

A: 20
L: 1 + 35 + 4 + 3 = 43
T: 30
S: 0

With the Hare method, S is eliminated for the second ballot:

A: 20
L: 1 + 20 + 4 + 3 = 43
T: 30

A is eliminated from the second ballot. The third ballot results are:

T: 30

L: $1 + 35 + 20 + 4 + 3 = 63$

Lillehammer (L) is the winner of the final vote. No, this does not violate the monotonicity criterion because the winning candidate from the first election Lillehammer is also the winner in this vote.

38. **a.** There is a total of 89 votes, so it would take $\frac{89 + 1}{2} = 45$ to win a majority. The winner of the plurality vote is Beijing.

   b. The results of the first ballot are:

Beijing:	32 votes
Berlin:	$3 + 6 = 9$ votes
Istanbul:	$5 + 2 = 7$ votes
Manchester:	$8 + 3 = 11$ votes
Sydney:	30 votes

Istanbul is the country with the fewest number of votes, so we eliminate Istanbul for the second ballot:

Beijing:	$32 + 5 = 37$ votes
Berlin:	$3 + 6 = 9$ votes
Manchester:	$8 + 3 + 2 = 13$ votes
Sydney:	30

The low vote belongs to Berlin, they are eliminated for the third round of voting.

Beijing:	$32 + 3 + 5 = 40$ votes
Manchester:	$8 + 2 + 3 = 13$ votes
Sydney:	$6 + 30 = 36$ votes

However, because of the disqualification and vote change, we have the following vote count:

Beijing:	40 votes
Manchester:	$13 - 2 = 11$ votes
Sydney:	$36 + 1 = 37$ votes

There is still no majority, so for this fourth round we eliminate Manchester from the voting.

Beijing:	$32 + 3 + 5 + 3 = 43$ votes
Sydney:	$8 + 6 + 30 + 2 = 46$ votes $-$ 1 disqualified; 45 votes

Sydney is declared with winner.

c. This example violates the monotonicity criterion because the winner of the original vote (Beijing) picked up votes in the process, but ending up losing the election.

39. a.

		B	L	I	M	S		B	L	I	M	S
										Total		
(BLIMS)	32(	5	4	3	2	1) =		160	128	96	64	32
(LBSIM)	3(	4	5	2	1	3) =		12	15	6	3	9
(IBLSM)	5(	4	3	5	1	2) =		20	15	25	5	10
(MSBLI)	8(	3	2	1	5	4) =		24	16	8	40	32
(LSBIM)	6(	3	5	2	1	4) =		18	30	12	6	24
(SBLMI)	30(	4	3	1	2	5) =		120	90	30	60	150
(IMSBL)	2(	2	1	5	4	3) =		4	2	10	8	6
(MBSLI)	3(	4	2	1	5	3) =		12	6	3	15	9
TOTAL:								370	302	190	201	272

Beijing (B) is the Borda count winner.

b. No, even though there was plenty of chance for the vote to change.

40. Here is the original vote:

Candidate	Party	Vote	Percent	Electoral College Vote
Harry Browne	Libertarian	386,024	00.37	0
Pat Buchanan	Reform	448,750	00.42	0
George W. Bush	Republican	50,456,167	47.88	271
Al Gore	Democrat	50,996,064	48.39	267
Ralph Nader	Green	2,864,810	02.72	0
14 others		238,300	00.23	0
TOTAL		105,390,115		538

Consider the results of a runoff election:

George W. Bush $50{,}456{,}167 + 448{,}750 + 0.20(2{,}864{,}810) + 0.50(238{,}300)$

$$= 51{,}597{,}029$$

Al Gore $50{,}996{,}064 + 386{,}024 + 0.80(2{,}864{,}810) + 0.50(238{,}300)$

$$= 53{,}793{,}086$$

The winner would be Al Gore.

41. a. The runoff election was between Chirac and Le Pen (the top two percentages).

b. Answers vary.

c. Answers vary. Notice that the abstention rate went way down. One possibility is that 9% more of the voters who abstained in the first vote went to the polls and voted for Chirac. Also notice that 63% (16% + 47%) voted for other candidates who were not entered in second ballot. Almost all of these changed their vote to Chirac. Finally, notice that even with all the shifting of votes, Le Pen stayed fairly static (17% to 18%). This would indicate that all other factions formed a coalition to get Chirac elected.

42. **a.** We find the one-on-one pairings:

A over B: 6; B over A: $5 + 4 = 9$

B wins

A over C: 6; C over A: $5 + 4 = 9$

C wins

B over C: $6 + 4 = 10$; C over B: 5

B wins

	A	B	C
A	–	B	C
B	B	–	B
C	C	B	–

Betty (B) is the Condorcet candidate.

b. There were 15 votes cast, so a majority is $\dfrac{15 + 1}{2} = 8$ votes. No person received a majority. Alice receives a plurality, which violates the Condorcet criterion.

43. Here is the Borda count:

A	B	C	A	B	C
6(3	2	1) =	18	12	6
5(1	2	3) =	5	10	15
4(1	3	2) =	4	12	8
TOTAL:			27	34	29

The Borda count winner is Betty (B). This does not violate the Condorcet criterion because both methods picked Betty.

44. Here is the count from the first ballot:

A: 6

B: 4

C: 5

Eliminate candidate B: Here is the count from the second ballot:

A: 6

C: $5 + 4 = 9$

Connie wins from the Hare method of counting. This does violate the Condorcet criterion because the Condorcet candidate is Betty.

45. A obtains 0 points, B 2 points, and C 1 one point. Betty (B) wins the pairwise comparison method. (See the solution for Problem 42 for details.) None of the conditions in Arrow's impossibility theorem are violated.

46.   **a.** There were 500 votes cast.

   b. Here is the outcome:

   A: 150

   B: $100 + 120 = 220$

   C: 130

   D: 0

A majority would be $\frac{500}{2} + 1 = 251$; no person received a majority. The plurality winner is Bate.

   c. We find the one-on-one pairings:

A over B: $130 + 150 = 280$

B over A: $100 + 120 = 220$; **A wins**

A over C: $120 + 150 = 270$

C over A: $100 + 130 = 230$; **A wins**

A over D: 150

	A	B	C	D
A	–	A	A	D
B	A	–	C	D
C	A	C	–	D
D	D	D	D	–

D over A: $100 + 120 + 130 = 350$; **D wins**

B with C: B: $100 + 120 = 220$; C: $130 + 150 = 280$; **C wins**

B with D: B: $100 + 120 = 220$; D: $130 + 150 = 280$; **D wins**

C with D: C: 130; D: $100 + 120 + 150 = 370$; **D wins**

Dave (D) is the Condorcet candidate.

47.   Here is the Borda count:

A	B	C	D			A	B	C	D
100(1	4	2	3)	=		100	400	200	300
120(2	4	1	3)	=		240	480	120	360
130(2	1	4	3)	=		260	130	520	390
150(4	1	2	3)	=		600	150	300	450
					TOTAL:	1,200	1,160	1,140	1,500

The Borda count winner is Dave (D). This does not violate any of the fairness criteria.

48.   A majority is $\frac{500}{2} + 1 = 251$. Here is the count from the first ballot:

   A: 150

   B: $100 + 120 = 220$

   C: 130

   D: 0

Eliminate candidate D: Here is the count from the second ballot:

> A: 150
> B: 100 + 120 = 220
> C: 130

Now, we eliminate C for the third ballot count:

> A: 130 + 150 = 280
> B: 100 + 120 = 220

A (Alberto) wins from the Hare method of counting. Violates the Condorcet criterion (Dave is the Condorcet candidate — see Problem 46c).

49. (See the solution to Problem 46c for details.)

> A wins over B; 1 point for A
> A wins over C; 1 point for A
> C wins over B; 1 point for C
> D wins over A; 1 point for D
> D wins over B; 1 point for D
> D wins over C; 1 point for D

A has 2 points; B, 0 point; C, 1 point; and D, 3 points. Dave (D) is the pairwise winner.

This does not agree with the plurality choice (Problem 46), but it does not violate any of the fairness criteria.

50. **a.** Here is the vote count:

> A: 5
> B: 3
> C: 3

No majority winner; the plurality winner is A.

  b. A over B: 5; B over A: 3 + 3 = 6; 1 point for B

  A over C: 5; C over A: 3 + 3 = 6; 1 point for C

  B over C: 5 + 3 = 8; C over B: 3; 1 point for B

  B is the winner using the pairwise comparison method.

  c. Yes, it is transitive; B → C and C → A; transitive property says B → A, which is the case.

51. **a.** A wins over B (33 to 7); A gets 1 point.

  A wins over C (25 to 15); A gets 1 point.

  C wins over B (21 to 19); C gets 1 point.

  C wins over D (25 to 15); C gets 1 point.

  D wins over A (21 to 19); D gets 1 point.

  D wins over B (21 to 19); D gets 1 point.

  There is no winner; A, C, and D each obtain 2 points.

   b. Tournaments: A/B → A; A/C → A; A/D → D; D wins this tournament.

                      A/B → A; A/D → D; D/C → C; C wins this tournament.

                      A/C → A; A/B → A; A/D → D; D wins this tournament.

                      A/C → A; A/D → D; D/B → D; D wins this tournament.

                      A/D → D; D/B → D; D/C → C; C wins this tournament.

                      A/D → D; D/C → C; C/B → C; C wins this tournament.

                      B/C → C; C/A → A; A/D → D; D wins this tournament.

                      B/C → C; C/D → C; C/A → A; A wins this tournament.

                      B/D → D; D/A → D; D/C → C; C wins this tournament.

                      B/D → D; D/C → C; C/A → A; A wins this tournament.

                      C/D → C; C/A → A; A/B → A; A wins this tournament.

                      C/D → C; C/B → C; C/A → A; A wins this tournament.

A, C, or D might win depending on the way they are paired.

   c. Yes, both of these methods violate the condition of decisiveness. Some means for breaking ties should be provided.

52. a. A wins over B (40 to 10); A gets 1 point.

        A wins over C (30 to 20); A gets 1 point.

        C wins over B (30 to 20); C gets 1 point.

        C wins over D (30 to 20); C gets 1 point.

        D wins over A (30 to 20); D gets 1 point.

        D wins over B (30 to 20); D gets 1 point.

        There is no winner; A, C, and D each obtain 2 points.

   b. Tournaments: A/B → A; A/C → A; A/D → D; D wins this tournament.

                      A/B → A; A/D → D; D/C → C; C wins this tournament.

                      A/C → A; A/B → A; A/D → D; D wins this tournament.

                      A/C → A; A/D → D; D/B → D; D wins this tournament.

                      A/D → D; D/B → D; D/C → C; C wins this tournament.

                      A/D → D; D/C → C; C/B → C; C wins this tournament.

                      B/C → C; C/A → A; A/D → D; D wins this tournament.

                      B/C → C; C/D → C; C/A → A; A wins this tournament.

                      B/D → D; D/A → D; D/C → C; C wins this tournament.

                      B/D → D; D/C → C; C/A → A; A wins this tournament.

$C/D \to C$; $C/A \to A$; $A/B \to A$; A wins this tournament.

$C/D \to C$; $C/B \to C$; $A/C \to A$; A wins this tournament.

The is no winner from the tournament method.

c. Yes, both of these methods violate the condition of decisiveness. Some means for breaking ties should be provided.

53. a. Here is the Borda count:

A	B	C	D		A	B	C	D
20(4	3	2	1)	=	80	60	40	20
20(2	4	3	1)	=	40	80	60	20
10(3	2	4	1)	=	30	20	40	10
TOTAL:					150	160	140	50

The Borda count winner is B.

b. Suppose we eliminate candidate C (who is not a winning candidate). The votes become:

(ABD)	(BAD)	(ABD)
20	20	10

A	B	D		A	B	D
20(3	2	1)	=	60	40	20
20(2	3	1)	=	40	60	20
10(3	2	1)	=	30	20	10
TOTAL:				130	120	50

The winning candidate is A, which violates the irrelevant alternative criterion.

54. a. There are 11 votes, so $\dfrac{11 + 1}{2} = 6$ is needed for a majority, so there is no majority winner. The plurality winner is C.

b. For a runoff between the second and third candidates we find:

(AB)	(BA)	(AB)
4	2	5

A: $4 + 5 = 9$ votes
B: 2 votes

A wins the runoff election.

c. If the election continues fairly, we have

(AC)	(AC)	(CA)
4	2	5

A: $4 + 2 = 6$ votes
C: 5 votes

A wins the election.

However if the supporters of C change their vote from (CAB) to (CBA). So here is what that (insincere) election will look like.

(ACB)   (BAC)   (CBA)

4   2   5

There is still no majority but the results of the runoff election between A and B is

(AB)   (BA)   (BA)

4   2   5

A: 4 votes

B: 2 + 5 votes

B wins the runoff election.

Now, between B and C we have:

(CB)   (BC)   (CB)

4   2   5

B: 2 votes

C: 4 + 5 = 9 votes

C wins the election ... and the C supporters will have accomplished this by voting insincerely.

55. a. There is a total of 13 votes, so $\frac{13 + 1}{2} = 7$ is needed for a majority, so we see there is no majority winner. There is a plurality winner, namely C (with 6 votes).

b. Here is the Borda count:

Outcome	A B C	A	B	C
(ACB), 2:	2(3 1 2) =	6	2	4
(BAC), 5:	5(2 3 1) =	10	15	5
(CBA), 4:	4(1 2 3) =	4	8	12
(CAB), 2	2(2 1 3) =	4	2	6
TOTAL:		24	27	27; There is no winner since B and C tie.

c. For the runoff election, eliminate the one with the most last-place votes:

A has 4, B has 4 and C has 5, so we eliminate C. The tally now looks like:

(AB)   (BA)   (BA)   (AB)   We can combine these: (AB)   (BA)

2   5   4   2   4   9

We see that B wins the runoff election.

d. The insincere preferences are:

(ACB)　(BAC)　(CBA)　(CBA)
　2　　　5　　　4　　　2

A: 2 votes; B: 5 votes: C: 6 votes; no majority winner. The candidate with the most last

place votes is: A: 6, B: 2; C: 5, so A is eliminated from the runoff. The tally now looks like:

(CB)　(BC)　　(CB)　　(CB)　We can combine these: (BC)　(CB)
　2　　　5　　　4　　　2　　　　　　　　　　　　　　5　　　8

C wins this runoff election.

Yes, they could change the outcome of the election by voting insincerely.

56.　**a.**　Here is the Borda count:

Outcome　　　　A　C　I　　A　　C　　I

(ACI), 10:　10(3　2　1) =　30　　20　　10

(CIA), 38:　38(1　3　2) =　38　114　76

(ICA), 52:　52(1　2　3) =　52　104　156

TOTAL:　　　　　　　　　　120　238　242; Choice I (grain subsidy) is the winner.

b.　If those who favor California change their vote to (CAI). Let's look at the Borda count for
this insincere vote.

Outcome　　　　A　C　I　　A　　C　　I

(ACI), 10:　10(3　2　1) =　30　　20　　10

(CAI), 38:　38(2　3　1) =　76　114　38　　　Insincere vote

(ICA), 52:　52(1　2　3) =　52　104　156

TOTAL:　　　　　　　　　　158　238　204; Choice C (freeway interchange) is the winner.

c.　The Iowa supporters could change their vote to (IAC). Let's look at the Borda count for this
double insincere vote.

Outcome　　　　A　C　I　　A　　C　　I

(ACI), 10:　10(3　2　1) =　30　　20　　10

(CAI), 38:　38(2　3　1) =　76　114　38　　　Insincere vote

(IAC), 52:　52(2　1　3) = 104　　52　156　　Second insincere vote

TOTAL:　　　　　　　　　　210　186　204; Choice A (bridge) is the winner.

It looks like they could keep it from California, but the change in vote would not be enough
to swing the election back to them.

Level 3, page 831

57. Here are the voters preferences

"Motion: The meeting time will change to evenings."

	Yes	No
Jane	x	
Linda	x	
Ann	x	
Melissa		x

It looks like it will pass, but Melissa is ingenuous. After the motion is on the floor, she offers an amendment: "I move that we change the meeting time to Saturday mornings."

	Yes	No
Jane		x
Linda	x	
Ann	x	
Melissa	x	

The amendment is passed with a 3 to 1 vote. Now, the vote on the amended motion is:

	Yes	No
Jane		x
Linda	x	
Ann	x	
Melissa		x

It is a tie vote, so it does not pass and the old rule stands.

58. Answers vary; see problems 50 or 60 for examples.

59. If the 90% vote is spread out evenly over the ten serious candidates, then a 10% vote for the radical candidate is theoretically enough to win the plurality vote.

Level 3 Problem Solving page 831

60. Consider the possible games:

a. A game with dice A and B is shown in the text.

$P(A$ winning$) = \frac{2}{3}$ and $P(B$ winning$)$ is $\frac{1}{3}$.

b. A game with dice A and C:

$C \mid A$	0	0	4	4	4	4	
2	(2, 0)	(2, 0)	(2, 4)	(2, 4)	(2, 4)	(2, 4)	
2	(2, 0)	(2, 0)	(2, 4)	(2, 4)	(2, 4)	(2, 4)	← A wins 16 out of 36
2	(2, 0)	(2, 0)	(2, 4)	(2, 4)	(2, 4)	(2, 4)	
2	(2, 0)	(2, 0)	(2, 4)	(2, 4)	(2, 4)	(2, 4)	
6	(6, 0)	(6, 0)	(6, 4)	(6, 4)	(6, 4)	(6, 4)	← C wins 20 out of 36
6	(6, 0)	(6, 0)	(6, 4)	(6, 4)	(6, 4)	(6, 4)	

$P(A \text{ winning}) = \frac{4}{9}$ and $P(C \text{ winning}) = \frac{5}{9}$.

c. A game with dice A and D:

$D \mid A$	0	0	4	4	4	4	
1	(1, 0)	(1, 0)	(1, 4)	(1, 4)	(1, 4)	(1, 4)	
1	(1, 0)	(1, 0)	(1, 4)	(1, 4)	(1, 4)	(1, 4)	← A wins 12 out of 36
1	(1, 0)	(1, 0)	(1, 4)	(1, 4)	(1, 4)	(1, 4)	
5	(5, 0)	(5, 0)	(5, 4)	(5, 4)	(5, 4)	(5, 4)	
5	(5, 0)	(5, 0)	(5, 4)	(5, 4)	(5, 4)	(5, 4)	← D wins 24 out of 36
5	(5, 0)	(5, 0)	(5, 4)	(5, 4)	(5, 4)	(5, 4)	

$P(A \text{ winning}) = \frac{1}{3}$ and $P(D \text{ winning}) = \frac{2}{3}$.

d. A game with dice B and C

$C \mid B$	3	3	3	3	3	3	
2	(2, 3)	(2, 3)	(2, 3)	(2, 3)	(2, 3)	(2, 3)	
2	(2, 3)	(2, 3)	(2, 3)	(2, 3)	(2, 3)	(2, 3)	← B wins 24 out of 36
2	(2, 3)	(2, 3)	(2, 3)	(2, 3)	(2, 3)	(2, 3)	
2	(2, 3)	(2, 3)	(2, 3)	(2, 3)	(2, 3)	(2, 3)	
6	(6, 3)	(6, 3)	(6, 3)	(6, 3)	(6, 3)	(6, 3)	← C wins 12 out of 36
6	(6, 3)	(6, 3)	(6, 3)	(6, 3)	(6, 3)	(6, 3)	

$P(B \text{ winning}) = \frac{2}{3}$ and $P(C \text{ winning}) = \frac{1}{3}$.

e. A game with dice B and D:

$D \mid B$	3	3	3	3	3	3	
1	(1, 3)	(1, 3)	(1, 3)	(1, 3)	(1, 3)	(1, 3)	
1	(1, 3)	(1, 3)	(1, 3)	(1, 3)	(1, 3)	(1, 3)	← B wins 18 out of 36
1	(1, 3)	(1, 3)	(1, 3)	(1, 3)	(1, 3)	(1, 3)	
5	(5, 3)	(5, 3)	(5, 3)	(5, 3)	(5, 3)	(5, 3)	
5	(5, 3)	(5, 3)	(5, 3)	(5, 3)	(5, 3)	(5, 3)	← C wins 18 out of 36
5	(5, 3)	(5, 3)	(5, 3)	(5, 3)	(5, 3)	(5, 3)	

$P(B \text{ winning}) = \frac{1}{2}$ and $P(D \text{ winning}) = \frac{1}{2}$.

f. Finally, a game with dice C and D:

$D \mid C$	2	2	2	2	6	6	
1	(1, 2)	(1, 2)	(1, 2)	(1, 2)	(1, 6)	(1, 6)	
1	(1, 2)	(1, 2)	(1, 2)	(1, 2)	(1, 6)	(1, 6)	← C wins 24 out of 36
1	(1, 2)	(1, 2)	(1, 2)	(1, 2)	(1, 6)	(1, 6)	
5	(5, 2)	(5, 2)	(5, 2)	(5, 2)	(5, 6)	(5, 6)	
5	(5, 2)	(5, 2)	(5, 2)	(5, 2)	(5, 6)	(5, 6)	← D wins 12 out of 36
5	(5, 2)	(5, 2)	(5, 2)	(5, 2)	(5, 6)	(5, 6)	

$P(C \text{ winning}) = \frac{24}{36} = \frac{2}{3}$ and $P(D \text{ winning}) = \frac{12}{36} = \frac{1}{3}$.

We now summarize these probabilities:

	Opponent Chooses			
	A	B	C	D
A	–	$\frac{2}{3}$	$\frac{4}{9}$	$\frac{1}{3}$
B	$\frac{1}{3}$	–	$\frac{2}{3}$	$\frac{1}{2}$
C	$\frac{5}{9}$	$\frac{1}{3}$	–	$\frac{2}{3}$
D	$\frac{2}{3}$	$\frac{1}{2}$	$\frac{1}{3}$	–

My choices: (rows A, B, C, D)

This means I would want to choose second.

If my opponent chooses die A, then I pick die D.

If my opponent chooses die B, then I pick die A.

If my opponent chooses die C, then I pick die B.

If my opponent chooses die D, then I pick die C.

If I make these choices, then my probability of winning is $\frac{2}{3}$. The game violates the transitive property.

Transparencies 32-34 can be used with Problem 60. I have found this to be an interesting classroom activity. Students enjoy playing this dice game, and it will help the students to remember the transitive property.

16.3 Apportionment, page 831

Level 1, page 847

1. Answers vary; see Table 16.9.

2. Give any two numbers a and b, the *arithmetic mean* is $\frac{a+b}{2}$, and the *geometric mean* is $\sqrt{ab}$.

3. Adams' plan favors the smaller states and Hamilton's and Jefferson's plans favor the larger states. You might also argue that Webster's and HH's plans also favor one or the other, but they are more equitable since they both round up sometimes and down other times.

4. The *quota rule* states that the number assigned to each represented unit must be either the standard quota rounded down to the nearest integer, or the standard quota rounded up to the nearest integer.

5. If you round down, then lower the standard divisor to find the modified quotients.

6. If you round up, then raise the standard divisor to find the modified quotients.

7. **a.** 3; 4 **b.** 3.5 **c.** 3.46 **d.** 4; 4   **8. a.** 1; 2 **b.** 1.5 **c.** 1.41 **d.** 1; 1   **9. a.** 1; 2 **b.** 1.5 **c.** 1.41 **d.** 1; 2   **10. a.** 3; 4 **b.** 3.5 **c.** 3.46 **d.** 3; 4   **11. a.** 2; 3 **b.** 2.5 **c.** 2.45 **d.** 2; 3   **12. a.** 2; 3 **b.** 2.5 **c.** 2.45 **d.** 3; 3   **13. a.** 1,695; 1,696 **b.** 1,695.5

c. 1,695.50   **d.** 1,695; 1,695   **14. a.** 1,695; 1,696   **b.** 1,695.5   **c.** 1,695.50   **d.** 1,696; 1,696

15. 6,500   **16.** 16,875   **17.** 126   **18.** 77.14   **19.** 120,833.33   **20.** 743,333.33

21. 184,000   **22.** 276,888.89

	Year	d	Manhattan	Bronx	Brooklyn	Queens	Staten Island
23.	1800	10,125	6.02	0.20	0.59	0.69	0.49
24.	1840	87,125	5.92	0.09	1.60	0.22	0.17
25.	1900	429,750	4.30	0.47	2.72	0.36	0.16
26.	1940	931,750	2.03	1.50	2.90	1.39	0.19
27.	1990	915,500	1.63	1.32	2.51	2.13	0.41
28.	2000	1,000,875	1.54	1.33	2.46	2.23	0.44

In Problems 29-32, the modified divisors, D, may vary.

29. $d = 11,600$ q $D = 10,000$ Q *Round Down*

		q		Q	Round Down
1st Precinct	3.02		3.50	3
2nd Precinct	1.81		2.10	2
3rd Precinct	1.03		1.20	1
4th Precinct	4.14		4.80	4
TOTAL				10

a. 11,600
b. 3.02, 1.81, 1.03, 4.14
c. $3 + 1 + 1 + 4 = 9$
d. 10,000

30. $d = 9,666.67$ q $D = 8,100$ Q *Round Down*

		q		Q	Round Down
1st Precinct	3.62		4.32	4
2nd Precinct	2.17		2.59	2
3rd Precinct	1.24		1.48	1
4th Precinct	4.97		5.93	5
TOTAL				12

a. 9,666.67
b. 3.62, 2.17, 1.24, 4.97
c. $3 + 2 + 1 + 4 = 10$
d. 8,100

31. $d = 80,000$ q $D = 65,000$ Q *Round Down*

		q		Q	Round Down
1st Precinct	1.69		2.08	2
2nd Precinct	2.89		3.55	3
3rd Precinct	1.48		1.82	1
4th Precinct	3.95		4.86	4
TOTAL				10

a. 80,000
b. 1.69, 2.89, 1.48, 3.95
c. $1 + 2 + 1 + 3 = 7$
d. 65,000

32. $d = 66,666.67$ q $D = 59,000$ Q *Round Down*

		q		Q	Round Down
1st Precinct	2.03		2.29	2
2nd Precinct	3.47		3.92	3
3rd Precinct	1.77		2.00	2
4th Precinct	4.74		5.36	5
TOTAL				12

a. 66,666.67
b. 2.03, 3.47, 1.77, 4.74
c. $2 + 3 + 1 + 4 = 10$
d. 59,000

In Problems 33-36, the modified divisors, D, may vary.

33. $d = 11{,}600$ q $D = 13{,}000$ Q *Round Up*

1st Precinct	3.02	2.69	3
2nd Precinct	1.81	1.62	2
3rd Precinct	1.03	0.92	1
4th Precinct	4.14	3.69	4
TOTAL			10

a. 11,600
b. 3.02, 1.81, 1.03, 4.14
c. $4 + 2 + 2 + 4 = 12$
d. 13,000

34. $d = 9{,}666.67$ q $D = 12{,}000$ Q *Round Up*

1st Precinct	3.62	2.92	3
2nd Precinct	2.17	1.75	2
3rd Precinct	1.24	1.00	2
4th Precinct	4.97	4.00	5
TOTAL			12

a. 9,666.67
b. 3.62, 2.17, 1.24, 4.97
c. $4 + 3 + 2 + 5 = 14$
d. 12,000

35. $d = 80{,}000$ q $D = 110{,}000$ Q *Round Up*

1st Precinct	1.69	1.23	2
2nd Precinct	2.89	2.10	3
3rd Precinct	1.48	1.07	2
4th Precinct	3.95	2.87	3
TOTAL			10

a. 80,000
b. 1.69, 2.89, 1.48, 3.95
c. $2 + 3 + 2 + 4 = 11$
d. 110,000

36. $d = 66{,}666.67$ q $D = 78{,}000$ Q *Round Up*

1st Precinct	2.03	1.73	2
2nd Precinct	3.47	2.96	3
3rd Precinct	1.77	1.51	2
4th Precinct	4.74	4.05	5
TOTAL			12

a. 66,666.67
b. 2.03, 3.47, 1.77, 4.74
c. $3 + 4 + 2 + 5 = 14$
d. 78,000

Level 2 page 848

37. $d \approx 37{,}089.51$

Maine (MA) is $\dfrac{96{,}643}{d} \approx 2.61$

The revised number for Massachusetts (MS) is $475{,}199 - 96{,}643 = 378{,}556$, so that

$$\frac{378{,}556}{d} \approx 10.21$$

This means that the standard quotas for the 16 states are:

CT	DE	GA	KY	ME	MD	MA	NH	NJ	NY	NC	PA	RI	SC	VT	VA
6.41	1.59	2.23	1.99	2.61	8.62	10.21	3.83	4.97	9.17	10.65	11.69	1.86	6.72	2.30	20.16

38.

6	2	2	2	3	9	10	4	5	9	11	12	2	7	2	20

Total is 106.

39.

6	1	2	1	2	8	10	3	4	9	10	11	1	6	2	20

Total is 96.

40.

7	2	3	2	3	9	11	4	5	10	11	12	2	7	3	21

Total is 112.

41. $n = 10; \ d = \dfrac{25,000}{10} = 2,500$ $D = 3,125$

		q	lower quota	upper quota	Q	Adams' plan
North	8,700	3.48	3	4	2.78	3
South	5,600	2.24	2	3	1.79	2
East	7,200	2.88	2	3	2.30	3
West	3,500	1.40	1	2	1.12	2
TOTAL: 25,000			8	12		10

42. $n = 10; \ d = \dfrac{25,000}{10} = 2,500$ $D = 2,100$

		q	lower quota	upper quota	Q	Jefferson's plan
North	8,700	3.48	3	4	4.14	4
South	5,600	2.24	2	3	2.67	2
East	7,200	2.88	2	3	3.43	3
West	3,500	1.40	1	2	1.67	1
TOTAL: 25,000			8	12		10

43. $n = 10; \ d = \dfrac{25,000}{10} = 2,500$

		q	lower quota	Hamilton's plan
North	8,700	3.48	3	4 (#2)
South	5,600	2.24	2	2
East	7,200	2.88	2	3 (#1)
West	3,500	1.40	1	1
TOTAL: 25,000			8	10

44. $n = 10; \ d = \dfrac{25,000}{10} = 2,500$ $D = 2,400$

		q	round (nearest)	Q	Webster's plan
North	8,700	3.48	3	3.63	4
South	5,600	2.24	2	2.33	2
East	7,200	2.88	3	3.00	3
West	3,500	1.40	1	1.46	1
TOTAL: 25,000			9		10

45. $n = 10; \ d = \dfrac{25,000}{10} = 2,500$ $D = 2,500$ (no modification necessary)

			a	b		
		q	lower quota	upper quota	$\sqrt{ab}$	HH's plan
North	8,700	3.48	3	4	3.46	4
South	5,600	2.24	2	3	2.45	2
East	7,200	2.88	2	3	2.45	3
West	3,500	1.40	1	2	1.41	1
TOTAL: 25,000			8	12		10

46. $n = 26; \ d = \dfrac{62,000}{26} \approx 2,384.62$ $D = 2,590$

		q	lower quota	upper quota	Q	Adams' plan
North	18,200	7.63	7	8	7.03	8
South	12,900	5.41	5	6	4.98	5
East	17,600	7.38	7	8	6.80	7
West	13,300	5.58	5	6	5.14	6
TOTAL: 62,000			24	28		26

47. $n = 26; \; d = \dfrac{62,000}{26} \approx 2,384.62$ $D = 2,210$

	q	lower quota	upper quota	Q	Jefferson's plan
North 18,200	7.63	7	8	8.24	8
South 12,900	5.41	5	6	5.84	5
East 17,600	7.38	7	8	7.96	7
West 13,300	5.58	5	6	6.02	6
TOTAL: 62,000		24	28		26

48. $n = 26; \; d = \dfrac{62,000}{26} \approx 2,384.62$

	q	lower quota	Hamilton's plan
North 18,200	7.63	7	8 (#1)
South 12,900	5.41	5	5
East 17,600	7.38	7	7
West 13,300	5.58	5	6 (#2)
TOTAL: 62,000		24	26

49. $n = 26; \; d = \dfrac{62,000}{26} \approx 2,384.62$ $D = 2,384.62$ (no modification necessary)

	q	round nearest	Webster's plan
North 18,200	7.63	8	8
South 12,900	5.41	5	5
East 17,600	7.38	7	7
West 13,300	5.58	6	6
TOTAL: 62,000		26	26

50. $n = 26; \; d = \dfrac{62,000}{26} = 2,384.62$ $D = 2,384.62$ (no modification necessary)

		a	b		
	q	lower quota	upper quota	$\sqrt{ab}$	HH's plan
North 18,200	7.63	7	8	7.48	8
South 12,900	5.41	5	6	5.48	5
East 17,600	7.38	7	8	7.48	7
West 13,300	5.58	5	6	5.48	6
TOTAL: 62,000		24	28		26

51. $n = 16; \; d = \dfrac{62,000}{16} = 3,875$ $D = 4,425$

	q	lower quota	upper quota	Q	Adams' plan
North 18,200	4.70	4	5	4.11	5
South 12,900	3.33	3	4	2.92	3
East 17,600	4.54	4	5	3.98	4
West 13,300	3.43	3	4	3.01	4
TOTAL: 62,000		14	18		16

52. $n = 16; \; d = \dfrac{62,000}{16} = 3,875$ $D = 3,400$

	q	lower quota	upper quota	Q	Jefferson's plan
North 18,200	4.70	4	5	5.35	5
South 12,900	3.33	3	4	3.79	3
East 17,600	4.54	4	5	5.18	5
West 13,300	3.43	3	4	3.19	3
TOTAL: 62,000		14	18		16

53. $n = 16$; $d = \dfrac{62,000}{16} = 3,875$

		q	lower quota	Hamilton's plan
North	18,200	4.70	4	5 (#1)
South	12,900	3.33	3	3
East	17,600	4.54	4	5 (#2)
West	13,300	3.43	3	3
TOTAL:	62,000		14	16

54. $n = 16$; $d = \dfrac{62,000}{16} = 3,875$ $D = 3,875$ (no modification necessary)

		q	round nearest	Webster's plan
North	18,200	4.70	5	5
South	12,900	3.33	3	3
East	17,600	4.54	5	5
West	13,300	3.43	3	3
TOTAL:	62,000		16	16

55. $n = 16$; $d = \dfrac{62,000}{16} = 3,875$ $D = 3,875$ (no modification necessary)

		q	a lower quota	b upper quota	$\sqrt{ab}$	HH's plan
North	18,200	4.70	4	5	4.47	5
South	12,900	3.33	3	4	3.46	3
East	17,600	4.54	4	5	4.47	5
West	13,500	3.43	3	4	3.46	3
TOTAL:	62,000		14	18		16

Level 3 page 849

56. Standard divisor: $d = \dfrac{16,630,000}{475} \approx 35,010.53$ Modified divisor: $D = 35,300$

Region	Number	Std Quota	Lower Quota	Upper Quota	Modified Quota	Adams' Plan
N	1,820,000	51.98	51	52	51.56	52
NE	2,950,000	84.26	84	85	83.57	84
E	1,760,000	50.27	50	51	49.86	50
SE	1,980,000	56.55	56	57	56.09	57
S	1,200,000	34.28	34	35	33.99	34
SW	2,480,000	70.84	70	71	70.25	71
W	3,300,000	94.26	94	95	93.48	94
NW	1,140,000	32.56	32	33	32.29	33
TOTAL:	16,630,000		471	479		475

57. Standard divisor: $d = \dfrac{16,630,000}{475} \approx 35,010.53$ Modified divisor: $D = 34,720$

Region	Number	Std Quota	Lower Quota	Upper Quota	Modified Quota	Jefferson's Plan
N	1,820,000	51.98	51	52	52.42	52
NE	2,950,000	84.26	84	85	84.97	84
E	1,760,000	50.27	50	51	50.69	50
SE	1,980,000	56.55	56	57	57.03	57
S	1,200,000	34.28	34	35	34.56	34
SW	2,480,000	70.84	70	71	71.43	71
W	3,300,000	94.26	94	95	95.05	95
NW	1,140,000	32.56	32	33	32.83	32
TOTAL:	16,630,000		471	479		475

58. Standard divisor: $d = \dfrac{16,630,000}{475} \approx 35,010.53$

Region	Number	Std Quota	Lower Quota	Upper Quota	Webster's Plan
N	1,820,000	51.98	51	52	52 (#1)
NE	2,950,000	84.26	84	85	84
E	1,760,000	50.27	50	51	50
SE	1,980,000	56.55	56	57	57 (#4)
S	1,200,000	34.28	34	35	34
SW	2,480,000	70.84	70	71	71 (#2)
W	3,300,000	94.26	94	95	94
NW	1,140,000	32.56	32	33	33 (#3)
TOTAL:	16,630,000		471	479	475

59. Standard divisor: $d = \dfrac{16,630,000}{475} \approx 35,010.53$ $D = 35,010.53$ (no modification necessary)

Region	Number	Std Quota	Round	Webster's Plan
N	1,820,000	51.98	52	52
NE	2,950,000	84.26	84	84
E	1,760,000	50.27	50	50
SE	1,980,000	56.55	57	57
S	1,200,000	34.28	34	34
SW	2,480,000	70.84	71	71
W	3,300,000	94.26	94	94
NW	1,140,000	32.56	33	33
TOTAL:	16,630,000		475	475

60. Standard divisor: $d = \dfrac{16{,}630{,}000}{475} \approx 35{,}010.53$　　$D = 35{,}010.53$ (no modification necessary)

Region	Number	Std Quota	Lower Quota, a	Upper Quota, b	$\sqrt{ab}$	HH's Plan
N	1,820,000	51.98	51	52	51.498	52
NE	2,950,000	84.26	84	85	84.499	84
E	1,760,000	50.27	50	51	50.498	50
SE	1,980,000	56.55	56	57	56.498	57
S	1,200,000	34.28	34	35	34.496	34
SW	2,480,000	70.84	70	71	70.498	71
W	3,300,000	94.26	94	95	94.499	94
NW	1,140,000	32.56	32	33	32.496	33
TOTAL:	16,630,000		471	479		475

16.4 Apportionment Paradoxes, page 849

Level 1, page 854

1. An increase in the total number of items to be apportioned resulting in a loss of items for a group is called the *Alabama paradox*.　　**2.** When there is a fixed number of seats, a reapportionment may cause a state to lose a seat to another state, even though the percent increase in the population of the state that loses the seat is larger than the percent increase of the state that wins the seat. When this occurs, it is known as the *population paradox*.　　**3.** When a reapportionment of an increased number of seats causes a shift in the apportionment of the existing states, it is known as the *new states paradox*.　　**4.** Any apportionment plan that does not violate the quota rule, must produce paradoxes. And any apportionment method that does not produce paradoxes must violate the quota rule.

5. $n = 100$; $d = \dfrac{128{,}700}{100} = 1{,}287$　　　　　　$D = 1{,}330$

		q	lower quota	upper quota	Q	Adams' plan	
A	68,500	53.22	53	54	51.50	52	← Violates the quota rule
B	34,700	26.96	26	27	26.09	27	
C	16,000	12.43	12	13	12.03	13	
D	9,500	7.38	7	8	7.14	8	
TOTAL:	128,700		98	102		100	

6. $n = 100$; $d = \dfrac{1{,}287}{100} = 12.87$　　　　　　$D = 13.3$

		q	lower quota	upper quota	Q	Adams' plan	
A	685	53.22	53	54	51.50	52	← Violates the quota rule
B	347	26.96	26	27	26.09	27	
C	160	12.43	12	13	12.03	13	
D	95	7.38	7	8	7.14	8	
TOTAL:	1,287		98	102		100	

7. $n = 100$; $d = \dfrac{127{,}500}{100} = 1{,}275$ $D = 1{,}240$

		q	lower quota	upper quota	Q	Jefferson's plan	
A	68,500	53.73	53	54	55.24	55	← Violates the quota rule
B	34,700	27.22	27	28	27.98	27	
C	14,800	11.61	11	12	11.94	11	
D	9,500	7.45	7	8	7.66	7	
TOTAL:	127,500		98	102		100	

8. $n = 132$; $d = \dfrac{79{,}903}{132} = 605.33$ $D = 595$

		q	lower quota	upper quota	Q	Jefferson's plan	
A	17,179	28.38	28	29	28.87	28	
B	7,500	12.39	12	13	12.61	12	
C	49,400	81.61	81	82	83.03	83	← Violates the quota rule
D	5,824	9.62	9	10	9.79	9	
TOTAL:	79,903		130	134		132	

9. $n = 200$; $d = \dfrac{10{,}315}{200} = 51.575$ $D = 50.9$

		q	lower quota	upper quota	Q	Jefferson's plan	
A	1,100	21.3282	21	22	21.61	21	
B	1,100	21.3282	21	22	21.61	21	
C	1,515	29.3747	29	30	29.76	29	
D	4,590	88.9966	88	89	90.18	90	← Violates the quota rule
E	2,010	38.9724	38	39	39.49	39	
TOTAL:	10,315		197	202		200	

10. $n = 150$; $d = \dfrac{45{,}000}{150} = 300$ $D = 295$

		q	lower quota	upper quota	Q	Jefferson's plan	
A	1,700	5.67	5	6	5.76	5	
B	3,300	11.00	11	12	11.19	11	
C	7,000	23.33	23	24	23.73	23	
D	24,190	80.63	80	81	82.00	82	← Violates the quota rule
E	8,810	29.37	29	30	29.86	29	
TOTAL:	45,000		148	153		150	

11. $n = 246$; $d = \dfrac{2{,}724}{246} \approx 11.07$ $n = 247$; $d = \dfrac{2{,}724}{247} \approx 11.03$

		q	lower quota	Hamilton's plan	q	Hamilton's plan	
A	181	16.35	16	17 (#1)	16.41	16	← Alabama paradox
B	246	22.22	22	22	22.31	22	
C	812	73.33	73	73	73.63	74 (#2)	
D	1,485	134.11	134	134	134.65	135 (#1)	
TOTAL:	2,724		245	246		247	

12. $n = 45$; $d = \dfrac{2{,}055}{45} \approx 45.67$ $n = 46$; $d = \dfrac{2{,}055}{46} \approx 44.67$

		q	lower quota	Hamilton's plan	q	Hamilton's plan	
A	235	5.15	5	5	5.26	6 (#2)	
B	318	6.96	6	7 (#1)	7.12	7	
C	564	12.35	12	12	12.62	13 (#1)	
D	938	20.54	20	21 (#2)	20.9984	20	← Alabama paradox
TOTAL:	2,055		43	45		46	

13. $n = 50; d = \frac{2,076}{50} = 41.52$ $n = 51; d = \frac{2,076}{51} \approx 40.71$

		q	lower quota	Hamilton's plan	q	Hamilton's plan	
A	300	7.23	7	7	7.37	7	
B	301	7.25	7	8 (#1)	7.39	7	← Alabama paradox
C	340	8.19	8	8	8.35	8	
D	630	15.17	15	15	15.48	16 (#1)	
E	505	12.16	12	12	12.41	13 (#2)	
TOTAL:	2,076		49	50		51	

14. $n = 82; d = \frac{3,301}{82} \approx 40.26$ $n = 83; d = \frac{3,301}{83} \approx 39.77$

		q	lower quota	Hamilton's plan	q	Hamilton's plan	
A	300	7.45	7	8 (#3)	7.54	7	← Alabama paradox
B	700	17.39	17	17	17.60	18 (#2)	
C	800	19.87	19	20 (#1)	20.12	20	
D	800	19.87	19	20 (#1)	20.12	20	
E	701	17.41	17	17	17.63	18 (#1)	
TOTAL:	3,301		79	82		83	

15. $n = 11$

	Orig	Revised	q (orig)	H's plan	q (rev)	H's plan	Increase	% Increase	
A	55,200	61,100	1.64	1	1.584	2 (#2)	5,900	10.69%	
B	124,900	148,100	3.71	4 (#1)	3.840	4 (#1)	23,200	18.57%	
C	190,000	215,000	5.65	6 (#2)	5.575	5	25,000	13.16%	← Pop. paradox
TOT:	370,100	424,200		11		11			(greater than A)
d:	33,645.45	38,563.64							

16. $n = 13$

	Orig	Revised	q (orig)	H's plan	q (rev)	H's plan	Increase	% Increase	
A	90,000	98,000	2.65	2	2.60	3 (#2)	8,000	8.89%	
B	124,800	144,900	3.68	4 (#1)	3.84	4 (#1)	20,100	16.11%	
C	226,000	247,100	6.67	7 (#2)	6.56	6	21,100	9.34%	← Pop. paradox
TOT:	440,800	490,000		13		13			(greater than A)
d:	33,907.69	37,692.31							

17. $n = 13$

	Orig	Revised	q (orig)	H's plan	q (rev)	H's plan	Increase	% Increase	
A	89,950	97,950	2.65	2	2.60	3 (#2)	8,000	8.89%	
B	124,800	144,900	3.68	4 (#1)	3.84	4 (#1)	20,100	16.11%	
C	226,000	247,100	6.67	7 (#2)	6.56	6	21,100	9.34%	← Pop. paradox
TOT:	440,750	489,950		13		13			(larger than A)
d:	33,903.85	37,688.46							

18. $n = 100$

	Orig	Revised	q (orig)	H's plan	q (rev)	H's plan	Increase	% Increase	
A	7,510	7,650	7.51	8 (#2)	7.60	7	140	1.86%	← Pop. paradox
B	20,500	20,800	20.50	20	20.67	21 (#2)	300	1.46%	← Pop. paradox
C	72,000	72,200	71.99	72 (#1)	71.73	72 (#1)	200	0.28%	(both larger than C)
TOT:	100,010	100,650		100		100			
d:	1,000.10	1,006.50							

19. $n = 12$; $d \approx 16,999.58$ $n = 12 + 2 = 14$; $d = 17,302.5$

	Pop	q	H's plan	Pop	q	H's plan	
A	144,899	8.52	9 (#1)	144,899	8.37	8	← New states paradox
B	59,096	3.48	3	59,096	3.42	4 (#1)	
C				38,240	2.21	2	
TOT:	203,995		12	242,235		14	

20. $n = 16$; $d \approx 71,808.75$ $n = 16 + 1 = 17$; $d \approx 75,734.71$

	Pop	q	H's plan	Pop	q	H's plan	
A	394,990	5.501	6 (#1)	394,990	5.22	5	← New states paradox
B	753,950	10.499	10	753,950	9.96	10 (#1)	
C				138,550	1.83	2 (#2)	
TOT:	1,148,940		16	1,287,490		17	

21. $n = 50$; $d = 325,820$ $n = 50 + 4 = 54$; $d \approx 328,537.04$

	Pop	q	H's plan	Pop	q	H's plan	
A	7,000,500	21.49	21	7,000,500	21.31	21	
B	9,290,500	28.51	29 (#1)	9,290,500	28.28	28	← New states paradox
C				1,450,000	4.41	5 (#1)	
TOT:	16,291,000		50	17,741,000		54	

22. $n = 16$; $d = 23,062.5$ $n = 16 + 2 = 18$; $d \approx 24,333.33$

	Pop	q	H's plan	Pop	q	H's plan	
A	265,000	11.49	11	265,000	10.89	11 (#1)	
B	104,000	4.51	5 (#1)	104,000	4.27	4	← New states paradox
C				69,000	2.84	3 (#2)	
TOT:	369,000		16	438,000		18	

Level 2, page 855

23. a. $\dfrac{18,834}{300} = 62.78$ **b.** A: 199.43; B: 72.55; C: 12.93; D: 15.08   **c.** A: 199, 200; B: 72, 73; C: 12, 13; D: 15, 16   **d.** modified divisor: 62.4; A: 200; B: 72; C: 13; D: 15   **e.** No, it does not violate the quota rule because all modified quotas are within the parameters specified in part **c.**

24. a. $d = \dfrac{999}{45} = 22.2$ **b.** EA: 6.94; MC: 6.40; M: 7.43; P: 13.83; SR: 10.41   **c.** EA: 6, 7; MC: 6, 7; M: 7, 8; P: 13, 14; SR: 10, 11   **d.** EA: 7; MC: 6; M: 8; P: 14; SR: 10   **e.** $d = \dfrac{999}{46} \approx 21.72$; EA: 7; MC: 6; M: 8; P: 14; SR: 11   **f.** Yes, it does illustrate the Alabama paradox because an increase in videos resulted in a loss of one video for Montgomery.   **25. a.** $\dfrac{20,330}{100} = 203.3$ **b.** U (uptown): 83.52; D (downtown): 16.48   **c.** U: 84; D: 16   **d.** U: 83; D: 17; New: 12   **e.** Yes, this illustrates the new states paradox because there was a shifting of the apportionments of the existing districts.

26. **a.** With a 5% raise the salaries are (in thousands):

	Salary	With Raise	Lower Quota	Hamilton's plan
Employee #1:	43.100	45.255	45	$45,000
Employee #2:	42.150	44.258	44	$45,000 (#1)
Employee #3:	20.000	21.000	21	$21,000
TOTAL				$111,000

b. With a 6% raise the salaries are (in thousands):

	Salary	With Raise	Lower Quota	Hamilton's plan
Employee #1:	43.100	45.686	45	$46,000 (#1)
Employee #2:	42.150	44.679	44	$44,000
Employee #3:	20.000	21.200	21	$21,000
TOTAL				$111,000

c. This illustrates the Alabama paradox. An increase in the raise resulted in a loss for Employee #2.

Level 3 Problem Solving page 856

27. Yes; the best that child #1 can do is to slice off 1/3 of the cake. Anything less than that will result in a diminished slice for child #1. The best that child #2 can do is to divide the larger piece into two parts; anything less than that will result in a diminished slice for child #2. Child #3 will be faced with, at worst, three slices of equal size. If they are not equal size, then the other children did not make good cuts and child #3 will pick the largest piece.

28. Answers vary; suppose a wise person gives an old nag horse to the estate so there are 18 horses. Then $\frac{1}{2}(18) = 9$ horses; $\frac{1}{3}(18) = 6$ horses; $\frac{1}{9}(18) = 2$ horses. Since the eldest picked 9 horses, and the middle child 6 horses, and the youngest 2 horses, they picked a total of 17 horses. Certainly, none picked the old nag horse, which was now returned to the wise person. (See the first paragraph of the next problem.)

29.

Age	Adams	Jeff	Web
Eldest	9	8	9
Middle	6	5	6
Youngest	2	1	2
Total	17	14	17

We see that Adams' or Webster's plan gives the correct apportionment of the horses.

30. The average for the numbers 1, 2, 3, 4, $\cdots N$ is $\frac{N+1}{2}$, so for $N = 17$, the average ranking is 9. The youngest one must receive 2 horses whose ranks add up to 18 ($9 \times 2 = 18$). The middle one must receive 6 horses whose ranks add up to 54 ($9 \times 6 = 54$), and the eldest must receive 9 horses whose rank adds up to 81 ($9 \times 9 = 81$). The following table shows one possible solution.

Eldest:	2, 3, 4, 7, 11, 12, 13, 14, 15	Rank: 81, average rank is 9
Middle	5, 6, 8, 9, 10, 16	Rank: 54, average rank is 9
Youngest:	1, 17	Rank: 18, average rank is 9

Chapter 16 Review Questions, page 858

1. There is a total of 45 votes, so a majority is $\frac{45 + 1}{2} = 23$; there is no majority. The plurality vote goes to C.

2. There is a total of $3! = 6$ arrangements, and 3 are shown here, so there are 3 possibilities that received no votes.

3. Here are the rankings:

   A: 15; B: 12; C: 18

   Eliminate B for the second round of votes:

   A: $15 + 12 = 27$

   C: 18

   A wins the majority in this round, so A is the winner.

4. A over B: 15; B over A: $18 + 12 = 30$; B wins 1 point.

   A over C: $15 + 12 = 27$; C over A: 18; A wins 1 point.

   B over C: $15 + 12 = 27$; C over B: 18; B wins 1 point.

   A has 1 point, B has 2 points, and C has 0 points, so B is the winner.

5. (1) The majority criterion is not violated because there is no majority winner.

   (2) We look for a Condorcet candidate by finding the one-on-one pairings:

		A	B	C
A with B:	A	–	B	A
B wins				
A with C:	B	B	–	B
A wins	C	A	B	–

   B with C: **B wins**

   B is a Condorcet candidate. We see that the Hare method (Problem 3) violates the Condorcet criterion.

6. Here are the results (in percents):

A:	$22 + 23 = 45$
B:	$15 + 29 = 44$
C:	$7 + 4 = 11$

   Candidate A has a plurality.

7. We look for a Condorcet candidate by finding the one-on-one pairings:

A over B: $22 + 23 + 4 = 49$

	A	B	C
A	–	B	A
B	B	–	B
C	A	B	–

B over A: $15 + 29 + 7 = 51$; **B wins** A

A over C: $22 + 23 + 15 = 60$

C over A: $29 + 7 + 4 = 40$; **A wins** C

B over C: $22 + 15 + 29 = 66$

C over B: $23 + 7 + 4 = 34$; **B wins**

B is the Condorcet winner.

8. Three points for a first place vote, 2 points for a second place vote, and 1 point for a third place.

David Carr (Fresno State):	$34(3) + 60(2) + 58(1) = 280$
Eric Crouch (Nebraska):	$162(3) + 98(2) + 88(1) = 770$
Ken Dorsey (Miami):	$109(3) + 122(2) + 67(1) = 638$
Dwight Freeney (Syracuse):	$2(3) + 6(2) + 24(1) = 42$
Rex Grossman (Florida):	$137(3) + 105(2) + 87(1) = 708$
Joey Harrington (Oregon):	$54(3) + 68(2) + 66(1) = 364$
Bryant McKinnie (Miami):	$26(3) + 12(2) + 14(1) = 116$
Julius Peppers (North Carolina):	$2(3) + 10(2) + 15(1) = 41$
Antwaan Randle El (Indiana):	$46(3) + 39(2) + 51(1) = 267$
Roy Williams (Oklahoma):	$13(3) + 36(2) + 35(1) = 146$

The winner was Eric Crouch with 770 Borda points.

9. The votes are:

A: 7
B: 4
C: 5
D: 1

The least first-place votes were for D, so we delete D from the second round of votes, with these results:

A: 7
B: $4 + 1 = 5$
C: 5

There is a tie for last place. The Hare method fails to pick a winner unless we specify some way of breaking a tie. If we have a runoff between just B and C, we find:

B: $7 + 4 + 1 = 12$
C: 5

B is the winner, so for the next round we eliminate C, with these results:

A: $7 + 5 = 12$
B: $4 + 1 = 5$

We declare A the winner.

10. Here are the pairwise point counts.

A over B: $7 + 5 = 12$; B over A: $4 + 1 = 5$; A wins 1 point.

A over C: $7 + 1 = 8$; C over A: $5 + 4 = 9$; C wins 1 point.

A over D: $7 + 5 = 12$; D over A: $4 + 1 = 5$; A wins 1 point.

B over C: $7 + 4 + 1 = 12$; C over B: 5; B wins 1 point.

B over D: $5 + 4 = 9$; D over B: $7 + 1 = 8$; B wins 1 point.

C over D: $5 + 4 = 9$; D over C: $7 + 1 = 8$; C wins 1 point. The final point count is:

A, 2 points; B, 2 points; C, 2 points; D, 0 points. There is no pairwise winner without specifying a method of breaking a tie. If we have a runoff among A, B, and C, we find

A: 7
B: $4 + 1 = 5$
C: 5

We declare A the winner.

11. If B pulls out of the race, we have the following preferences:

(ADC)   (CAD)   (CDA)   (DAC)

7   5   4   1

The first vote is:

A: 7
C: $5 + 4 = 9$
D: 1

C has a majority, so C is the winner. If we compare this with the result of Problem 9 we see that this result does violate the irrelevant alternatives criterion.

12. The standard divisor is $\frac{790}{100} = 7.9$.

13. The standard quotas are EA: 11.39; MC: 27.22; M: 33.92; P: 16.84; SR: 10.63.

The lower and upper quotas are EA: 11, 12; MC: 27, 28; M: 33, 34; P: 16, 17   SR: 10, 11.

14. Adams' plan (modified divisor is 8.15): EA: 12; MC: 27; M: 33; P: 17; SR: 11.

15. Jefferson's plan (modified divisor is 7.67): EA: 11; MC: 28; M: 34; P: 17; SR: 10.

16. Hamilton's plan: EA: 11; MC: 27; M: $33 + 1 = 34$ (#1); P: $16 + 1 = 17$ (#2);
SR: $10 + 1 = 11$ (#3).

17. Webster's plan (modified divisor is 8): EA: 11; MC: 27; M: 34; P: 17; SR: 11.

18. HH plan: EA: 11; MC: 27; M: 34; P: 17; SR: 11.

19. No, not from the given information. We can see that the data does not violate the quota rule, and without more information, we cannot check to see whether any of the other paradoxes might be violated.

20. The sum of the number of violent crimes per 100 residents for the five areas is 62.68. Thus,

$d = \dfrac{62.68}{180} = 0.348\overline{2}.$ *Webster's Plan*

Downtown: $\dfrac{24.45}{d} \approx 70.2$ 70

Fairground: $\dfrac{10.04}{d} \approx 28.8$ 29

Columbus Square: $\dfrac{9.75}{d} \approx 28.0$ 28

Downtown West: $\dfrac{9.43}{d} \approx 27.1$ 27

Peabody: $\dfrac{9.01}{d} \approx 25.9$ 26

TOTAL: 180

Group Research Problems, page 860

G60. Answers vary.   **G61.** Answers vary.   **G62.** Answers vary.

Individual Research Problems, page 860

P16.1 to **P16.4** Answers vary.

CHAPTER 17
THE NATURE OF CALCULUS

17.1 What Is Calculus?, page 862

I use a definition of mathematics (Transparency 35) to introduce the idea of a mathematical model.

Level 1, page 870

1. limits, derivatives, and integrals   **2.** Answers vary; it involves a real world problem, abstraction to write the real world problem in mathematical terms. This step usually involves some assumptions. Then derive results and compare with the real world problem. This step may involve a modification of the assumptions and abstraction step. Mathematical models allow us to answer questions about real world problems.   **3.** Answers vary; theoretically, she never reaches the wall, but in reality she must hit the wall because the woman has physical dimensions.   **4.** Answers vary; theoretically, he never reaches second bases, but in reality he must reach second base because the baseball player has physical dimensions.   **5.** Answers vary; $\frac{1}{3}$ **6.** Answers vary; $6\frac{2}{3}$ **7.** Answers vary; 1

8. Answers vary; $\frac{3}{11}$ **9.** Answers vary; π

10.

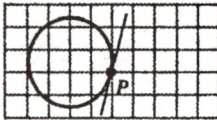

11.

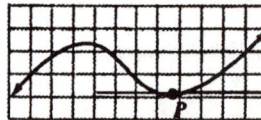

12.

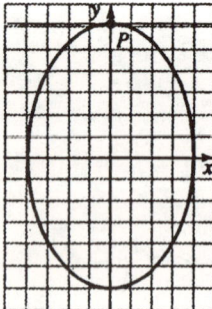

13.

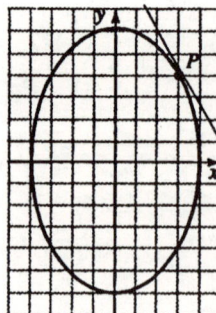

14. 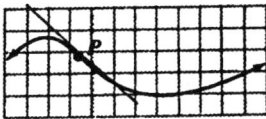   **15.**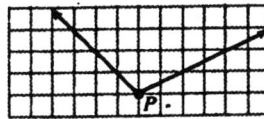

No single tangent line exists.

Level 2, page 871

Answers to Problems 16-27 may vary.

16. 2   **17.** $\frac{2}{3}$ **18.** 1   **19.** $\frac{1}{2}$ **20.** 0   **21.** $\frac{3}{2}$ **22.** 3   **23.** 20   **24.** 6 or 7   **25.** 4 or 5   **26.** 6

27. 3

Level 3 Problem Solving, page 871

28. a. 0.3984375   **b.** 0.365234375   **c.** Answers vary; $\frac{1}{3}$ **29.** Answers vary.   **30.** Answers vary.

17.2 Limits, page 872

Level 1, page 876

1. Answers vary; the limit of a sequence means that successive terms of a sequence are getting closer and closer to the number called the limit.   **2.** Answers vary; use a table of values, a graph, or a limit to infinity theorem.   **3.** 0, 2, 0, 2, 0   **4.** $-\frac{1}{8}, \frac{1}{16}, -\frac{1}{32}, \frac{1}{64}, -\frac{1}{128}$ **5.** $\frac{4}{3}, \frac{7}{4}, 2, \frac{13}{6}, \frac{16}{7}$

6. $0, \frac{1}{3}, \frac{1}{2}, \frac{3}{5}, \frac{2}{3}$ **7.** $\lim\limits_{n \to \infty} \dfrac{8,000n}{n+1} = \lim\limits_{n \to \infty} \dfrac{8,000}{1 + \frac{1}{n}} = 8,000$ **8.** $\lim\limits_{n \to \infty} \dfrac{8,000}{n-1} = 0$

9. $\lim\limits_{n \to \infty} \dfrac{2n+1}{3n-4} = \lim\limits_{n \to \infty} \dfrac{2 + \frac{1}{n}}{3 - \frac{4}{n}} = \dfrac{2}{3}$ **10.** $\lim\limits_{n \to \infty} \dfrac{4 - 7n}{8 + n} = \lim\limits_{n \to \infty} \dfrac{\frac{4}{n} - 7}{\frac{8}{n} + 1} = -7$

11. $\lim\limits_{n \to \infty} \dfrac{5n+8}{n} = \lim\limits_{n \to \infty} \dfrac{5 + \frac{8}{n}}{1} = 5$ **12.** $\lim\limits_{n \to \infty} \dfrac{5n}{n+7} = \lim\limits_{n \to \infty} \dfrac{5}{1 + \frac{7}{n}} = 5$

13. $\lim\limits_{n \to \infty} \dfrac{8n^2 + 800n + 5,000}{2n^2 - 1,000n + 2} = \lim\limits_{n \to \infty} \dfrac{8 + \frac{800}{n} + \frac{5,000}{n^2}}{2 - \frac{1,000}{n} + \frac{2}{n^2}} = \dfrac{8}{2} = 4$

14. $\lim\limits_{n \to \infty} \dfrac{100n + 7,000}{n^2 - n - 1} = \lim\limits_{n \to \infty} \dfrac{\frac{100}{n} + \frac{7,000}{n^2}}{1 - \frac{1}{n} - \frac{1}{n^2}} = \dfrac{0}{1} = 0$

15. $\lim\limits_{n \to \infty} \dfrac{8n^2 + 6n + 4,000}{n^3 + 1} = \lim\limits_{n \to \infty} \dfrac{\frac{8}{n} + \frac{6}{n^2} + \frac{4,000}{n^3}}{1 + \frac{1}{n^3}} = \dfrac{0}{1} = 0$

15. $\displaystyle\lim_{n\to\infty} \frac{8n^2 + 6n + 4{,}000}{n^3 + 1} = \lim_{n\to\infty} \frac{\frac{8}{n} + \frac{6}{n^2} + \frac{4{,}000}{n^3}}{1 + \frac{1}{n^3}} = \frac{0}{1} = 0$

16. $\displaystyle\lim_{n\to\infty} \frac{n^3 - 6n^2 + 85}{2n^3 - 5n + 170} = \lim_{n\to\infty} \frac{1 - \frac{6}{n} + \frac{85}{n^2}}{2 - \frac{5}{n^2} + \frac{170}{n^2}} = \frac{1}{2}$

Level 2, page 876

17. $\displaystyle\lim_{n\to\infty} \frac{1}{3n} = 0$ **18.** $\displaystyle\lim_{n\to\infty} \frac{1}{2^n} = 0$ **19.** 0.7 **20.** 6

21. $\displaystyle\lim_{n\to\infty} \frac{3n^2 - 7n + 2}{5n^3 + 9n^2} = \lim_{n\to\infty} \frac{\frac{3}{n^2} - \frac{7}{n^3} + \frac{2}{n^4}}{5 + \frac{9}{n^2}} = \frac{0 + 0 + 0}{5 + 0} = 0$

22. $\displaystyle\lim_{n\to\infty} \frac{4n^4 + 10n - 1}{9n^3 - 2n^2 - 7 + 3} = \lim_{n\to\infty} \frac{4 + \frac{10}{n^3} - \frac{1}{n^4}}{\frac{9}{n} - \frac{2}{n^2} - \frac{7}{n^3} + \frac{3}{n}} = \infty$

23. $\displaystyle\lim_{n\to\infty} \frac{2n^4 + 5n^2 - 6}{3n + 8} = \lim_{n\to\infty} \frac{2 + \frac{5}{n^2} - \frac{6}{n^4}}{\frac{3}{n^3} + \frac{8}{n^4}} = \infty$

24. $\displaystyle\lim_{n\to\infty} \frac{10n^3 + 13}{7n^2 - n + 2} = \lim_{n\to\infty} \frac{10 + \frac{13}{n^3}}{\frac{7}{n} - \frac{1}{n^2} + \frac{2}{n^3}} = \infty$

25. $\displaystyle\lim_{n\to\infty} \frac{15 + 9n - 6n^2}{2n^2 - 4n - 1} = \lim_{n\to\infty} \frac{\frac{15}{n^2} + \frac{9}{n} - 6}{2 - \frac{4}{n} - \frac{1}{n^2}} = \frac{0 + 0 - 6}{2 - 0 - 0} = -3$

26. $\displaystyle\lim_{n\to\infty} \frac{n^4 + 5n^3 + 8n^2 - 4n + 12}{2n^4 - 7n^2 + 3n} = \lim_{n\to\infty} \frac{1 + \frac{5}{n} + \frac{8}{n^2} - \frac{4}{n^3} + \frac{12}{n^4}}{2 - \frac{7}{n^2} + \frac{3}{n^3}} = \frac{1 + 0 + 0 - 0 + 0}{2 - 0 + 0} = \frac{1}{2}$

27. $\displaystyle\lim_{n\to\infty} \frac{-21n^3 + 52}{-7n^3 + n^2 + 20n - 9} = \lim_{n\to\infty} \frac{-21 + \frac{52}{n^3}}{-7 + \frac{1}{n} + \frac{20}{n^2} - \frac{9}{n^3}} = \frac{-21 + 0}{-7 + 0 + 0 - 0} = 3$

28. $\displaystyle\lim_{n\to\infty} \frac{12n^5 + 7n^4 - 3n^2 + 2n}{n^5 + 8n^3 + 14} = \lim_{n\to\infty} \frac{12 + \frac{7}{n} - \frac{7}{n^3} + \frac{2}{n^4}}{1 + \frac{8}{n^2} + \frac{14}{n^5}} = \frac{12 + 0 - 0 + 0}{1 + 0 + 0} = 12$

Level 3 Problem Solving, page 876

29. The sequence is 12, 6, 3, $\frac{3}{2}$, $\frac{3}{4}$; there is 0.75 mg of the drug present. At the end of n hours, there is $24(\frac{1}{2})^n$ mg present.

30. a. Answers vary; look at the illustration.

b. $\frac{1}{1} = 1.0000$; $\frac{2}{1} = 2.0000$; $\frac{3}{2} = 1.5000$; $\frac{5}{3} \approx 1.6667$; $\frac{8}{5} = 1.6000$;

$\frac{13}{8} = 1.6250$; $\frac{21}{13} \approx 1.6154$; $\frac{34}{21} \approx 1.6190$; $\frac{55}{34} \approx 1.6176$; $\frac{89}{55} \approx 1.6182$

c. $\dfrac{a_{n+1}}{a_n} = \dfrac{a_{n-1} + a_n}{a_n} = \dfrac{a_{n-1}}{a_n} + 1$

Let $L = \dfrac{a_{n+1}}{a_n}$, so that $\dfrac{1}{L} = \dfrac{1}{\frac{a_{n+1}}{a_n}} = \dfrac{a_n}{a_{n+1}}$; since L is the limit, this

relationship must be the same as $\dfrac{a_{n-1}}{a_n}$. This means that L must satisfy the
following relationship:

$$L = 1 + \frac{1}{L}$$
$$L^2 = L + 1$$
$$L^2 - L - 1 = 0$$
$$L = \frac{1 \pm \sqrt{(-1)^2 - 4(1)(-1)}}{2(1)} \qquad \textit{Reject negative value.}$$
$$= \frac{1 + \sqrt{5}}{2}$$

17.3 Derivatives, page 877

Level 1, page 886

1. a. AVERAGE $= \dfrac{80 - 40}{8 - 1} \approx 5.71$ **b.** AVERAGE $= \dfrac{100 - 40}{5 - 1} = 15$ **c.** AVERAGE $= \dfrac{80 - 40}{2 - 1} = 40$

d. It looks like the slope of the tangent line has a rise of 4 squares and a run of 1 square for a slope of

40. **2. a.** AVERAGE $= \dfrac{2{,}000 - 1{,}000}{800 - 100} \approx 1.42$ **b.** AVERAGE $= \dfrac{1{,}750 - 1{,}000}{500 - 100} = 1.875$

c. AVERAGE $= \dfrac{1{,}500 - 1{,}000}{300 - 100} = 2.5$ **d.** It looks like the slope of the tangent line has a rise of 6

squares and a run of 3 squares for a slope of 5. **3. a.** AVERAGE $= \dfrac{26.7 - 0}{6{:}36 - 6{:}09} = \dfrac{26.7}{27/60} \approx 59$ mph

b. AVERAGE $= \dfrac{35 - 26.7}{7{:}03 - 6{:}36} = \dfrac{8.3}{27/60} \approx 18$ mph **c.** AVERAGE $= \dfrac{50 - 35}{7{:}28 - 7{:}03} = \dfrac{15}{25/60} = 36$ mph

d. AVERAGE $= \dfrac{50 - 0}{7{:}28 - 6{:}09} = \dfrac{50}{79/60} \approx 38$ mph **4. a.** AVERAGE $= \dfrac{530 - 572}{2001 - 1997} \approx -10.5$ pt/yr

b. AVERAGE $= \dfrac{530 - 560}{2001 - 1998} = -10$ pt/yr **c.** AVERAGE $= \dfrac{530 - 548}{2001 - 1999} \approx -9$ pt/yr

d. AVERAGE $= \dfrac{530 - 540}{2001 - 2000} \approx -10$ pt/yr

5. Let \$ be trillions of dollars. **a.** AVERAGE $= \dfrac{7.6360 - 0.5153}{1996 - 1960} \approx 0.198$ \$/yr

b. AVERAGE $= \dfrac{7.6360 - 1.0155}{1996 - 1970} \approx 0.255$ \$/yr

d. AVERAGE $= \dfrac{7.6360 - 5.5461}{1996 - 1990} \approx 0.348$ \$/yr

e. AVERAGE $= \dfrac{7.6360 - 7.2654}{1996 - 1995} \approx 0.371$ \$/yr

f. It is changing at the rate of \$371 billion per year.

6.

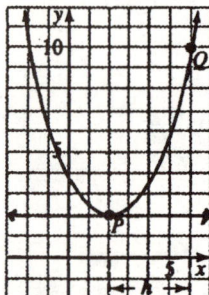

slope is 0

7.

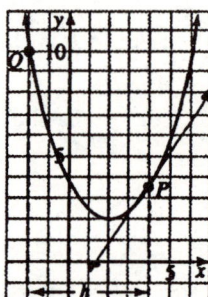

slope is 3/2

8.

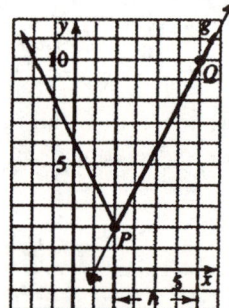

slope is 2

9.

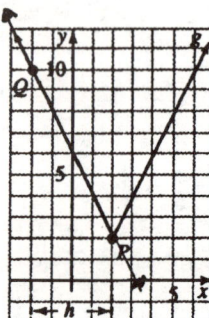

slope is -2

10.

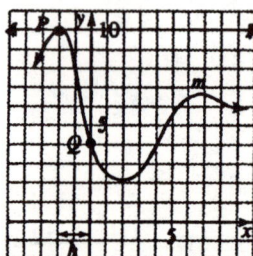

slope is 0

11.

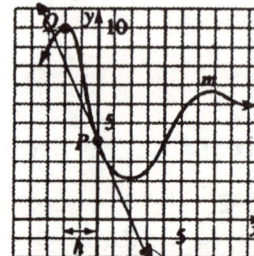

slope is -2

Level 2, page 887

12. a. $\dfrac{f(x + h) - f(x)}{h} = \dfrac{f(2) - f(x)}{2 - (-3)}$

$= \dfrac{-2 - 13}{5}$

$= -3$

b. $\displaystyle\lim_{h \to 0} \dfrac{f(x + h) - f(x)}{h} = \lim_{h \to 0} \dfrac{f(-3 + h) - f(-3)}{h}$

$= \displaystyle\lim_{h \to 0} \dfrac{4 - 3(-3 + h) - 13}{h}$

$= \displaystyle\lim_{h \to 0} \dfrac{4 + 9 - 3h - 13}{h}$

$= \displaystyle\lim_{h \to 0} (-3)$

$= -3$

13. a. $\dfrac{f(x+h)-f(x)}{h}=\dfrac{f(3)-f(-3)}{3-(-3)}$

$=\dfrac{5-5}{6}$

$=0$

b. $\displaystyle\lim_{h\to 0}\dfrac{f(x+h)-f(x)}{h}=\lim_{h\to 0}\dfrac{f(-3+h)-f(-3)}{h}$

$=\displaystyle\lim_{h\to 0}\dfrac{5-5}{h}$

$=\displaystyle\lim_{h\to 0}0$

$=0$

14. a. $\dfrac{f(x+h)-f(x)}{h}=\dfrac{f(3)-f(1)}{3-1}$

$=\dfrac{27-3}{2}$

$=12$

b. $\displaystyle\lim_{h\to 0}\dfrac{f(x+h)-f(x)}{h}=\lim_{h\to 0}\dfrac{f(1+h)-f(1)}{h}$

$=\displaystyle\lim_{h\to 0}\dfrac{3(1+h)^2-3(1)^2}{h}$

$=\displaystyle\lim_{h\to 0}\dfrac{3+6h+3h^2-3}{h}$

$=\displaystyle\lim_{h\to 0}\dfrac{h(6+3h)}{h}$

$=\displaystyle\lim_{h\to 0}(6+3h)$

$=6$

15. a. $\dfrac{f(x+h)-f(x)}{h}=\dfrac{f(9)-f(4)}{9-4}$

$=\dfrac{[-2(9)^2+9+4]-[-2(4)^2+4+4]}{5}$

$=\dfrac{-149-(-24)}{5}$

$=\dfrac{-125}{5}$

$=-25$

b. $\displaystyle\lim_{h\to 0}\dfrac{f(x+h)-f(x)}{h}=\lim_{h\to 0}\dfrac{[-2(4+h)^2+(4+h)+4]-[-24]}{h}$

$=\displaystyle\lim_{h\to 0}\dfrac{-32-16h-2h^2+4+h+4+24}{h}$

$=\displaystyle\lim_{h\to 0}\dfrac{-15h-2h^2}{h}$

$=\displaystyle\lim_{h\to 0}(-15-2h)$

$=-15$

16. $f(x)=\frac{1}{2}x^2;\ f'(x)=\frac{1}{2}(2)x^{2-1}=x$

17. $f(x)=\frac{1}{3}x^3;\ f'(x)=\frac{1}{3}(3)x^{3-1}=x^2$

18. $f(x)=e^{1.2x};\ f'(x)=1.2e^{1.2x}$

19. $f(x)=e^{-6x};\ f'(x)=-6e^{-6x}$

20. $f(x)=25-250x;\ f'(x)=0-250=-250$

21. $f(x) = 3 + 2x - 3x^2 = 0 + 2 - 6x = 2 - 6x$

22. $y' = 10x$; at $x = -3$, $m = 30$, so at $(-3, 45)$ the equation of the tangent line is

$$y - 45 = 30(x + 3)$$

$$x - y + 45 = 0$$

23. $y' = 4x$; at $x = 4$, $m = 16$, so at $(4, 32)$ the equation of the tangent line is

$$y - 32 = 16(x - 4)$$

$$16x - y - 32 = 0$$

24. $y' = -5$; at $(-2, 14)$ the equation of the tangent line is

$$y - 14 = -5(x + 2)$$

$$5x + y - 4 = 0$$

25. $y' = 6x + 4$; at $x = 0$, $m = 4$, so at $(0, 0)$ the tangent line is

$$y = 4x$$

$$4x - y = 0$$

26. $\displaystyle\lim_{h \to 0} \frac{16(2 + h)^2 - 16(2)^2}{h} = \lim_{h \to 0} \frac{64 + 64h + 16h^2 - 64}{h} = \lim_{h \to 0} (64 + 16h) = 64$

27. a. The height of the tower is found when $t = 0$: $h(0) = -16(0)^2 + 96(0) + 176 = 176$

   b. $v(t) = h'(t) = -32t + 96$

28. a. $P(30) - P(20) = 600 - 600 = 0$; there is no increase in profit.

   b. $P(25) - P(20) = 625 - 600 = 25$; profit increases by $25,000.

   c. $P(21) - P(20) = 609 - 600 = 9$; profit increases by $9,000.

   d. $\displaystyle\lim_{h \to 0} \frac{P(20 + h) - P(20)}{h} = \lim_{h \to 0} \frac{50(20 + h) - (20 + h)^2 - [50(20) - 20^2]}{h}$

$$= \lim_{h \to 0} \frac{1{,}000 + 50h - 400 - 40h - h^2 - 1{,}000 + 400}{h}$$

$$= \lim_{h \to 0} \frac{10h - h^2}{h}$$

$$= \lim_{h \to 0} (10 - h) = 10$$

   This is $10,000/item.

29.   **a.** $\text{AVERAGE} = \dfrac{C(200) - C(100)}{200 - 100} = \dfrac{1{,}180{,}000 - 290{,}000}{100} = 8{,}900$; average is $8,900/item.

     b. $\text{AVERAGE} = \dfrac{C(110) - C(100)}{110 - 100} = \dfrac{352{,}000 - 290{,}000}{10} = 6{,}200$; average is $6,200/item.

     c. $\text{AVERAGE} = \dfrac{C(101) - C(100)}{101 - 100} = \dfrac{295{,}930 - 290{,}000}{1} = 5{,}930$; average is $5,930/item.

     d. $\text{AVERAGE} = \dfrac{C(x + h) - C(h)}{h} = \dfrac{30(x + h)^2 - 100(x + h) - [30x^2 - 100x]}{h}$

$$= \dfrac{30x^2 + 60xh + 30h^2 - 100x - 100h - 30x^2 + 100x}{h}$$

$$= \dfrac{60xh + 30h^2 - 100h}{h}$$

$$= 60x - 100 + 30h$$

        This is in dollars/item. In particular, if $x = 100$, then this is $5{,}900 + 30h$.

Level 3 page 888

30.   **a.** $\displaystyle\lim_{h \to 0} \dfrac{N(t + h) - N(t)}{h} = \lim_{h \to 0} \dfrac{2(t + h)^2 - 200(t + h) + 1{,}000 - [2t^2 - 200t + 1{,}000]}{h}$

$$= \lim_{h \to 0} \dfrac{2t^2 + 4th + 2h^2 - 200t - 200h + 1{,}000 - 2t^2 + 200t - 1{,}000}{h}$$

$$= \lim_{h \to 0} \dfrac{4th + 2h^2 - 200h}{h}$$

$$= \lim_{h \to 0} (4t - 200 + 2h)$$

$$= 4t - 200$$

     b. When $t = 3$; the instantaneous rate of change is $4(3) - 200 - 2(3) = -194$

     c. The beginning of this experiment is when $t = 0$; $4(0) - 200 - 2(3) = -206$

17.4   Integrals, page 888

Level 1, page 895

1. The figure is a rectangle, so we use the formula $A = \ell w$ where $w = 5$ and $\ell = x$. Thus, $A(x) = 5x$, so $A(8) = 5(8) = 40$.   **2.** The figure is a rectangle, so we use the formula $A = \ell w$ where $w = 8.3$ and $\ell = x$. Thus, $A(x) = 8.3x$.   **3.** The figure is a triangle, so we use the formula $A = \frac{1}{2}bh$, where $b = x$ and $h = 3x$. Thus, $A(x) = \frac{1}{2}x(3x) = \frac{3}{2}x^2$, so $A(4) = \frac{3}{2}(4)^2 = 24$.   **4.** This figure is a triangle, so we use the formula $A = \frac{1}{2}bh$, where $b = x$ and $h = 2x$. Thus, $A(x) = \frac{1}{2}(x)(2x) = x^2$.

5. This is a trapezoid; the area of a trapezoid is $A = \frac{1}{2}h(b_1 + b_2)$: $A(x) = \frac{1}{2}(3)(2 + 11) = \frac{39}{2}$

6. This is a trapezoid; the area of a trapezoid is $A = \frac{1}{2}h(b_1 + b_2)$: $A(x) = \frac{1}{2}x(5 + x + 5) = 5x + \frac{1}{2}x^2$

7. $\int 6\, dx = 6x + C$ **8.** $\int 3\, dx = 3x + C$ **9.** $\int (x + 5)\, dx = \frac{1}{2}x^2 + 5x + C$

10. $\int (x + 6)\, dx = \frac{1}{2}x^2 + 6x + C$ **11.** $\int (3x + 4)\, dx = \frac{3}{2}x^2 + 4x + C$

12. $\int (2x + 5)\, dx = x^2 + 5x + C$

Level 2, page 895

13. Answers vary. For a given function f, we define the derivative of f at x, denoted by $f'(x)$ to be
$$f'(x) = \lim_{h \to 0} \frac{f(x + h) - f(x)}{h}$$ provided this limit exits.

14. Answers vary. Let f be a function defined over the interval $[a, b]$. Then the definite integral of f over the interval is denoted by $\int_a^b f(x)\, dx$ and is the net change of an antiderivative of f over that interval. Thus, if $F(x)$ is an antiderivative of $f(x)$, then $\int_a^b f(x)\, dx = F(x)\big|_b^a = F(b) - F(a)$.

15. Answers vary. The area function, $A(t)$, is the area bounded below by the x-axis, above by a function $y = f(x)$, on the left by the y-axis, and on the right by the vertical line $y = t$.

16. Answers vary. The definite integral is a number, the indefinite integral is a function.

17. If $f(x) = \frac{1}{2}x^2$, then $f'(x) = x$, so $\frac{1}{2}x^2$ is an antiderivative of x.

18. If $f(x) = \frac{1}{2}x^2 - 5$, then $f'(x) = x$, so $\frac{1}{2}x^2 - 5$ is an antiderivative of x.

19. If $f(x) = 2x^3$, then $f'(x) = 6x^2$, so $2x^3$ is an antiderivative of $6x^2$.

20. If $f(x) = 2x^3 - 3$, then $f'(x) = 6x^2$, so $2x^3 - 3$ is an antiderivative of $6x^2$.

21. $\int_1^9 x^2\, dx = \frac{1}{3}x^3\big|_1^9 = 242\frac{2}{3}$ **22.** $\int_2^5 x^2\, dx = \frac{1}{3}x^3\big|_2^5 = 39$ **23.** $\int_3^8 3x\, dx = \frac{3}{2}x^2\big|_3^8 = 82.5$

24. $\int_1^4 (2x + 1)\, dx = (x^2 + x)\big|_1^4 = 18$ **25.** $\int_0^2 e^{0.5x}\, dx = 2e^{0.5x}\big|_0^2 = 2e - 2$

26. $\int_{-2}^3 (9 - x^2)\, dx = (9x - \frac{1}{3}x^3)\big|_{-1}^3 = \frac{80}{3}$ **27.** $\int_1^5 (1 + 6x)\, dx = (x + 3x^2)\big|_1^5 = 76$

28. $\int_0^2 (e^x + e^2)\, dx = (e^x + e^2 x)\big|_0^2 = e^2 + 2e^2 - 1 = 3e^2 - 1$

Level 3, page 895

29. $\int_0^{10} 200 e^{0.5t}\, dt = 200(2) e^{0.5t}\Big|_0^{10} = 400 e^5 - 400 \approx 58{,}965.3$

Answers vary; this represents the increase in the number of bacteria after 10 hours (or net change).

Level 3 Problem Solving, page 895

30. $18 = 21 e^{-0.01t}$

$\dfrac{18}{21} = e^{-0.01t}$

$-0.01t = \ln \dfrac{18}{21}$

$t = -100 \ln \dfrac{18}{21} \approx 15.4$

$\int_0^{15.4} (21 e^{-0.01t} - 18)\, dt \approx 22.5287$

Answers vary; this represents the total profit.

Chapter 17 Review Questions, page 897

1. $\displaystyle\lim_{n\to\infty} \frac{1}{n} = 0$ **2.** $\displaystyle\lim_{n\to\infty} \frac{3n^4 + 20}{7n^4} = \lim_{n\to\infty} \frac{\dfrac{3n^4}{n^4} + \dfrac{20}{n^4}}{\dfrac{7n^4}{n^4}} = \lim_{n\to\infty} \frac{3 + \dfrac{20}{n^4}}{7} = \frac{3}{7}$

3. $\displaystyle\lim_{n\to\infty} (2n + 3) = \infty$ **4.** $\displaystyle\lim_{n\to\infty} \left(1 + \frac{1}{n}\right)^n$

Construct a table of values:

$n = 1$;	2
$n = 10$;	2.5937424601
$n = 10^3$;	2.71692393224
$n = 10^{10}$;	2.71828182832

It looks like the limit is e. In fact, this limit is often given as the *definition* of e.

5. The series illustrates the idea of a tangent line ("instantaneous growth rate"). The first shows 5 measurements, the second 20 measurements, and the third 40 measurements.

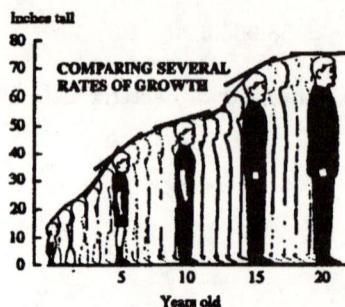

COMPARING SEVERAL RATES OF GROWTH

Inches tall / Years old

Each point on the curve has a tangent which indicates rate of growth at that point. The steepest tangent shown here occurs at about age three: the boy's growth was most rapid then.

Growth Rate at an Instant

The lines AE, AD, AC and AB above show average growth rates for successively smaller periods of time. But for the instant A, the growth rate is shown by the tangent at A.

Note the last one measures the growth rate at an instant. The closer the measurements are together, the better the approximation at a particular point.

6. The main ideas of calculus are limits, derivatives, and integrals.

(1) $\lim\limits_{n \to \infty} a_n = L$ means that the sequence a_n becomes closer and closer to the number L as n becomes larger and larger.

(2) The derivative illustrates the idea of a tangent line.

(3) The integral is used to find the area under a curve.

Answers vary.

7.
$$\lim_{h \to 0} \frac{f(x + h) - f(x)}{h} = \lim_{h \to 0} \frac{[6 - 4(x + h)^2] - [6 - 4x^2]}{h}$$
$$= \lim_{h \to 0} \frac{6 - 4x^2 - 8xh - 4h^2 - 6 + 4x^2}{h}$$
$$= \lim_{h \to 0} \frac{-8xh - 4h^2}{h}$$
$$= \lim_{h \to 0} (-8x - 4h) = -8x$$

8. $\int_{-1}^{0} (x - 4x^3)\, dx = (\frac{1}{2}x^2 - x^4)\Big|_{-1}^{0} = \frac{1}{2}$

9. $\int_{1}^{2} e^x\, dx = e^2 - e \approx 4.67$

10. Total amount used is $\int_0^t 32.4e^{6x/125}\,dx = 675(e^{6t/125} - 1)$

Solve

$$670 = 675(e^{6t/125} - 1)$$

$$\frac{670}{675} = e^{6t/125} - 1$$

$$\frac{1{,}345}{675} = e^{6t/125}$$

$$\frac{6t}{125} = \ln\left(\frac{1{,}345}{675}\right)$$

$$t = \frac{125}{6}\ln\left(\frac{1{,}345}{675}\right)$$

$$\approx 14.363$$

This says that the oil reserves will be depleted in 2014.

Epilogue Problem Set, page E1

Level 1, page E18

1. Answers vary.   **2.** Answers vary.   **3. a.** social science   **b.** natural science   **c.** natural science

d. humanities   **e.** social science   **4. a.** natural science   **b.** humanities   **c.** social science

d. natural science   **e.** natural science   **5. a.** social science   **b.** natural science   **c.** social science

d. social science   **e.** natural science   **f.** natural science   **6. a.** humanities   **b.** social science

c. social science   **d.** social science   **e.** social science   **f.** social science   **7.** all natural science

8. a. humanities   **b.** social science   **c.** natural science   **d.** natural science   **e.** social sciences

f. natural sciences   **9.** 1.46×10^8 **10.** 3.2998×10^{13} **11.** Venus and Neptune have the smallest

eccentricity and therefore have the most circular orbits.   **12.** $2 \times 2 = 4$ for the first pair;

$4^{23} \approx 7.04 \times 10^{13}$ **13.** Since eggs are three dimensional, $s^3 = 8$, so $s = \sqrt[3]{3}$; $2\sqrt[3]{3} \approx 2.89$. We

expect the ostrich egg to weigh 2.89 oz.

14. $\dfrac{\text{SMALL EGGS}}{\text{JUMBO EGGS}} = \dfrac{34 \text{ lb/case}}{56 \text{ lb/case}} \approx 0.61$ which is the scaling factor for one jumbo egg; or if you want

the scaling factor for a small egg, find the reciprocal which is 1.65.

15. Consider $(p + q)^2$ where $p = 0.48$, $q = 0.55$:

     genotype $SS = (0.55)^2 = 0.3025$

     genotype wS or $Sw = 2(0.55)(0.45) = 0.4950$

     genotype $ww = (0.45)^2 = 0.2025$

16. Consider $(p + q)^2$ where $p = 0.52$, $q = 0.48$:

     genotype $SS = (0.52)^2 = 0.2704$

     genotype wS or $Sw = 2(0.52)(0.48) = 0.4992$

     genotype $ww = (0.48)^2 = 0.2304$

17. Since this is measuring area, the scaling factor is $s^2 = 2,000$ or $s \approx 44.721$, so the area of the

unmagnified cell is $20/44.721 \approx 0.45 \text{ cm}^2$.

Level 2, page E18

18. a. further   **b.**

T=9
X=400.92954   Y=4.0295449

X₁ᴛ🔲(63cos 45)T
Y₁ᴛ🔲(63sin 45)T−
4.9T^2
    Tmin=0
    Tmax=10
    Tstep=.1
Xmin=0   Ymin=-25
Xmax=400 Ymax=150
Xscl=50 Yscl=25

Conjecture was correct; almost 50 ft further.

19.

T=11.3
X=361.6   Y=.62857202

X₁ᴛ🔲(64cos 60)T
Y₁ᴛ🔲(64sin 60)T−
4.9T^2
    Tmin=0
    Tmax=12
    Tstep=.1
Xmin=0   Ymin=-25
Xmax=400 Ymax=200
Xscl=50 Yscl=25

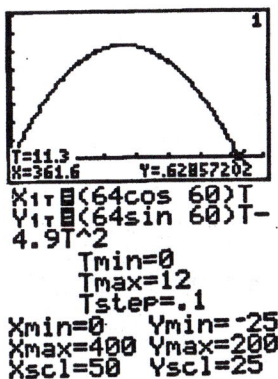

20. a. $a = \sqrt{4.5837 \times 10^{15}} \approx 67{,}703{,}028$

$b = \sqrt{4.5835 \times 10^{15}} \approx 67{,}701{,}551$

Since a and b are almost the same, we note the ellipse is almost circular.

$$\epsilon = \sqrt{1 - \frac{b^2}{a^2}} \approx 0.0066$$

b. Let each unit be $\sqrt{10^{15}}$.

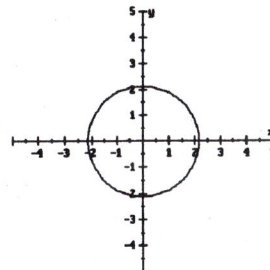

21. a. $a = \sqrt{2.015 \times 10^{16}} \approx 141{,}950{,}696$

$b = \sqrt{1.995 \times 10^{16}} \approx 141{,}244{,}469$

Since a and b are almost the same, we note the ellipse is almost circular.

$$\epsilon = \sqrt{1 - \frac{b^2}{a^2}} \approx 0.0996$$

b. Let each unit be $\sqrt{10^{16}} = 10^8$.

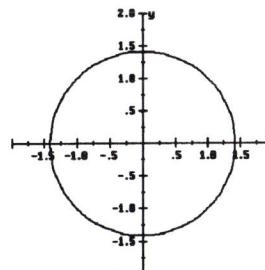

Page 407

22. $6 \cdot 5 \cdot 4 \cdot 3 = 360$

23. $4! = 24$

24. **a.**

	A	B	C	D
a	(4, 1)	(2, 1)	(1, 2)	(3, 1)
b	(4, 3)	(3, 2)	(1, 1)	(2, 2)
c	(2, 4)	(1, 6)	(3, 4)	(4, 5)
d	(1, 5)	(2, 5)	(4, 5)	(3, 6)
e	(1, 6)	(2, 4)	(3, 3)	(4, 3)
f	(3, 2)	(1, 3)	(4, 6)	(2, 4)

    b. Answers vary.

	A	B	C	D			A	B	C	D	
a	**[(4, 1)]**	(2, 1)	(1, 2)	(3, 1)	unstable	*a*	(4, 1)	**[(2, 1)]**	(1, 2)	(3, 1)	stable
b	(4, 3)	**[(3, 2)]**	(1, 1)	(2, 2)		*b*	(4, 3)	(3, 2)	**[(1, 1)]**	(2, 2)	
c	(2, 4)	(1, 6)	**[(3, 4)]**	(4, 5)		*c*	(2, 4)	(1, 6)	(3, 4)	(4, 5)	
d	(1, 5)	(2, 5)	(4, 5)	**[(3, 6)]**		*d*	(1, 5)	(2, 5)	(4, 5)	(3, 6)	
e	(1, 6)	(2, 4)	(3, 3)	(4, 3)		*e*	(1, 6)	(2, 4)	(3, 3)	**[(4, 3)]**	
f	(3, 2)	(1, 3)	(4, 6)	(2, 4)		*f*	**[(3, 2)]**	(1, 3)	(4, 6)	(2, 4)	

25. **a.**

	A	B	C	D
a	(1, 2)	(3, 1)	(4, 1)	(2, 2)
b	(2, 3)	(4, 4)	(3, 2)	(1, 1)
c	(2, 4)	(3, 3)	(4, 4)	(1, 3)
d	(3, 1)	(2, 2)	(4, 3)	(1, 4)

    b. Answers vary.

	A	B	C	D			A	B	C	D	
a	(1, 2)	(3, 1)	(4, 1)	**[(2, 2)]**	unstable	*a*	**[(1, 2)]**	(3, 1)	(4, 1)	(2, 2)	stable
b	(2, 3)	(4, 4)	**[(3, 2)]**	(1, 1)		*b*	(2, 3)	(4, 4)	(3, 2)	**[(1, 1)]**	
c	(2, 4)	**[(3, 3)]**	(4, 4)	(1, 3)		*c*	(2, 4)	(3, 3)	**[(4, 4)]**	(1, 3)	
d	**[(3, 1)]**	(2, 2)	(4, 3)	(1, 4)		*d*	(3, 1)	**[(2, 2)]**	(4, 3)	(1, 4)	

26.

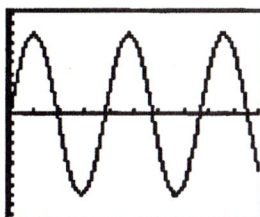

```
Y₁≣8sin (360*261
.626X)
Xmin=0    Ymin=-10
Xmax=.01  Ymax=10
Xscl=.001 Yscl=1
```

27.

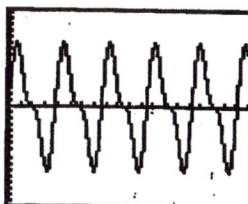

```
Y₁≣8sin (360*261
.626X)+4sin (720
*261.626X)
Xmin=0    Ymin=-15
Xmax=.02  Ymax=15
Xscl=.001 Yscl=1
```

28. There are $3! = 6$ possibilities.
The matrix of their stated choices is:

	A	B	C
a	(2, 1)	(3, 2)	(1, 1)
b	(2, 2)	(3, 1)	(1, 3)
c	(3, 3)	(2, 3)	(1, 2)

Here are the six possibilities:

	A	B	C
a	$\boxed{(2, 1)}$	(3, 2)	(1, 1)
b	(2, 2)	$\boxed{(3, 1)}$	(1, 3)
c	(3, 3)	(2, 3)	$\boxed{(1, 2)}$

a prefers C and C prefers a; not stable

	A	B	C
a	$\boxed{(2, 1)}$	(3, 2)	(1, 1)
b	(2, 2)	(3, 1)	$\boxed{(1, 3)}$
c	(3, 3)	$\boxed{(2, 3)}$	(1, 2)

a prefers C and C prefers a; not stable

	A	B	C
a	(2, 1)	$\boxed{(3, 2)}$	(1, 1)
b	$\boxed{(2, 2)}$	(3, 1)	(1, 3)
c	(3, 3)	(2, 3)	$\boxed{(1, 2)}$

a prefers C and C prefers a; not stable

Page 409

	A	B	C
a	(2, 1)	[(3, 2)]	(1, 1)
b	(2, 2)	(3, 1)	[(1, 3)]
c	[(3, 3)]	(2, 3)	(1, 2)

a prefers *C* and *C* prefers *a*; not stable

	A	B	C
a	(2, 1)	(3, 2)	[(1, 1)]
b	[(2, 2)]	(3, 1)	(1, 3)
c	(3, 3)	[(2, 3)]	(1, 2)

stable

	A	B	C
a	(2, 1)	(3, 2)	[(1, 1)]
b	(2, 2)	[(3, 1)]	(1, 3)
c	[(3, 3)]	(2, 3)	(1, 2)

b prefers *A*, and *A* prefers *b*; not stable

29. There are $3! = 6$ possibilities.

The matrix of their stated choices is:

	A	B	C
a	(3, 1)	(1, 3)	(2, 2)
b	(2, 2)	(3, 1)	(1, 3)
c	(1, 3)	(2, 2)	(3, 1)

Here are the six possibilities:

	A	B	C
a	[(3, 1)]	(1, 3)	(2, 2)
b	(2, 2)	[(3, 1)]	(1, 3)
c	(1, 3)	(2, 2)	[(3, 1)]

stable; each school has its first choice

	A	B	C
a	[(3, 1)]	(1, 3)	(2, 2)
b	(2, 2)	(3, 1)	[(1, 3)]
c	(1, 3)	[(2, 2)]	(3, 1)

a would rather be with *C*, and *C* with *a*; unstable

	A	B	C
a	(3, 1)	$\boxed{(1, 3)}$	(2, 2)
b	$\boxed{(2, 2)}$	(3, 1)	(1, 3)
c	(1, 3)	(2, 2)	$\boxed{(3, 1)}$

c would rather be with *B*, and *B* with *c*; unstable

	A	B	C
a	(3, 1)	$\boxed{(1, 3)}$	(2, 2)
b	(2, 2)	(3, 1)	$\boxed{(1, 3)}$
c	$\boxed{(1, 3)}$	(2, 2)	(3, 1)

stable; each person has his or her first choice

	A	B	C
a	(3, 1)	(1, 3)	$\boxed{(2, 2)}$
b	$\boxed{(2, 2)}$	(3, 1)	(1, 3)
c	(1, 3)	$\boxed{(2, 2)}$	(3, 1)

a prefers *B*, but *B* is happier with *c*

b prefers *C*, but *C* is happier with *a*

c prefers *A*, but *A* is happier with *b*

stable

	A	B	C
a	(3, 1)	(1, 3)	$\boxed{(2, 2)}$
b	(2, 2)	$\boxed{(3, 1)}$	(1, 3)
c	$\boxed{(1, 3)}$	(2, 2)	(3, 1)

b would rather be with *A* and *A* with *b*; unstable

30. Let $\frac{x^2}{a^2} + \frac{y^2}{b^2} = 1$ be the equation of the orbit of the moon. Let c be the distance from the center to the focus. Then,

$$\begin{cases} a + c = 378{,}000 \\ a - c = 199{,}000 \end{cases}$$

Solving, we find $a = 288{,}500$ and $c = 89{,}500$; $\epsilon = \frac{c}{a} = \frac{89{,}500}{288{,}500} \approx 0.31$

THE NATURE OF MATHEMATICS

MATH ANXIETY

Math Anxiety Bill of Rights*
by Sandra L. Davis

1. I have the right to learn at my own pace and not feel put down or stupid if I'm slower than someone else.
2. I have the right to ask whatever questions I have.
3. I have the right to need extra help.
4. I have the right to ask a teacher or TA for help.
5. I have the right to say I don't understand.
6. I have the right not to understand.
7. I have the right to feel good about myself regardless of my abilities in math.
8. I have the right not to base my self-worth on my math skills.
9. I have the right to view myself as capable of learning math.
10. I have the right to evaluate my math instructors and how they teach.
11. I have the right to relax.
12. I have the right to be treated as a competent adult.
13. I have the right to dislike math.
14. I have the right to define success in my own terms.

*From *Overcoming Math Anxiety,* by Sheila Tobias, pp. 236–237.

GUIDELINES FOR PROBLEM SOLVING

FIRST: UNDERSTAND THE PROBLEM

SECOND: DEVISE A PLAN

THIRD: CARRY OUT THE PLAN

FOURTH: LOOK BACK

1.   Look for a simpler **related problem**

2.   Work backward

3.   Work forward

4.   Narrow the condition

5.   Widen the condition

6.   Seek a counterexample

7.   Guess and test

8.   Divide and conquer

9.   Change the conceptual mode

MAP OF SAN FRANCISCO

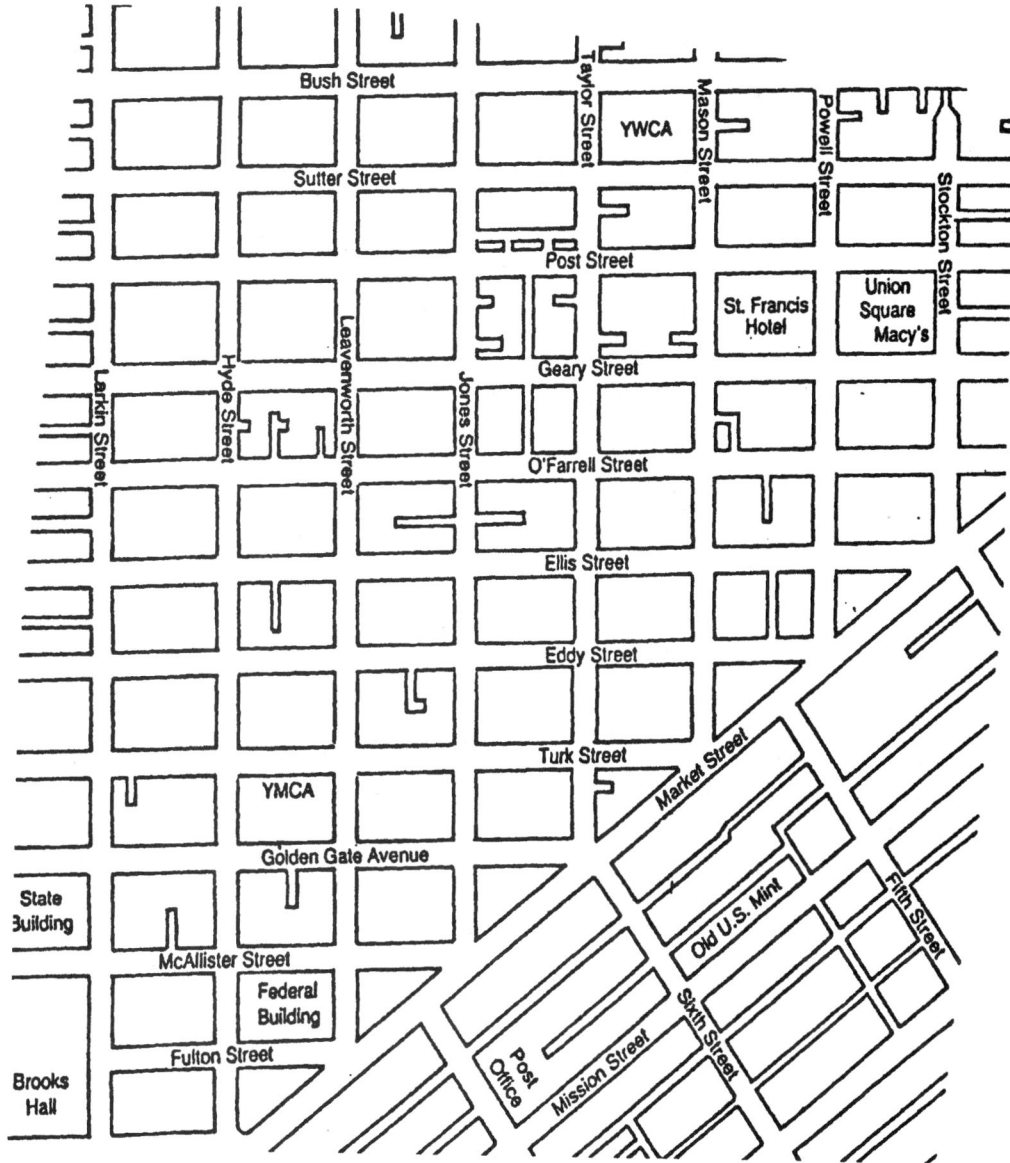

PASCAL'S TRIANGLE

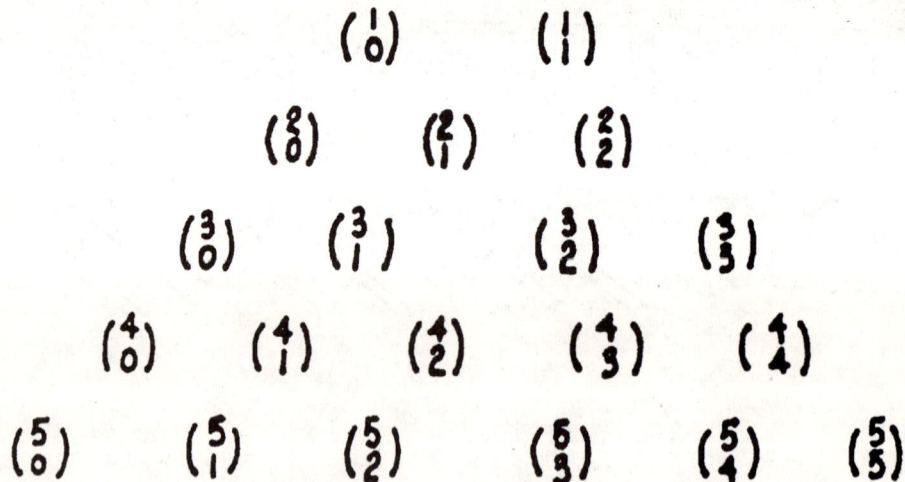

PATTERNS IN MULTIPLICATION

NINE

$$1 \times 9 = 9$$
$$2 \times 9 = 18$$
$$3 \times 9 = 27$$
$$4 \times 9 = 36$$
$$5 \times 9 = 45$$
$$6 \times 9 = 54$$
$$7 \times 9 = 63$$
$$8 \times 9 = 72$$
$$9 \times 9 = 81$$
$$10 \times 9 = 90$$

USE PATTERNS TO SIMPLIFY:

$$\frac{(999{,}999{,}999)(999{,}999{,}999)}{1+2+3+4+5+6+7+8+9+8+7+6+5+4+3+2+1}$$

EXTRA! EXTRA!

ENTIRE WORLD POPULATION MOVES TO FLORIDA

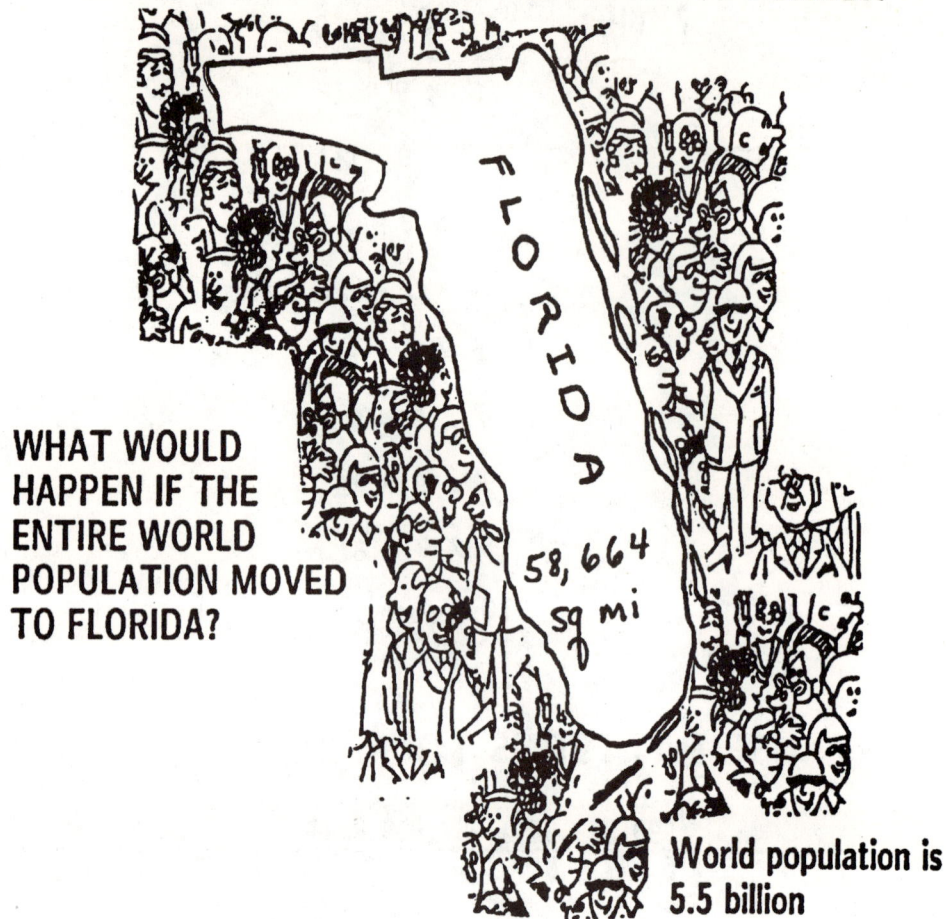

WHAT WOULD
HAPPEN IF THE
ENTIRE WORLD
POPULATION MOVED
TO FLORIDA?

FLORIDA

58,664 sq mi

World population is
5.5 billion

How much space would a family of four persons have?

A. 12 sq in   (oh no!)

B. 12 sq ft   (standing room only)

C. 120 sq ft   (a small room)

D. 1,200 sq ft   (a typical home)

E. 12,000 sq ft (a grand estate)

TRUTH OR CONSEQUENCES

FUNDAMENTAL OPERATORS

p	q	p∧q AND	p∨q OR	~p NOT
T	T	T	T	F
T	F	F	T	F
F	T	F	T	T
F	F	F	F	T

CONDITIONAL

p	q	p→q IMPLIES
T	T	T
T	F	F
F	T	T
F	F	T

ADDITIONAL OPERATORS

p	q	either...or	neither..nor	unless	because	no p is q
T	T	F	F	T	T	F
T	F	T	F	T	F	T
F	T	T	F	T	F	T
F	F	F	T	F	F	T

NUMERATION SYSTEMS AND PLACE VALUE

HOW MANY?

a. **Hindu Arabic** b. **Egyptian**

c. **Babylonian** d. **Roman**

Base	Place Value				
two	2^5	2^4	2^3	2^2	$2^1 = 2$ $2^0 = 1$
three	3^5	3^4	3^3	3^2	$3^1 = 3$ $3^0 = 1$
four	4^5	4^4	4^3	4^2	$4^1 = 4$ $4^0 = 1$
five	5^5	5^4	5^3	5^2	$5^1 = 5$ $5^0 = 1$
ten	10^5	10^4	10^3	10^2	$10^1 = 10$ $10^0 = 1$
twelve	12^5	12^4	12^3	12^2	$12^1 = 12$ $12^0 = 1$

A HUMAN COMPUTER

TWO STATES

SWITCHING STATE

COUNTER: 001100_{two}

overflow

↑ ↑
Impulse Impulse

ADD: 111_{two}

overflow

↑ ↑ ↑
Impulse Impulse Impulse

RAILROAD PROBLEM

A SINGLE RAILROAD TRACK IS LAID ONE MILE OVER LEVEL GROUND. IT IS FIRMLY SECURED AT THE ENDS SO THAT THEY CAN NOT MOVE. IF IN THE HEAT OF THE DAY, THE TRACK EXPANDS ONE INCH OVER ITS LENGTH AND ARCS UP ABOVE THE GROUND, THEN HOW HIGH IS THE ARC AT ITS CENTER?

HIGH ENOUGH TO:

- A. SLIP A SHEET OF PAPER UNDER?
- B. SLIP YOUR HAND UNDER?
- C. CRAWL UNDER?
- D. WALK UNDER?
- E. DRIVE A LOCOMOTIVE UNDER?

RAILROAD PROBLEM SOLUTION

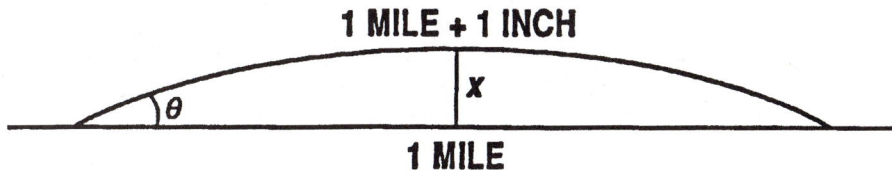

1 MILE + 1 INCH

θ

x

1 MILE

1/2 MILE + 1/2 INCH

x

1/2 MILE

A triangle may be used to approximate x, since θ is so small (approximately 0.03°)

$$(\tfrac{1}{2} \text{ mile} + \tfrac{1}{2} \text{ inch})^2 = x^2 + (\tfrac{1}{2}\text{mile})^2$$

$$(31{,}680.5)^2 = x^2 + (31{,}680)^2 \text{ in inches}$$

$$x \approx 177.989466$$

$$\approx 14.8 \text{ feet}$$

CASEY JONES CAN GO RIGHT ON UNDER!!!

SYMMETRIES OF A SQUARE

LETTER	DESCRIPTION	RESULT
A	90° clockwise rotation	3 4 / 2 1
B	180° clockwise rotation	2 3 / 1 4
C	270° clockwise rotation	1 2 / 4 3
D	360° clockwise rotation	4 1 / 3 2
E	Flip about a horizontal line	3' 2' / 4' 1'
F	Flip about a vertical line	1' 4' / 2' 3'
G	Flip about upper to lower diagonal	4' 3' / 1' 2'
H	Flip about lower to upper diagonal	2' 1' / 3' 4'

	A	B	C	D	E	F	G	H
A	B	C	D	A	H	G	E	F
B	C	D	A	B	F	E	H	G
C	D	A	B	C	G	H	F	E
D	A	B	C	D	E	F	G	H
E	G	F	H	E	D	B	A	C
F	H	E	G	F	B	D	C	A
G	F	H	E	G	C	A	D	B
H	E	G	F	H	A	C	B	D

FIELD OF DREAMS

A <u>FIELD</u> is a set F, with two operations, + and × satisfying the following properties for any elements a, b, and c of F.

ADDITION	MULTIPLICATION

1. CLOSURE

$(a + b)$ is an element of F

2. CLOSURE

ab is an element of F

3. COMMUTATIVE

$a + b = b + a$

4. COMMUTATIVE

$ab = ba$

5. ASSOCIATIVE

$(a + b) + c = a + (b + c)$

6. ASSOCIATIVE

$(ab)c = a(bc)$

7. IDENTITY

There exists a real number 0 so that

$0 + a = a + 0 = a$

8. IDENTITY

There exists a real number 1 so that

$1 \times a = a \times 1 = a$

9. INVERSE

For each a in F there is a unique number $-a$ in F number in F so that

$a + (-a) = (-a) + a = 0$

10. INVERSE

For each nonzero a in F there is a unique number $1/a$ in F so that

$a \times \frac{1}{a} = \frac{1}{a} \times a = 1$

11. DISTRIBUTIVE FOR MULTIPLICATION OVER ADDITION

$a \times (b + c) = a \times b + a \times c$

THE FBI HAS BED BUGS

OLD OR YOUNG WOMAN?

OLD WOMAN

YOUNG WOMAN

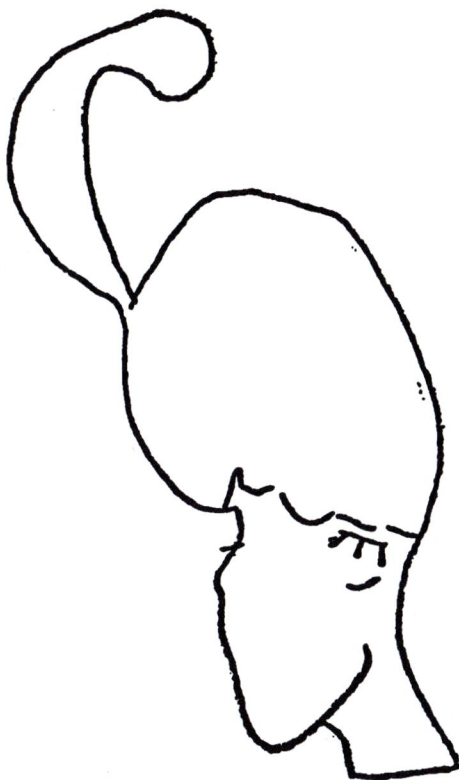

STEEL BAND PROBLEM

SUPPOSE THERE WERE A STEEL BAND FITTING TIGHTLY AROUND THE EQUATOR OF THE EARTH. NOW SUPPOSE THAT YOU REMOVE IT AND CUT IT AT ONE PLACE, THEN SPLICE IN AN ADDITIONAL PIECE 10 FEET LONG.

IF YOU NOW REPLACE IT ON THE EQUATOR, IT WILL FIT MORE LOOSELY, WOULD IT NOT? ASSUMING THAT THE SLACK IS UNIFORM ALL THE WAY AROUND THE EQUATOR, THE BAND WOULD BE LOOSE ENOUGH TO:

A. WALK UNDER?
B. CRAWL UNDER?
C. SLIP YOUR HAND UNDER?
D. SLIP A SHEET OF PAPER UNDER?
E. NOT EVEN GET THE SHEET OF PAPER UNDER?

STEEL BAND PROBLEM SOLUTION

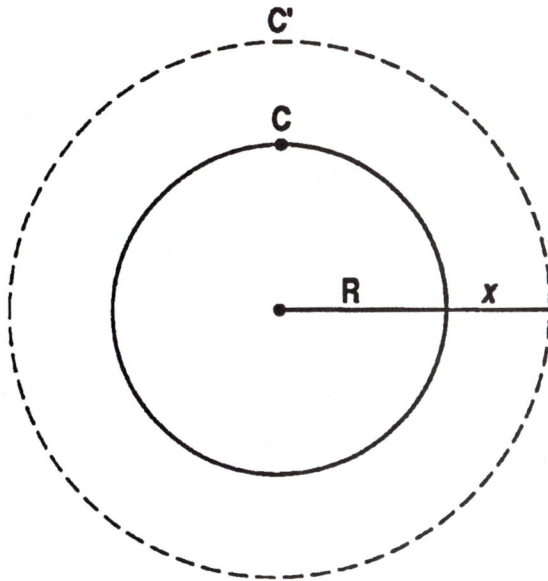

$$C = 2\pi R$$
$$C' = 2\pi(R + x)$$
$$C' - C = 2\pi(R + x) - 2\pi R$$
$$10 = 2\pi R + 2\pi x - 2\pi R$$
$$10 = 2\pi x$$
$$\frac{10}{2\pi} = x$$
$$x \approx 1.6$$

The band would allow a uniform distance of 1.6 ft. all the way around the equator.

(ENUF' TO CRAWL!)

Moreover, wouldn't it be the same around the moon? Or...a basketball?

Note the result is independent of R!!!

EXTRA SQUARE INCH PROBLEM

CONSIDER THE FOLLOWING 8 in. BY 8 in. SQUARE.

THE AREA OF THIS SQUARE IS $8 \times 8 = 64$ in.^{2}

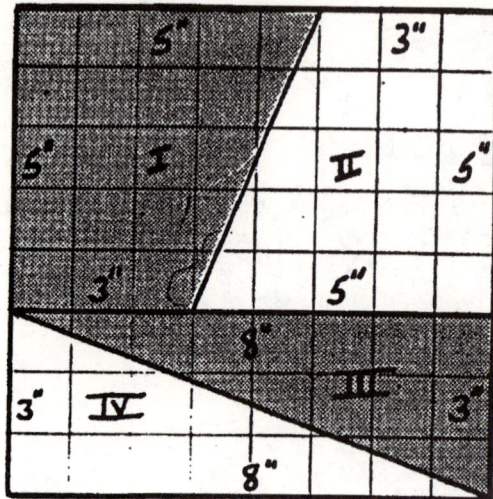

Cut this square into four pieces marked I, II, III, IV, and rearrange the pieces as shown at the right.

THE AREA OF THIS RECTANGLE IS

$5 \times 13 = 65$ in.^{2}

WHERE DID THE EXTRA SQUARE INCH COME FROM?

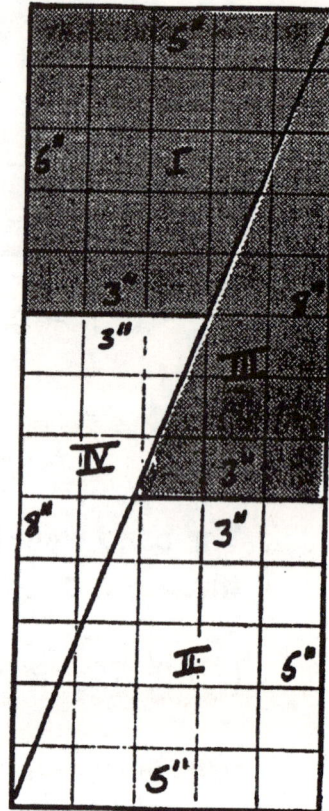

TIME IS MONEY

FINANCIAL VARIABLES

P   for present value (or principal)
A   for future value
I   for interest
r   for annual percentage rate
t   for time (in years)
m   amount of periodic payment
n   number of times compounded per year

COMPOUND INTEREST

Find the future value for $1 invested at 100% interest for 1 year;
i.e. Find A for P = 1, r = 1, t = 1

Number of periods	Formula	Amount
annually, $n = 1$	$(1 + \frac{1}{1})^1$	$2.00
semiannually, $n = 2$	$(1 + \frac{1}{2})^2$	$2.25
quarterly, $n = 4$	$(1 + \frac{1}{4})^4$	$2.44
monthly, $n = 12$	$(1 + \frac{1}{12})^{12}$	$2.61
daily, $n = 360$	$(1 + \frac{1}{360})^{360}$	$2.715 \approx $2.72
$n = 10,000$		$2.718145926
$n = 1,000,000$		$2.718280469
$n = 100,000,000$		$2.718281828

Define $e = \lim_{n \to \infty} (1 + \frac{1}{n})^n \approx 2.718281828$

ARE YOU A GENIUS?

PROBLEM 1

A, D, G, J, ...

PROBLEM 2

1, 3, 6, 10, ...

PROBLEM 3

1, 1, 2, 3, 5, ...

PROBLEM 4

21, 20, 18, 15, 11, ...

PROBLEM 5

8, 6, 7, 5, 6, 4, ...

PROBLEM 6

40, 35, 34, 29, 28, 23, ...

From *Mensa* test reprinted by permission by Mensa, 50 E. 42 St., New York, NY 10017

ELEVEN PUZZLE

PICK ANY TWO INTEGERS BETWEEN –5 AND 5.

ADD THESE TWO NUMBERS TO FILL IN SPACE #3

ADD #2 AND #3 TO FILL IN #4

CONTINUE UNTIL YOU HAVE FILLED IN TEN DIGITS.

WHAT IS THE SUM OF THESE 10 NUMBERS?

FIBONACCI'S DELIGHT

PICK ANY TWO NUMBERS: (1) _____ n

(2) _____ m

ADD TO OBTAIN A THIRD: (3) _____ $n + m$

CONTINUE: (4) _____ $n + 2m$

(5) _____ $2n + 3m$

(6) _____ $3n + 5m$

(7) _____ $5n + 8m$

(8) _____ $8n + 13m$

(9) _____ $13n + 21m$

(10) _____ $21n + 34m$

ADD THE ENTIRE COLUMN: _____ $55n + 88m$

NOTICE: $55n + 88m = 11(5n + 8m)$ This is 11(7th no.)

SAMPLE SPACE

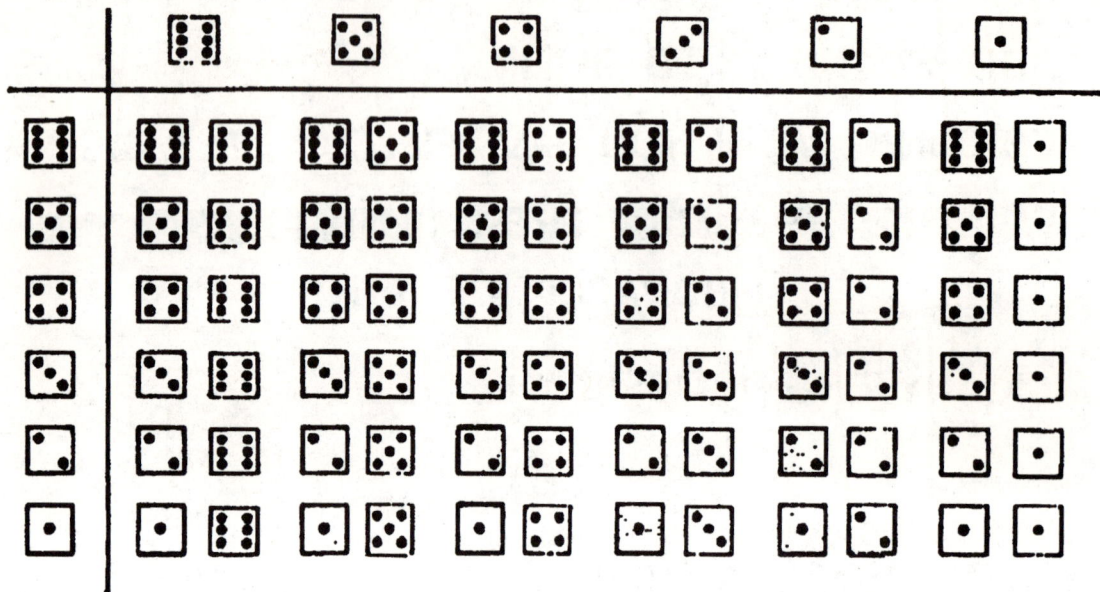

A DECK OF CARDS

Hearts (red cards)

Spades (black cards)

Diamonds (red cards)

Clubs (black cards)

The Nature of Mathematics, 10th Edition

Suppose that your friend George shows you three cards. One card is white on both sides, one is black on both sides, and the last is black on one side and white on the other. He mixes the cards and tells you to select one at random and place it on the table. Suppose the upper side turns out to be black. It is not the white-white card; it must be either the black-black or black-white card.

"Thus," says George, "I'll bet you $1 that the other side is black." Would you play?

Perhaps you hesitate. Now George says he feels generous. You need to pay him only 75¢ if you lose, and he will still pay you $1 if he loses.

Would you play now?

There are two cars built in Sweden. Before you buy theirs, drive ours.

When people who know cars think about Swedish cars, they think of them as being strong and durable. And conquering some of the toughest driving conditions in the world.

But, unfortunately, when most people think about buying a Swedish car, the one they think about usually isn't ours. (Even though ours doesn't cost any more.)

Ours is the SAAB 99E. It's strong and durable. But it's also a lot different from their car.

Our car has Front-Wheel Drive for better traction, stability and handling.

It has a 1.85 liter, fuel-injected, 4-cylinder, overhead cam engine as standard in every car. 4-speed transmission is standard too. Or you can get a 3-speed automatic (optional).

Our car has four-wheel disc brakes and a dual-diagonal braking system so you stop straight and fast every time.

It has a wide stance. (About 55 inches.) So it rides and handles like a sports car.

Outside, our car is smaller than a lot of "small" cars. 172" overall length, 57" overall width.

Inside, our car has bucket seats up front and a full five feet across in the back so you can easily accommodate five adults.

It also has more headroom than a Rolls Royce and more room from the brake pedal to the back seat than a Mercedes 280. And it has factory air conditioning as an option.

There are a lot of other things that make our car different from their car. Like roll cage construction and a special "hot seat" for cold winter days.

So before you buy their car, stop by your nearest SAAB dealer and drive our car. The SAAB 99E. We think you'll buy it instead of theirs.

SAAB 99E

Phone 800-243-6000 toll-free for the name and location of the SAAB dealer nearest you. In Connecticut, call 1-800-942-0655.

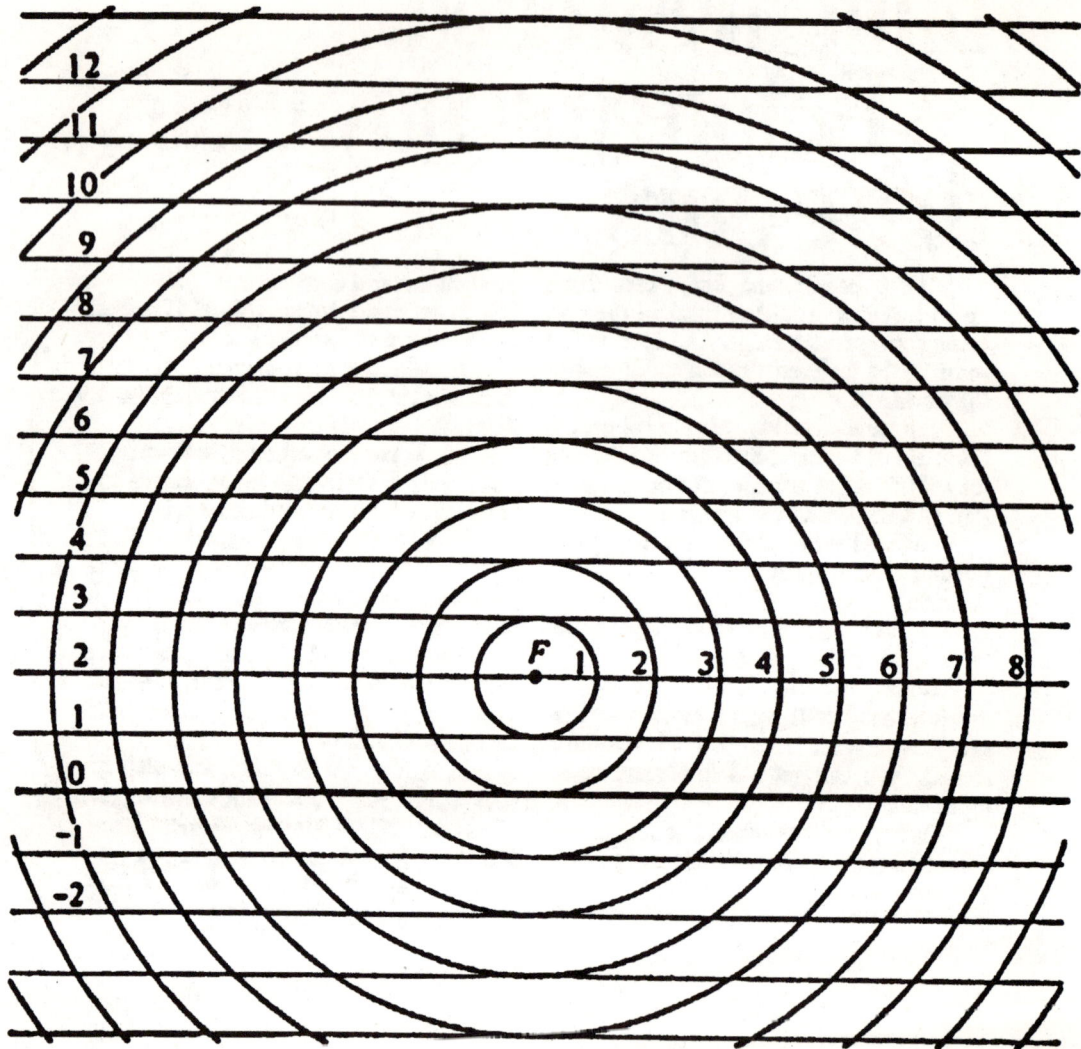

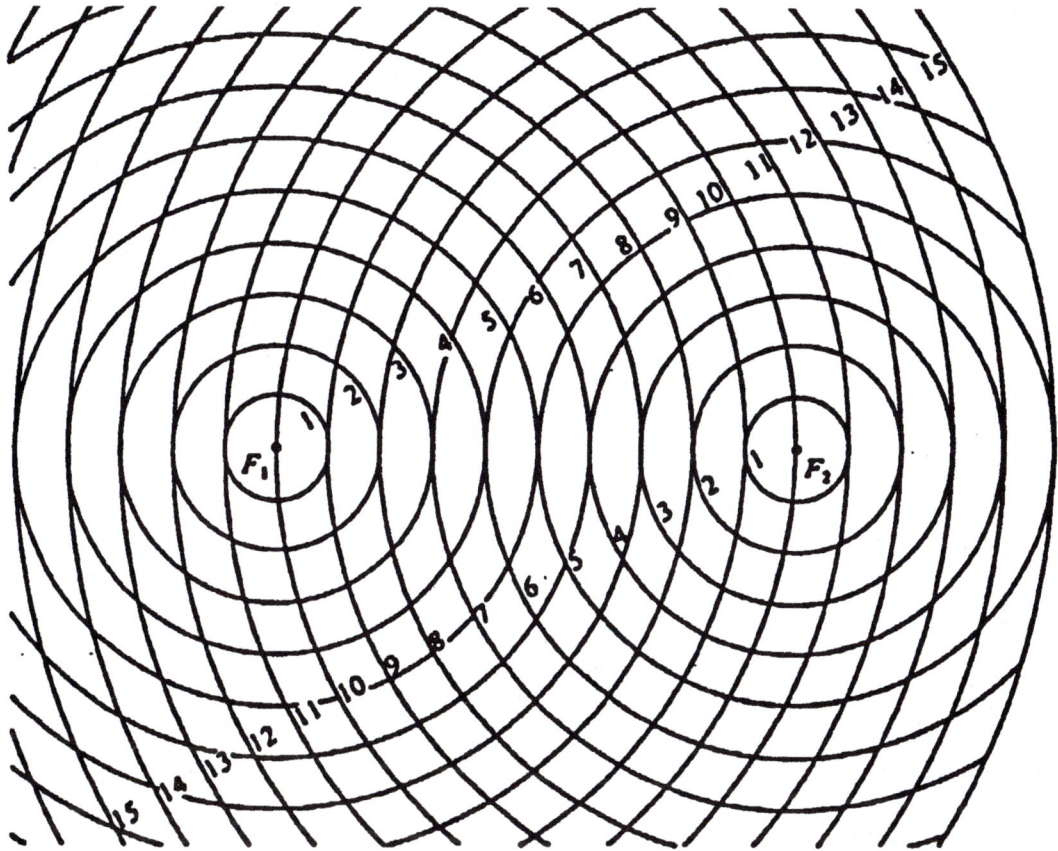

SANTA ROSA STREET PROBLEM

On Saturday evenings, a favorite pastime of students is to cruise the marked streets in Santa Rosa.

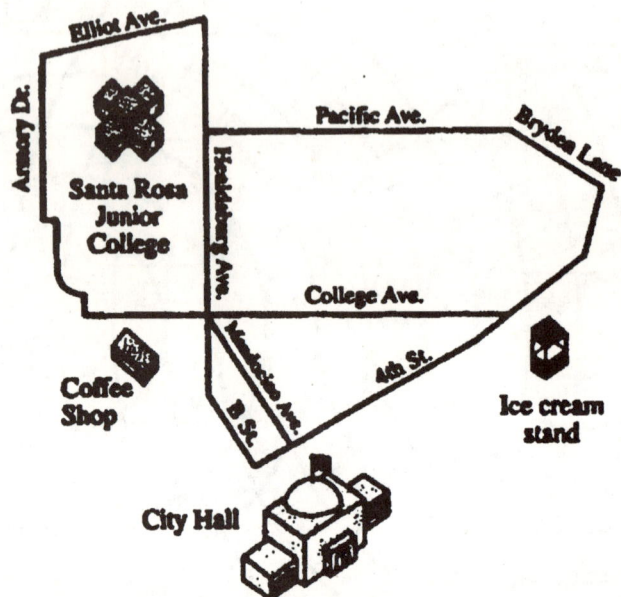

IS IT POSSIBLE TO CHOOSE A ROUTE SO THAT ALL OF THE PERMITTED STREETS ARE TRAVELED EXACTLY ONCE?

The intersections are identified by the following buildings.

SRJC (Mendocino and Pacific Avenues)

Coffee Shop (Mendocino, College, and Healdsburg Avenues)

City Hall (Mendocino and Fourth Streets)

Fresh Freeze (College and Fourth Streets)

THE GAME OF WIN

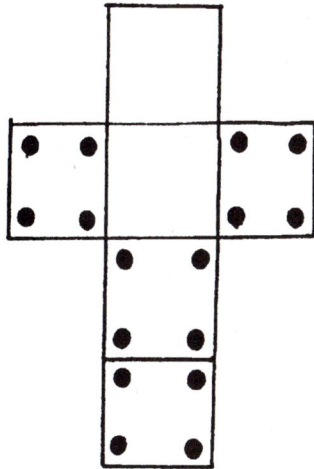

A

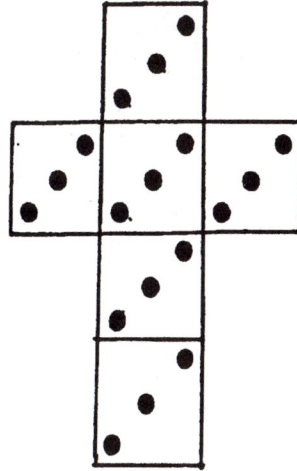

B

C

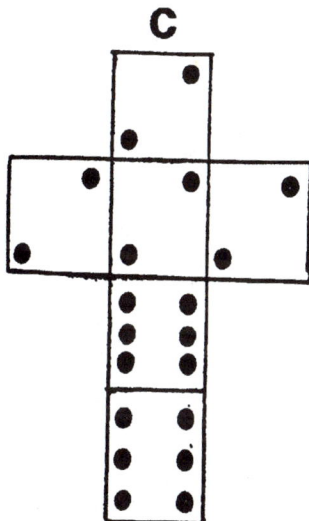

D

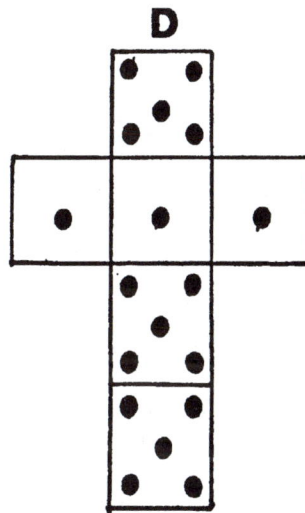

SAMPLE SPACE FOR DICE CHOICES

A\B

	3	3	3	3	3	3
0	B	B	B	B	B	B
0	B	B	B	B	B	B
4	A	A	A	A	A	A
4	A	A	A	A	A	A
4	A	A	A	A	A	A
4	A	A	A	A	A	A

A\C

	2	2	2	2	6	6
0	C	C	C	C	C	C
0	C	C	C	C	C	C
4	A	A	A	A	C	C
4	A	A	A	A	C	C
4	A	A	A	A	C	C
4	A	A	A	A	C	C

B\C

	2	2	2	2	6	6
3	B	B	B	B	C	C
3	B	B	B	B	C	C
3	B	B	B	B	C	C
3	B	B	B	B	C	C
3	B	B	B	B	C	C
3	B	B	B	B	C	C

B\D

	5	5	5	1	1	1
3	D	D	D	B	B	B
3	D	D	D	B	B	B
3	D	D	D	B	B	B
3	D	D	D	B	B	B
3	D	D	D	B	B	B
3	D	D	D	B	B	B

A\D

	5	5	5	1	1	1
0	D	D	D	D	D	D
0	D	D	D	D	D	D
4	D	D	D	A	A	A
4	D	D	D	A	A	A
4	D	D	D	A	A	A
4	D	D	D	A	A	A

C\D

	5	5	5	1	1	1
2	D	D	D	C	C	C
2	D	D	D	C	C	C
2	D	D	D	C	C	C
2	D	D	D	C	C	C
6	C	C	C	C	C	C
6	C	C	C	C	C	C

OPPONENT'S CHOICE MY CHOICE PROBABILITY(WIN)

OPPONENT'S CHOICE	MY CHOICE	PROBABILITY(WIN)
	B	1/3
A	C	5/9
	D	2/3 **
	A	2/3 **
B	C	1/3
	D	1/2
	A	4/9
C	B	2/3 **
	D	1/3
	A	1/3
D	B	1/2
	C	2/3 **

CONCLUSION:

A BEATS B
B BEATS C
C BEATS D
D BEATS A

NOTE: DICE ARE NOT COMMUTATIVE

IF I CHOOSE SECOND, THE PROBABILITY OF MY WINNING IS 2/3.

MATHEMATICAL MODEL

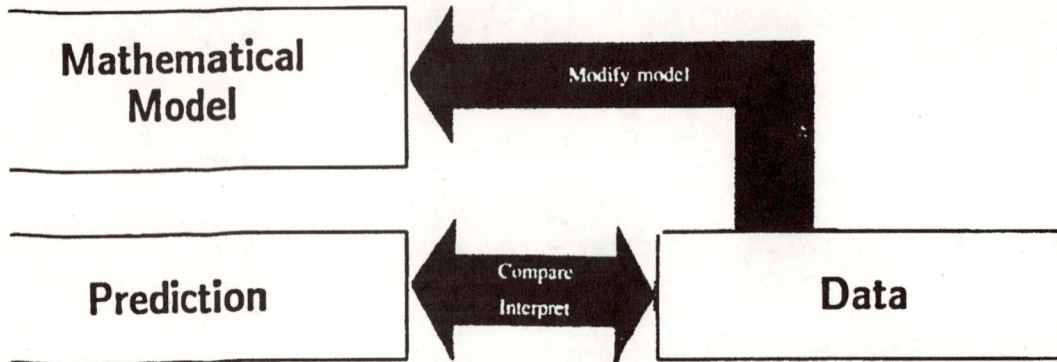

Mathematical Model	← Modify model

Prediction ← Compare Interpret → Data

math·e·mat·ics \'math-ə-'mat-iks \ <u>n pl but usu sing in constr</u> **1**: the science of numbers and their operations, interrelations, combinations, generalizations, and abstractions and of space configurations and their structure, measurement, transformations, and generalizations **2**: a branch of, operation in, or use of mathematics

This is what it is.